MICROWAVE PHOTONICS

Microwave Photonics

From Components to Applications and Systems

edited by

Anne Vilcot
IMEP - INPG, France

Béatrice Cabon
IMEP - INPG, France

and

Jean Chazelas
Thales Airborne System, France

SPRINGER SCIENCE+BUSINESS MEDIA, LLC

A C.I.P. Catalogue record for this book is available from the Library of Congress.

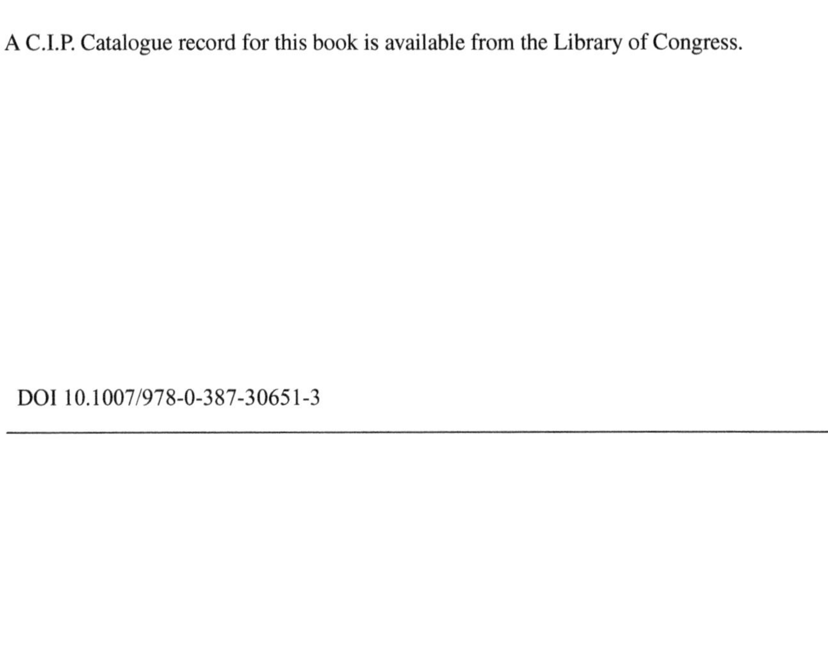

DOI 10.1007/978-0-387-30651-3

Printed on acid-free paper

TABLE OF CONTENTS

PREFACE

Microwave-Photonics can be defined as the study of interactions between optical signals and electrical signals in the frequency range of microwave and millimeter-wave.

This field has been growing very rapidly over the last two decades and recent development of wideband technologies has resulted in sophisticated devices that are now under commercial development.

This book covers all aspects of the field of microwave photonics, including components, modeling methods, circuits, optical processing of microwave signals and broadband systems for a wide variety of applications, such as telecommunications, radar and fiber-wireless systems. New optical architectures for antenna systems are also reviewed in the applications.

The strength of the microwave-photonics area has been demonstrated by the number of papers published in this field, in the literature and topical meetings, workshops and summer schools in the past years.

In particular, in 1998,1999 and 2000, three summer-schools/tutorials "OMW, Interactions between microwave and optics" were held in Autrans (France, 38). This book is a selection of about 40 papers presented during the OMW series by international experts of microwave-photonics. The very latest developments in microwave-photonics components, systems and applications are presented by specialists of over ten organizations in Europe, the United States and Israel.

This manual is targeted at Research and Development, engineers in industry, postgraduate students, and all professionals interested in the use of photonics in microwave and millimeter-wave wide band applications.

ACKNOWLEDGEMENTS

We wish to thank all speakers and participants to the summer-schools OMW (Interactions between Microwaves and Optics), who contributed to the success of the events, and who made it possible the publication of this manual.

Nota Bene : The adresses indicated for each author are those at the time of the corresponding summer-school.

GENERAL INTRODUCTION

Microwave photonics is an emerging technology that may be described as the unification of microwave and photonic techniques for applications such as fibre delivery of mm-waves (e.g. in fibre radio) and high-speed fibre-optic links.

While wireless systems offer mobility, optical fiber communications provide the massive bandwidth that fuelled the huge demand in internet traffic. These two nowadays complementary technologies may converge in Microwave Photonics.

By the example of the number of Microwave Photonics Conferences and Workshops organised in the last ten years, it is satisfying to see how Microwave Photonics has become a flourishing business area and an exciting field for interdisciplinary research.

The editors of this manual have tried to produce a book in the field of Microwave Photonics reflecting the most exciting recent developments in Microwave Photonics. About 40 contributions to the OMW summer-schools have been selected and are published here. Special thanks are due to the large team of authors, internationally renowned and who have worked hard to make OMW a success. We would like to thank them again for their support of this manual.

The full range of activities in microwave photonics is covered in this book, from novel devices to systems experiments in broad-band radio and antennas.

The book is divided into five general chapters.

The first chapter is directed to enabling wide band components for the generation, amplification and high speed detection of microwave photonics signals.

The second chapter moves then on electronics for optics with special constraints in optoelectronic microwave monolithic integrated circuits.

Then, modeling methods for optoelectronics are descibed in the third chapter.

We then move to applications of microwave photonics with two last and long chapters addressing optical links, photonic techniques for microwave signal processing and beamforming for microwave phased array antennas. Microwave and millimeterwave systems are reviewed for telecommunications as well as wireless systems. We have also included in the last part digital converters, optoelectronic processors, and optoelectronics for the terahertz frequency range.

We hope that this book will provide a useful tool that gives the state of the art of microwave photonics.

Since research and industrial exploitation in this field is progressing very rapidly, the reader will find information that has probably evolved in the meantime.

CHAPTER 1 : MICROWAVE PHOTONICS COMPONENTS

1. INTRODUCTION

Transmission of analogue microwave/millimeterwave and high-speed digital signals is an enabling technology that has many applications in our life on the basis of the modern Information Society.

Very wide-bandwidth microwave photonic components are essential components because they are now needed in optical fiber communication systems with a data rate exceeding 40 Gb/s per channel. Microwave Photonics should allow as well the development of ultra wide band remote front-end RF transponder.

This chapter reports on enabling microwave photonics technologies, devices and components including lasers, modulators, semiconductor optical amplifiers, phototransistors and photodetectors.

The first three groupings of this chapter describe the very wide-bandwidth lasers and external optical modulators that are now essential components for high-speed transmission. The necessary optical amplification is also presented.

The last part reports on high speed photodetection. The photodetector is a key component in optical transmission and optical measurement systems. Devices of high efficiency are especially important as systems become faster and a wide variety of photodetectors and phototransistors are presented in this chapter.

2. FAST LASER SOURCES

2.1. Fast Lasers Sources

F. Deborgies
Thomson-CSF / Laboratoire Central de Recherches
91404 Orsay, France
E-mail: deborgies@lcr.thomson-csf.com

2.1.1. Laser Basics

What is a laser? It is the acronym for Light Amplification by the Stimulated Emission of Radiation or practically speaking a source of coherent light. The first laser operation was demonstrated in 1960, with a ruby laser (red light). In its basic principles, the operation of a laser is very similar to an electrical oscillator which is depicted in figure1.a. In order, to obtain oscillation, several conditions have to be meet: the gain in the amplifier must be greater than the combined losses in the feedback loop and from the output while the phase in the round trip must be a multiple of 2π. In that case oscillations will build up from the noise and grow until the amplifier saturates. Since the gain and the phase of the amplifier are frequency dependant, only a limited number of discrete frequencies can be generated by the oscillator. These are called the resonance frequencies.

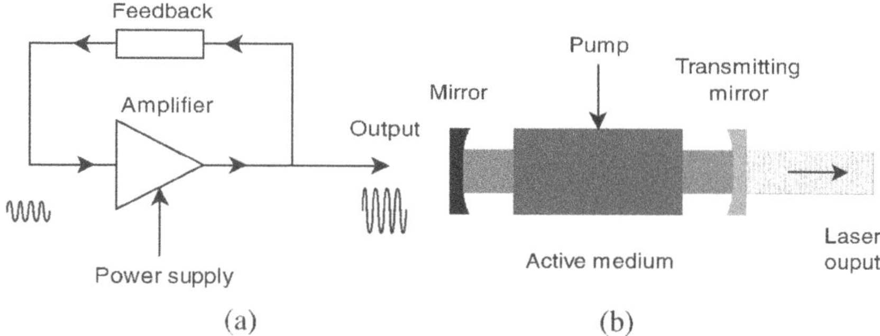

Figure 1 : Principle operation of an electrical oscillator (a) and a laser (b).

Similarly in a laser (see figure 1.b), the gain is provided by the active medium which can be a gas, a solid, a liquid or a semiconductor, while the feedback is obtained with mirrors, one of them being semi-transparent in order to couple light out of the cavity. Light amplification is obtained by exciting or pumping the medium so that more atoms are in the upper energy level than in the lower level (see figure 2). This operation is called "population inversion", while pumping can be achieved electrically (DC or RF) or optically (flashlamp or laser). Most lasers are four level systems (see figure 2), the only exception being the ruby and erbium lasers which are three level lasers. The latter can also be seen as four level lasers in which the transition to the ground state is infinitely fast. The lasing wavelength is directly related to the transition energy between the "lasing states" (thick lines in figure 2). The table below lists some of the most common lasers apart from semiconductor lasers which will be detailed in the following.

Laser	KrF	Argon	He-Ne	Ruby	Nd^{3+}:YAG	Erbium	CO_2
Wavelength	UV	515 nm	632.8 nm	694.3 nm	1.064 μm	1.55 μm	10.6 μm

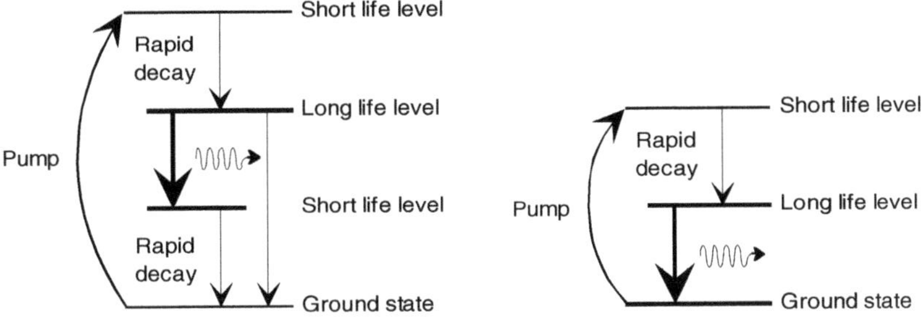

Figure 2 : Energy levels for 4 and 3 level lasers.

2.1.2. Semiconductor Lasers

The first semiconductor (SC) laser was demonstrated in 1962 nearly simultaneously by IBM, GE and MIT Lincoln Lab. However this was achieved at a temperature of 77 K and the first CW laser at room temperature was obtained by Bell Labs in 1973. For SC lasers, the active medium is obviously the SC material and the "lasing transition" energy is related to the bandgap energy as shown in the table below.

SC Material	Ge	Si	InP	GaAs
Bandgap (eV)	0.66	1.11	1.35	1.42
Wavelength (μm)	1.88	1.15	0.92	0.87

However, not any kind of SC can be used to design a laser. As a matter of fact, there are two types of semiconductors: the direct gap SC and the indirect gap. In a direct gap SC such as GaAs or InP, the absorption of a photon with enough energy (greater than the bandgap) leads to a generation of an electron-hole pair in the material as shown in figure 3.a. Similarly, the recombination of an electron-hole pair results in the spontaneous emission of a photon with the corresponding energy (see figure 3.b). Furthermore, an electron-hole pair recombination can be "stimulated" by an incoming photon. In that case, the created photon is identical to the original one; this process called "stimulated emission" is responsible for the light amplification in lasers.

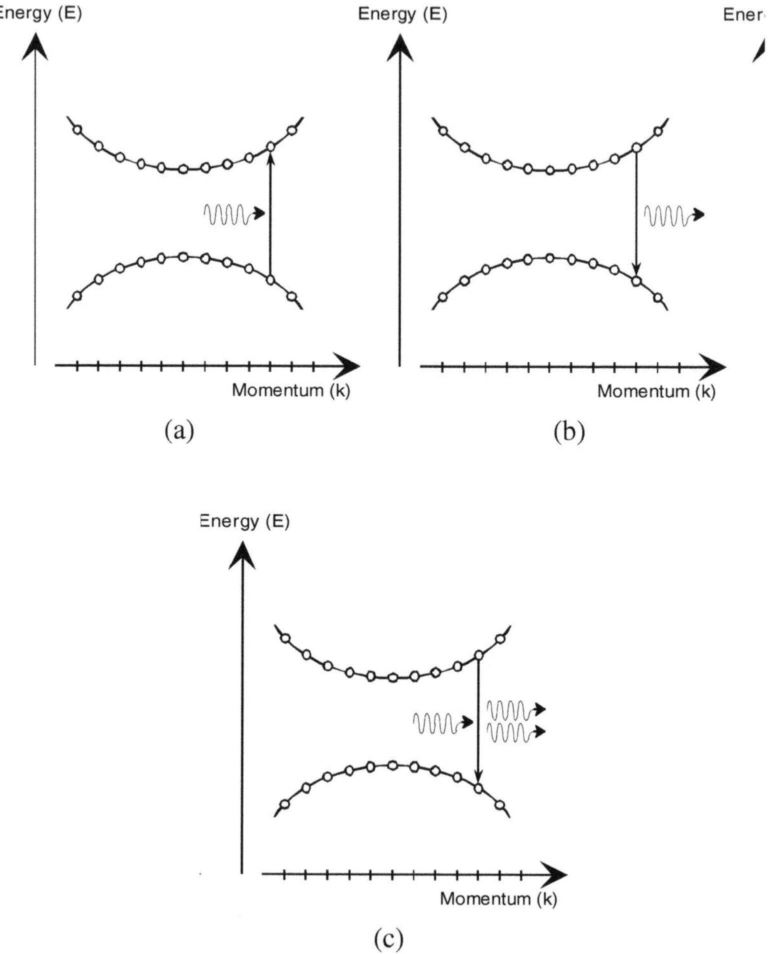

Figure 3 : Photon absorption (a), photon emission (b) and stimulated emission (c) in a direct gap SC.

In indirect gap semiconductors such as Si, Ge or GaP, the picture is more complicated. As for any kind of semiconductors, photon emission or absorption is conditioned by two basic rules: conservation of energy (E) and conservation of momentum (k). A transition from the bottom of the conduction band to the top of the valence band (dashed line in figure 4.a) where most of the electrons and holes respectively are likely to be, should in principle lead to the emission of a photon. However, the photon cannot carry the necessary momentum in order to comply with conservation rules. Hence one phonon which carries little energy and large momentum is simultaneously needed (dotted line in figure 3.a). Since this three body process (electron-hole pair, photon and phonon) has a low probability to occur, this means in practise that indirect gap SC are very poor light emitters. However, this does not mean that indirect gap SC cannot detect light. The absorption of a photon leads to a vertical transition from the top of the valence band to a higher energy level for the electron (see figure 4.b). The electron can then go down to the bottom of the conduction band by a process named "thermalisation" in which it transfers little energy and large momentum to successive phonons (dotted lines in figure 4.b). Since this is a sequential process, it does not have a low probability of occurrence as for the three body process involved in the photon emission. This explains why silicon is widely used for photodetectors.

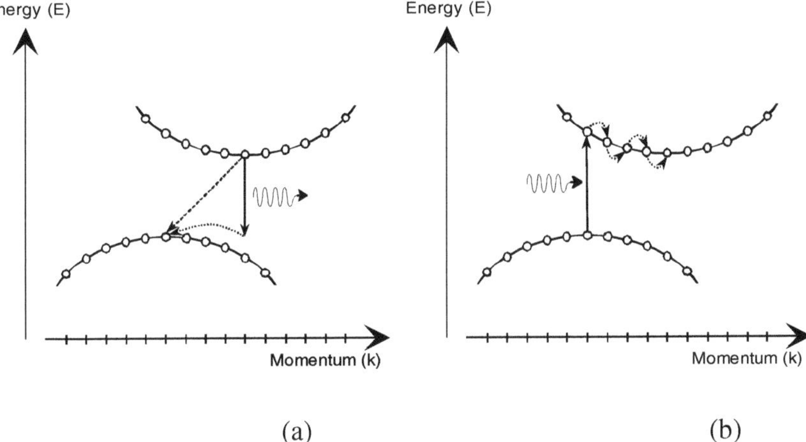

(a) (b)

Figure 4 : Photon emission (a) and photon absorption (b) in an indirect gap SC.

SC lasers have a very wide wavelength coverage from less than 500 nm to more than 5 μm. Telecom-munications have for long been the major thrust behind the development of SC lasers. As a matter of fact, SC lasers are the ideal light sources for fibre communications. The emergence of low loss fibres (see figure 5.a) has lead to the realisation of GaAs laser in the 800 nm window. Later the improvement of fibre shifted the wavelength towards 1.3 μm where the dispersion of standard fibre is minimal and then to 1.5 μm where the losses are the lowest. Further improvement of the fibre is now intrinsically limited by the Rayleigh scatterings due to inhomogeneities of the refractive index of the silica (dashed line on the lower part of the spectrum in figure 5.a) and the infrared absorption of the material in the upper part of the spectrum. Water contamination revealed by the presence of two OH peaks in the attenuation of the fibre is another limitation but extrinsic to the fibre itself.

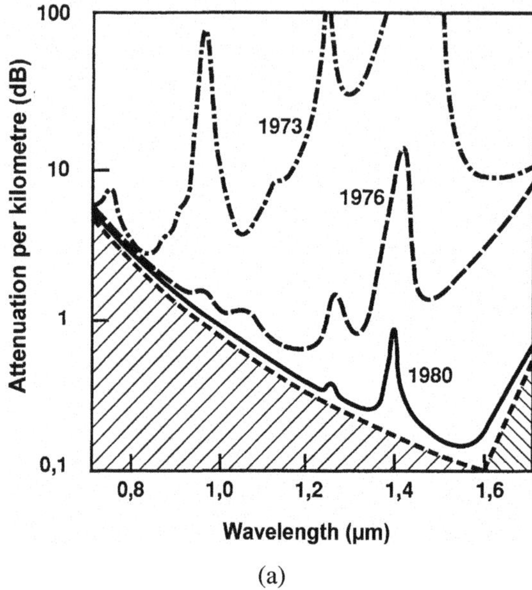

(a)

Telecom. Window	*Wavelength (µm)*	*SC Material*	*Fibre Specificity*
1	*0.8*	*GaAs*	-
2	*1.3*	*InP*	*Minimum dispersion*
3	*1.5*	*InP*	*Minimum losses*

(b)

Figure 5 : Attenuation of fibre versus wavelength and time (a) and the corresponding telecom windows (b).

Other SC wavelengths of interest are listed in the table below.

Wavelength (nm)	650	780	808	980	1017	1480
Power	low	low	high	high	high	high
Application	DVD player	CD player	YAG pump	EDF A pump	PDF A pump	EDF A pump

The simplest structure for a SC laser is the Fabry-Perot (FP) structure in which the active medium is a forward biased PN junction and the cavity is made from the partly reflecting cleaved facets (see figure 6.a). This buried ridge structure (BRS) is grown on an N-doped InP substrate and the active layer (the ridge itself) is made of P-doped GaInAsP. Proton implantation is realised on both sides of the ridge to reduce the leakage current. The output power of a laser is a function of the bias current (see figure 6.b): below the threshold current the power is negligible while it increases rapidly above before saturating. This increase rate called "slope efficiency" of the laser. Since both the threshold current and the efficiency are temperature, most of the time the laser chip has to be temperature controlled in order to have stable performances whatever the operating conditions.

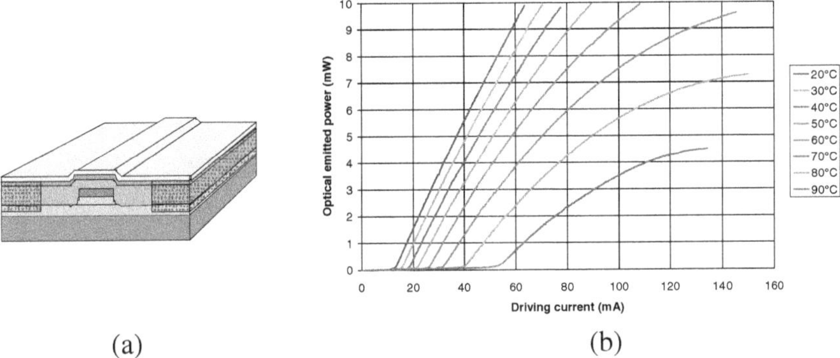

Figure 6 : Fabry-Perot BRS laser structure (a) and typical SC laser output power vs. bias current.

FB lasers are multimode lasers (see figure 7) and the number of modes is dependent on the spectral width of the SC material as well as the cavity length. It is a limiting factor in most telecom application because mode competition in the laser leads to excess noise while the combination of the fibre dispersion and the FP broad linewidth reduce the maximum transmission distance.

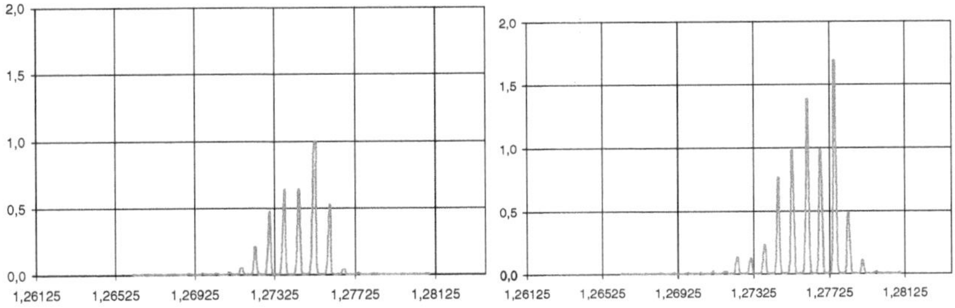

Figure 7 : Typical spectra of a Fabry-Perot laser at two operating currents.

In order to answer the telecom needs for longer distances, singlemode SC lasers have been developed to overcome the limitations of FP lasers. There are various types of singlemode SC lasers and among them the most common are the DFB and DBR lasers (see figure 8.a). In a DBR, the cavity is realised by adding Bragg gratings which act as wavelength sensitive mirrors, at both ends of the structure while in a DFB, the grating is on top of the active zone. This leads to a singlemode operation of the laser (see figure 8.b) which is simply evaluated by the unwanted mode suppression or side-mode suppression ratio (SMSR) which can be better than 40 dB in good lasers. Singlemode laser consequently have improved noise performances (better RIN) as well as a narrow linewidth (usually less than 10 MHz). But they are more sensitive to optical feedback and need an optical isolator to operate properly.

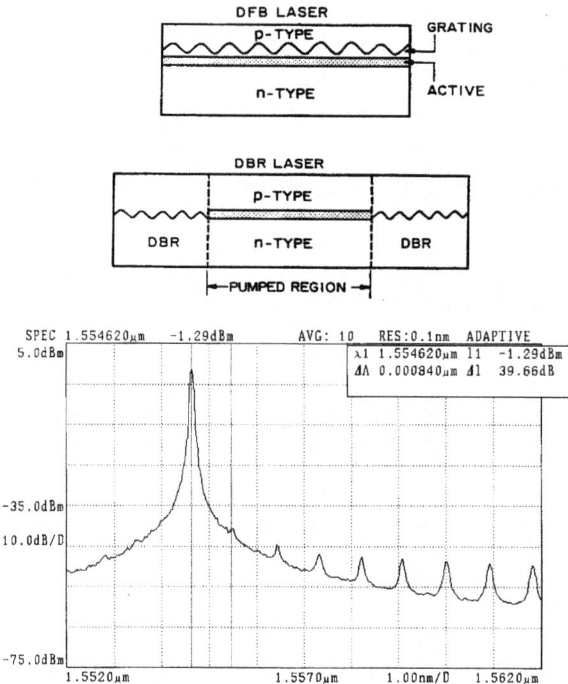

Figure 8 : DFB and DBR laser structures with a typical singlemode spectrum.

Amplitude modulation is mostly used in optical links. It can be either direct modulation of the drive current of a SC laser or external modulation of a CW source with an external modulator. The main advantage of the direct is its simplicity. External modulation although more complex to implement, offers the optimum choice since the modulator and the source (which can be a SC laser) can be selected independently for best overall performances. One of the main limiting factor of directly modulating a laser is the chirp or spectral broadening due to the fact that the emitted wavelength is not only dependent on the temperature but also on the bias current.

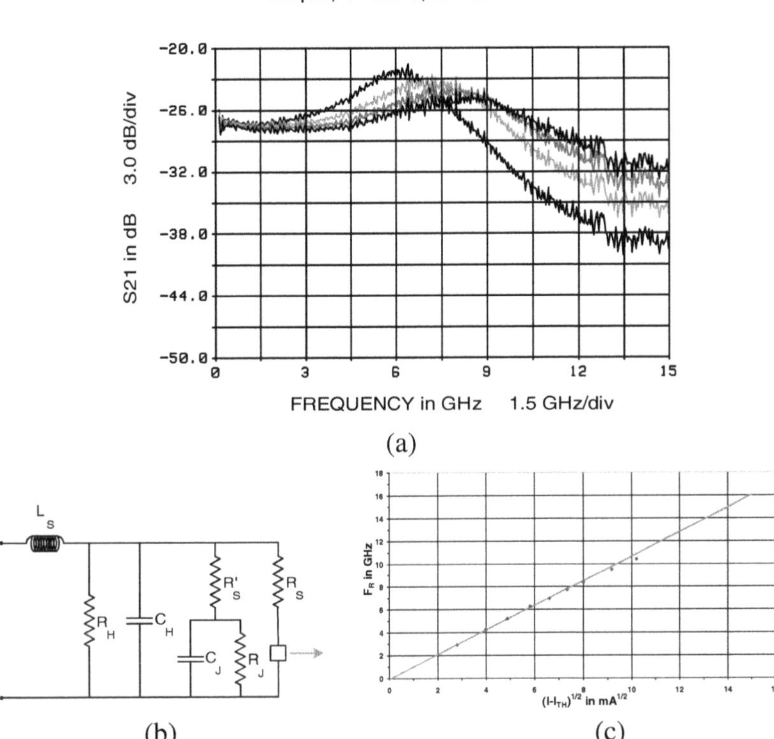

Figure 9 : Frequency response of a high speed SC laser at different bias currents (a), its physical equivalent circuit (b) and its resonant frequency dependence with bias current (c).

The best SC lasers have a 3 dB cut-off frequency in the 25-30 GHz range depending on the wavelength. A typical frequency response of a high speed SC laser is given in figure 9.a. It has a second order low-pass filter shape and depends on the bias current. As a matter of fact, the resonant frequency is proportional to the square root of the emitted optical power as shown in the plot of figure 9.c. A physical equivalent circuit of the laser is depicted in figure 9.b: C_H is the capacitance of the proton implanted region while R_H is the associated resistance, C_J is the PN junction capacitance and R_J the associated resistance, R'_S is the series resistance to this junction while R_S is the series resistance to the active zone. Some typical values are few pF for the capacitances, several kΩ for R_H, infinity for R_J and few ohms for R_S and R'_S.

Noise is also an important parameter for lasers. It is critical for many high performance analogue short range links where the excess noise of the laser is predominant. Lasers are characterised by the relative intensity noise (RIN) which is the ratio of the average rms optical noise over mean optical power after detection. RIN is very much dependent on the structure of the laser and is also related to the intrinsic frequency response of the laser (see figure 10).

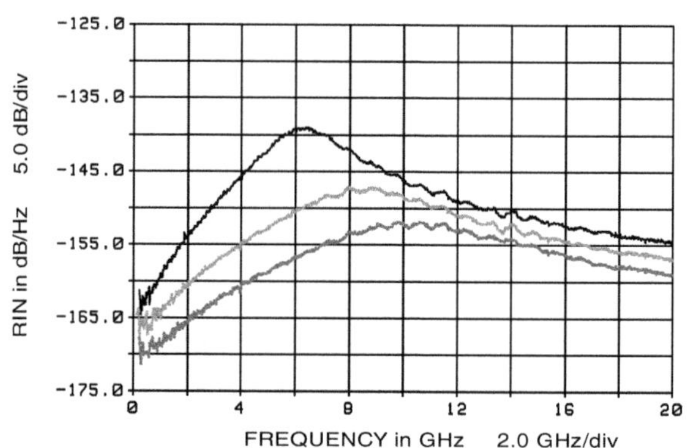

Figure 10 : RIN of a high speed SC laser at different bias currents.

Finally in demanding applications such as CATV, the linearity of the laser is more than crucial. If the laser is not perfectly linear, distortion leads to intermodulation products as shown in figure 11.a. The most critical are the products which are always located nearby the signals i.e. the odd order products starting from the third (2*F2-F1 and 2*F1-F2). From this one can define the spurious free dynamic range (SFDR in figure 11) which is the domain in which the system does not introduce third order product terms. It is expressed in dBc.$Hz^{2/3}$. For linearised systems in which usually the third order terms are greatly suppressed, the limiting unwanted terms are the fifth order terms (3*F2-2*F1 and 3*F1-2*F2). In that case, the SFDR is in dB.$Hz^{4/5}$. For some specific applications (single tone) the dynamic range is defined as the difference between the 1 dB compression point and the noise (D1 in figure 11): this is always larger than the SFDR.

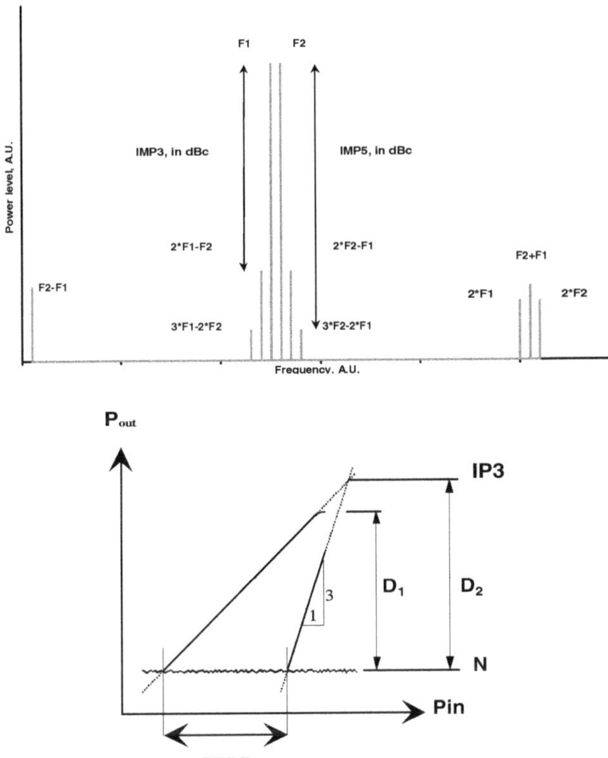

Figure 11 : Effect of non-linearities in a system (a), and definitions of the major dynamic range values (b).

2.1.3. High Frequency Laser Packaging

The packaging of high speed lasers is also of great importance for a laser transmitter since it should not degrade the intrinsic performance of the chip while providing all the necessary interfaces to the external world as well as a protection to the device. The submodule approach can meet all these requirements to frequencies over more than 20 GHz. The transmitter is build around a SC laser (Fabry-Perot or DFB, at 1.3 μm or 1.5 μm) which is mounted on a diamond heatsink for a good thermal dissipation. This subassembly is brazed on a metallic carrier or corner plate on which an hybrid circuit is also mounted (see figure 12.a). A tilted feedback photodiode (power monitoring) and a thermistor (temperature control) are placed onto the circuit (usually alumina). The laser is connected to the RF signal through a 50 Ω microstrip line and a matching resistor for wide band operation while DC biased through a bias circuit.

A critical factor is the coupling efficiency between the laser and the fibre which is directly related to the overall insertion losses. A lensed fibre provides a good coupling but at the expenses of the tolerances on the position of the fibre. This is all the more difficult in harsh environments often found in military or space applications. In order to obtain a good stability a patented YAG welding technique has been developed. The lensed fibre is attached into a fibre carrier which slides on the corner plate. The carrier is dynamically moved to obtain the optimum position and then welded (see figure 12.b). The same technique is used for adding an isolator to the submodule by simply replacing the lensed fibre by an association of a lens, an isolator and a collimated fibre.

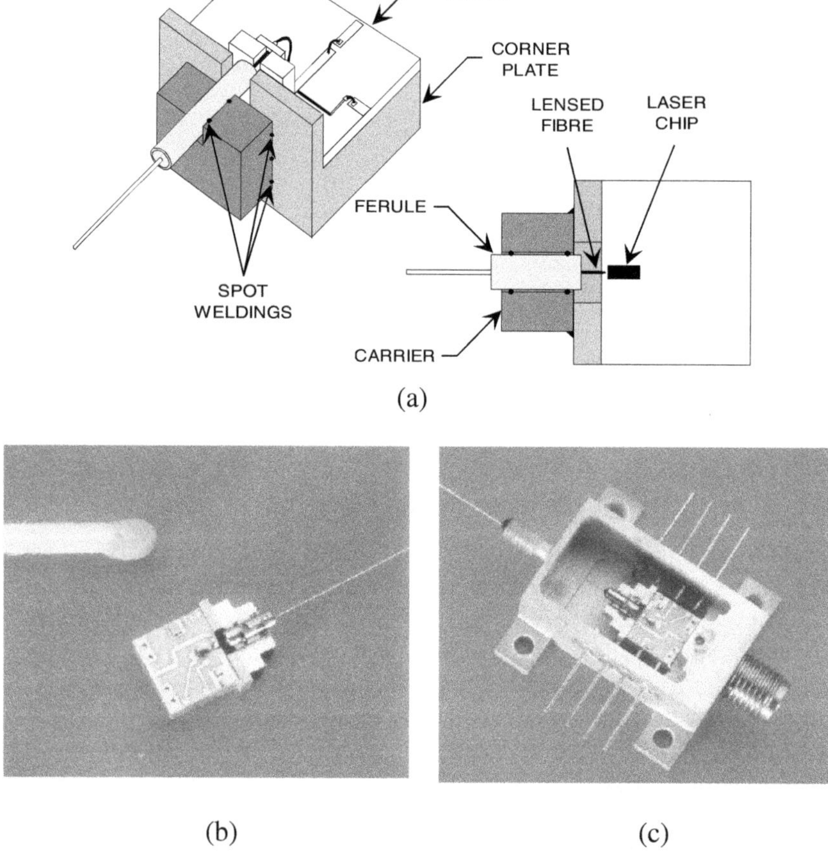

Figure 12 : Submodule principles (a), laser submodule (b) and laser module (c).

If necessary, the same submodule can be included in a small package which contains a military class thermo-cooler and provides a connectorized input (see figure 12.c). Tightness is classically obtained with glass beads for the RF and electrical inputs/outputs and silica/glass/metal feedthrough for the fibre pigtail while the cover is YAG welded in an atmosphere of nitrogen or argon. The coupling scheme has been validated by submitting the module to more than 100 thermal cycles (one cycle consists of 30 minutes at -50°C and +125°C respectively) without any degradation of the coupling ratio.

Figure 13.b shows the equivalent circuit of a mounted SC laser extracted form the S_{11} measurement (see figure 13.a). This equivalent circuit is very close to the physical equivalent circuit mentioned earlier. Of those elements, only the series inductance (related to the wire), the series resistance and parallel capacitance of the laser are of interest. Furthermore, on a well designed laser, the capacitance do not play a significant role and the mounted laser can be seen, in the first order, as a pure resistance of few ohms in series with an inductance. It is the reason why a simple series resistance gives a good wide band 50 Ω match. However, this matching is done at the expenses of increased microwave losses for the optical link based on this kind of transmitter: for a 5 Ω laser, the energy lost in the 45 Ω series resistance translates into a 10 dB additional loss of the link. Passive reactive matching can reduce the losses by replacing the resistance with the equivalent to an impedance transformer, but is limited in terms of bandwidth. This trade-off (bandwidth vs. losses) can be overcome by active matching with an MMIC such as the impedance transformer (50 Ω to 10 Ω) shown in figure 13.c.

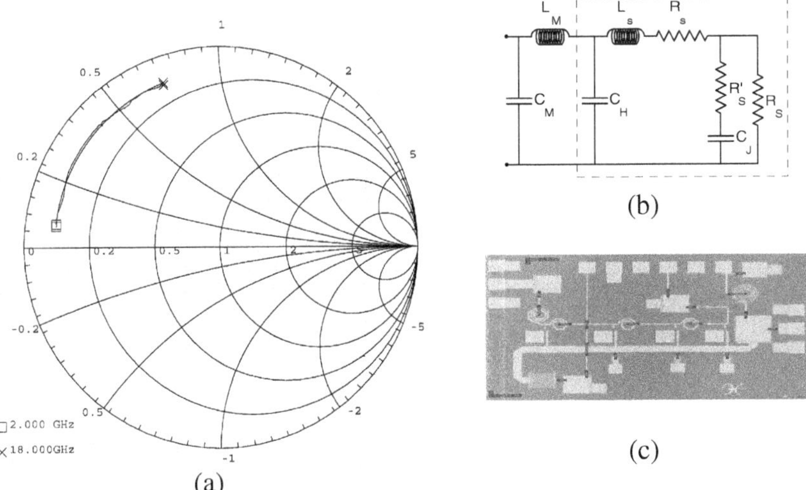

Figure 13 : Smith chart of laser impedance as a function of frequency (a), equivalent circuit of mounted laser chip (b) and active impedance transformer MMIC(c).

2.1.4 Alternative High-Speed Semiconductor Lasers

Many new laser structure have been proposed to overcome the frequency limitation of standard SC lasers such as Fabry-Perot or DFB. Among those, the most common ones are the mode-locked SC laser, the dual mode laser and the harmonic laser.

The mode-locked laser can cover very high frequencies but has a narrow locking band and is sensitive to temperature. The dual mode laser has essentially the same advantages and drawbacks as the mode-locked laser except that it has also a low dispersion penalty. Finally the harmonic generation is a simple solution which can, also go to high frequencies but with a limited efficiency while the frequency increases.

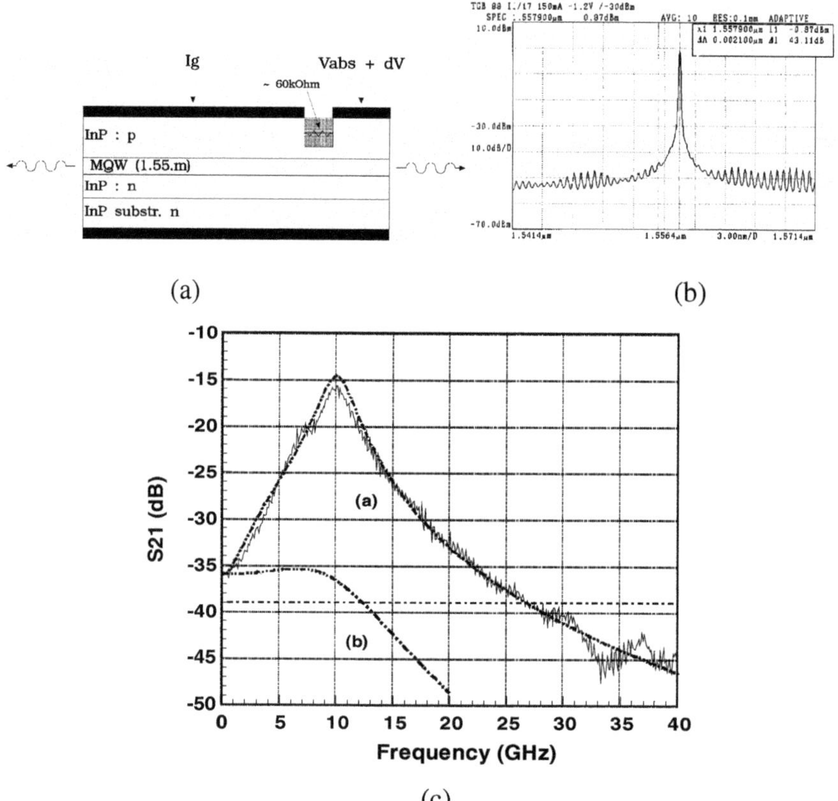

Figure 14 : Two electrode laser structure (a), spectrum (b) and frequency response (c).

The two-electrode laser could be an interesting alternative. It is based on a standard DFB structure (see figure 14.a) which has been slightly modified in order to add a small electrode which acts as a voltage controlled saturable absorber. Similarly to a DFB laser, it is a singlemode laser (see figure 14.b) but it behaviour concerning its frequency response is rather different. As a matter of fact the resonance is greatly enhanced compared to an identical structure with a single electrode (see curve (a) and curve (b) in figure 14.c) while the slope ,at higher frequencies, is now a first order slope instead of second order for standard lasers. This enables to reach a 30 GHz cut-off frequency with a structure which was optimised as a standard DFB laser, for a typical cut-off around 12 GHz. Furthermore, because of the first order slope, this laser can be operated at 40 GHz with only a few decibels of additional losses. By refining the single electrode DFB structure, it should be possible to reach a cut-off frequency of more than 40 GHz with a two-electrode configuration.

2.1.5 Acknowledgements

This work would not have been possible without the active participation of my co-workers of the Central Research Laboratory of Thomson-CSF and some other colleagues in other parts of the company.

Support from DGA, France Telecom and the European Community is also acknowledged.

2.2. Tunable/Selectable Sources

F. Brillouet
Alcatel Alsthom Recherche
Unité Composants Photoniques, Route de Nozay, 91 460 Marcoussis, FRANCE
e-mail : Francois.Brillouet@aar.alcatel-alsthom.fr

2.2.1. Introduction

Due to the very rapidly increasing traffic demand, there has been a recent trend of operators to upgrade the already installed fiber transmission capacity launching different wavelengths on the same fiber, opening the way to a Wavelength Division Multiplexing (WDM) concept, quickly evolving toward a dense multiplexing (DWDM). In the same movement, progress in intrinsic transmission performances leads to an increased temporal bit rate from 2.5 Gb/s to 10 Gb/s (Time Division Multiplexing, TDM) so that typical system presently installed have the following configuration : 10 Gb/s , 100 channels with 50 GHz spacing. The spectral range was recently limited to 30nm, corresponding to the gain spectrum of an Erbium Doped Fiber Amplifier (EDFA). Increased fiber performances open presently a continuum wavelength allocation spectrum in the 1.3-1.6 μm.

2.2.2. WDM Transmission and Routing

The WDM evolution was initially concerning a *point-to-point* transmission where different wavelengths at the emission are multiplexed on a same fiber, and after propagation optically de-multiplexed before electric detection (figure 15 a).

In a second step, new functions are added with the possibility to add and drop specific wavelength along the transmission line (Figure 15 b). These optical function are bit rate transparent, and the optical switching element only requires a specific band pass filter.

A third complexity level allowed by WDM approach is for a multiple-to-multiple points transmission, between which Optical Cross Connects (OXCN) are inserted, allowing routing and switching functions without electrical demultiplexing a high bit rate of data (Figure 15 c).

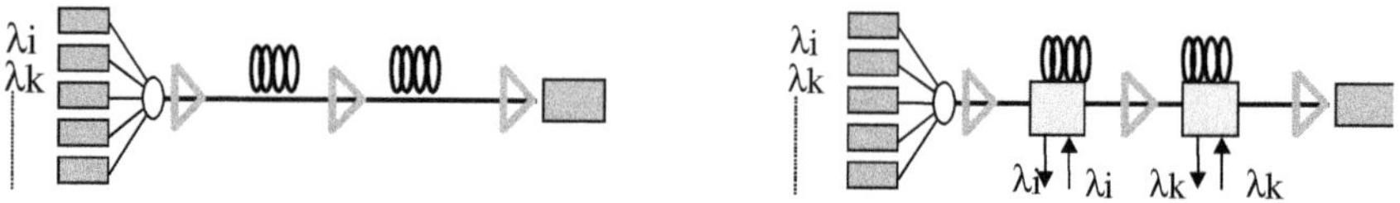

Figure 15a : Point-to-point transmission

Figure 15b : Transmission with add-and-drop functions

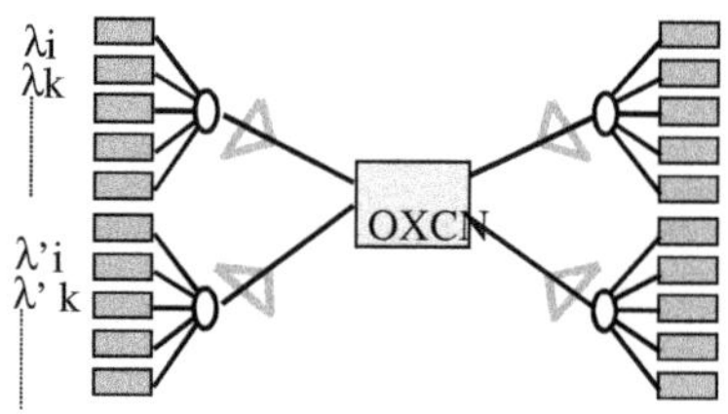

Figure 15c : Transmission with optical cross connect function

If the WDM approach is developing in parallel with the TDM, both have their respective *advantages and limitations:*

- TDM :

• the technology of electronic multiplexing is well mastered up to 10 Gb/s, but leads to an increased difficulty/cost for increasingly high bit rate multiplexing : the next step, from 10 to 40 Gb/s is still a major issue for electronic drivers, receivers and electronic processing before demultiplexing; new technologies based in SiGe, GaAs and even InP materials are the object of an intense competition between development laboratories.

• the fiber propagation, however, presents an increasing difficulty when increasing the basic bit rate that directly slows down the interest of TDM vs WDM: if B is the basic bit rate, the transmission performances on a standard fiber scales as B^{-1} for the propagation distance and B^{-2} or the Polarisation Mode Dispersion (PMD) characteristics of the fiber. As an example, the typical transmission distance on a standard fiber at 2.5 Gb/s is in the range of 1000 km with an external modulator, and only 150 km at 10Gb/s. However, the transmission distance at a given bit rate can be improved if some dispersion compensation is provided by specific fiber section. The PMD performances of a fiber are directly linked to the residual ellipticity of the fiber and if new fiber characteristics fit the 10Gb/s propagation requirements, the characteristics of 5 years old installed fibers often prevent any 10 Gb/s propagation.

- WDM :

• WDM offers the complementary advantages of the above mentioned TDM limitations, that is mainly a simpler electronic technology, and a wider fiber propagation tolerance.

• however, the wavelength emitter stability is now the critical requirement for dense WDM propagation, and other issues like inter-channel cross-talk and Raman amplification will also be key elements to optimise a WDM transmission.

• cost improvements is also a major issue for WDM terminals, as long as the number of channels is increased. In this respect, the emitter source which is assigned to a given normalised ITU wavelength has to be designed specifically for each channel. It is now well accepted that a significant cost improvement would occur if each specific ITU source could be replaced by a source that would be identical for a given number of ITU channels.

2.2.3. Need for Tunable/Selectable Sources

Wavelength tunable or selectable source (TS) has been one of the important target of research laboratories since few years with different potential applications in WDM systems. A tunable source is a single chip with a quasi-continuous wavelength selection through a specific section electrical drive (figure 16 a) : different wavelengths can be successively emitted from a single chip with the same characteristics as those of a single wavelength emitter (currently realised with a Distributed Feedback - DFB- laser).

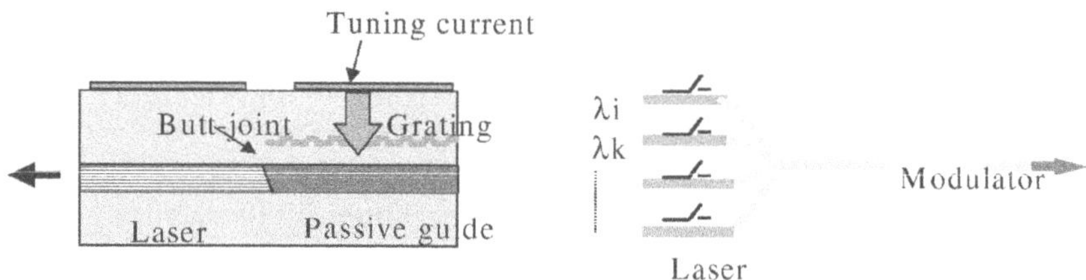

Figure 16 a : Tunable DBR laser. *Figure 16 b : Selectable source.*

On the other hand, the selectable sources are designed with a typical source array combined in a single waveguide, each addressing a given ITU wavelength, and electrically selected.(Figure 16 b).

The main applications for these sources are the following :

- it can be used as a spare source : in case of failure occurring on a single wavelength source, the TS source will be tuned to the failing wavelength source, and will replace all failing source that are in its wavelength range. As an example, the availability of a 20 channels TS source does reduces from 40 DFB sources to 2 TS sources the required number of spare sources.
- it can be used as a 1:N protection, to secure any of the N emitted wavelengths in a short time delay (typically few microseconds)
- it can be used in placed of single wavelength DFB sources, replacing N different emitters by the same one with a wavelength selection facility. In this last case, there is an important cost issue, and an operating lifetime identical to a standard single mode source (15 years).

2.2.4. Tunable Sources

The wavelength tunability is obtained by changing the effective index of the tuning section. Two main effects can be used : *the temperature and the current injection.*

- As an illustration of the first one, a three section DFB - with a different Bragg wavelength filter in each one- is realised, and biasing two sections at the transparency and the third one above to reach the laser threshold condition, each wavelength segment is successively turned-on, and within each one a tunability of 5 nm is reached through a 50°C temperature cycling. When combining two elements in parallel, a 34nm tunability is reached [1].
- Most of other approaches use the current injection to prevent any excessive lifetime device degradation due to the significant increase of operating temperature. The basic approach is represented by the DBR laser structure in which the Bragg section is separated from the

active section. The tunability is obtained by a current injection through the Bragg section, and a maximum tunability of 17 nm has been reached [2], (Figure 17). In this case a quasi-continuous tuning is obtained through a 80 mA current injection. Specific current injection values are required to reach both a given ITU wavelength and a maximum value of the spectral side mode suppression ratio (SMSR).

- An improved version of the DBR in term of tuning range is realised using specific grating structures on each side of the active structure, and extensive tuning range of more than 100 nm can be reach with a careful adjustment of two or three tuning currents [3,4] .

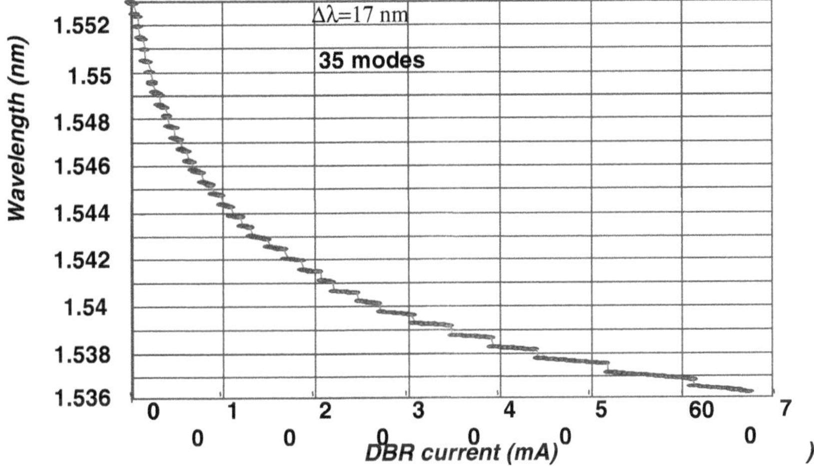

Figure 17 : Tunability range of a DBR laser.

The tunable sources present two main issues related to the emitted wavelength and the SMSR value. Both parameters are suspected to drift all along the operating time and in each case a simple process has to be found to reassess the right values. The most critical case occurs in the improved DBR versions [3,4] where no specific related function connects the set of emitted wavelengths and the set of tuning currents. On the other hand in the simple DBR approach, if any wavelength drift is detected, a single parameter (temperature or current) allows to control the ITU wavelength through a monotonic response.

2.2.5. Selectable Sources

As mentioned above, the selectable source structure is based on a laser array, each addressing a predetermined wavelength. No tuning is required, but, as the wavelength are preselected, the ITU compatible wavelength precision and stability are required.

Two main families can be distinguished : the integrated version and the hybrid one. In the first one, all the functions are integrated on a single InP chip (amplification, wavelength inscription, and optical combiner) resulting in very compact sources with a simple fiber-pigtail assembly.

A typical example of the integrated version is provided by a 6 DFB lasers combined in a single waveguide output and integrated with a modulator [5] with a very good wavelength accuracy of 1.8 Å with respect to an ITU grid (figure 18).

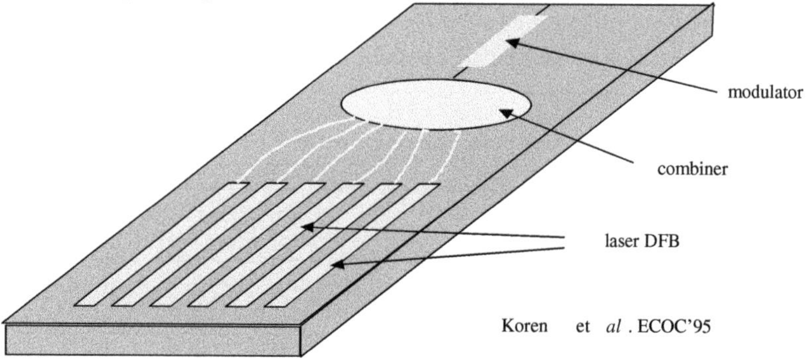

Figure 18 : Integrated DFBs.

However, the intrinsic effective index dispersion from laser to laser prevents any further wavelength accuracy improvement to reach a compatibility with the present very narrow channel spacing of 50 GHz (0.4 Å) of the ITU grid. The only solution to overcome these limits is to use a specific temperature tuning for each laser through a Peltier cooler control. In addition all combining function will degrade the output power of the selectable laser to almost one order of magnitude compared with a single DFB source. Integration of an additional amplification function can partly compensate this additional loss.

The hybrid version allows to separate the wavelength selection and the laser function. The efficiency of a such approach is to design the source so that the InP chip properties are wavelength transparent in a wide range (typ. 30 nm), and the specific wavelength selection is realised by UV Bragg grating inscription in a fiber or in Silica waveguide, on which the wavelength can be adjusted with a high precision by a post control process step.

In the figure 19, a laser stripe is integrated on InP in an integrated Mach-Zehnder modulator, and lasing effect is realised trough an on-chip mirror which have 30 nm bandwidth and external fiber or silica waveguide grating with an ITU pitch [6].

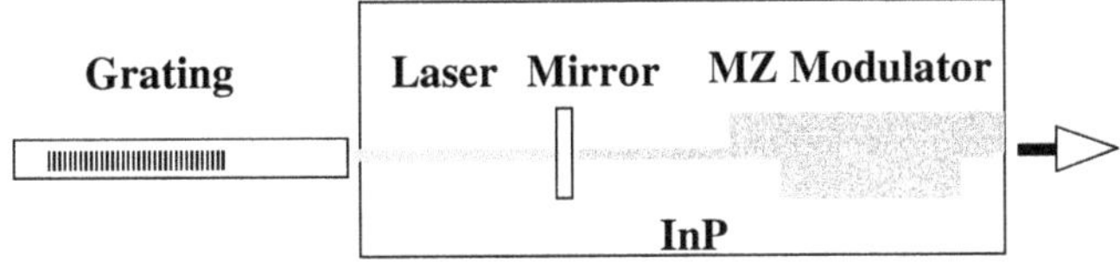

Figure 19 : Hybrid Version of a selectable source integrated with a modulator.

In this configuration, the same InP chip can be used to provide any ITU wavelengths defined by external grating. In addition, this DBR structure is a 100% single mode yield (lower, in the DFB case), a very critical value when a great number of lasers are put in parallel on the same chip. On the other hand, one of the issues of the hybrid approach is that the SMSR stability is dependent on the whole external cavity stability and on the residual parasitic optical feedback at the interface.

2.2.6. Conclusions

Though the research activity on these topics is 10 years old, a highly performance, reliable and easy-to-handle product is still to come. With the WDM network development, the wavelength spacing is decreasing quickly, reaching now 50 GHz, hereby requiring an increasing wavelength stability : most of the DFB WDM sources are now equipped with a wavelength locker.

If the hybrid selectable source appears simpler to master because of an in- plane function separation that could be independently optimised, the power output is limited to the 0dBm range, and, in addition the single mode stability of the hybrid structure is still an issue.

In parallel, the tunable source, more sophisticated, is a promising structure if the long-term InP parameters drifts can be controlled through feed-back loops.

2.3. Transverse Mode, Patterns and Polarization Behavior in VCSELs

J. G. McInerney
Optronics Ireland/Physics Department,
National University of Ireland, University College, Cork, Ireland
and Optical Sciences Center, University of Arizona, Tucson, AZ 85721, USA
E-mail : mcinerney@ucc.ie

Abstract

Vertical cavity surface-emitting semiconductor lasers provide solutions for many engineering applications and fundamental scientific investigations. Knowledge of the the transverse field and polarisation properties is often essential, and in many cases it is highly desirable to select a single predetermined transverse mode and polarization state. Here we review recent research in characterising, modelling and controlling transverse modes and polarisation effects in vertical cavity lasers.

2.3.1. Introduction

During the past decade, vertical cavity surface-emitting semiconductor lasers (VCSELs) provide solutions for a variety of engineering applications including optical data interconnects in free space and fiber arrays, laser printing, displays and sensor arrays, smart pixels and optical backplanes. They are also fascinating scientific vehicles for studying microcavity physics, nonlinear optics and spatio-temporal complexity at very large Fresnel numbers. For all these applications knowledge of the the transverse field and polarisation properties is essential, and in many cases it is highly desirable to select a single predetermined transverse mode and polarization state. In this paper we review recent research in characterising, modelling and controlling transverse modes and polarisation effects in VCSELs. Section 2.3.2 describes recent numerical modelling based on self-consistent solution of optical field, carrier and lattice temperature equations and results which indicate how to optimise structures and pumping for fundamental transverse mode operation. Section 2.3.3 describes experimental measurements of transverse mode spectra from various VCSEL types including single mode devices. Section 2.3.4 discusses transverse mode and polarization control. Section 2.3.5 describes injection locking measurements and modelling for mode selection, and Section 2.3.6 contains discussion and conclusions.

2.3.2. Theory of Transverse Mode Competition

2.3.2.1. Optical Electric Field

A general vector Maxwell theory of transverse modes in VCSELs will be complicated and is currently under development, but a simplified scalar theory provides physical intuition and appreciation of spectral and far-field data. Other authors have generated self-consistent carrier and field distributions [7] as well as detailed models for temperature distributions [8]. Here we analyse VCSEL modes including the light field, carriers and lattice temperature. An even more detailed approach, required for ultrafast dynamics and description of phenomena occurring over broad spectral bandwidths, would be to include the carrier plasma temperature, using many-body semiconductor physics [9] or a phenomenological description of carrier temperature [10]. Here we begin with the 3D Helmholtz equation for the scalar electric field components E

$$\nabla^2 E + k^2 E = 0 \tag{1}$$

which in cylindrical coordinates becomes

$$\frac{\partial^2 E}{\partial r^2} + \frac{1}{r}\frac{\partial E}{\partial r} + \frac{1}{r^2}\frac{\partial^2 E}{\partial \phi^2} + \frac{\partial^2 E}{\partial z^2} + k^2 E = 0 \tag{2}$$

with $k = k_0 n$ the propagation constant in the material and $n(r)$ the radially varying refractive index. Using the cylindrical symmetry to express these components as

$$E(r,\phi,z) = \psi(r)\exp(im\phi)\exp(i\beta_z z) \tag{3}$$

where m is an integer, we obtain the following equation for the radial field dependence:

$$\frac{d^2\psi}{dr^2} + \frac{1}{r}\frac{d\psi}{dr} + (q^2 - \frac{m^2}{r^2})\psi = 0 \tag{4}$$

The lateral wavenumber q is given by $q^2 = k^2 - \beta_z^2$. For each m this equation has a spectrum of eigenvalues q_{mn} determined by the boundary conditions on the fields as given by the index distribution $n(r)$. The boundary conditions will govern the existence of modes of even symmetry with $\partial\psi / \partial r\big|_{r=0} = 0$ and odd modes with $\psi(r=0) = 0$.

The longitudinal eigenvalues eigenvalues β_z are determined by the laser cavity modes, and the resultant wave propagation constant k should correspond to a frequency close to the gain peak. In the simplest approximation, applicable to air-post or oxide-apertured devices, $n(r)$ is a top-hat function with a small enough diameter and large enough steps that thermal and carrier effects can be neglected; then we write the usual Gauss-Laguerre modes and the problem is relatively simple. However, in large aperture devices (> 20 μm) and in gain-guided lasers of any size, fully self-consistent modelling between field, carriers and temperature is required. The background index n_r in the presence of gain guiding becomes

$$n = n_r + i(g / 2k_0) \tag{5}$$

with carrier- and temperature-induced changes

$$\Delta n_r(r) = \frac{\partial n_r}{\partial N}\Delta N(r) + \frac{\partial n_r}{\partial T}\Delta T(r) \tag{6}$$

with [11]

$$\frac{\partial n_r}{\partial N} = -e^2 / 2n_0\varepsilon_0\omega^2 m^* \text{ and } \frac{\partial n_r}{\partial T} = 4 \times 10^{-4}\ \mathrm{K}^{-1} \tag{7}$$

The distributions $\Delta N(r)$ and $\Delta T(r)$ are obtained from the carrier diffusion and thermal conduction equations respectively.

2.3.2.2. Carrier Diffusion

The radial diffusion equation assumes uniform carrier density along the z axis

$$D_n \frac{1}{r}\frac{\partial}{\partial r}\left(r\frac{\partial N(r)}{\partial r}\right) - \frac{N(r)}{\tau_s} - \frac{g(N(r))P_a\left|E_t(r)\right|^2}{\hbar\omega} + \frac{J(r)}{ed} = 0 \tag{8}$$

with the average optical power P_a and normalized intensity distribution $|E_t(r)|^2$ given by

$$P_a = \frac{1}{\pi s^2}\iint_{active}\frac{1}{2}cn_r\varepsilon_0\left|\psi_t(r)\right|^2 dA \tag{9}$$

and

$$\left|E_t(r)\right|^2 = \frac{\left|\psi_t(r)\right|^2}{\frac{1}{2s}\int_{-\infty}^{\infty}\left|\psi_t(r)\right|^2 dr} \tag{10}$$

with s the active region radius. The current spreads differently inside and outside the active area bounded by $r = s$:

$J(r) = J_0\exp[(r-s)/r_1], r \le s$

and

$$= J_0\exp[-(r-s)/r_2], r \ge s \tag{11}$$

2.3.2.3. Thermal Conduction

Heat flow from the top and sides of the laser into the air is neglected, hence we determine 2-D heat flow through the bottom metal contact into the heat sink which is at temperature T_0. The conduction equation may be written

$$\frac{1}{k_i}\frac{\partial T}{\partial t} = \frac{1}{r}\frac{\partial}{\partial r}\left(r\frac{\partial T}{\partial r}\right) + \frac{1}{\lambda_i}Q_i(r,t) \tag{12}$$

with boundary conditions $T(z=0)=T_0$ and $\partial T/\partial z\big|_{z=h}=\partial T/\partial r\big|_{r=0}=\partial T/\partial r\big|_{r=s}=0$. Q_i is the thermal energy density, λ_i the conductivity and k_i the diffusivity of the ith layer, and h is the height of the laser. Following [8], the solution of (12) is obtained using Green's function methods as

$$T(r,z,t)=T_0+\frac{4}{hs^2J_0^2(L_ns)}\sum_{m=1}^{\infty}\sum_{n=1}^{\infty}A_{mn}\sin(K_mz)J_0(L_nr) \qquad (13)$$

with

$$A_{mn}=\int_0^t\int_0^h\int_0^s\frac{k_i}{\lambda_i}\sin(K_mz)J_0(L_nr)rQ(r,z)\exp[k_i(K_m^2+L_n^2)(u-t)]dudrdz \qquad (14)$$

where the eigenvalues are $K_m=(2m-1)\pi/2h, m=1,2,3\ldots$ and $L_n=\mu_n/s, n=1,2,3\ldots$ J_0 and J_1 are the usual Bessel functions of the first kind, and μ_n are the roots of $J_1(L_ns)=0$. The energy densities $Qi(r,t)$ are due to Joule heating

$$Q_i(r)=J^2(r)\rho_i \qquad (15)$$

with additional nonradiative recombination and radiative transfer in the active region represented by

$$Q_a=\frac{V(r)(1-\eta_{sp}f_{sp})}{d}[J_{th}+(J(r)-J_{th})(1-\eta_{\text{int}})] \qquad (16)$$

where η_{sp} and η_{int} are the internal quantum efficiencies for spontaneous and stimulated emission respectively, $V(r)$ the junction voltage (determined as in [8]), f_{sp} the (geometrical) fraction of spontaneous emission escaping from the active region, and J_{th} the spatially averaged threshold current density

$$J_{th}=\frac{1}{\pi s^2}\int_{-\infty}^{\infty}d\theta\int_{-\infty}^{\infty}2\pi rJ_{th}(r)dr \qquad (17)$$

2.3.2.4. Numerical Analysis

Calculation of the field and carrier profiles was based on finite difference algorithms and the fields obtained self-consistently for the particular case where the slowly varying transverse lasing field comprises the zero- and first-order modes:

$$\psi_t(r)=\psi_0(r)+\psi_1(r) \qquad (18)$$

after which an initial temperature profile was found using (13). At first, near threshold, we set the average power $P_a=0$ and estimate the current density J_0, then iterate the following procedure: solve diffusion equation, find temperature profile, find index profile, solve field equation, check for self-consistency and whether the mode gain equals the loss. The parameters used in the calculations are summarised in table 1. In the following sections we describe the effects of three practical design parameters on the competition between the fundamental and first-order transverse modes, for gain-guided VCSELs.

2.3.2.5. Variation of Current Density

The near field intensity, temperature, carrier density and refractive index profiles for the gain-guided VCSEL in figure 20 are shown in figure 21 for injection currents I = 1.1, 1.3 and 1.5 I_{th}. The device design parameters are s = 5 μm (*ie* 10 μm aperture diameter), inside and outside current spreading parameters r_1 = 80 μm and r_2 = 10 μm respectively. The data show clearly the onset of the first-order mode due to spatial hole burning at 1.3 times threshold. The time required to establish transverse mode equilibrium is limited by the onset of thermal lensing: this time is estimated to be 0.5-2.5 μs depending on structural details and heat sinking. Paradoxically any attempts to reduce thermal dissipation in the VCSEL will prolong this time delay since thermal lensing will then occur mode gradually. These conclusions are in broad agreement with experiments [12-14]. Note that the conflict between the gain, carrier and temperature effects is initially resolved in favor of the carriers, resulting in a depressed refractive index on center, *ie* a defocusing profile. Only at higher current densities does thermal lensing dominate.

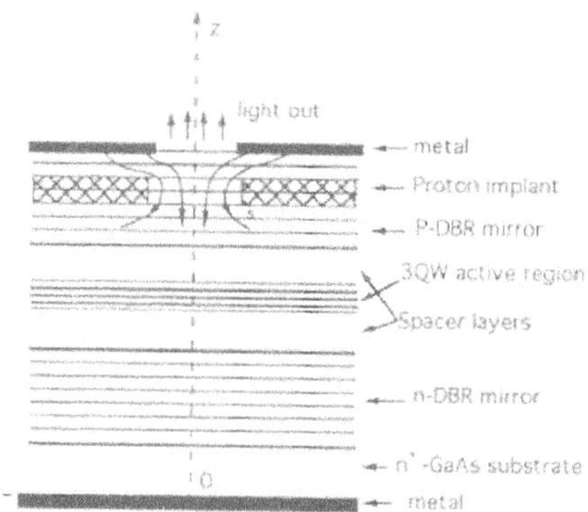

Figure 20 : Schematic of gain-guided VCSEL used in calculations and measurements.

TABLE I
PARAMETERS FOR THE STUDIED VCSEL'S

Parameter	Symbol	Value
Electron diffusion coefficient	D_n	50 cm^2/s
Carrier lifetime	τ_s	2×10^{-9} s
Gain coefficient	a	2.5×10^{-16} cm^2
Carrier concentration for transparency	N_{tr}	1.5×10^{18} cm^{-3}
Internal loss	α	46 cm^{-1}
Reflectivity of the end mirror	R	0.99
GaAs refractive index		3.655
AlAs refractive index		3.178
Effective mass of electron of GaAs	m_n	0.067 m_0
Laser frequency	ν	3.482×10^{14} 1/s

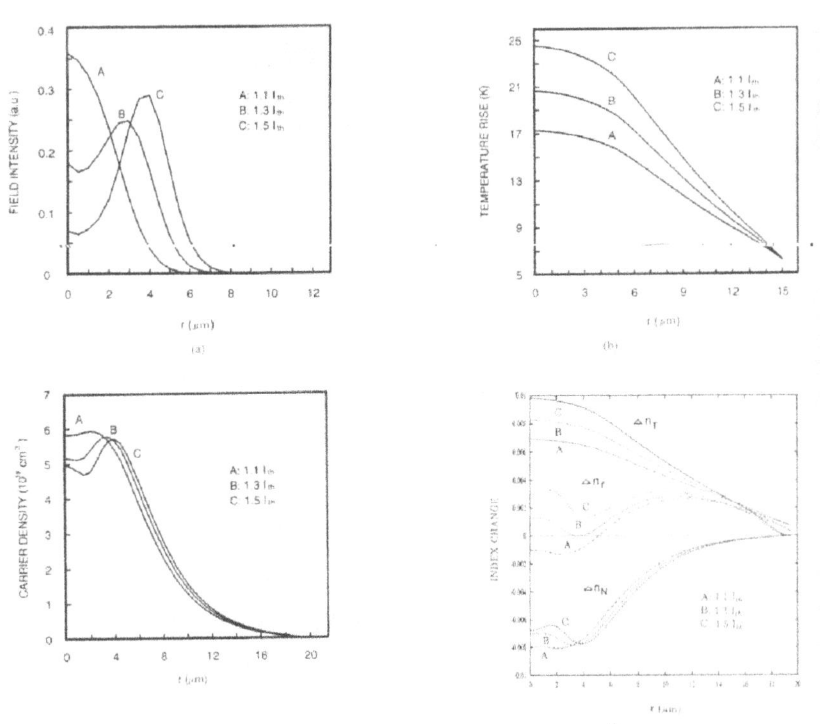

Figure 21 : Current density, temperature, carrier density, optical field and refractive index profiles for $I = 1.3\ I_{th}$, $s = 5\ \mu m$, $r_1 = 80\ \mu m$ and three different values of the external current spreading parameter r_2 = 2, 6 and 10 μm. In (d), Δn_T is the thermal contribution, Δn_N the carrier contribution and Δnr the net index change.

2.3.2.6. Influence of Current Spreading in Implanted Region

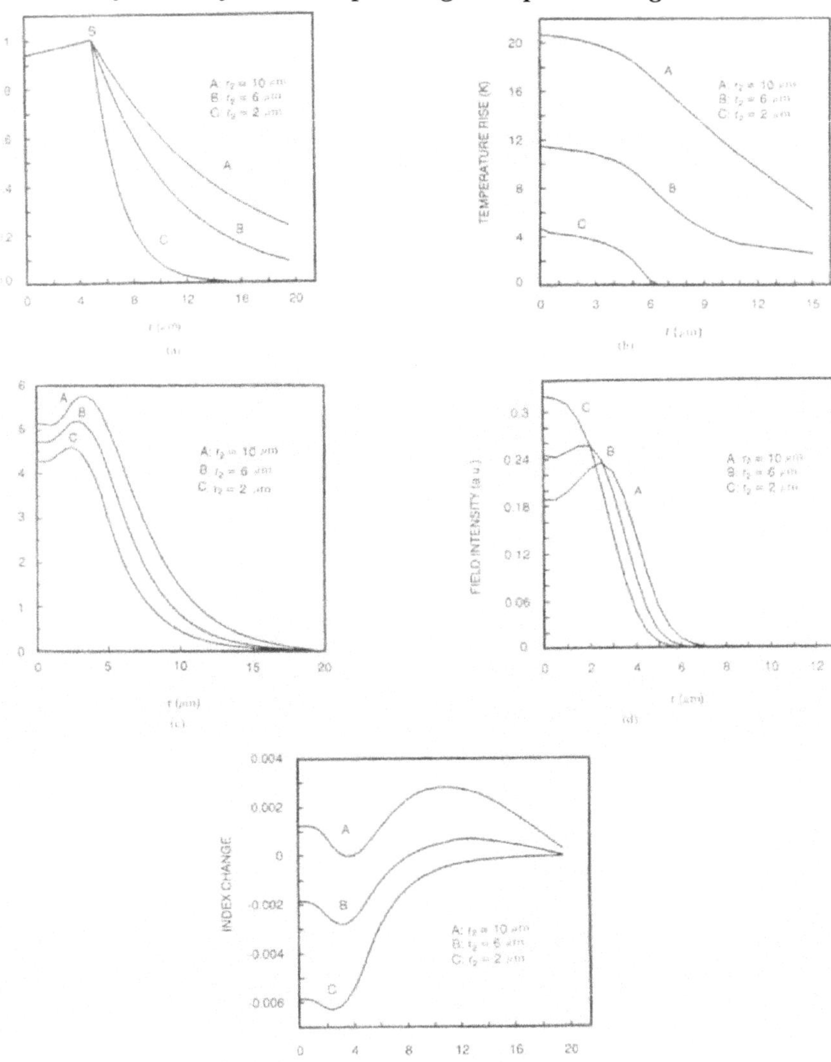

Figure 22 : Current, temperature, carrier density and index profiles for various degrees of external current spreading.

Figure 22 shows simulations of the current density, temperature, carrier density, optical field and refractive index profiles for $I = 1.3\ I_{th}$, $s = 5$ μm, $r_1 = 80$ μm and three different values of the external current spreading parameter $r_2 = 2$, 6 and 10 μm. Decreasing r_2 corresponds to increasing the resistivity of the implanted and annealed region surrounding the active area. If r_2 is too large then obviously carrier leakage becomes excessive. Smaller

r_2 gives less thermal dissipation and tends to promote fundamental transverse mode operation, giving a strong negative index change which is established slowly (several μs).

2.3.2.7. Effects of Aperture Size

Figure 23 shows the temperature, carrier density, index and intensity profiles for various values of s = 5, 7.5 and 10 μm, with different values of the internal current spreading parameter r_1 to offset changes in carrier density uniformity in the active region. Larger aperture devices have larger and wider temperature and carrier density profiles with correspondingly stronger spatial hole burning.

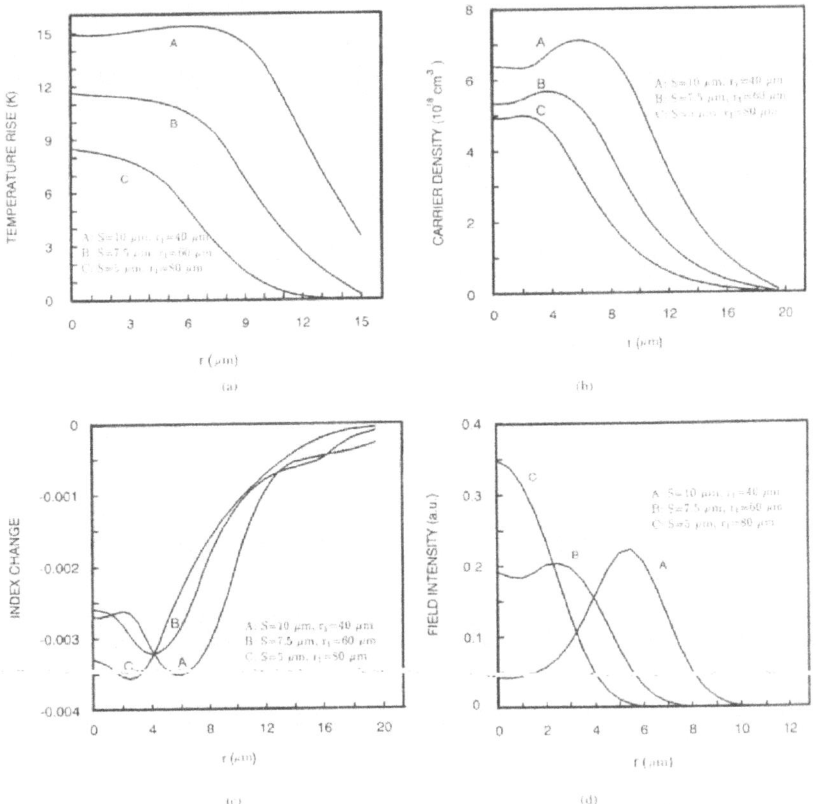

Figure 23 : Temperature, carrier, index and intensity profiles for I = 1.3 I_{th}, r_2 = *4 μm and various* r_2, *s*.

Negative lensing effects near the axis vary significantly with aperture size, so that efforts to control mode structure in gain-guided VCSELs by aperturing alone should meet with limited success. Due to thermal effects larger devices will tend to operate in high-order transverse modes above ~1.2 I_{th}, in agreement with experiments [14-16].

2.3.2.8. Theoretical Conclusions and Discussion

We have presented a detailed and self-consistent theoretical model for examining transverse mode competition in gain-guided VCSELs and have examined the straightforward case of the fundamental-first order mode competition near threshold. It is clear that the variations in index due to gain, carriers and temperature are complicated and difficult to control. Only when a strong built-in index step is included, as for air-post or oxide-apertured VCSELs, will large scale transverse mode stability be achieved. We note in passing that both air-post and gain-guided VCSELs suffer abnormally large mode losses when the device diameter is decreased significantly below 10 μm. Our conclusions point to the clear superiority of oxide confined devices for most applications and operating conditions.

2.3.3. Measured Transverse Mode Spectra

In this section we describe measurements of CW transverse mode spectra from gain-guided VCSELs operated well above threshold, showing high-order Gauss-Laguerre and Gauss-Hermite modes. This result is important in that, although Gauss-Laguerre modes are the eigenmodes of the empty cavity, it is not obvious that they should also be the natural basis in the presence of the nonlinear semiconductor gain medium. The devices [17] were proton-implanted, with a high-Q cavity formed by epitaxial DBR mirrors, the active region comprised four 10-nm GaAs quantum wells clad by AlGaAs spacers [18] and was designed for operation at 850 nm. They emitted through circular windows ranging from 10-25 μm in diameter.Because of their very short cavity length L = 1.8 μm, the VCSELs always operated in a single longitudinal mode, but their high nominal Fresnel numbers $F = s^2/\lambda L$ (from 50-200 for the samples tested) caused large numbers of transverse modes to be excited [19]. Only at very low currents, up to 1.3 times threshold, was the emission predominantly in the fundamental mode. Under these conditions the spectral linewidths were typically 50-60 MHz and relative intensity noise (RIN) was -140 to -150 dB/Hz. The experimental arrangement used for spectral and polarization measurements is shown in figure 24. A Si CCD camera was used to observe the near-field intensity profiles directly, via a polarizer and through a pair of etalons for spectral resolution. Another beam train went to an optical spectrum analyzer and scanning Fabry-Perot interferometer. The total output power and the polarization-resolved power were also measured.

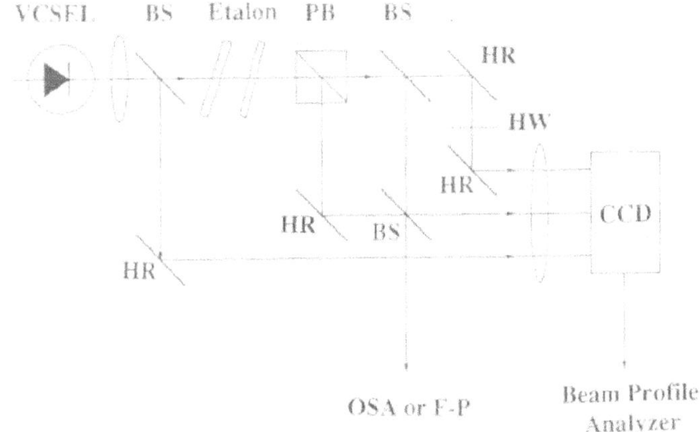

Figure 24 : Experimental arrangement for CW transverse mode and polarization studies.

2.3.3.1. Polarization Behavior

The 20 VCSELs tested tended to emit preferentially in two orthogonal directions, <011> and <011> (denoted in this paper as S and P respectively), consistent with other observations [20,21]. Two typical types of behavior were observed, as shown in figure 25: S- and P-polarized emission coexisted in most devices with comparable powers up to thermal shutdown at about five times threshold, with small anti-correlated changes in their relative powers as the current was varied (Fig 25(a)). In a sizeable minority of devices emission was initially polarized along one characteristic direction (in Fig. 25(d) the P-state) but switched suddenly and entirely with increasing current to the orthogonal state. Some devices produced slow self-pulsations at this polarization transition point. In general, the two orthogonal polarization states corresponded to different transverse patterns and emission frequencies [19]. This would result in mode beating effects at several GHz, causing difficulties in certain telecommunication or data interconnection applications. Quite apart from the problems they will cause in polarization-sensitive situations, it is clear that polarization instabilities can produce other deleterious effects including self-pulsations, LI kinks, mode partition fluctuations, excess RIN, modulation and beam pointing errors [21-27].

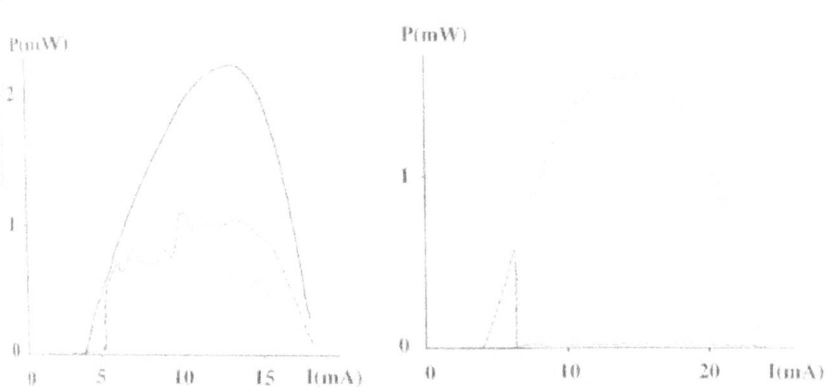

Figure 25 : Polarization resolved light-current characteristics for different VCSELs.

2.3.3.2. Transverse Mode Spectra

Measurements were performed on 15 and 25 μm window devices. Figure 26 depicts total frequency spectra of the smaller devices at different currents, showing the onset of modes up to 4th order. The transverse mode spacing shifts from 185 GHz at 10.3 mA to 240 GHz at 15.3 mA. Some modes had fine structure (see *eg* the first-order mode at 10.3 mA which has a doublet spaced by several GHz) due to polarization dispersion. These patterns were then imaged by the CCD camera and broken into individual mode images by inserting etalons in front of the CCD.

Figure 27 illustrates the near field images of the components of the spectra in figure 26. The first-order mode doublet appears to be the classic "donut" shape, made up of an equal mixture of TEM_{01} and TEM_{10}. Unlike most appearances of this object, neither bistability nor vortex-like behavior have been observed, presumably because of the frequency degeneracy. We note that this degeneracy can be removed by injection locking both components to an external reference [28] whereupon vortices can be observed.

The 25 μm aperture lasers tested had threshold currents of 10-12 mA. Figure 28 shows the P-polarized transverse mode spectrum from such a device at 42 mA (3.5 times threshold) indicating modes up to 10th order. Again we imaged the total and spectrally-resolved near fields for this spectrum; some of the results are shown in figure 29. Figure 30 shows the differences between 5th-order S- and P-polarized modes at 3.1 times threshold. Very high order modes were obtained at 4-5 times threshold, near thermal shutdown. Although the Fresnel number for these lasers is

>100 according to the simple formula $F = s^2 / \lambda L$, there we saw no modes of order higher than 14, probably because of thermal distortion of the wavefront which reduced the effective *F*. Interestingly, both Gauss-Laguerre and Gauss-Hermite modes coexist in these devices, the former because of the cavity geometry and the latter presumably because of the existence of linear strain patterns. Experiments carried out with equivalent optically pumped devices have shown mostly Gauss-Laguerre modes, as expected in the absence of strain induced by the fabrication of electrically pumped lasers.

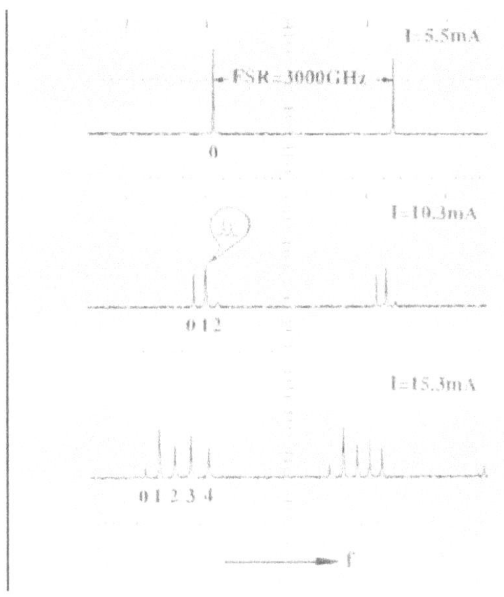

Figure 26 : frequency spectra of 15 mm VCSEL at various currents, showing transverse mode behavior.

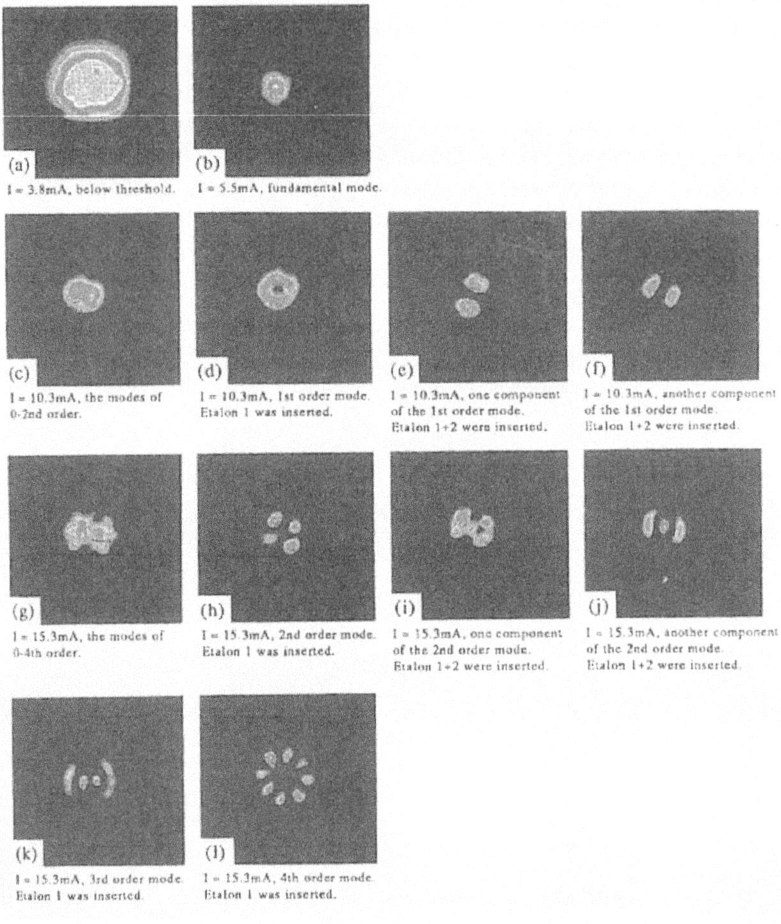

(a)
I = 3.8mA, below threshold.
(b)
I = 5.5mA, fundamental mode.
(c)
I = 10.3mA, the modes of 0-2nd order.
(d)
I = 10.3mA, 1st order mode. Etalon 1 was inserted.
(e)
I = 10.3mA, one component of the 1st order mode. Etalon 1+2 were inserted.
(f)
I = 10.3mA, another component of the 1st order mode. Etalon 1+2 were inserted.
(g)
I = 15.3mA, the modes of 0-4th order.
(h)
I = 15.3mA, 2nd order mode. Etalon 1 was inserted.
(i)
I = 15.3mA, one component of the 2nd order mode. Etalon 1+2 were inserted.
(j)
I = 15.3mA, another component of the 2nd order mode. Etalon 1+2 were inserted.
(k)
I = 15.3mA, 3rd order mode. Etalon 1 was inserted.
(l)
I = 15.3mA, 4th order mode. Etalon 1 was inserted.

Figure 27 : transverse mode patterns, P-polarization, 15 μm VCSEL.

FSR=780GHz
0 1 2 3 4 5 6 7 8 9 10
f

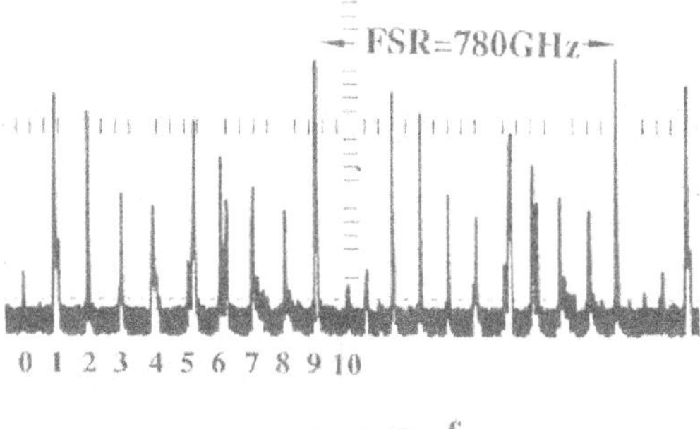

Figure 28 : P-polarized tranverse mode spectrum.

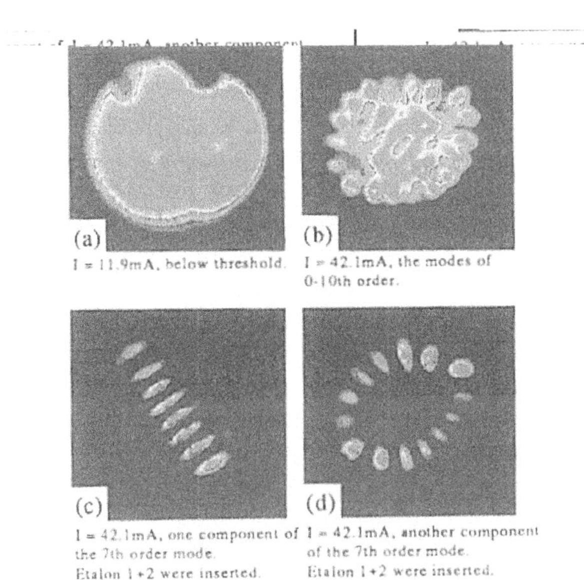

Figure 29 : Transverse patterns of P-polarized emission from 25 μm square VCSEL at various currents.

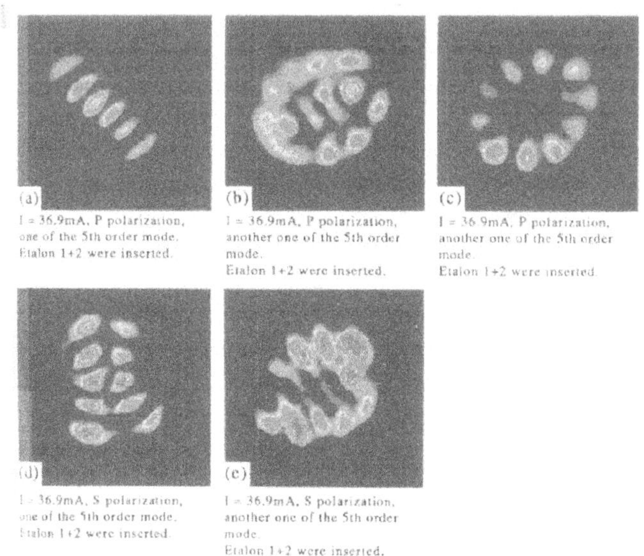

Figure 30 : S and P polarized transverse patterns under same driving conditions.

2.3.4. Control and Selection of Transverse Modes

2.3.4.1. Transverse Mode Control Strategies

Much recent work has been directed to the problem of selecting single transverse modes, preferably the fundamental one, in VCSELs. Introduction of apertures [18,29] is at best partially successful, while the use of strongly dispersive external cavities [30] is undesirable in practical applications. The use of passive antiguiding regions [21] is imaginative and promising but at the cost of significant fabrication difficulty including epitaxial regrowth. Such buried heterostructure type VCSELs may in future be implemented using techniques such as impurity-induced disordering [31] provided that carrier leakage problems can be overcome. Hybrid DBR designs incuding implantation, contact apertures and dielectric top mirrors [33] are also promising. Finally, we note that VCSELs less than 10 μm in diameter provide better single mode performance than larger devices, even though their threshold current densities are much larger - especially for air-post or gain-guided devices with peripheral optical losses and nonradiative recombination.

For practical applications it is best to consider exploiting the intrinsic dispersive nature of the VCSEL cavity [32]. It has already been found [34] that detuning of the cavity mode relative to the gain peak can influence the transverse mode spectra. We propose that the combination of a highly selective cavity - incorporating tapered oxide apertures to provide strong lateral field discrimination, with a narrow gain spectrum and the usual high-Q longitudinal structure - can constrain the eigenmodes so that only one mode can oscillate over a wide range. The cost would be sensitivity to temperature and process variations, but these tradeoffs are open to study. Such an approach has recently produced robust single mode operation [35] but the mode in question was of higher order and hence produced multiple off-axis lobes. It is clear that the presence of an oxide defined intracavity aperture does more than simply constraining the current: there are clear waveguiding effects which need to be exploited by placement and profiling of single or multiple apertures for optimum mode control . Another fertile area of investigation is to consider the use of nonlinearities such as self-phase modulation or saturable absorption (which tend to select patterns even in uniform unbounded media [36]) and propagation effects such as diffractive coupling or Talbot imaging in discriminating between transverse modes. There is still much fundamental investigation to be done, and for this purpose injection locking is a useful technique to examine the detailed physical conditions for achievement of single mode selection [19,28]: some typical results will be presented in the next section.

2.3.4.2. Polarization Dynamics and Control

Gain-guided VCSELs grown on <100> GaAs substrates tend to lase with preferential linear polarizations along the <011> and $\langle 01\bar{1}\rangle$ (or $\langle 1\bar{1}0\rangle$) directions [19-21], with significant variations occurring between individual devices. The degree of birefringence - and hence polarization selection - may be gauged by the frequency difference between these eigenmodes. In normal devices this is a few GHz. Index-guided devices show similar behavior [27]. The most successful techniques for control and selection of VCSEL polarization have been anisotropic pumping distributions as in non-circular current apertures [38-40] or non-circular air posts [41,42], application or variation of stress by global or local temperature changes or by making holes in the wafer adjacent to the VCSEL [43-47], producing anisotropic gain or loss by forming metal gratings on the top layer [48] or by growth on misoriented substrates [49-53], and using the frequency difference between polarizations to discriminate using an external cavity [54]. Polarization dynamics including switching and bistability have also been observed [40,55-57], with characteristic speeds limited to ~100 MHz by thermal or electrical parasitics. These strategies can produce stronger polarization birefringence, ~10 GHz. Although some devices may show polarization selection ratios in excess of 20 dB when operated CW, under strong modulation (modulation depth close to unity) this ratio can reduce to as low as 2-3 dB. Devices grown on misoriented substrates or with non-circular cavities tend to have 15-20 dB polarization ratios even under large signal modulation, an important consideration in systems applications.

The theoretical picture of polarization selection is not quite as advanced, mostly because of lack of a complete vector Maxwell treatment and the fact that the experiments have relied heavily on non-ideal factors such as stress-induced birefringence. Current theoretical pictures have treated the natural circular birefringence of the semiconductor band structure and elasto-optic effects in real devices [58-62]. Closer interaction between theory and experiments will be crucial.

2.3.5. Summary and Conclusions

For many scientific and engineering applications of vertical cavity lasers control of the transverse field and polarization properties is essential, and in many cases it is highly desirable to select a single predetermined transverse mode and polarization state. Recent approaches to transverse mode selection include careful aperturing, external cavities, and passive antiguiding regions. Promising avenues include buried heterostructures, hybrid DBR designs and exploiting transverse and longitudinal cavity tuning effects. We have described recent theoretical modeling based on self-consistent computation of the optical field, carriers, refractive index and temperature. Experimental data on transverse mode imaging and their stabilisation by injection locking have also been described.

Conventional VCSEL outputs tend to be polarized preferentially along certain orthogonal directions, with slight frequency differences between these eigenmodes. Recent approaches to polarization selection include anisotropic pumping and cavity geometries in non-circular apertures or waveguides, application of stress, modifying the wafer surface, defining gratings on the top layer, external cavity dispersion and growth on misoriented substrates. The latter is probably the most promising approach. Phase-coupled or injection-locked arrays may also stabilize the polarization states of individual devices, although the collective dynamics of large arrays may be complicated. Theoretical models of VCSEL polarization dynamics are in their infancy and are likely to improve substantially in close connection with experiments. Much work remains to be done.

2.4. Mode Locked Microchip Lasers for the Generation of Low Noise Millimeter Wave Carriers

P. R. Herczfeld
Center for Microwave-Lightwave Engineering,
Drexel University Philadelphia, PA, USA
E-mail : pherczfe@ece.Drexel.edu

Abstract

This presentation is concerned with the generation of high fidelity microwave and millimeter wave signals in solid state lasers, and their utilization in optically fed wireless systems.

2.4.1. Introduction

The demand for broad band wireless services implies the use of higher and higher frequency bands. The future trend is to apply a millimeter-wave carrier frequency for wireless access networks. The optical distribution of the millimeter wave carrier, concurrently with the data signals, to the nodes of a cellular system is a favored solution. This requires high speed fiberoptic networks operating in the microwave and millimeter wave regions with good noise figure and high dynamic range. Key to the design and implementation of proficient millimeter wave over fiber networks are: high performance optical transmitters at the central stations, low cost base stations and a practical network topology. The distribution network considered here, shown in figure 31, comprises of a central station that is connected to dispersed microcells by a star network. Within each microcell there are several picocells which are linked to the central station by an open ring fiberoptic network. To avoid interference, the i-th picocell has its own distinct millimeter wave carrier, f_{pi}, carrying several channels with a total bandwidth of B_{pi}.

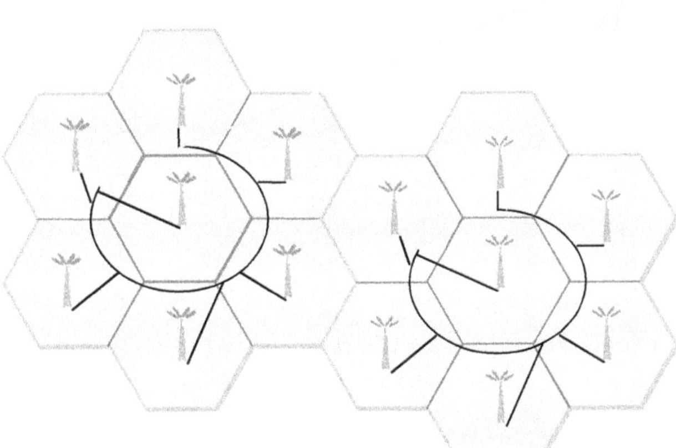

Each microcell is divided into several picocells with individual millimeter wave carrier, f_{pi} . The central station will generate the optical and the millimeter wave carriers.

Figure 31 : Network topology for optically fed millimeter wave wireless communications.

The functional system is depicted in figure 32. The most critical component is the high performance optical transmitter with microwave/millimeter wave carrier.

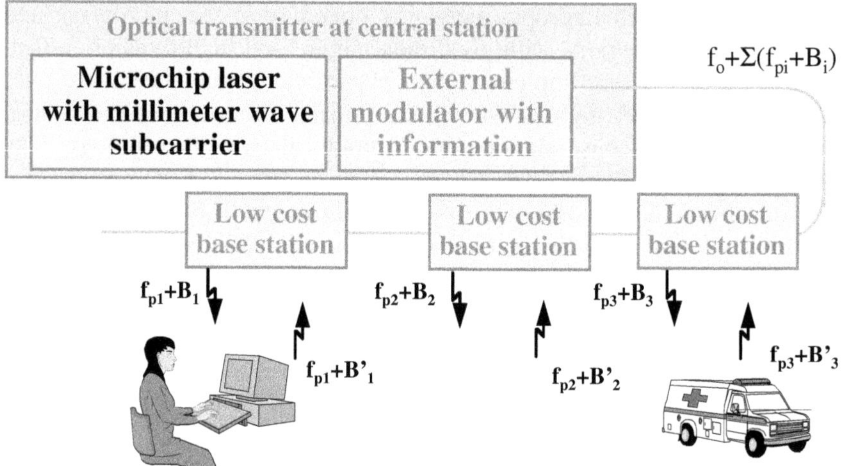

Figure 32 : Fiberoptic access to wireless communications. In this figure f_o is the optical carrier, f_{pi} is the millimeter wave carrier of the i-th picocell and B_i is its information bandwidth.

2.4.2. Optical Transmitter

The optical transmitter, shown in more detail in figure 33, consists of two part: the microchip laser and the modulator. The mode-locked microchip laser generates the optical carrier (f_o) as well as the millimeter wave carrier (f_m). The subcarrier for the i-th picocell (f_{si}) and the information signal (B_i) is superimposed by an external modulator. Note, for the i-th picocell: $f_{pi}=f_m+f_{si}$. The microchip laser, the focus of the research, must provide for the subcarrier with high modulation index and low amplitude and phase noise.

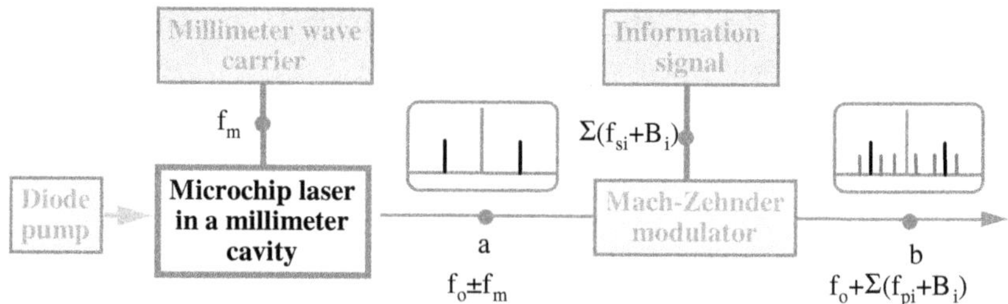

Figure 33 : Block diagram of the optical transmitter.

The principal ideas regarding the design were: i. employ a low noise, efficient, diode pumped solid state laser as source, ii. use active mode-locking to generate the microwave or millimeter wave carrier, and iii. fully integrate the laser with the microwave subsystem.

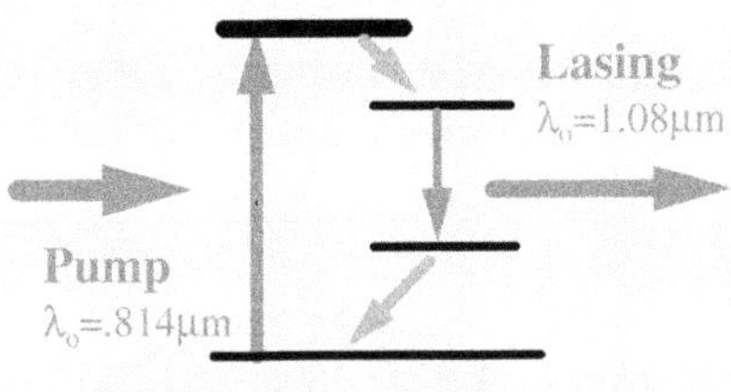

- The host, $LiNbO_3$, is an excellent electrooptic material that can be easily modulated
- The Nd doping provides for efficient lasing

- The laser is pumped by a diode at 814 mm
- The gain profile is centered around 1,08 µm and has a width of approximately 120 GHzt

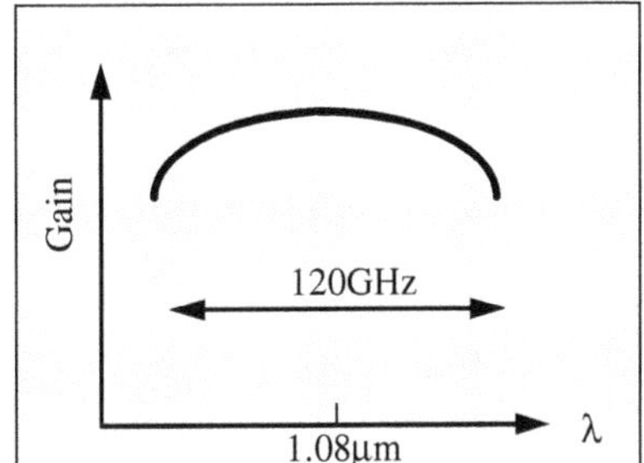

Figure 34 : Lasing in Nd:LiNbO₃.

To accomplish this a $Nd:LiNbO_3$ crystal, with mirrors deposited at the ends, served as the gain medium (optical source), the millimeter wave subcarrier was produced by mode-locking, and the laser was embedded into a microwave reentry cavity for the injection locking process. Figure 34, depicts the relevant electronic transitions and gain bandwidth of the laser. The length of the crystal (i.e. the round trip time) determines the free spectral range or mode structure. For this experiment the mode separations was 20GHz. To obtain a clean, low noise signal, the modes must be locked by an applied field (see figure 36). To lock the modes the laser was placed in the high field region of a cavity, as shown in figure 37.

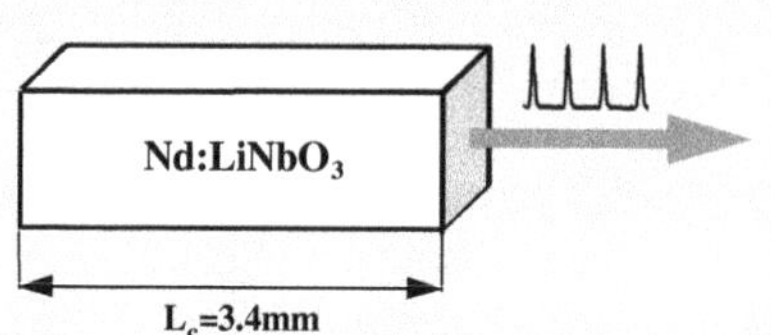

- The lenght of the optical cavity, L_c, determines the roundtrip time, T_{rt}= $2nL_c/c=f_m$
- The roundtrip time defines the free spectral range, or mode spacing

- For the present design L_c=3.4mm, which yields a roundtrip time of 50ps and a mode spacing of 20 GHz
- There are six longitudinal modes produced
- To generate a low noise output the modes must be locked

Gain

1.08µm λ

Figure 35 : Mode structure in Nd:LiNbO$_3$ laser.

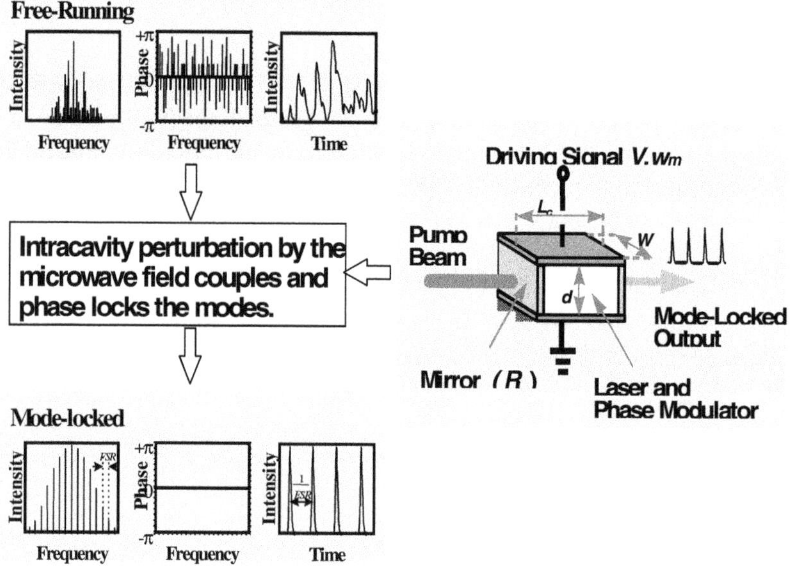

Figure 36 : Mode-locking concept.

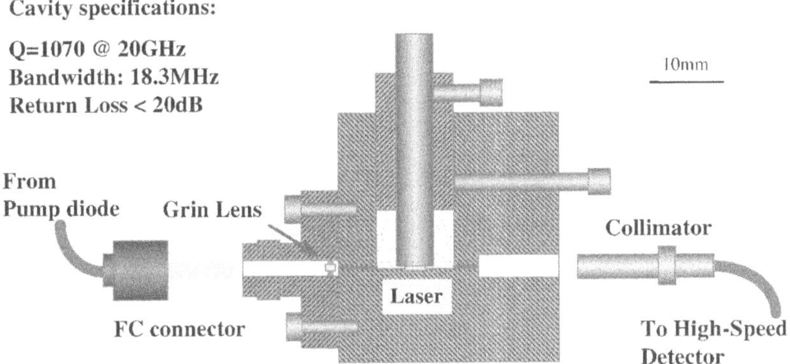

Figure 37 : Reentrant microwave cavity.

On the next few figures we describe the performance of the microchip laser. The experimental setup for the time, frequency and optical domain characterization of the device is shown in figure 38.

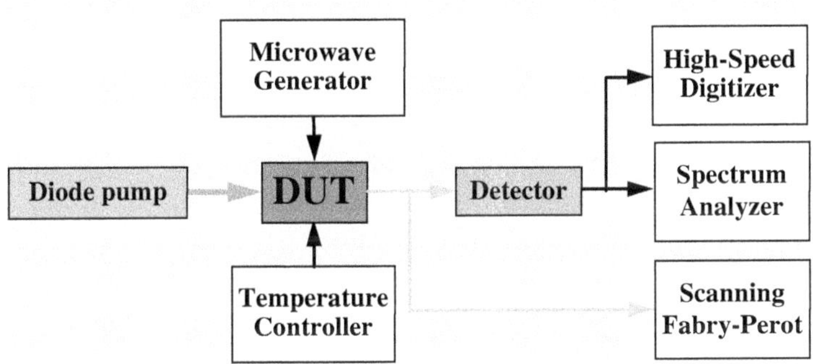

Figure 38 : Experimental setup.

The microwave domain characterization of the mode locked laser, namely the quality of the millimeter wave signal is shown in figure 39. The most important result, the measured phase noise is depicted in figure 40.

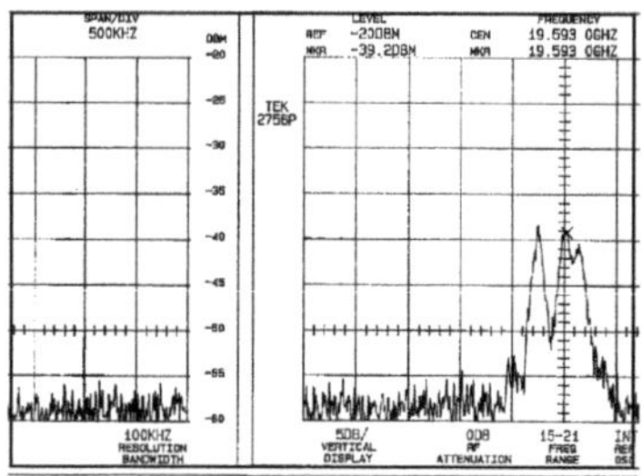

Free-running, no microwave input, self mode-locking
Modulation index: 24%
Spectrum unstable

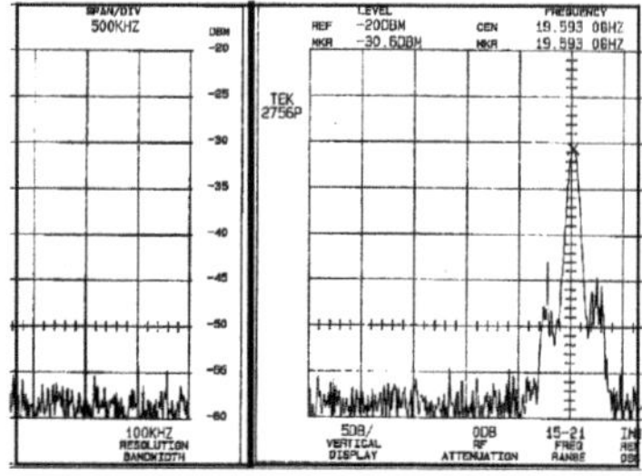

12.6 dBm microwave input
Modulation index: 95.6%
Spectrum very stable
Microwave bandwidth: 84kHz

Figure 39 : The free running and mode locked laser output.

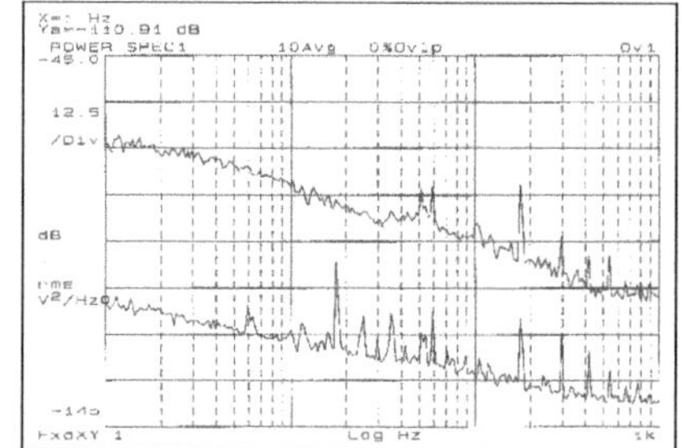

<u>Residual SSB phase noise:</u>
Lower trace: background
Upper trace: device
-110dBc/Hz @ 1KHz offset

<u>Amplitude noise (not shown):</u>
RIN @ 20GHz: <-150dB/Hz

Figure 40 : The phase noise of the mode locked laser. The lower trace represents the residual (system) noise and the upper trace is the laser noise.

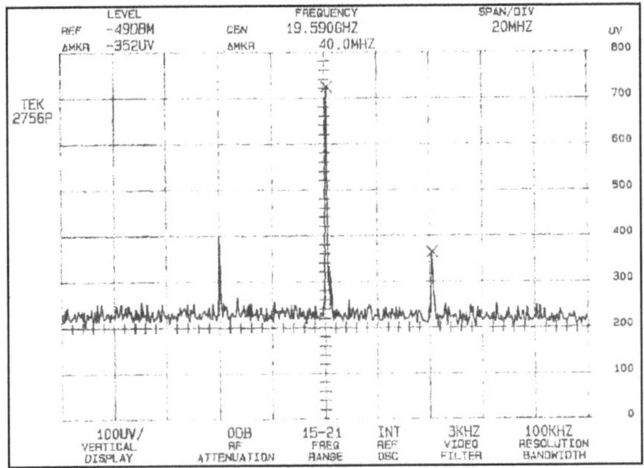

Figure 41 : 20 GHz subcarrier with 40MHz side band on the optical carrier.

Table 2 summarizes the performance of the prototype optical transmitter.

	Measured
Material	**Nd:LiNbO$_3$**
Size (wxhxl) in mm	**1x1x3.4**
Lasing wavelength	**1.08µm**
Linewidth (opt.)	**<25KHz**
Efficiency	**18%**
Millimeter wave freq.	**20 & 40GHz**
Lock. power (min/max)	**6/12dBm**
Bandwidth (microw.)	**~80KHz**
Modulation index	**97%**
RIN	**<-150dB/Hz**
Phase noise @1KHz	**<-110dBc/Hz**
Temp. stability	**.01°C**
Output power*	**35mW**

*limited by pumping power

Table 2 : Summary of transmitter performance.

2.4.3. Base Station

Next we consider the base station. The incoming optical signal is detected by an optical sensor. We prefer an HBT detector because it can provide for high frequency detection with gain, and it can be integrated with other MMIC components. The detected signal is split by a branch coupler. In the transmit arm a filter selects the appropriate millimeter wave signal, f_{pj} and it's information bandwidth, B_j, for this (j-th) picocell. The signal is amplified and transmitted. There is no need for the generation of the millimeter wave carrier. On the receiver side the detected signal is amplified, and down converted. The original signal from the central station is filtered and amplified to provide the necessary millimeter wave signal for the down conversion. A inexpensive laser operating at lower frequencies is used for the upstream signal transmission. The base station, as seen, can use simple, low cost MMIC circuitry.

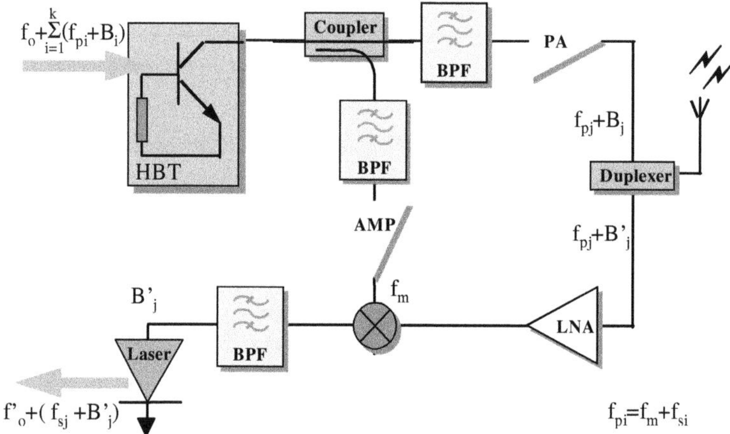

Figure 42 : Base station with HBT optical detector, MMIC components and low cost laser.

2.4.4. Summary of Results

The simultaneous generation of an optical carrier and a microwave/mm-wave subcarrier by a compact mode-locked microchip laser was demonstrated. Specifically, a clean, low noise millimeter wave signal was generated. High quality optical transmitter permitted the design of a low cost, MMIC based base station.

2.4.5. Future Efforts

Current efforts include the design and fabrication of an Erbium doped $LiNbO_3$ laser @ 1.55μm (see table 3). We are also working on a microchip amplifier and a Fabry-Perot modulator. Finally, the microchip laser in combination with an optical filter can be used as an optical domain microwave to millimeter wave multiplier. The concept is depicted in figure 43. Mode locking the laser at 20 GHz and then suppressing the internal modes frequency multiplication can be achieved.

	Measured	New design
Material	**Nd:LiNbO$_3$**	**Er:LiNbO$_3$**
Size (wxhxl) in mm	**1x1x3.4**	**.5x.5x3.4**
Lasing wavelength	**1.08μm**	**1.55μm**
Linewidth (opt.)	**<25KHz**	**<25KHz**
Efficiency	**18%**	**>25%**
Millimeter wave freq	**20 & 40GHz**	**20 & 40GHz**
Lock. power (min/max)	**6/12dBm**	**3/6dBm**
Bandwidth (mw.)	**~80KHz**	**~80KHz**
Modulation index	**97%**	**97%**
RIN	**<-150dB/Hz**	**<-150dB/Hz**
Phase noise @1KHz	**<-110dBc/Hz**	**<-110dBc/Hz**
Temp. stability	**.01°C**	**.01°C**
Output power	**35mW**	**>250mW**

Table 3 : Actual performance of Nd:liNbO$_3$ laser and predicted performance of Er:LiNbO$_3$ laser under test.

On the long term we envision a compact optical transmitter, as shown in figure 44. It consists of a microchip laser, as described above. It will also contain a solid state optical amplifier, a filter and a Fabry-Perot modulator. The amplifier and the laser is optically pumped. The function of the filter is to reduce the number of modes and thereby provide for multiplication. Fewer modes also reduce the potential of chromatic dispersion. The Fabry-Perot modulator, using Nd or Er doped LiNbO$_3$, can be optically pumped to produce gain, which increases the finesse. This implies system gain and improved linearity. It is significant to point out that there is no dc bias required, and only to microwave inputs are needed; for the mode locking and for the modulation.

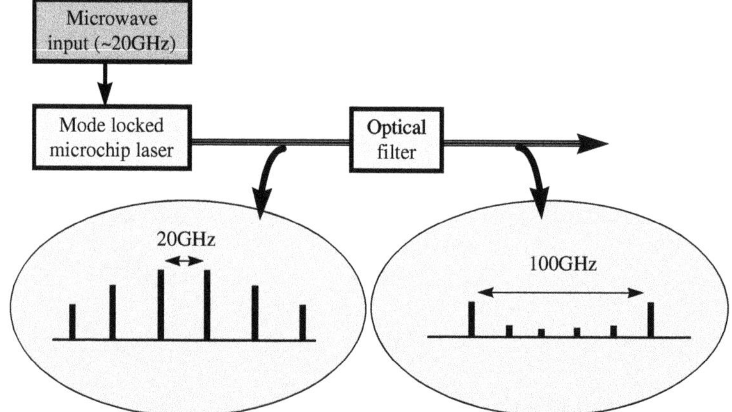

Figure 43 : Optical domain millimeter wave multiplier.

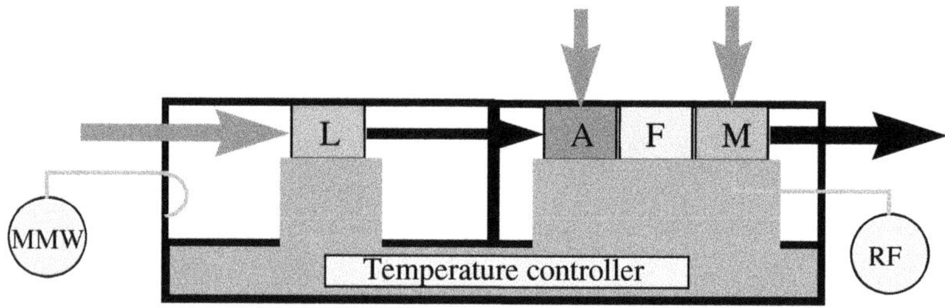

	Laser	Amplifier	Filter	Modulator
MMW (20-100GHz)	<10dBm	—	—	—
RF (0.1 - 1Ghz)	—	—	—	20dBm
Pump @ 814nm	150mW	100mW	—	100mW
DC (tuning)	—	20V	—	20V
Temperature control	.01°C	.01°C	.1°C	?
Size mm^3	.2 - .8	.2 - .8	.1 - .2	.2 - .8

Figure 44 : Compact optical transmitter configuration.

2.4.6. Conclusion

The optical generation and transmission of high fidelity millimeter wave signals over fiber is promising. New applications in communications, radar and remote sensing using this technique are expected.

2.4.7. Acknowledgement

I wish to acknowledge Dr.s Amarildo Vieira and Tibor Berceli for their valuable input to this work.

3. SEMICONDUCTORS OPTICAL AMPLIFIERS

J.C. Simon
France Telecom BD/CNET DTD/RTO
Technopole Anticipa, 2 av. Pierre Marzin
F22307 Lannion, France
E-mail : jc.simon@cnet.francetelecom.fr

3.1. Introduction

Semiconductor Optical Amplifiers (SOA) are now commercially available, with rather attractive features in both 1300 and 1550 nm wavelength windows: fiber-to-fiber gains ranging from 25 to 30 dB, polarisation sensitivity below 1 dB, saturation output powers up to 13 dBm, noise figures around 6-7 dB and a large optical bandwidth of 50 nm. In addition, they could be integrated on small chips with many other InP based components, and fabrication technology is quite compatible with mass production, a key step towards low prices. So, why can't we find any of these devices in optical networks ?

There are several reasons for this situation. One of them can be found if we briefly turn back to the late eighties, when the Erbium doped fiber amplifier (EDFA) suddenly came into the optical amplifier R&D field, with its totally polarisation insensitive 35 dB gain, and unmeasurable non-linear distorsion levels for signal bandwidths greater than a few MHz. At that time, the SOA was yielding internal gains of about 25 dB, with fiber insertion losses around 10 dB, polarisation sensitivities greater than 5 dB, and such high levels of non-linear distorsions that it was not thinkable of using this device elsewhere than in a lab. It was thus clear that R&D effort would not be as huge as in the most promising field of EDFAs, and that it would be a difficult challenge to solve all these problems. But in spite of these hard conditions, considerable progress has been achieved during the last few years. We will now review some basic features of these devices, such as gain, polarisation sensitivity, noise figure, non-linear distorsions and finally discuss some promising applications in all optical signal processing.

3.2. Gain and Polarisation Sensitivity

Regarding basic principles, the structure of an SOA is quite similar to that of a laser diode : differential gain coefficient, current and photon confinement, Auger recombination, etc... Early devices were simply commercially available diodes with additional anti-reflection coatings on facets. Very quickly, researchers realized that this could not work : the device behaved like an active Fabry-Perot cavity, which cavity gain is given by :

$$G_c = \frac{T_1 T_2 G_i}{(1-r)^2 + 4r\sin^2\frac{\varphi}{2}},$$

with :

$$r = \sqrt{R_1 R_2} G_i, \qquad G_i = \exp\left[\Gamma\left(g - \alpha_{IVBA}\right) - \alpha_s\right] L,$$

$$T_{1,2} = 1 - R_{1,2}, \qquad \varphi = 2\beta_0 L$$

where R_1,R_2 are modal reflectivities of coated facets, r is a "resonance parameter", equal to the internal gain G_i times the average facet reflectivity, T_1 and T_2 the facet transmission coefficients, φ is the round trip phase shift, Γ is the filling factor, α_{IVBA} is the intervalence band absorption coefficient, α_s is the scattering losses coefficient. The cavity gain spectrum thus showed a wavelength dependent « ripple » defined as :

$$R = \left(\frac{1+r}{1-r}\right)^2$$

With an average reflectivity ten times smaller than G_i^{-1} , the gain ripple is almost 2 dB high, which is not acceptable for most of system applications. For a typical internal gain of 30 dB, the average reflectivity should be less than 10^{-5}, which cannot be reproducibly obtained without a combination of tilted waveguide axis (7 to 10 degrees) and antireflection coatings.

Another very critical point of early devices was polarisation sensitivity of gain, which arises from the active waveguide geometry which was a very flat stripe, in which the confinement factor of TM mode is significantly smaller than that of TE mode. Also, effective refractive indices are different, which adds an additional contribution to the polarisation sensitivity, when ripple is not negligible, because the TE and

TM ripples are not in-phase. Neglecting this last contribution, the gain difference is given (in dB) by :

$$\Delta G_i = \left(1 - \frac{\Gamma_{TM}}{\Gamma_{TE}}\right)\left(G_{TE} + 4.34\alpha_s L\right)$$

For an internal gain of 30 dB, the filling factor difference should be kept below 3 percent for a 1 dB gain difference.

In order to improve this point, different approaches have been investigated. One technique consisted of playing with the shape of the waveguide: by making an almost square shape buried active layer, polarisation sensitivities below 1 dB for chip gains greater than 30 dB were obtained. Also, with appropriately designed ridge waveguide structures, similar results were achieved. However, reproducibility seems to be a problem, because of very tight tolerances on stripe dimensions. Presently, a quite different but promising approach consists of inducing some amount of strain in the active material, in order to create a material gain birefringence which compensates for the difference in confinement factors. Various Multiple Quantum Well structures with the right mixture of tensile and compressive strain in wells and/or barriers has led to polarisation sensitivities of about 1 dB in the 1500 nm window, and to a record result of 0.3 dB in the 1300 nm window. Very recently, the concept of a small tensile strain in a bulk material active layer has been reported at OAA'96, allowing process tolerant fabrication of high gain amplifiers with polarisation sensitivities below 1 dB. However, despite this impressive progress, these figures are not yet challenging the EDFA polarisation « insensitivity ».

Insertion losses have considerably been improved. In the past, very low coupling losses to fibers could be achieved in the lab with high precision translation stages, but it was another story to keep good results for packaged devices. The basic reason was the very small size of the amplifier guided mode (about ten times smaller than that of the fiber), which implied unmanageable positioning tolerances. The only way to overcome this difficulty was to increase the spot size of the amplifier mode. This is now obtained with integrated adiabatic mode size converters of different kinds. Typical insertion losses for packaged devices have now dropped to about 2-3 dB per facet, and they should still decrease. As a matter of fact, the main motivation for lower coupling losses is to reduce noise figure, and to increase the output power.

3.3.2. Noise Figure

In an optical amplifier, noise arises from amplified spontaneous emission (ASE). If a photodetector is placed at the amplifier output, mixing of amplified signal and ASE gives rise to noise beats in excess of shot noise terms. It has now become a common way to define an « electrical » noise figure $NF_{electr.}$ as the ratio of input to output SNR, yielding :

$$NF_{electr} = \frac{1}{\eta G} + 2n_{sp}\frac{G_i - 1}{G_i}$$

where n_{sp} is the population inversion parameter including non resonant loss contributions, and η is detector quantum efficiency, and all other noise contributions except for signal-spontaneous emission beat noise have been neglected.

Although commonly used, this noise figure definition is not quite satisfactory, as it does not depend *intrinsically* on the amplifier characteristics alone. This is why people prefer now to use an « optical » noise figure definition, in analogy with the one used by the « microwave » community, and which consist of *neglecting any beat noise contribution* : only average signal and ASE powers are considered, and the quantum noise contribution. This « optical » noise figure is given by :

$$NF_{optic} = \frac{1}{G} + n_{sp}\frac{G_i - 1}{G_i}$$

We notice that when the gain is very high, the optical noise figure is equal to n_{sp}, while it is equal to $2n_{sp}$ in the electrical definition. The factor of 2 difference simply arises from the fact that beat noise terms are neglected in the optical definition.

An important point not to be neglected in the noise figure, is the contribution of coupling losses of the incoming signal to the amplifier. For input and output coupling losses C_1, C_2, respectively, the noise figure writes :

$$NF_{optic} = \frac{1}{C_1 G C_2} + \frac{n_{sp}}{C_1}\frac{G_i - 1}{G_i}$$

Now, let's focus on the specific case of a SOA, for which the optical noise figure is given by :

$$N_{sp} = K'\frac{1}{1 - \exp\frac{E - \Delta F}{k\Theta}} \times \frac{1}{1 - \frac{\alpha_s}{g_{em}}} \times \xi\frac{G_i - 1}{G_i} = K' n_{sp} \rho_{nr} \xi \frac{G_i - 1}{G_i}$$

with : $\xi = \dfrac{1 + R_1 G_i}{1 - R_1}$

-ΔF is the electron-hole quasi-Fermi level difference.

-α_s is the non-resonant losses attenuation coefficient (scattering, intervalence band absorption, etc...)

-R_1 is input facet reflectivity

- The first factor (K') only occurs for gain-guiding amplifiers: it is linked to the Peterman *K factor*. However in most of index guided state-of-the-art amplifiers, K'=1.
- The second factor is the population inversion parameter n_{sp}, which is simply the ratio of spontaneous emission transition rate per unity frequency to the stimulated emission rate. As the quasi-Fermi level difference increases with carrier density, best results are obtained with devices operating at high carrier densities.
- The third factor (ρ_{nr}), arises from non-resonant losses (i.e. losses due to other contributions than stimulated absorption between lower and upper laser levels). This contribution can be as high as 1.5 to 2 dB when intervalence-band absorption (IVBA) is significant. Strained multiquantum well materials seem to be attractive owing to a lower IVBA.
- The last factor ξ arises from increased ASE contribution due to backward ASE which reflects on the input facet and then adds to co-propagative ASE. $R_1 G_i << 1$ is a prerequisite for low noise amplifier.

3.4. Non-linear Distorsions

The non-linear regime appears when the stimulated emission rate induced by the signal instantaneous intensity dominates the spontaneous emission rate. As it can be seen from the set of equations below, describing signal propagation through an SOA, the carrier density, and thus the gain, depend on the signal intensity :

$$\frac{\partial I}{\partial \tau} = (\Gamma g - \alpha) I$$

$$\frac{\partial \Phi}{\partial z} = -\frac{1}{2}\Gamma g \alpha_H$$

$$\frac{\partial n}{\partial \tau} = \frac{J}{ed} - \frac{n}{\tau_c} - g\frac{I}{h\nu}$$

$$g = a(n - n_0)$$

where :

$\tau = t - \frac{z}{v_g}$ pulse reference frame time

α_H phase - amplitude coupling factor

J, g current density and gain coefficient

τ_c carrier lifetime

a differential gain coefficient

$I_{sat} = \frac{h\nu}{a\tau_c}$ saturation Intensity

This non-linear regime is characterized by the saturation intensity I_{sat}. If the signal intensity is constant, as with FSK or PSK modulation formats, the saturation regime does not significantly disturb the amplified signal, since the only consequence is a constant gain compression. But if the signal intensity is not constant, as with an AM modulated signal, or if several wavelength multiplexed channels are amplified, then more or less severe signal distorsions can happen. Actually, two cases have to be considered :

1) If the signal power evolves « slowly », i.e. on a time scale much longer than the carrier lifetime, then the population inversion instantaneously follows signal variations, and so does the gain : strong distorsions occur (harmonic distorsions, intermodulation products in multichannel systems, etc..).
2) If the signal intensity evolves « rapidly » , then the population inversion cannot follow signal variations : there are thus negligible distorsions for signal frequencies higher than the reciprocal of the carrier lifetime. Actually, the SOA (or any optical amplifier) behaves much better for very high signal frequencies. Unfortunately, as long as linear amplification is concerned, gain saturated SOAs yield strong distorsions for signal frequencies up to about 10 GHz, as the carrier lifetime ranges between 100 and 300 ps.

Let's consider the amplification of an AM modulated optical carrier. If the input optical power writes as :

$$P_{in} = P(1 + m\cos\omega t)$$

then second (third) order distortion ratio *IMD2* (*IMD3*), defined as the ratio of the output second (third) harmonic power to the fundamental tone power, are approximately given by :

$$IMD2 = \frac{m^2}{4}\left(\frac{\ln 2}{1+(\omega\tau_c)^2}\right)^2\left(\frac{GP_{in}}{P_{sat}}\right)^2$$

$$IMD3 = \frac{m^4}{64}\left(\frac{\ln 2}{1+(\omega\tau_c)^2}\right)^4\left(\frac{GP_{in}}{P_{sat}}\right)^4$$

It can be seen that the third order harmonic power is generally much weaker than the second order one. For analog CATV applications, requirements generally preclude using SOAs for high optical outputs.

Now, the probably most recent and significant advance is the considerable reduction of non-linear distorsions in SOAs. The very simple concept of gain clamping by laser oscillation in a SOA has been successfully experimented in different labs. The principle consists of pinning the population inversion by forcing the amplifier to oscillate at a wavelength located far away from the useful spectral gain window. Based on this principle, a gain-clamped amplifier operating in the 1300 nm window was reported at OAA'96, with analog CATV -grade linearity for output power levels higher than 13 dBm.

3.5. Applications for SOA Gain Non-Linearities in Signal Processing

During the last few years, there has been a considerable interest for SOA non-linearities, but now in a positive way : it has been realized that these non-linearities could be exploited for numerous applications : wavelength conversion, optical gating and sampling, clock recovery, phase conjugation, all-optical regeneration are some examples of signal processing applications for SOAs. There are roughly two kinds of physical effects involved in these applications : gain saturation, involving interband relaxation mechanisms on a time scale of tens to hundreds of picoseconds, with very high power efficiencies, and intraband relaxations, with much shorter relaxation times (< 1 ps). The first effect is more generally used for optical gating (wavelength conversion, non-linear optical gates for signal regeneration) up to 40 Gbit/s, while the second effect is used in broadband four wave mixing (FWM) for phase conjugation or optical sampling. Recent experiments show that data driven gates operating at a 100 Gbit/s rate will be possible very soon.

3.6. Conclusion

As far as linear amplification is concerned, it will not be easy for SOAs to compete with EDFAs, as general purpose linear amplifiers. It is clear that the future for SOAs will be in the field of high speed optical signal processing, because of their high efficiency, compactness, and compatibility with large scale integration.

4. FAST MODULATORS

4.1. Fast Modulators

M.Varasi
Alenia Research Dpt – Roma, Italy
E-Mail : varasi@alenia.finmeccanica.it

4.1.1. Introduction

External modulation offers advantages over direct laser diode modulation of optical radiation in fiber optic systems, mainly in terms of bandwidth and linearity range [77,78]. This drove the development of the fabrication technologies of integrated optical modulators at a very high maturity level. A broad range of solutions for different system requirements are now realised exploiting $LiNbO_3$ based technologies. Phase, amplitude and frequency of the optical carrier can be modulated over a very broad frequency range. The fabrication technologies for $LiNbO_3$ integrated optical modulators will be presented focusing on the Thermal Annealed Proton Exchange (TAPE) process for the optical waveguide fabrication.

The integrated electro optical amplitude modulator will be discussed in detail starting from the modelling approaches, to the very high frequency (>20 GHz) and high linearity configurations. The overcoming of the bandwidth limitations imposed by the difference of the propagation velocities between the microwave modulating signal and the optical carrier, is the task for the very high fequency modulators. The research of the best compromise between bandwidth and modulation efficiency leads the choice between the periodic electrodes for phase reversal approaches and the velocity matched travelling wave configurations. Transmission of analog signals in CATV [79] and antenna remoting [80,81] applications demands high dynamic range electro-optic modulators (EOM). A review of different methods will be presented to reduce EOM non linearities, and a comparison will be carried out in terms of third order intermodulation products, sensitivities and bandwidth for the cascade MZ modulator. Integrated optical circuit for frequency modulation of the optical carrier will be also presented because of their application in coherent system architectures [82,83].

4.1.2. Waveguide Technology

Two technologies are used for the industrial fabrication of optical circuits in $LiNbO_3$: the thermal Diffusion of Titanium (TiD) and the Proton Exchange followed by Thermal Annealing (TAPE) [84-87]. The TAPE technology in X-cut $LiNbO_3$ has been developed and is industrially used in our laboratory. The X-cut has been preferred rather than the Z-cut because of the higher thermal stability, reduced in the Z-cut mainly by the pyroelectric effect, and lower DC drift, the variation vs temperature of the DC bias requested to electro-optically induce a constant phase delay. The TAPE solution has been motivated by the following main reasons:

- the resistance to the optical radiation damage is increased of 3-4 orders of magnitude by the proton exchange;
- only the TE polarised modes are guided in TAPE waveguides because only the extraordinary refractive index is increased by proton exchange, that allowing a very effective polarisation filtering function;
- the lower temperature and higher semplicity of the TAPE process contribute to reduce the fabrication costs.

The proton exchange is realised by the immersion of the substrate in a melt of Benzoic Acid diluted by Lithium Benzoate (1%) at 235°C. The exchange is carried out through a thin film (sputtered SiO_2) mask patterned to be open in the waveguide region. The ion exchange $LiNbO_3 + (1-x)H^+ \leftrightarrow Li_xH_{1-x}NbO_3 + (1-x)Li^+$ is buffered by the presence of the Lithium ions supplied by the Lithium Benzoate salt, reducing exchange speed and percentage and allowing better control of the process. A step wise waveguide results from the Proton Exchange, which depth is linearly proportional to the square root of the exchange time and in which the increase of the extraordinary refractive index is determined by the percentage "x" of the ion exchange in the crystal. A qualitative correlation between the two parameters is shown in figure 45[88].

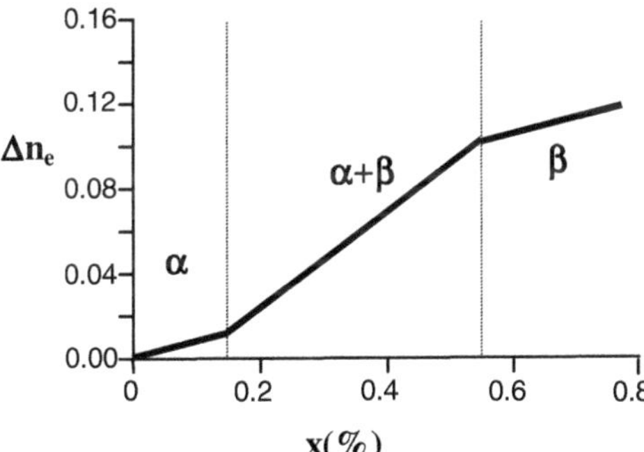

Figure 45 : Extraordinary refractive index modification vs H^+-Li^+ exchange percentage.

A subsequent thermal annealing at 400°C is performed in order to obtain a diffused refractive index profile and reduce the Proton concentration in the crystal. The diffused profile allows a better optical field matching with the fiber in order to reduce the coupling losses. The reduction of the H^+-Li^+ exchange percentage below 0.12 allows to obtain a single rombohedric (α) phase close to that of the unexchanged $LiNbO_3$, in which the properties of the crystal, and in particular the electro-optic efficiency, are very close to those of the virgin crystal.

4.1.3. Phase/Amplitude Modulation

The conventional approach to the amplitude modulation by $LiNbO_3$ integrated optical circuits adopts the Mach-Zehnder inteferometric configuration [89], schematically shown in figure 46a. The relative phase delay of the radiation in the two arms is electro-optically induced by push pull electrodes configurations (figure 46b), in which the two optical waveguides are placed in the gap between the central electrode and the two common external electrode. This configuration allows the modulating electric field to be parallel to the Z crystal axis and to exploit the highest electro-optical coefficient $r_{33} \approx 31$ pm/V.

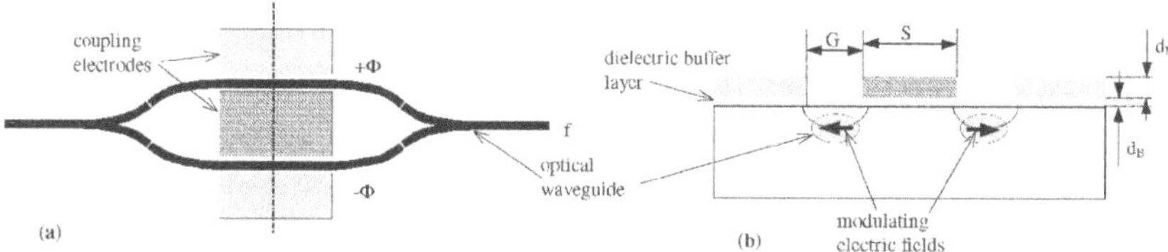

Figure 46 : Schematic view of the Mach-Zehnder modulator.

The system matrix equation of this optical circuit is simply: $\begin{pmatrix} f \\ 0 \end{pmatrix} = \frac{1}{2}\begin{pmatrix} 1 & 1 \\ 0 & 0 \end{pmatrix} \cdot \begin{pmatrix} e^{i\Phi} & 0 \\ 0 & e^{-i\Phi} \end{pmatrix} \cdot \begin{pmatrix} 1 & 0 \\ 1 & 0 \end{pmatrix} \cdot \begin{pmatrix} 1 \\ 0 \end{pmatrix}$ and the modulator transfer function $F=f^*f$ is: $F=\cos^2(\Phi)$.

Travelling Wave (TW) configuration are usually adopted to overcome the RC bandwidth limitations imposed by the simple capacitive coupling [90,91], typically at 2÷3 GHz. In this approach the electrode structure is essentially a coplanar microstrip transmission line in which the RF modulating signal propagates in the same direction as the optical radiation. The TW electrodes adopt a dielectric buffer layer, typically 100-200 nm sputtered SiO_2 [92], to minimise the risk of optical losses by metal absorption, and 2-3 μm thick Gold metallic layers. The resulting structure is then optimised for the electro-optical coupling efficiency, described by the voltage V_π defined as the voltage by which the modulator transfer function moves from its minimum to the maximum value: $V_\pi = G\lambda/(2n_e^3 r_{33}\Gamma_s L)$, where L is the length of the electro-optical coupling region and Γ_s is the overlapping integral between the optical mode distribution and the modulating electric field. The characteristic impedance of the coplanar microstrip is around 16-24 Ω for typical G/S values. Even if the TW configuration is adopted the difference between the propagation speeds of the optical ($n_{opt} \approx 2$) and RF ($n_{RF} \approx 4$) signals causes bandwidth limitations. The resulting cut off frequency f_c is given by the following approximate relation: $f_c \approx 1.4c/(n_{RF}-n_{opt})\pi L$. A lot of different approaches have been proposed to overcome this limitation, but in particular the phase reversal solutions [93,94] has been investigated despite the narrow bandwidth characteristics. Anyway the phase reversal configuration, suitable for the Z-cut substrates, does not well adapt to the X-cut substrates. A solution for X-cut substrates can be obtained by matching the two propagation velocities. This can be obtained in different ways [95,96], but the easiest and chipest approach adopts a modified coplanar microstrip configuration in which the increase of both the buffer and metallic layer thicknesses, and the increase of the ratio G/S allow to reduce the RF signal propagation speed and the matching of the characteristics impedance to the 50 Ω of the external lines. In figure 47 an example of a velocity matched configuration is shown.

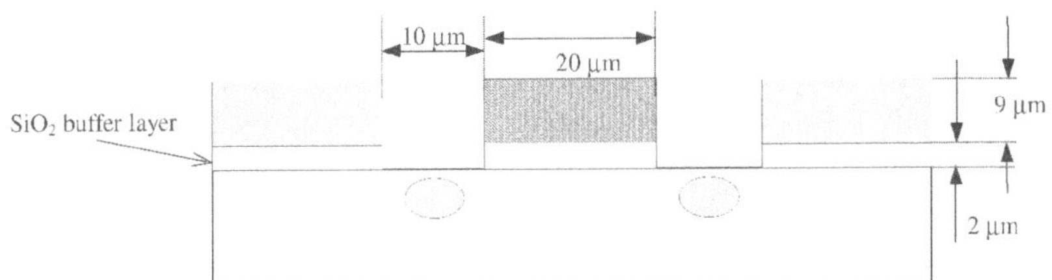

Figure 47 : Coplanar microstrip section for velocity matched configuration.

The drawback of the reduced electro-optical coupling efficiency, caused by the increased distance between the optical waveguide and the electrodes, is compensated by the possibility to increase the coupling length L up to the limit imposed by the RF losses in the coplanar structure, and by the matching of the characteristic impedance, allowing more efficient use of the RF input power. In practice the optimisation of the structure takes account of the real bandwidth requirement and then of the acceptable velocity mismatch in order to maximise the electro-optical coupling efficiency. Adopting this approach it is possible to realise velocity matched modulators the electro-optic efficincy of which is higher than that obtained by the conventional configuration for bandwidths higher than 5-6 GHz.

4.1.4. Linearity and Linearisation of the Modulator

The dynamic range of the system is a key feature of an analog optical link. Because the sinusoidal nature of the MZ modulator trasfer function causes non linearities and in consequence dynamic range limitations, it is necessary to adopt advanced modulator architectures improving the linearity of the transfer function. Many different solution have been suggested to reduce the modulator non linearities, ranging from signal predistortion [97,98] and dual polarisation techniques[99], to optical or RF feedback [100,101] and parallel/cascade MZ configurations [102-105].

While the predistortion approaches require complex RF electronics, only the last offer the opportunity to integrate all the components in the same optical circuit with single input and output. An example of cascaded MZ circuit will be discussed in the following because some technological advantages can be obtained adopting these configurations: minimum number of electrodes, maximum optical efficiency, minimum RF crosstalk, two useful linearised outputs. The operating principle of all these configurations is to adjust a couple of configuration parameters in order to have the spurious signals cancelled or significantly reduced.

Because the dynamic performances of the modulator are defined vs the noise level of the system, it is necessary to consider the modulator included in a test fiber optic link. This link (figure 48) has been defined as the simplest as possible including state of the art components working a at 1550 nm: the laser source, characterised by its output power (P_L=20 dBm) and the RIN (-165 dBc), the detector, characterised by the responsivity (η_D=0.75 A/W) and output impedance (R_D=50 Ω), and the modulator characterised by the input impedance (R_M=50 Ω), the electrooptic efficiency (V_π =10 volts), the transmission loss (Γ_M=-5 dB) and its transfer function (F). The transfer function of the modulator F is defined as the ratio between the optical powers at the output and at the input of the modulator, excluding the optical transmission losses. The noise bandwidth of the system has been conventionally set at 1 Hz.

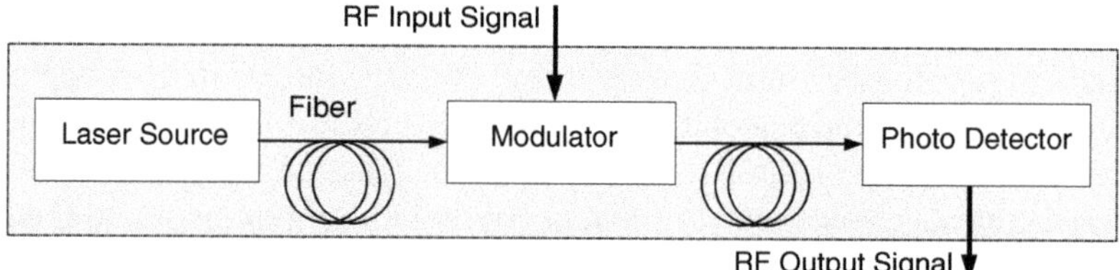

Figure 48 : The reference link.

Before to proceed in the system analysis it is necessary to define the dynamic range concept . In the case of very narrow instantaneous bandwidth signals, those that can be considered single tone signals, the value of the dynamic range can be limited by thedeviation from the linearity of the system transfer function at the fundamental frequency, or by the generation of spurious harmonic signals. In the more general case of wide instantaneous bandwidth the Intermodulation Products (IP) contribute with the harmonics to the limitation of the dynamic range.

In the first case if the width of the tunability band of the signal is less than one octave, the dynamic range is limited only by the deviation from the linearity, and a Linear Dynamic Range (LDR) can be defined as the ratio between the input power (P_{in}^{CP}) at which the output signals deviates of n dB, typically 3 dB, from the straight line response, and the input power (P_{in}^{N}) at which the output signal equals the noise level.

$LDR = P_{in}^{CP} / P_{in}^{N}$

The P_{in}^{CP} level is usually named "-n dB compression point". When the tunability band is wider than one octave, the input power (P_{in}^{Ha}) at which the harmonics level at the output is higher than the noise floor can be less then the P_{in}^{CP}, and then the dynamic range is limited by these spurious frequency components. In this case it is conventional to define the Spurious Free Dynamic Range (SFDR) as the ratio between the input power (P_{in}^{Ha}) at which the output harmonics signals equal the noise level and P_{in}^{N}.

$SFDR = P_{in}^{Ha} / P_{in}^{N}$

The situation change in the case of wide instantaneous bandwidth signals for which the Intermodulation Products are close to carrier and, independently from the width of the tunability band, they limit the dynamic range. In this case the SFDR is defined as follow: $SFDR = P_{in}^{IMP} / P_{in}^{N}$

where P_{in}^{IMP} is the input signal power at which the power of the IP product equals the noise floor in output signal.

It is then evident that in order to increase the dynamic range of the system it is necessary increase the $P_{in}^{IMP,Ha}$ and reduce the P_{in}^{N} acting on the modulator characteristics.

The modulator transfer function F is a periodic function of the variable Φ, defined as: $\Phi=\pi V/V_{\pi}$, where V is the voltage induced by the input signal in the modulator coupling electrodes: $V = \sqrt{2R_M P_{in}}$, where P_{in} is the RF input power. The output signal is given by the following expression:

$$P_{out} = i_D^2 \frac{R_D}{2} = \frac{R_D}{2}(P_L \cdot \eta_D \cdot \Gamma_M \cdot F)^2$$

In the ideal case of linear response of the modulator its transfer function can be defined as:

$$P_{out}^{Lin} = \left\{\left(P_L \cdot \eta_D \cdot \Gamma_M \cdot \eta_M\right)^2 R_D R_M\right\} P_{in}$$

where η_M is the small signal modulator efficiency, also named "modulator sensitivity": $\eta_M = \left.\frac{\partial F}{\partial V}\right|_{V=0}$

The input power at the -n dB compression point will be the P_{in}^{CP} solution of the following equation:

$$\frac{F^2(P_{in}^{CP})}{2 \cdot \eta_M^2 \cdot P_{in}^{CP}} = 10^{-n/10}$$

For the definition of the SFDR it is necessary to identify the noise level of the system. It includes the following contributions:

- RIN induced noise $N_{RIN} = RIN \cdot R_L \cdot (\eta_D \cdot P_L \cdot \Gamma_M \cdot F_{DC})^2 / 2$
- shot noise $N_{shot} = e \cdot R_L \cdot \eta_D \cdot P_L \cdot \Gamma_M \cdot F_{DC}$
- thermal noise $N_{th} = kT$

where F_{DC}=F(V=0), and e is the electron charge.

The two most relevant contributions to the noise floor, the RIN induced and the shot noises, can be expressed in terms of Equivalent Input Noise (EIN), defined as the power of the input signal by which an output signal is induced equal to the corresponding noise level. The following expressions are obtained:

$$EIN_{RIN} = \frac{RIN}{2 \cdot R_M}\left(\frac{F_{DC}}{\eta_M}\right)^2 \quad EIN_{shot} = \frac{e \cdot F_{DC}}{\eta_D \cdot P_L \cdot R_M \cdot \Gamma_M}\left(\frac{1}{\eta_M}\right)^2$$

These relations indicate that the ratio between the value of F_{DC} and the modulator sensitivity strongly influences the noise floor level, and then the dynamic range. It is worthwhile to observe that during the optimisation of the modulator performances it is at the same time convenient to reduce the IMPs and harmonics product and reduce the noise level, in order to increase the dynamic range. In this direction the polarisation of the modulators far from the quadrature could offer the advantage to reduce F_{DC}.

In the more general case the performance analysis of the modulator must consider a wideband signal in order to include the intermodulation products in the model as limiting factors of the dynamic range. To this aim a simple two tone test signal can be used in the model, in which the two tone have the same amplitude:

$$V(t) = V_0 \cdot [\sin(\omega_a t) + \sin(\omega_b t)]$$

The approach to the analysis of the amplitude various spectral components of the RF output signal exploits the serie expansion of the modulator transfer function in Bessel function of the first type. In practice the transfer function F is separated in its DC and time dependent components: $F = F_{DC} + F_{RF}(t)$

Being the time dependent component of F a periodic function of V(t), it can be expressed as a linear combination of sine and cosine functions of the argument V(t), an then expanded in series of Bessel functions, allowing an easy spectrum analysis vs input RF power.

4.1.4.1. Simple MZ Modulator (MZ)

A simple MZ modulator is considered first in order to have a comparison term (figure 49). The MZ circuit is DC biased at the quadrature point, the Y junction is perfectly balanced at -3 dB and the value of the coupling angle of the output directional coupler is π/4, to obtain a 3 dB coupling. The phase shift angle Φ is expressed in terms of external parameters by the following relation :

$$\Phi = \frac{\pi}{2V_\pi}\left(V_{RF}(t) + V_{DC}\right) = \Phi_{RF} + \Phi_{DC}$$

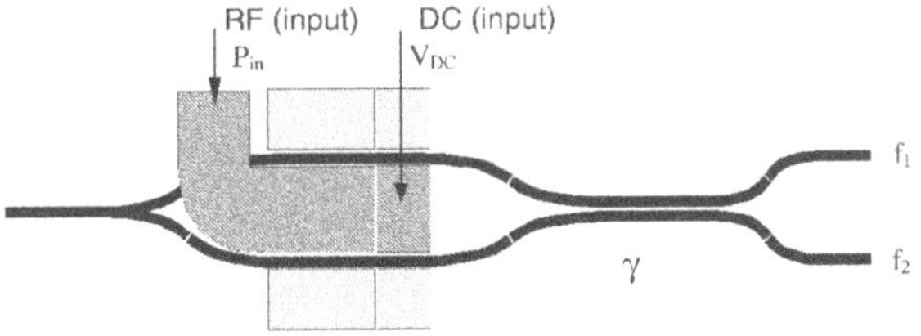

Figure 49 : Schematic view of the MZ modulator.

In order to perform a tolerance analysis of the system the parameters Φ_{DC} and γ have been considered as variables. With this configuration the transfer function (F=f*f) of one of the two outputs is given by the following relations: $F_{DC} = 0.5 + a_c$ $\qquad F_{RF} = a_a \sin(\Phi_{RF}) + a_c \cos(\Phi_{RF}) - a_c$

Where: $a_c = -\frac{1}{2}\sin(2\gamma)\cdot\sin(\Phi_{DC})$, $a_s = -\frac{1}{2}\sin(2\gamma)\cdot\cos(\Phi_{DC})$ and the modulator sensitivity is: $\eta_M = a_s$

The series expansion of the time dependent transfer function in Bessel functions of the first type, allow us to obtain the power levels of the various frequency components of the output signal spectrum vs Φ_{RF} (figure 50) and then the SFDR.

The quadrature polarisation of the MZ allows the exctinction of all the even harmonics and IMPs. The dynamic range is then limited at 116 dB/Hz$^{2/3}$ by the 3rd order IMP, being the 2/3 power correlation with the bandwidth induced by the cubic power dependence of these IMP.

The sensitivity of the dynamic range vs the change of the configuration parameters γ and Φ_{DC} is shown in the figure 51. With Φ_{DC} constant and at the design value 0, the dynamic range is weakly influenced by quite large variations of the value of the coupling angle of the directional coupler. The situation is very different when γ is mantained constant at its best configuration value (π/4), and Φ_{DC} is variable in a quite narrow range (±1.8°). The value of the dynamic range remains stable at the maximum value (116 dB) until the 3rd order IMP prevails over the 2nd harmonic. But as soon the value of Φ_{DC} exceeds the range ±0.018° the value of the 2nd harmonic is over the value of the 3rd order IMP and the value of the dynamic range rapidly decrease.

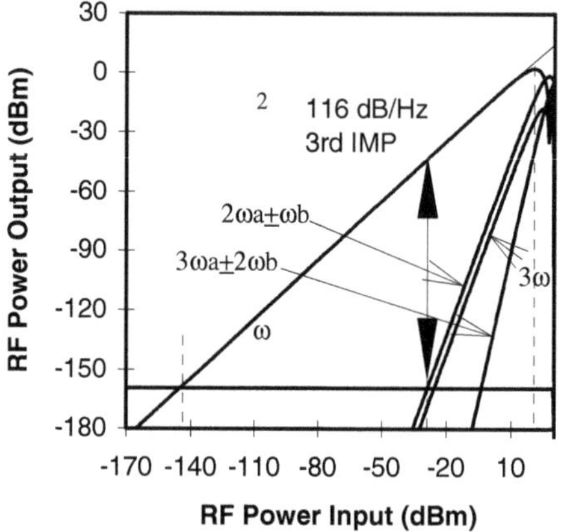

Figure 50 : RF power outputs vs RF power input for single MZ modulator.

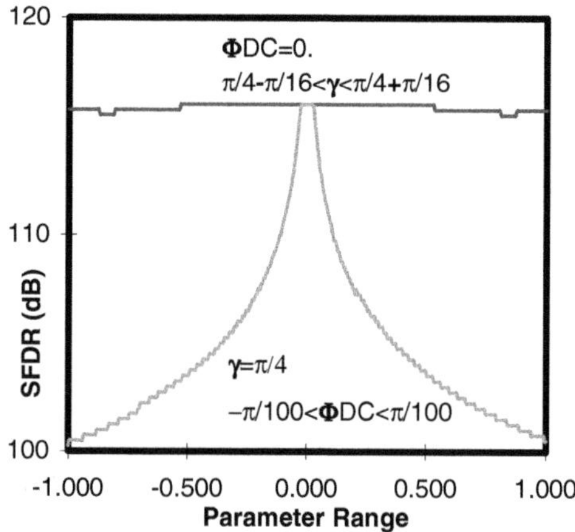

Figure 51 : SFDR vs configuration parameters for MZ modulator.

4.1.4.2. Cascade Quadrature Interferometric Modulator (CQIM)

This integrated optical circuits is based on the cascade of two identical MZ modulators polarised at their quadrature points, see figure 52. Two identical directional couplers are placed between the two MZ and at the end of the second MZ. The coupling angle γ of the directional couplers and the ratio $k=V_1/V_2$ between the two signal voltages induced by the RF signal delivered at the modulators, are considered as configuration parameters to be adjusted to optimise the system performances. In all the cases the basic principle is to identify those configuration parameter values allowing the cancellation of the main contributions to the dynamic range limits.

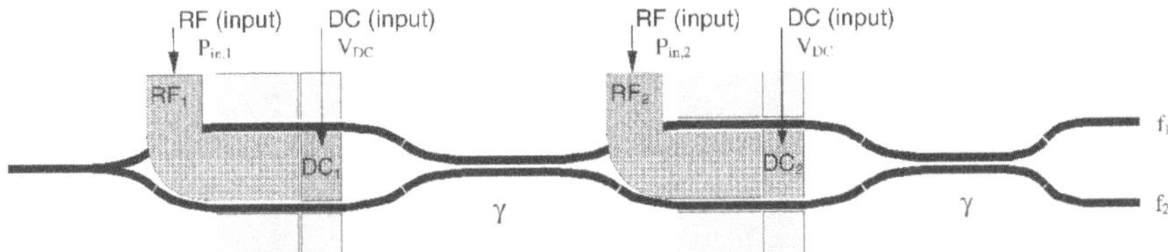

Figure 52 : Schematic view of the CQIM modulator.

With this configuration the transfer function of one of the two outputs is given by the following relation:

$$F_{DC} = 0.5 \qquad F_{RF} = a_1 \sin(\Phi_{RF}) + a_2 \sin((k+1)\Phi_{RF}) + a_3 \sin((k-1)\Phi_{RF})$$

Where: $a_1 = \sin(2\gamma)\cdot\cos(2\gamma)$, $a_2 = \sin(2\gamma)\cdot\cos^2(\gamma)$, $a_3 = \sin(2\gamma)\cdot\sin^2(\gamma)$

The modulator sensitivity is given by:

$$\eta_M = -\frac{\pi}{2V_\pi}\left[a_1 + (k+1)a_2 + (k-1)a_3\right]$$

The quadrature biasing of the modulators results in the extinction of all the even harmonics and IMPs.

An analytical approach to the modulator optimisation has been carried out. In this approach the small signal approximation has been adopted to obtain an extimation of the coefficient of the 3rd order IMP:

$$F^{RF}_{3rd-IMP} \approx (a_1 + a_2(k+1) + a_3(k-1))\cdot \text{const.}$$

Then the values of the configuration parameters have been identified satisfying both the following conditions:

$$F^{RF}_{3rd-IMP} = 0 \qquad \eta_M = \text{max imum}$$

Solving this system the following values of the configuration parameters have been obtained: γ=0.4702 and k=-0.498. With this values the modulator sensitivity (η_M) is 0.55.

Adopting these values for the configuration parameters, the analysis of the transfer function has been carried out vs the RF input power. The results are shown in figure 53, in terms of various frequency components of the output signal spectrum vs RF input power. The SFDR is again limited by the 3rd order IMP but its value is now 133.5 dB/Hz$^{4/5}$, 17.5 dB higher than that obtained with the single MZ modulator. Being nulled the cubic components of the 3rd order IMP vs the input power, still the higher odd order components, neglected in the approximation of $F^{RF}_{3rd-IMP}$, remain to limit the SFDR. The 4/5 power dependence of the SFDR vs the bandwidth is caused by the 5th power dependence of the IMPs now limiting the SFDR.

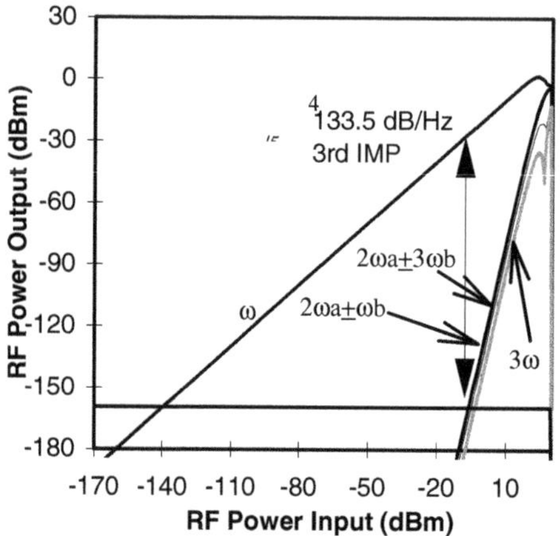

Figure 53 : RF power outputs vs RF power input for CQIM modulator as resulted from analytical optimisation.

Further improvement of the dynamic range has been obtained performing a numerical optimisation of the systems adjusting the vales of the configuration parameters around the values obtained by the analytical optimisation. As results of this numerical optimisation the following values have been obtained: : γ=0.4804 and k=-0.4759.

And the amplitudes of the various frequency components of the output signal spectrum vs RF input power are shown in figure 54. An improvement of 3 dB has been obtained, adjusting the parameters in order to place the singularity of the curve describing the 3rd order IMP vs the input power in a position such that the discontinuity of the curve is just at the noise floor level. This singularity is originated by the exact cancellation of the IMP products, that occurs only with a single set of configuration parameters and input power values.

Adopting this optimised configuration a tolerance analysis has been carried out considering ±2.5% variations of the configuration parameters. The results of this calculations are shown in the figure 55. The stability of the configuration is very critical. Variation of configuration parameters of less than 1% far from the optimised values results in a decrease of the dynamic range of 12-16 dB.

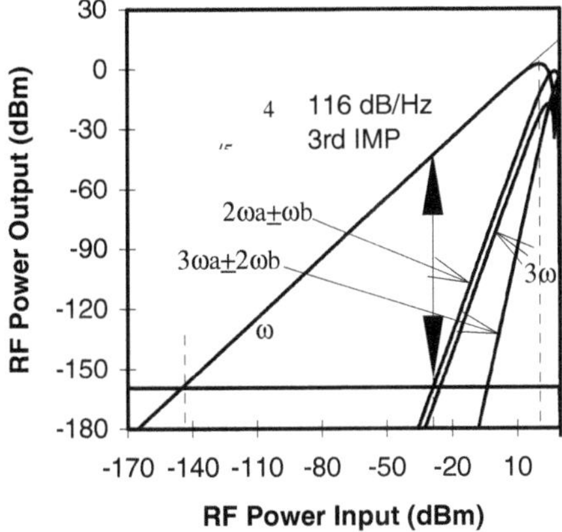

Figure 54 : RF power outputs vs P_{in} for CQIM modulator after numerical optimisation.

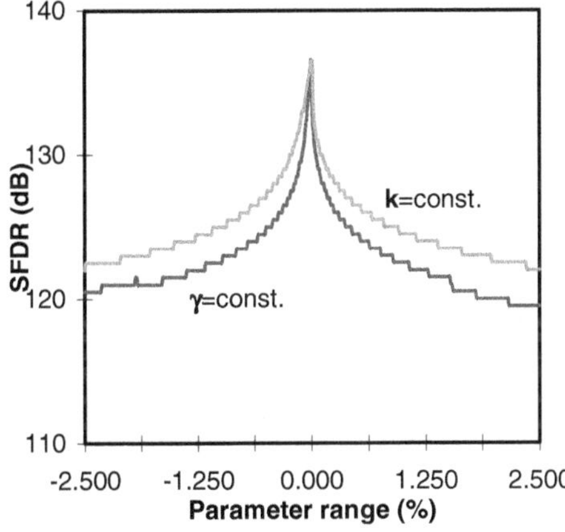

Figure 55 : SFDR vs configuration parameters variations for the CQIM modulator numerically optimised.

In conclusion significant improvements of the SFDR dynamic range can be obtained up to 20 dB adopting cascade MZ configurations. The risk is that the cost of the increased circuit complexity is vanified because the performances could be rapidly degradated to those obtained by the simple MZ modulator. It is then necessary to work at technological level in order to set up processes by which highly stable electro-optical integrated circuits can be fabricated, and identify very effective closed loop active stabilisation strategies.

4.1.5. Frequency Modulation

Various frequency modulation approaches of the optical carrier offer interesting solutions in many significant applications, such as: linearisation and dynamic range improvements, very long optical fiber transmission, processing and distribution in coherent optical architectures.

Considering the linearisation problem discussed in the previous paragraph, the modulation in the frequency domain can be effective to extend the linear dynamic range of the transmission system by adopting a Double Sideband Suppressed Carrier (DSSC) modulation approach in conjuction with optical amplification [106]. The frequency spectrum at the output of a conventional MZ amplitude modulator at its quadrature point consists of the carrier and the sidebands at ±nω from the carrier. Since only a few % modulation index can be used in order to maintain all the spurious products sufficiently low, the high carrier level is not efficiently utilised and it causes high noise floor level at the output. In the DSSC approach the carrier is suppressed and only the residual spectrum is optically amplified. The carrier is then added back (see figure 56) after the optical amplification so that, overall, the modulation index is effectively increased up to 100% without deteriorating the signal linearity. The suppression of the carrier can be easily obtained adopting the configuration shown in figure 56 in which the modulator in integrated in an arm of a MZ at the second arm of which a phase modulator allow to adjust the interference for the carrier cancellation. The feedback form the first optical detector allow to mantain the right phase sfhift minimising the DC carrier induced signal.

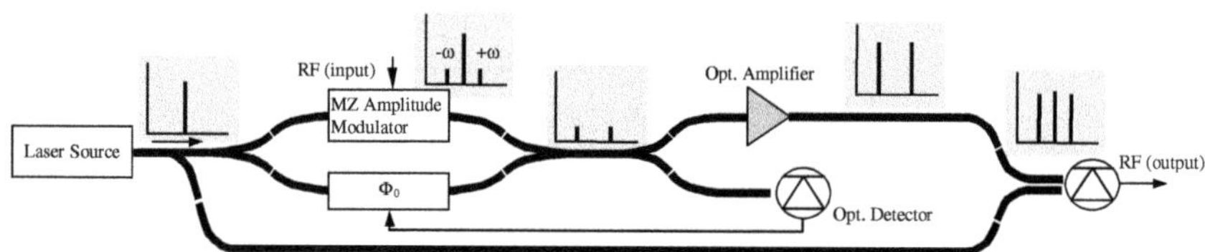

Figure 56 : DSSC configuration.

Another example in which the modulation in the frequency domain offer significant advantages is given by the signal fading in very lonk fiber optic link. In a conventional AM approach the two sidebands around the carrier propagate in the fiber with not the same velocity. Because the information at the output of the fiber is given by the beat between these signals and the carrier, the differential phase delay due to the chromatic dispersion in long fiber links generates interference and periodic cancellation of the information along the fiber. The suppression of one of the two sidebands results in a significant advantage in reducing this periodic fading. A very simple way to do that is to use a Single Side Band

(SSB) modulation approach [107,108], in which no interference between the two sidebands can occur. An optical circuit configuration allowing SSB modulation is shown in the figure 57, in which the second arm of a MZ modulator is driven with a T/4 delayed signal. The two modulator outputs can be used as input of two fiber links.

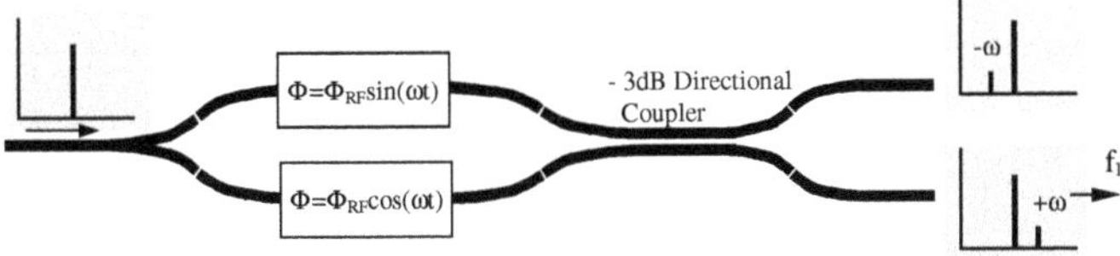

Figure 57 : SSB configuration.

The previous two examples suggest the desiderability of efficient and broadband SSB modulators or Optical Frequency Shifters (OFS). This desiderability extends over a very wide range of applications, in coherent, or heterodyne, optical transmission or processing systems, high resolution spectroscopy and photochemistry. Great efforts have been applied to the investigation of coherent architectures for the processing and distribution of microwave signals in phased array active antennas. Acousto optical interactions in Bragg regime have been exploited to obtain frequency shifting, but there is a frequency limitation at 3-4 GHz obtainable with bulk wave anisotropic interactions in $LiNbO_3$. In order to overcome this frequency limitation and improve the integration level, two approaches can be adopted both exploiting electro-optical coupling in integrated optical circuits. The first based on the cancellation of the unwanted frequency content of the optical spectrum adopting two or four arms interferometric architectures [109,110]. The second exploting the electro-optically induced mode coupling between the odd and even modes of a coupled waveguides structure [111].

A very simple approach is possible exploting again the circuit shown in the figure 57, the upper output of which gives an optical spectrum described by the following serie expansion in Bessel functions:

$$f_1 = \frac{1}{2}\left\{(1-j)J_0(\Phi_{RF}) + \sum_1^{\infty} {}_n J_n(\Phi_{RF})\left[((-1)^n + (-j)^{n+1})e^{jn\omega} + (1+(-j)^{n+1})e^{-jn\omega}\right]\right\}$$

Where it is possible to verify the cancellation of the +ω component. Being the amplitudes of the carrier and of the -ω sideband respectively proportional to $J_0(\Phi_{RF})$ and $J_1(\Phi_{RF})$ the ratio between the two components can be made maximum with cancellation of the carrier that occurs at $\Phi_{RF} = 2.41\text{rad}$. The resulting frequency spectrum, shown in the figure 58 a, still contains residual higher order frequency components at -4.6 dB from the fundamental, even if the carrier and the +ω sideband have been suppressed. A compromise can be adopted in order to reduce these spurious components, driving the modulator at $\Phi_{RF} = 1.84\text{rad}$ and allowing the carrier level to increase at the same level of the higher order spectral components. The result of this compromise is shown in figure 58 b, all the spurious spectral components are now at less than -8.3 dB, including the residual carrier.

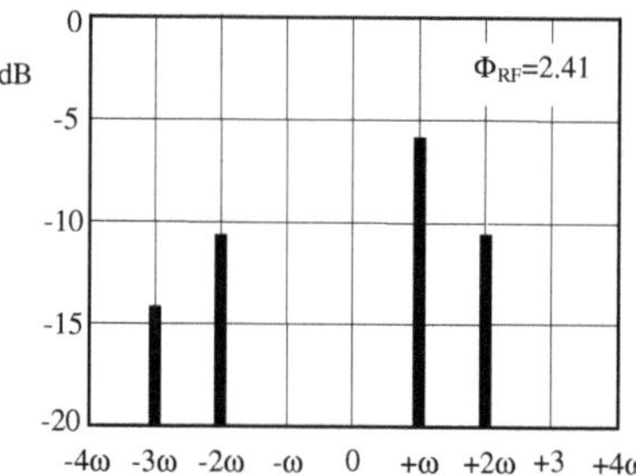

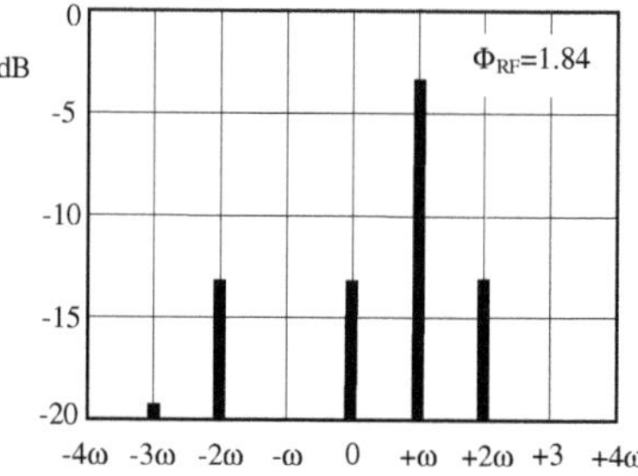

Figure 58 a/b : Optical power spectrum of SSB with two different Φ_{RF}.

To further reduce the spurious spectral components and still suppress the carrier, a four arm interferometer configuration can be adopted as shown in figure 59. The various parameters involved in this configuration allow to adopt different solutions. In one of these solutions the relative optical phase shifts introduced in each arm are 0, π/2, π and 3/2π, starting from the first to the fourth arm, and the relative fase shifts introduced in the RF signals are 3/2π, π, π/2 and 0 respectively. The best performance of this configuration is achieved when all the modulators are driven at $\Phi_{RF} = 1.84\text{rad}$, in this situation all the spurious spectral components of the output optical signal are at less than -18 dB from the fundamental.

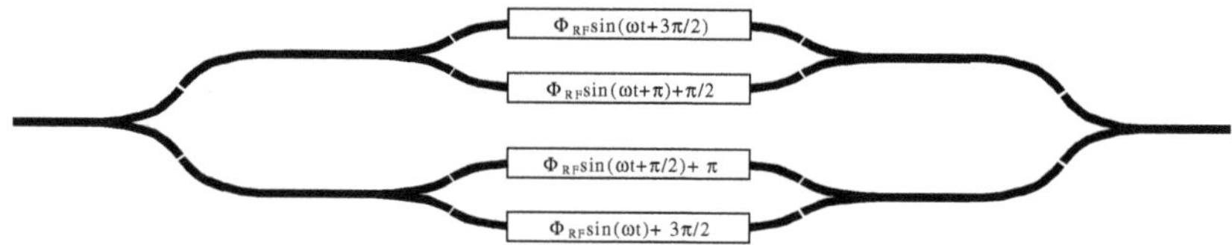

Figure 59 : Four arms SSB.

A completely different approach has been suggested by B.Desormiere et al. [61] to achieve the optical frequency shifting. The device exploits the electro-optically induced coupling between the two supermodes of a Directional Coupler (DC). The input radiation is equally splitted into the two arms of a directional coupler (figure 60), and only the symmetric supermode of the DC is then excited, but the perturbation induced by the RF signal causes the coupling between the symmetric and antisymmetric DC supermodes. In this mode coupling the following “phase matching” and “energy conservation” conditions are satisfied:

$$\frac{2\pi n_s}{\lambda_s} - \frac{2\pi n_a}{\lambda_a} = \frac{2\pi n_{RF}}{c} f_{RF} \quad h\omega_a - h\omega_s = 2\pi h f_{RF}$$

The frequency of the antisymmetric mode product of the mode coupling is shifted by the frequency f_{RF} of the modulating RF signal. The MZ placed at the output of the DC introduce a differential phase delay between the two arms resulting in the extinction of the symmetric mode and the constructing interference of the antisymmetric mode at its output. In conclusion the at the output of the devices only the frequency shifted radiation will be delivered.

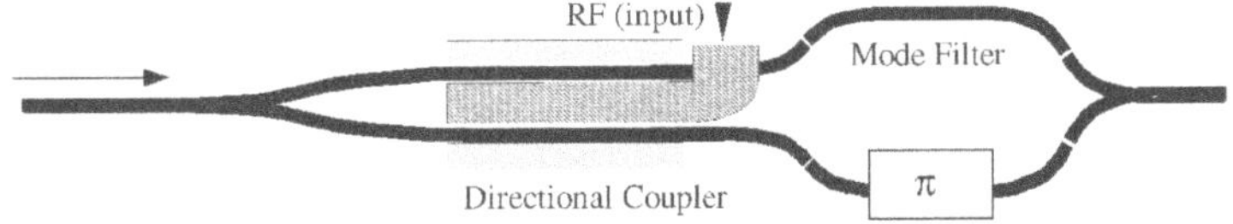

Figure 60 : Schematic view of the Mode Coupling OFS.

4.1.6. Conclusions

The velocity matched TW electrode configurations allow to obtain very wide bandwidth modulators with electro-optical efficiency increasingly higher than those obtained by the standard configurations as the bandwidth increase.

Significant improvements of the SFDR dynamic range can be obtained, up to 20 dB, adopting cascade MZ configurations. But the critical stability of the performances vs the configuration parameters, imposes the necessity to improve the fabrication technologies in order to obtain highly stable electro-optical integrated circuits, and to investigate very effective closed loop active stabilisation strategies.

Both the SSB and the mode coupling OFS offer the possibility to modulate the frequency of the optical carrier and then to exploit the opportunities of the coherent signal processing.

4.2. Electroabsorption Modulators and Photo-oscillators for Conversion of Optics to Millimeterwaves

C. Minot
France Telecom, CNET-DTD, Laboratoire de Bagneux
BP 107, 196 avenue Henri Ravera, 92225 Bagneux, France
E-mail : christophe.minot@cnet.francetelecom.fr

4.2.1. Introduction

The rapid development of wireless communications and the growing demand for broadband services has recently raised up new investigations of millimetre-wave (mm-wave) devices and systems. With progressive investments, Local Multipoint Distribution Systems (LMDS) offer rapidly deployed alternatives to the traditional cabled networks. They are being implemented mostly in the Ka band (26.5-40 GHz). In the V band (50-75 GHz), the 60 GHz frequency is also of particular interest because mm-waves are strongly absorbed by the atmosphere at this wavelength. It is then possible to design picocellular mobile communication systems based on numerous mm-wave radiolinks, in which frequency reuse is easy, available bandwidth large and electromagnetic power low. Since services are supplied by external, generally optical, distribution networks, low-cost interfaces have to be developped in order to connect both subsystems.

The concept of hooking microwave terminals on optical fiber networks is well-known [112]. More specifically, we consider optically-fed mm-wave radiolinks inserted in a broadband optical distribution network, which is supposed to optically deliver the radio signal (i.e. the carrier and the data) from control stations. We focus on some of the optoelectronic components required in such "radio over fibre" systems, a simplified picture of which is given in figure 61. This architecture centralises the generation of the mm-wave signals in the control stations, which is expected to be cost-effective and allow easy maintenance. The optical carrier is emitted by a CW laser and modulated by a distinct device, which operates a conversion from mm-waves to optics and gives very good control of the optically transmitted data. In the base stations, the radio signal is converted from optics to the mm-waves as efficiently as possible, in order to avoid expensive amplification stages, before emission in free space. It is clear then that the modulators and the optical to mm-waves transducers require special attention in the design of the system. The former must not give rise to additional noise and have to be highly reliable. The latter must be cost-effective, since they are present in each base station, and may be key devices to demonstrate the economical advantage of the system. In the following, we briefly describe from a designer point of view two devices which have been developed in order to fulfil such objectives: *i)* the electroabsorption modulator and a sophisticated Photonic Integrated Circuit (PIC) derived from it, the integrated single-sideband lightwave source, as a reliable modulated optical signal generator *ii)* the superlattice photo-oscillator, as a simple and powerful frequency locked photo-receiver.

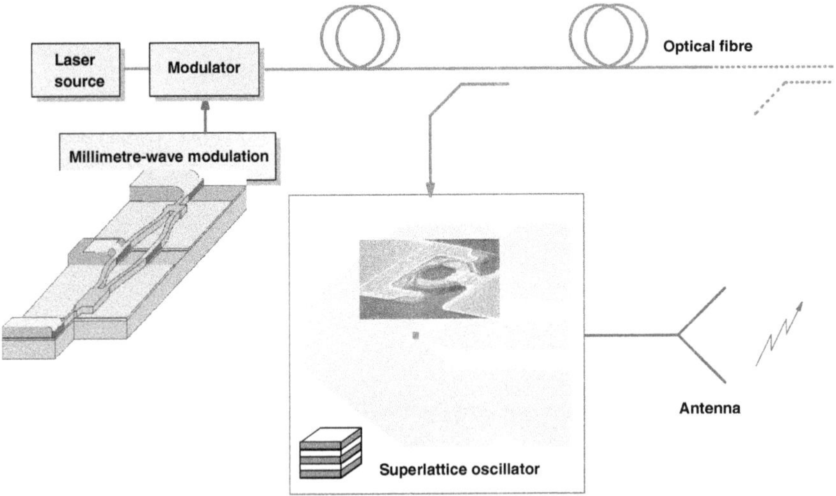

Figure 61 : Down-link in a hybrid fibre-radio system, with a single-sideband source (schematic view by courtesy of E. Vergnol) in the control station and a superlattice photo-oscillator in the base station.

4.2.2. Electroabsorption Modulator

In electroabsorption modulators in linear regime, the light input P_{in} and output P_{out} powers are proportionnal and the transmitted power depends on an applied voltage or electric field. As a result, we can distinguish between the high transmission on-state and the low transmission off-state, and define the extinction ratio:

$$R_{ext} = 10\log\left(\frac{P_{out}^{off}}{P_{out}^{on}}\right) = -4.34\left[\Gamma\alpha(F_{off}) - \Gamma\alpha(F_{on})\right]L \qquad (19)$$

and the on-state losses:

$$A_{ext} = 10\log\left(\frac{P_{out}^{on}}{P_{in}^{on}}\right) = -4.34\Gamma\alpha(F_{on})L \qquad (20)$$

where Γ is the confinement factor of the optical mode in the absorbing region, α the absorption coefficient, F the internal electric field and L the device length (figure 62). In general, the design constraints impose an upper bound on the extinction ratio and a lower bound on the on-state losses [113]. As a result the contrast, i.e. the ratio: $\frac{R_{ext}}{A_{ext}} = \frac{\Delta\alpha}{\alpha_0}$, must be larger than a limit value, e.g. $\frac{\Delta\alpha}{\alpha_0} > 3.3$ for A_{ext}>-3dB and R_{ext}<-10dB.

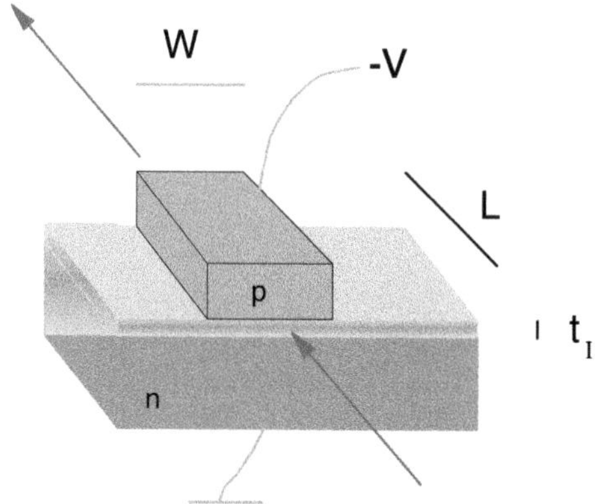

Figure 62 : Ridge waveguide and p-i-n diode structure of the electroabsorption modulators.

A second design constraint can be defined from the modulator cut-off frequency:

$$\nu_{3dB} = \frac{1}{\pi R_L C} \quad \text{with} \quad C = \frac{\varepsilon WL}{t_I} \tag{21}$$

where R_L is the load resistance, C the device capacitance, ε the dielectric constant, W the device width and t_I the thickness of the region where the internal field develops (figure 62). Then, using the drive voltage $\Delta V = t_I \Delta F$, a figure of merit which does not depend on the geometrical dimensions L and t_I can be written as:

$$\frac{\nu_{3dB}}{\Delta V} = \frac{4.34}{\pi \varepsilon R_L R_{ext}} \; \frac{\Gamma}{W} \; \frac{\Delta\alpha}{\Delta F} \tag{22}$$

This quantity must be maximised to get the largest cut-off frequency with the smallest drive voltage. It is expressed as the product of three terms: the first one reflects external constraints, the second one the geometrical parameters and the third one optical properties of the electroabsorption material. W can be minimised as far as the technology allows and Γ maximised by adjusting the thickness of the absorbing material (smaller than t_I). Eventually, the quantity $\frac{\Delta\alpha}{\Delta F}$ must be made as large as possible thanks to effective electro-optic materials. It has been demonstrated that electroabsorption modulators based on the Quantum Confined Stark Effect (QCSE) in quantum wells can exhibit better figures of merit than similar devices based on the bulk materials[114].

In order to design a modulator, figure 63 shows typical variations of the material parameters $\left|\frac{\Delta\alpha}{\alpha_0}\right|$ and $\frac{\Delta\alpha}{\Delta F}$ as a function of wavelength owing to the QCSE. In the region of interest, $\left|\frac{\Delta\alpha}{\alpha_0}\right|$ increases and $\frac{\Delta\alpha}{\Delta F}$ decreases with wavelength, so that the most favourable wavelength λ_C can be determined from the criterion on $\frac{\Delta\alpha}{\alpha_0}$ only, which also gives α, $\Delta\alpha$ and ΔF. Then, the length L is obtained from the required extinction ratio (19) and W from technological limitations. The thickness of the absorbing material is adjusted to preserve monomode propagation with maximum confinement factor Γ. Finally, once t_I is related to the cut-off frequency through (21), the drive voltage can be derived from (22). Experimental transmission curves are shown in figure 64 at different wavelengths.

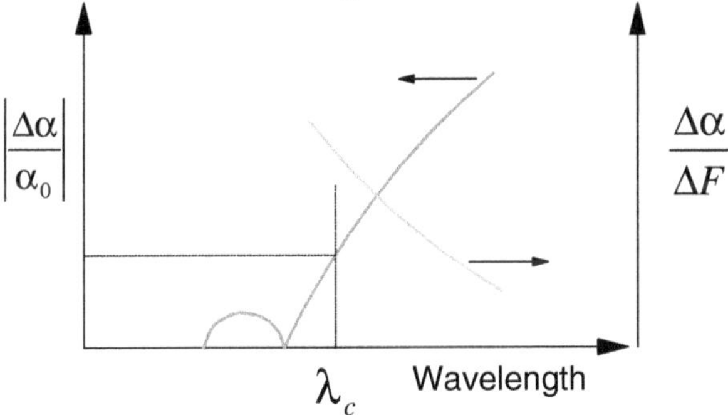

Figure 63 : Typical wavelength dependence of the absorption contrast and figure of merit in QCSE electrooptic materials.

Several PICs have been developed in which one or two electroabsorption modulators are integrated with other optoelectronic or photonic functions: tandem of modulators separated by a semiconductor optical amplifier [115], DFB laser and modulator[116], single-sideband (SSB) modulated source [117] (a sophisticated circuit which includes a DFB laser, two passive multimode interferometers, two modulators and an amplifier. A schematic view can be seen in figure 61). All the devices and circuits are grown on InP and make use of the InGaAsP quaternary material. The circuits are based on the "identical active layer" approach in which the same active layer is used for the modulators and the amplifiers or DFB lasers (the grating of the latter is detuned to longer wavelength in order to ensure compatibility). The SSB source is a very promising device for radio over fibre systems: in addition to its compactness, stability and flexibility, it eliminates the dispersion problems encountered with traditional modulated sources [118].

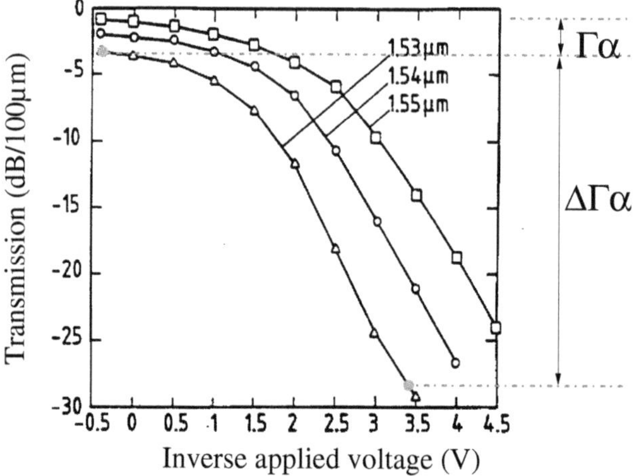

Figure 64 : Transmission as a function of applied voltage in a GaInAsP QCSE electroabsoption modulator at several wavelengths near 1.55 μm (by courtesy of E. Vergnol and F. Devaux).

4.2.3. Superlattrice Photo-Oscillator

Optical injection locking of negative differential conductance (NDC) oscillators is an interesting technique to simultaneously generate the powerful mm-wave carrier and recover the numerical data, provided the latter are encoded through frequency or phase keying. A 1A/W sensitivity 50 Ω loaded photodiode only generates ~ -33 dBm electrical power from -10 dBm incident optical power, whereas a very simple optically locked NDC oscillator can supply power gain in the conversion from optics to mm-waves. This may be particularly important in the millimetre range where photodiodes and transistor amplifiers are expensive. Since most of optical communication networks operate at 1.3 or 1.55 μm and the usual transferred electron devices and Gunn diodes are not optically sensitive at these wavelengths, new NDC devices have been investigated [118], the optical and electronic properties of which can be tailored according to specific requirements. Their active region consist of a semiconductor superlattice, in which the very design of the layer stack and the formation of new energy bands (the so-called "minibands") give rise to negative differential velocity and non-linear transport properties. With the materials GaInAs/AlInAs lattice-matched to InP, the superlattice bandgap can be tuned into the adequate near-infrared wavelength range.

In the samples which have been studied, weakly doped superlattices ($\sim 1\mathrm{x}10^{16}$ cm^{-3}) are sandwiched between n^+ contact layers (figure 65). They are processed in a passivated mesa technology. Although requiring very careful alignment to obtain very small area devices, all steps use only conventional optical lithography.

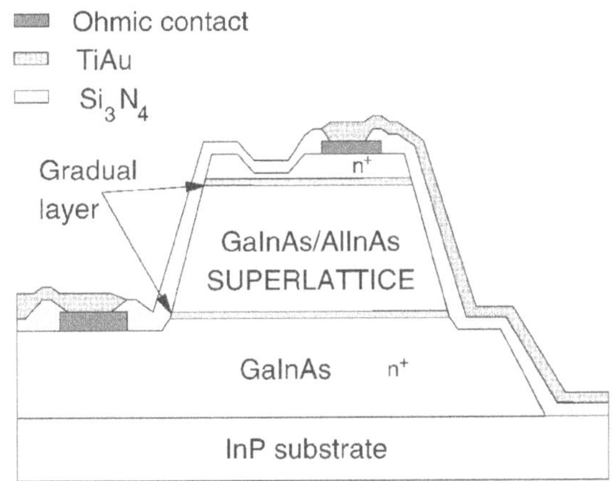

Figure 65 : Superlattice diode structure and technological process.

Figure 66 shows the module of the reflection coefficient (S_{11} in a one-port S-parameter measurement) in a superlattice sample with a moderately wide miniband (Δ=56 meV), as a function of frequency from 0 to 65 GHz and for successive values of the applied voltage. When the conductance is negative, the reflection gain is larger than unity. At -1V, the gain cut-off frequency can be extrapolated linearly to ~75 GHz. The results can be accounted for by a simple admittance model for NDC devices, in which two main parameters determine the resonant behaviour of the gain: the resonance frequency is approximately given by the inverse of the electron transit time T through the superlattice, and the resonance amplitude by the ratio of that transit time over the dielectric relaxation time T/τ ($\tau=\varepsilon/\sigma$ where ε is the dielectric constant and σ the differential conductivity). The shift of the resonance to lower frequency when bias increases reflects the diminution of the electron velocity (T increases), in very good agreement with usual models of electron transport in superlattice minibands [119,120]. As a result, T~1/f and τ, together with the wavelength λ, are the relevant parameters to design a superlattice for an oscillator at frequency f, through their relation to the structure and the filling of the miniband: the transit time depends on the overall thickness of the superlattice and the electron velocity (an increasing function of miniband width), the dielectric relaxation time depends on the superlattice doping and the differential velocity, the bandgap depends on the thickness and height of the wells and barriers. However, under illumination, the exact values of the gain at a given frequency cannot be precisely predicted in the present state of our understanding.

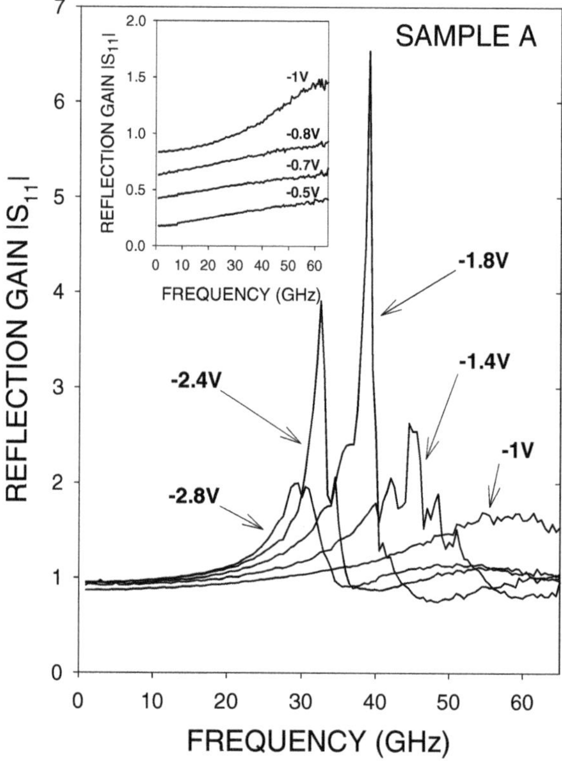

Figure 66 : Reflection gain in a GaInAs/AlInAs superlattice oscillator structure as a function of applied voltage.

The design of the resonant circuit which is connected to the active chip also deserves very careful attention, for the oscillator output power and locking bandwidth can be significantly improved by an adequate choice of the circuit impedance and (external) quality factor Q. Such problems have been discussed extensively by K. Kurokawa [121] in terms of the amplitude dependent device impedance on the one hand, and dephasing between oscillator and locking signal amplitudes on the other hand.

Similar concepts have been applied to optical locking of NDC oscillators to get an expression for the locking bandwidth $\Delta\omega$:

$$\frac{\Delta\omega}{\omega} = \frac{1}{\omega T}\frac{1}{Q}\sqrt{\frac{P_i}{P_0}} \tag{23}$$

where $\omega=2\pi f$, P_i and P_0 are the injected locking signal and free-running oscillator powers respectively. Since in general $\omega T \sim 2\pi$, (23) shows that optical locking is slightly less efficient than electrical locking in NDC oscillators, due to their photoconductive behaviour.

The question arises whether a locked oscillator is able to recover FSK or PSK (Frequency or Phase Shift Keying) encoded data, since its output certainly does not follow the locking signal linearly nor instantaneously. Actually, data transmission experiments have been successfully carried out[7] with an electrically locked oscillator near 40 GHz and DPSK data at 40 Mbits/s ($BER<10^{-8}$). Optical locking experiments have also been performed at 20 and 38 GHz [122,123], and optical locking bandwidths up to 120 MHz demonstrated.

4.2.4. Conclusion

Electroabsorption modulators are very intensively employed in high bit rate optical communication systems, so that they are able to operate in the mm-wave range. As waveguided devices, they can benefit from the advanced photonic technologies. Thus they offer efficient answers to the needs of new hybrid fibre-radio concepts involving mm-waves. Contrarily, the locked NDC photo-oscillator approach is much less mature and is permanently threatened by the continuous advance of transistor technology to higher cut-off frequencies. Nevertheless, its simplicity makes it very attractive in the mm-wave range and calls for further investigations on its basic electronic mechanisms and limitations.

Acknowledgements

The author is very grateful to Eric Vergnol, Abderrahim Ramdane and Fabrice Devaux whose assistance has been very helpful to present the main issues on electroabsorption modulators. He also wants to thank all those who have contributed to the development of both devices discussed here, in Bagneux and Lannion, and made possible a lecture at the OMW summer school.

5. HIGH SPEED PHOTODETECTION

5.1. Microwave Optical Interaction Devices

D. Jäger
Fachgebiet Optoelktronik, Sonderforschungsbereich 254
Gerhard-Mercator-Universität Duisburg
D-47048 Duisburg, Germany
E-mail : jaeger@optorisc.uni-duisburg.de

Abstract

In this paper the fundamental concepts of ultrafast microwave photonic devices based upon the interaction of propagating microwaves and optical signal beams are discussed. Such travelling wave optoelectronic devices utilizing, for example, microstrip or coplanar transmission lines as electrical waveguides exhibit cut-off frequencies not limited by the usual RC time constant. As a result, a high bandwidth together with improved efficiency and power capabilities are obtained. In particular, travelling wave photodetectors, waveguide and electrooptical modulators, optical switches and microstrip laser diodes are discussed.

5.1.1. Introduction

The realisation of ultrafast photonic devices operating in the microwave regime is expected to play a key role in developing future high-speed and high-capacity lightwave systems. The electrical cut-off frequency of photonic devices is usually limited by internal physical time constants and additionally by the device structure and the external circuitry. In order to achieve operation up to millimeterwave frequencies, the device dimensions may not exeed a few μm to get a capacitance far below 1pF leading with to a characteristic impedance of 50Ω to a cut-off frequency of merely 3GHz. On the other hand, an electrical contact size of about 100 μm can reach the order of a quarter wavelength already at 10GHz when slow mode effects occur [124]. In that case, the device properties depend on travelling wave effects, and no RC time constant can be defined in the usual way. In contrast, wave propagation effects have necessarily to be included in the simulation, modelling and fabrication of such components [125-133].

In this paper microwave photonic devices are presented, that utilize microstrip or coplanar transmission lines as electrical waveguides. The metallization is formed according to well known microwave techniques, and the input resistance is determined by the characteristic impedance of the coplanar waveguide. The light is also guided using conventional optical waveguides and the optoelectronic conversion takes place via a microwave-optical interaction process in space and time domain.

5.1.2. High Speed Optical Links

Figure 67 shows the key building blocks of a high-speed optical link replacing standard metallic transmission lines such as coaxial cables or rectangular waveguides. The advantages of such an optical link are a result of the extremely low propagation losses of an optical wave in a fiber (or even in free space) as compared to the attenuation in a metallic connection medium. The optical link further requires high-speed and efficient electrical-optical power converters on both sides.

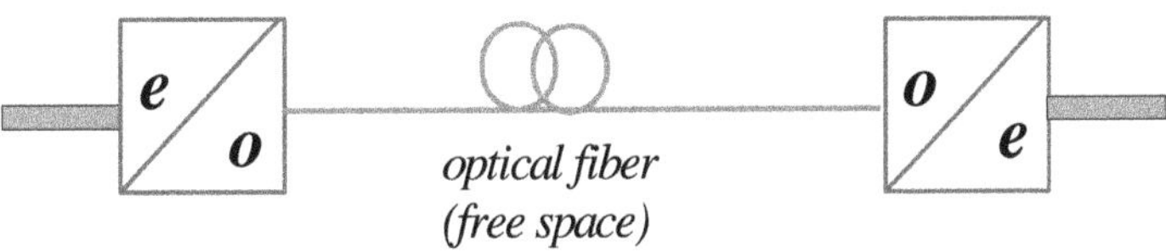

Figure 67 : The basic concept of an optical interconnection replacing a usual microwave connection via a coaxial cable or a waveguide.

5.1.3. Microwave Optical Interaction Devices

In figure 68, basic concepts of microwave optical interaction devices used as optical-electrical or electrical-optical converters are sketched. Figure 68(a) shows an optoelectronic "photodetector", converting optical power into microwave power, whereas figure 68(b) shows electrical-optical converters such as laser diodes (LD) or LEDs. Figure 68(c) and (d) represent 3-port devices, i.e. modulators where an external optical or electrical power supply is additionally used. Such devices are generally a kind of optoelectronic (hybrid) transistors exhibiting amplification or switching capabilities.

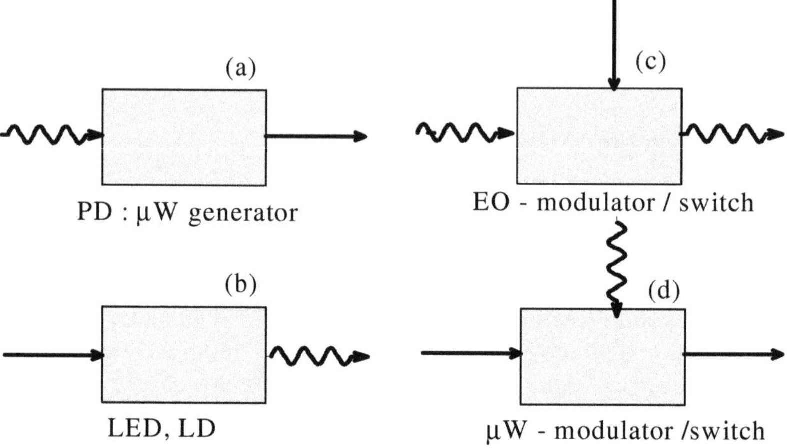

Figure 68 : Principle interaction mechanisms in microwave optical interaction devices (see text).

5.1.4. Travelling Wave Devices

The general outline of a travelling-wave optoelectronic device is sketched in figure 69.

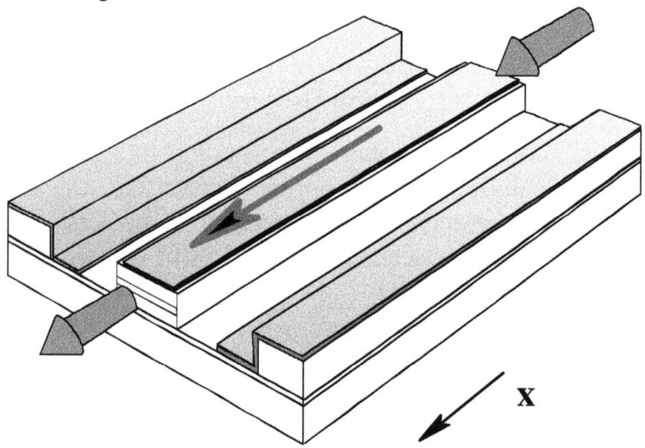

Figure 69 : Sketch of a microwave optical interaction device using travelling wave effects.

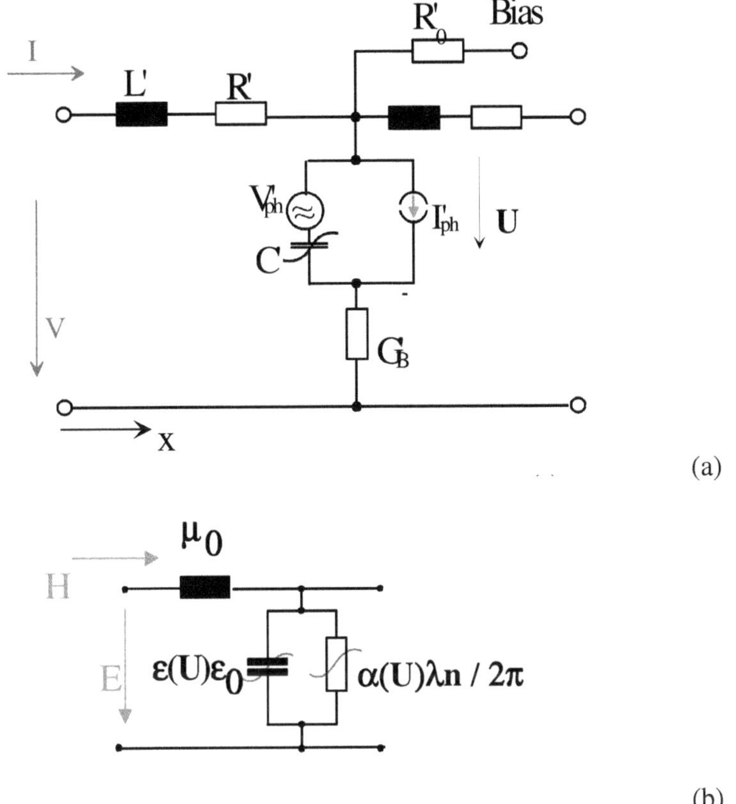

Figure 70 : Equivalent circuits for electrical (a) and optical (b) wave propagation in the photonic device of figure 69.

Electrically, the metallic contacts are used as microwave or millimeterwave transmission lines. The electrical wave propagation is now mainly determined by the multilayered semiconductor substrate material, where the cross section in most cases is that of a pn-, Schottky-, or pin-diode [129].

Because the magnetic field is not influenced by the conductivity of the layers, slow wave properties will arise, where the slowing factor can be as large as 20 to 100 [124,135]. Optically, the layer structure is used as a waveguide for the propagation of light. Here different cases can be distinguished: The optical input energy can be absorbed to generate a microwave signal or the optical beam can be modulated by an electrical, i.e. a microwave signal. The resulting devices are called travelling-wave (TW) photodetector and modulator [126,134-137]. In case of a laser diode, light is generated and the optical output is controlled by the microwave signal [138]. Note, that a vertical/oblique illumination (photodetector), transmission or reflection (modulator) and emission of light (vertical cavity surface emitting laser) can also lead to travelling wave effects provided that the extension of the optical beam in x-direction exceedes a quarter of the microwave wavelength, approximately.

In Figure 70(a) the equivalent circuit for electrical wave propagation on a coplanar transmission line on layered media is shown. Note that C' and G' are nonlinear elements controlled by the properties of a depletion layer. In case of a photodetector I'_{ph} is an impressed current source per unit length describing the opto-electric conversion, and here $I'_{ph} = I_{ph}(x,t)$ is also a wave due to the propagation of light. The circuit in figure 70(b) describes the optical domain where the optical losses lead to the value of $I'ph$ .of figure 70(b). In case of a travelling wave modulator, $I'ph=0$ in figure 70(a), and the voltage drop can be used to calculate the modulation effect via the electrooptical properties of the active layer, see for example [136]. In case of laser diodes, the nonlinear G' of figure 70(a) gives rise to an "optical" current source, i.e. generation of photons.

In summary, travelling-wave (TW) optoelectronic devices can be described by electrical and optical equivalent circuits, where the interaction is given by elements with a parametric space and time dependence. The efficiency of TW-devices depends critically on the degree of phase matching between the optical and microwave signal. Optimum conversion efficiencies are only achieved under phase matching conditions.

5.1.5. TW-Photodetector

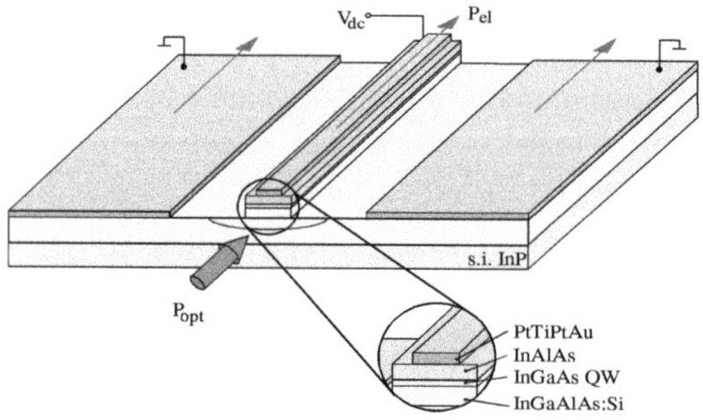

Figure 71 : Sketch of the travelling-wave photodetector. Popt and Pel are the input optical and output electrical powers, respectively.

Figure 71 shows the sketch of a TW-photodetector. Here the quaternary semiconductor together with the InAlAs cladding layers form the optical waveguide. The InGaAs quantum well is used as an absorbing film where the optical attenuation is due to a leakage effect and the losses can be controlled by several geometrical parameters. Such a photodetector can easily be used to generate microwave power when two optical beams with different frequencies are propagating down the line. As a result of wave mixing effect in the heterodyne photodetector a microwave signal is generated, the frequency of which is given by the difference of the optical frequencies.

From numerical simulations it is concluded that in case of phase matching the microwave amplitude increases monotonically with distance x and the microwave output signal becomes a maximum. This is also obvious from an analytical solution as derived from a simplified equivalent circuit model :

$$u(z) = \frac{Z\hat{I}_{ph}(z=0)}{2} e^{-\gamma_{el} L} \frac{e^{-\gamma_{opt} L} - e^{\gamma_{el} L}}{\gamma_{opt} + \gamma_{el}} e^{\gamma_{el} z}$$

$$+ \frac{Z\hat{I}_{ph}(z=0)}{2} \frac{-e^{-\gamma_{opt} z} + e^{\gamma_{el} z}}{\gamma_{opt} + \gamma_{el}} + \frac{e^{-\gamma_{opt} z} - e^{-\gamma_{el} z}}{\gamma_{opt} - \gamma_{el}}$$

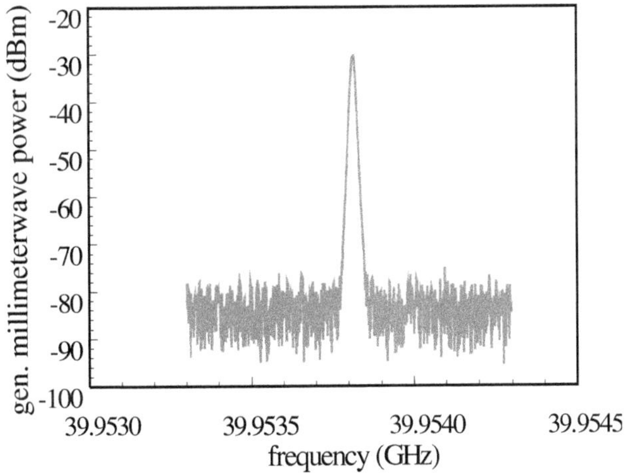

Figure 72 : Measured spectrum.

The device of figure 71 has been measured at optical wavelength of $\lambda = 1.3$ μm and $\lambda = 1.55$ μm. Figure 72 shows the spectrum and figure 73 the measured frequency response.

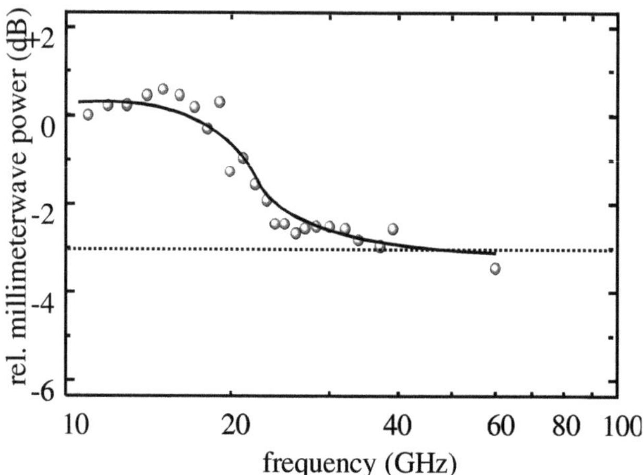

Figure 73 : Frequency responce of a TW photodetector

5.1.6. TW-Laser Diode

High speed waveguide laser diodes are today fabricated with a structure as shown in figure 69. Again the coplanar metallization leads to cut-off frequencies well above 20 GHz [142]. In such a laser the center conductor length varies typically between 100 μm and about 200 μm and due to slow wave effects the metallic contact may exhibit an inhomogeneous voltage distribution. Preliminary results showing the effect of travelling microwave signals have been published recently [143]. It is therefore foreseen, that a layout using microwave propagation effects may lead to further enchancement of the bandwidth of laser diodes.

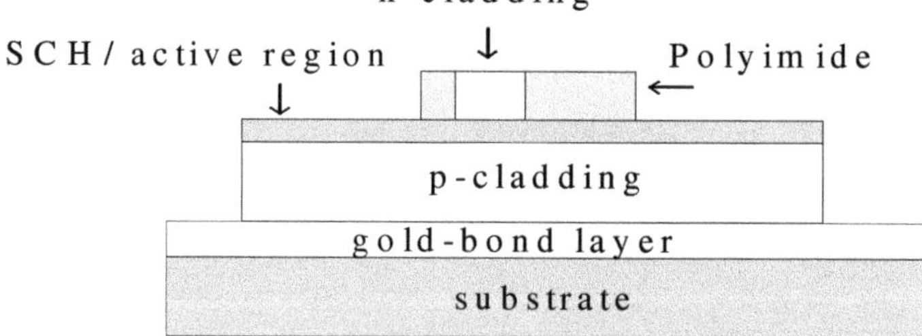

Figure 74 : Cross section of a microstrip laser diode [143].

In figure 74, the cross section of a recently proposed and so called microstrip laser diode is presented [143]. In this case the metallization structure is that of a microstrip line.

5.1.6. TW-Modulator

In figure 75 a TW - waveguide modulator is sketched. Experimentally, an electroabsorption (EA) modulator has been investigated using the quantum confined Stark effect (QCSE) in strained InGaAs/AlGaAs MQW waveguide structures [136]. Experimental results of the electrical bandwidth measured in a common 50 Ω system reveal a cut-off frequency in excess of 70 GHz due to an optimum impedance and phase matching.

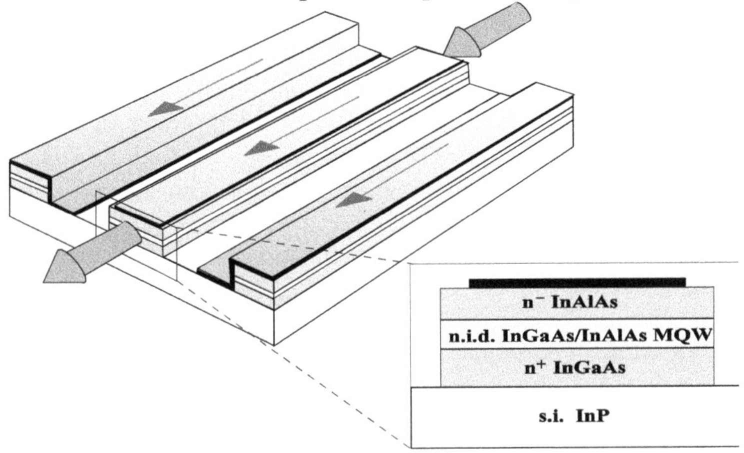

Figure 75 : Sketch of coplanar waveguide modulator.

At this time it should be noted that an EA modulator (EAM) can also be used as a photodetector because the physical mechanism of the QCSE is based upon the generation of electron-hole pairs. As a result, the EAM combines photodetector and modulator behaviour. We call such a device an electroabsorption transceiver (EAT) element. We have studied such an EAT for operation at 1.3μm wavelength. The device consists of a slightly Silicon doped lattice matched InAlAs top cladding layer and a highly Silicon doped lattice matched InAlAs bottom cladding layer. The active region is formed by 20 n.i.d quantum wells with a thickness of 7.7nm each. By implementing 1% tensile strain in the InGaAs quantum wells and 1% compressive strain in the InAlAs barrier layers polarization insensitive operation is achieved, which avoids expensive measures for polarization control within any system. For operation up to the (M)MW regime we used a hybrid coplanar microstrip configuration for the metallization of the modulator. In a former experiment we have demonstrated millimeter wave operation of up to 70GHz using such nin-EA-modulators with hybrid coplanar-microstrip metallization.

5.1.7. Other Microwave Optical Interaction Devices

Today there is a huge class of microwave optical interaction devices used and proposed for different applications:

- switches: such as interferometric or total internal reflection switches and so called digital optical switches
- optically controlled microwave devices: transistors, such as HEMTs or HBTs, diodes, mixers, phase shifters, filters
- signal processing elements: electrooptical elements resembling the well known acoustooptical devices, optoelectronic signal generators and pulse compressors, optocoupler based flip-flops and bistable or switching elements with memory, SEED (self-electrooptic effect device) elements for switching and logic operations, nonlinear vertical Bragg reflectors, etc

It is foreseen, that further improvement and ideas will lead to highly interesting devices

5.1.8. Technical Applications

As a result of recent advances in microwave photonic devices several technical application areas of using fiber optic links instead of metallic transmission lines are under discussion.

In phased array antenna systems, optical links are used to interconnect the antenna array with the central control station where an optical control is also discussed today.

Fiber wireless systems on the other hand are candidates for the distribution of radio and cable TV or even for bidirectional mobile telecommunication. Here the picocells are optically connected to the base station and indoor as well as outdoor applications are under development. Similarly, optical interconnects are in progress for wireless PC networkas for multimedia techniques.

Optical measurements of microwave signals for EMC applications or in integrated circuits is another area of using optical links here for high.speed sensing of electrical signal.

5.1.9. Conclusion

It is shown by various examples that coplanar optoelectronic devices can meet the current requirements for ultra-high-speed operation. In particular, TW-photodetectors, modulators and laser diodes are not limited by the usual RC time constants. Instead, microwave properties determine the bandwidth and the input resistance ia given by the characteristic impedance. As a result, travelling wave devices are much more flexible with respect to design parameters and provide a layout ideally suited for further monolithic integration in optical MMICs. Simulation and modelling of the devices can be carried out by using equivalent circuits for the optical and electrical domain, where the interaction can be considered by parametrically controlled elements. Today, there is already a huge amount of microwave optical interaction devices which are suitable for high-speed optical links in different application areas.

5.2. The GaAs MESFET as an Optical Detector

A. Madjar* , A. Paollela+ , P.R. Herczfeld+
* Technion - Israel Institute of Technology and RAFAEL,
Haifa, Israel, E-mail : asher@ee.technion.ac.il
+ Drexel University, ECE Department, Philadelphia, PA, USA.

Abstract

The GaAs MESFET is a light sensitive device. For normal microwave applications this is considered an undesirable property, and device manufacturers attempt to reduce it. In the last decade researchers discovered ways of utilizing this effect. In this paper we present briefly the optical detection properties of the MESFET and describe some typical applications.

5.2.1. Introduction

Gallium Arsenide is a light sensitive material. When illuminated by light with photon energy greater than the bandgap ($h\upsilon > E_g = 1.41$ eV) each absorbed photon generates an electron-hole pair. The optically generated charge carriers alter the material properties and are responsible for the optical effects in GaAs devices. The GaAs MESFET is an important microwave device and serves as the building block for MMICs. For normal microwave applications the light sensitivity is very undesirable, and device manufacturers try very hard to minimize this effect. However, by the end of the seventies it was clear that the light sensitivity of MESFETs can be utilized favorably, and during the eighties many applications have emerged.

In this review paper we begin by outlining the most important potential applications of light interaction with MESFETs in section 5.2.2. The physical photodetection processes are explained in section 5.2.3. The response to constant illumination is presented in section 5.2.4, and the modulated light effects are explained in section 5.2.5.

5.2.2. Applications

The light sensitivity of GaAs MESFET and its applications have been investigated quite thoroughly for more than a decade. In this section we present the main potential applications and explain their importance.

Optical port on MMIC - Since the MESFET is a building block of MMICs it is reasonable to utilize it as an on-chip optical port. This can be useful mostly for optically fed phased arrays, where the microwave signal is distributed to the radiating elements by use of optical fibers. In this application the MESFET can serve as an integral optical detector in the T/R module.

Tuning of MESFET oscillators - By direct illumination of the MESFET, which serves as the active device of a microwave oscillator, it is possible to tune the oscillator's frequency. This application has been demonstrated by several researchers. Already in 1979 Moncrief([169]) has demonstrated successfully such an optical tuning of a 12 GHz oscillator. He achieved tuning range of 400 MHz with 100 mw of optical power. Generally, the achievable optical tuning range of MESFET oscillators is in the order of few percent (see also [170,171]).

Injection locking of MESFET oscillators - Optical injection locking of MESFET oscillators is achieved by illuminating the device with modulated light at a frequency very close to the oscillation frequency. Such a technique can be very useful in optically fed phased array antennas to distribute the reference signal. Initial observations of direct optical injection locking were reported by De Salles and Forrest ([172]) at 2.35 GHz, which achieved a locking range of 5 MHz. Similar experiment was conducted by Buck and Cross ([173]) and a modelling technique was presented by Warren et al. ([174]). All the reported experiments have demonstrated a very small locking range (a fraction of a percent), which is attributed to poor coupling of the light into the active region of the device and also to the relatively poor response of the MESFET to modulated light at microwave frequencies (section 5 below).

Control of MESFET amplifiers and phase shifters - The gain of amplifiers and the phase of phase shifters can be controlled optically by illuminating the MESFET with varying light intensity. By proper design the gain or phase are monotonically increasing functions of the light intensity. Very effective gain control has been reported by several authors, and tuning ranges as large as 25-30 db were achieved ([175,176]). Complete optical phase control (360 degrees) was demonstrated for a 6 bit phase shifter by Jemison et al ([177]).

Optical switching - Direct and indirect switching of MESFETs by pulsed illumination can be very useful for many applications, including optically controlled T/R modules for phased arrays. Direct switching is achieved by use of the photovoltaic effects in the MESFET (section 5 below). Indirect optical switching was reported by Paollela et al. ([178]).

Optical and optoelectronic mixing - The MESFET can be used as an optical detector and a mixer simultaneously by illuminating it with an optical carrier modulated by the RF information. If the MESFET is self-oscillating, the IF output is extracted at the drain; otherwise, the MESFET has to be fed by an RF local oscillator to achieve the same result. This type of optical receiver was reported by Rauscher et al. ([179]). A different type of mixing is achieved by illuminating the MESFET from two different lasers simultaneously. The two optical signals mix in the MESFET and the resulting difference frequency, which is in the microwave or millimeterwave range is extracted. This can be used as an alternate method to injection lock oscillators or for generation of microwave signals. This approach was demonstrated by Goldberg et al. ([180]), Fetterman et al ([181]) and Ni et al ([182]).

5.2.3. Optical Detection Mechanisms

During the last decade a large number of researchers have investigated the photodetection properties of the MESFET. Most of the published work was experimental, and demonstrated the effectiveness of light detection in MESFETs. The first known study on the effects of light on the DC characteristics was performed by Gaffuil et al. [183]. Mizuno [184] conducted an experimental study of DC optical response and microwave scattering parameters of the MESFET as function of the biasing conditions and light intensity. Gautier et al.[185] measured the effect of optical illumination on the MESFET both at DC and at microwave frequencies for several biasing conditions. Simons et al.[186,187] reported extensive measurements of the optical response of MESFETs and HEMTs both at DC and at microwave frequencies. Madjar et al.[188] have identified photoavalanche effects in MESFETs, which can be utilized to increase the optical response. An experimental and theoretical MESFET characterization, with emphasis on the photovoltaic effect, was carried out by De Salles [189]. He also investigated the photoresponse as a function of light intensity and performed a preliminary study of the backgating effect and optically induced substrate current. Darling [190] developed a perturbation analysis that accounts for the photoconductive effect under low level illumination. Recently an in-depth experimental and theoretical investigation of the optical response of MESFETs has been performed by us, and the resulting theoretical model and experimental results are presented in [191].

The physical processes responsible for the light sensitivity of the MESFET are presented in detail in [191]. In this section we present briefly the nature of these photodetection mechanisms. The various current components induced in the illuminated MESFET are depicted in figure 76. In general, the current components can be attributed to either photoconductive or photovoltaic effects. Both effects exist in the device and should be considered.

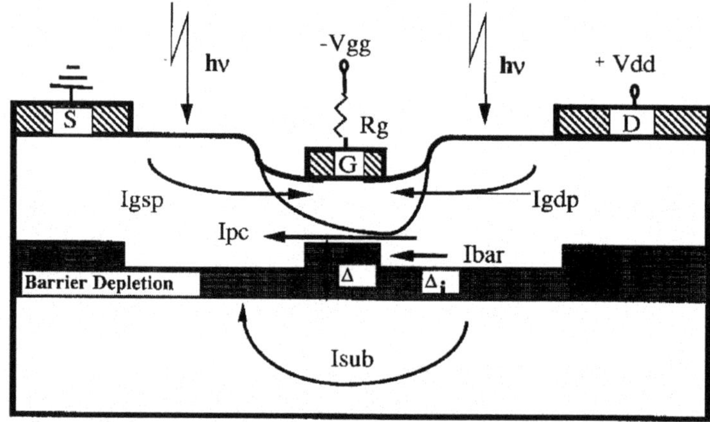

Figure 76 : Cross sectional view of MESFET demonstrating optically induced current components.

The device cross section area in figure 76 is divided into five regions. In each region a different photodetection mechanism is occurring:

Region 1 is the illuminated portion of the gate depletion region. The electron-hole pairs optically generated here are swept by the large electric field and contribute to the optically induced gate current ($I_{gate}=I_{gsp}+I_{gdp}$). The holes are swept to the gate, while the electrons generated on the source side are swept to the source (and contribute to I_{gsp}) and those generated on the drain side are swept to the drain (and contribute to I_{gdp}).

Region 2 is the illuminated portion of the channel. The optically generated carriers here establish an accumulation of excess carriers just like in any photoconductor. This excess charge distribution contributes to two current components: (a) photoconductive drain current - I_{pc}; and (b) gate current - I_{gate} (due to diffusion of holes from the channel to the gate depletion region).

Region 3 is the illuminated inter-electrode area. The optically generated charge carriers here increase the conductivity and thus decrease the parasitic resistances of these regions. Since the epitaxial layer is quite heavily doped the excess carrier concentration in this region is relatively small, and its effect is usually negligible.

Region 4 is the depletion region of the barrier junction existing between the substrate and the epitaxial layer. This barrier is created due to the large doping step between the two materials. Electrons diffuse from the heavily doped epilayer to the semi-insulating substrate, thus leaving behind a depleted region. The diffused electrons reside in the substrate very close to the junction since they are electrically attracted to the ionized donors in the epilayer. The optically generated electron-hole pairs here are swept by the high electric field (electrons to the epilayer and holes to the substrate), and thus establish an optically induced (vertical) current between the substrate and the epilayer. This current, which flows via the junction and the large substrate resistance, creates a photovoltage across the barrier junction and effectively reduces the potential barrier and the physical height of the barrier region(D). The decrease in the barrier height is equivalent to an increase of the channel height (see figure 76), namely, an increase in the drain current (denoted I_{bar}). This effect has been recognized in the past, however the first thorough investigation of this effect and its quantitative contribution to the photoresponse of the MESFET is presented in [191]. We have named this the internal photovoltaic effect.

Region 5 is the illuminated portion of the substrate. The optically generated charge carriers here constitute the substrate drain current (denoted I_{sub}). This contribution to the drain current is possible due to the optically induced decrease of the barrier, as explained in the previous paragraph. Without the barrier decrease the excess carriers in the substrate cannot enter the epilayer.

5.2.4. Constant Illumination

In this section we consider the MESFET's response to constant illumination taking into account all the physical processes described in the previous section. The optical response under constant illumination includes the following measurable effects: (a) increase of gate current, (b) increase of drain current and (c) change in the microwave scattering parameters. All of these effects have been observed and documented (i. e. [183-190]).

Gate current - In the dark the gate current is the reverse saturation current of the gate junction and is usually negligible. Under constant illumination the gate current consists of optically generated holes in region 1, which are swept to the gate by the strong electric field, and holes which diffuse from region 2 (figure 76). The derivation of the expression for the gate current is presented in [191] (Eq. 24). The gate current is relatively small (microamp range). It is a linear function of the absorbed optical power density and the illuminated area of the gate depletion region. The illuminated area can be increased by increasing both V_{ds} (drain to source voltage) and V_{sg} (source to gate voltage), because this causes further extension of the depletion region beyond the gate metalization.

Drain current - The drain current, which is the major optical response consists of several components, and can be written as

$$I_{drain}=I_{pc}+I_{pvi}+I_{gdp}+I_{pvx} \quad (24)$$

where I_{pc} is the photoconductive current, $I_{pvi}=I_{bar}+I_{sub}$ is due to the internal photovoltaic effect, I_{gdp} is the gate current contribution from the drain side and I_{pvx} is due to the external photovoltaic effect.

I_{pc} originates from region 2 in figure 76 and is calculated by solving the generation/recombination/diffusion continuity equation. The expression for I_{pc} is presented in Eq. 24 of [191]. This photoconductive current contribution is a linear function of the optical power density and for typical microwave MESFETs it turns out to be extremely small (submicroamp). The main reason is that this current is proportional to the cube of the channel height, which is very small for microwave devices.

The internal photovoltaic effect is explained in the previous section. Complete theoretical analysis and the expressions for I_{bar} and I_{sub} are presented in [191]. Generally, this effect is of great importance and constitutes the largest photodetection response - tens of milliamps. $I_{bar}=g_m V_{ph}$ is the main contribution, where g_m is the transconductance and V_{ph} is the optically induced photovoltage across the barrier. V_{ph} is approximately a logarithmic function of the absorbed optical power density. For large optical power V_{ph} approaches the value of the built-in potential of the barrier junction (~0.8 volt), which limits the maximum photoresponse due to this effect.

The external photovoltaic effect is manifested when a large external resistor is present at the gate circuit. In this case, the optically induced gate current when flowing via the external resistor creates a photovoltage, which tends to increase the gate to source voltage, and thus increases the drain current. For large external resistor values this can be a very large effect. In fact, it is possible to optically switch the device between cutoff and saturation. This effect was characterized by several researchers (i. e. [189],[191]). Exact expressions for this contribution are presented in [191]. Generally, this contribution is proportional to the induced photovoltage across the gate junction. For low optical power there exists a linear relationship between the optical power and the photovoltage, however for large optical power the gate junction becomes forward biased, and the photovoltage becomes a logarithmic function of the optical power. The upper limit is reached for large optical power or large resistor value, which forward bias the gate junction, and cause the channel to reach its maximum height (epilayer thickness).

To sum up, photovoltaic effects dominate the MESFET's response to constant illumination. These effects are of a compressive nature, compared to the linear relationship for the photoconductive effect. This behavior is depicted in figure 77, which displays the drain current photoresponse as a function of optical power for a FUJITSU MESFET. The figure is taken from [191], and it shows the response for both no gate resistor (external photovoltaic effect missing) and a 1 megohm resistor in the gate. The compressive nature of the response is clearly visible in figure 77.

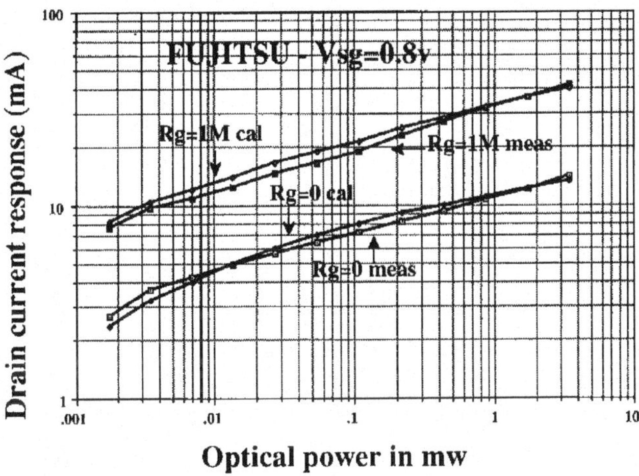

Figure 77 : Drain current photoresponce of FUJITSU MESFET vs. Optical power under constant illumination.

Scattering parameters - The change in the scattering parameters of the MESFET under constant illumination has been measured and documented in several publications (i. e. [184-187,189]). There is a complete agreement between all researchers that the main reason to the change in the S parameters is due to optically induced change in the bias point. As is well known, microwave MESFETs are quite sensitive to bias. The optically induced change in the drain current and gate voltage as outlined above is therefore associated with a change in the scattering parameters. Observations show that the main change is in S_{21}, namely, the gain parameter, which is reasonable, because the transconductance g_m is quite sensitive to bias.

5.2.5. Modulated Illumination

In this section we look at the photoresponse of the MESFET under non-constant illumination, namely, the light intensity is changing with time. The most common modulation types are: (a) small signal sinusoidal modulation for communication links and (b) large signal square wave modulation for switching applications. The response of the MESFET to modulated light is derived similarly to the constant illumination case, except that now the time constant associated with each one of the different physical mechanisms outlined in section 5.2.3 must be taken into account.

We have analyzed both of the above modulation types, and the complete analysis is being prepared for publication ([192-193]). Here we present the main results and their practical implications. In general, large time constants are associated with current components that are large at low frequencies. Thus, at high frequencies the relative magnitudes of the various current components are different than those at low frequencies. Furthermore, the relative magnitudes are strong functions of frequency. The following can be stated regarding the "speed" of the various contributions:

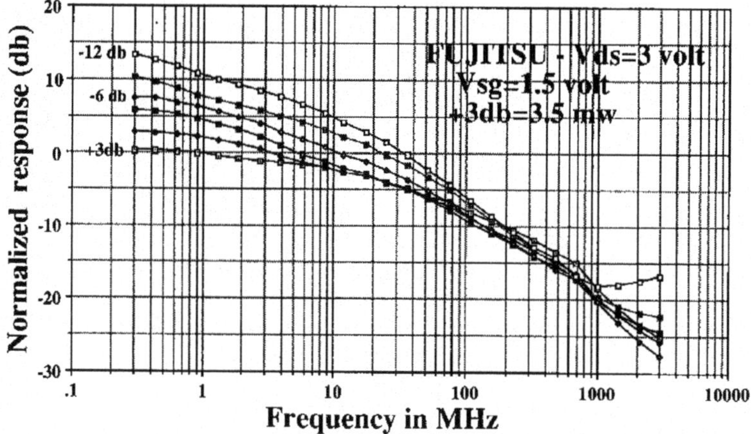

Figure 78 : Frequency response of FUJITSU MESFET for several optical power levels.

<u>Gate current</u> - The gate current is associated with a very small time constant attributed to the gate capacitance and load resistance, which is typically in the picosecond range.

<u>Photoconductive current</u> - Solution of the time-dependant continuity equation in the channel yields a time constant of around 100-500 times the characteristic frequency of the recombination process. Since the

recombination lifetime for GaAs is around 3-5 ns, the time constant associated with the photoconductive current is in the picosecond range.

Internal photovoltaic effect - The characteristic frequency associated with this effect is related to the RC time constant of the barrier junction. The unique feature is that this frequency is dependant on the optical power because both the junction capacitance and junction resistance are nonlinear. Generally, as the optical power increases the capacitance increases and the resistance decreases. It turns out that the net effect is an increase of the characteristic frequency with increasing optical power. Unfortunately, the low frequency gain is decreasing with the increase of optical power, so that the gain-bandwidth product is almost a constant. For typical microwave devices the characteristic frequency ranges from around several MHz for low optical power to around 200-500 MHz for large optical power. The frequency response is almost constant from DC up to the characteristic frequency, beyond which the response rolls off. The slope of the rolloff is less than 20 db/decade, and this is attributed to the deep level traps in the GaAs([192,194,195]). A typical frequency response plot for a FUJITSU MESFET is depicted in figure 78. Shown is the normalized response vs. frequency for several values of optical power. The dependance of gain and 3db cutoff frequency on optical power as well as the less than 20db/decade roloff slope are clearly visible.

External photovoltaic effect - This effect is associated with the gate circuit time constant. For typical microwave MESFETs the gate capacitance is very small (less than 1 pF), but the external gate resistance is very large (greater than 100 kohm); thus typically the time constant is around 1 microsecond. Since the gate capacitance is an increasing function of optical power so is the time constant, however, this is a weak dependance.

To sum up, the photovoltaic effects have a large but "slow" response. At microwave frequencies the external photovoltaic effect is attenuated very strongly and is practically negligible. The internal photovoltaic effect has usually larger cutoff frequency, which can be controlled by the optical power, however this effect is also very small at microwave frequencies. The gate current and photoconductive current are very small, but at high frequencies their magnitude exceeds the photovoltaic effect, because they are "fast" and do not roloff up to very high frequencies well into the microwave range. Therefore, at microwave frequencies the MESFET optical response is small, associated with the gate junction and resembles a photodiode. These facts explain the small locking range achieved by optical injection locking. For switching applications the large photovoltaic effects can be utilized, however, the rise and fall times are typically in the microsecond range, which means that the device is limited to switching rates of several MHz.

5.2.6. Conclusions

In this paper we have reviewed the phenomenon of light interaction with MESFET. The importance and applications of this photodetection have been explained, and the physical mechanisms were described. It was shown that the response to modulated light is quite different from the response to constant illumination. In general, the MESFET is a relatively "slow" photodetector, and its response decreases strongly with frequency. At low frequencies and DC the response is very large, and exhibits gain, but at microwave frequencies the response is small and similar to a photodiode. Despite the above, the MESFET is useful at high frequencies for applications such as injection locking and optical mixing, which cannot be obtained by photodiodes.

5.3. HBT Phototransistor as an Optic/Millimetre-wave Converter – Part I : the Device

C. Gonzalez
France Telecom, CNET-DTD, Laboratoire de Bagneux
BP 107, 196 avenue Henri Ravera, 92225 Bagneux, France
E-mail : carmen.gonzalez@cnet.francetelecom.fr

5.3.1. Introduction

HBT phototransistor (HPT) is the subject of intensive research as one of the most promising optic/millimetre-wave converter [196-198]. HPT has a structure similar to the heterojunction bipolar transistor and it can be seen as an HBT with a window area on the base side, for the optical input. Therefore, HPTs keep potentially the excellent frequency performances of HBTs in the mm-wave domain [199-201]. Also, by using the inherent non-linear properties of HBTs, the phototransistors can provide high optoelectronic mixing efficiency [197,202-203].

On the other hand, high speed fibre radio communication systems in the long wavelength regime (1.3 μm to 1.55 μm), require a large number of optical/radio frequency converters operating in the mm-wave band for signal radiation or distribution. This paper analyses the performances of InP/InGaAs HBT phototransistors as an optical/RF converter for this type of communication systems.

We focus on the performances of InP/InGaAs phototransistor as a direct photodetector and as an optoelectronic upconverting mixer to the millimetre-wave band. The noise performances are also described.

5.3.2. InP/InGaAs HBT Phototransistor

Figure 79 shows the schematic diagram of the cross-section of a n-p-n epitaxial phototransistor and its corresponding energy-band. This n-p-n structure consists of a wide band-gap (1.35 eV) n-type InP emitter, a p-type $In_{0.53}Ga_{0.47}As$ base, and a lightly doped n-type $In_{0.53}Ga_{0.47}As$ collector. In the long wavelength regime (1.3 to 1.55 μm), the absorbing semiconductor is $In_{0.53}Ga_{0.47}As$ with a gap energy of 0.75 eV. The phototransistor optical detection process can be explained in the following way : the input light at 1.55 μm is absorbed in the base and base-collector depletion regions which creates electron-hole pairs. The photogenerated holes are swept into the base and modify the base-emitter potential, which causes a large electron current to flow from the emitter into the base. Current amplification is achieved through the transistor gain mechanism. Similar to HBT structure, the function of the wide gap emitter is to increase the emitter injection efficiency, by preventing reverse injection of holes from the base into the emitter. The primary photocurrent I_{ph}, that is the photocurrent without amplification, is created by the absorbed light. I_{ph} is a function of the optical input power P_{opt} and of the device parameters. In this way, R is the reflection coefficient to the surface of the base layer, W_B is the thickness of the base layer, W_{BC} is the thickness of the depletion layer and L_p is the diffusion length of the holes in the base layer.

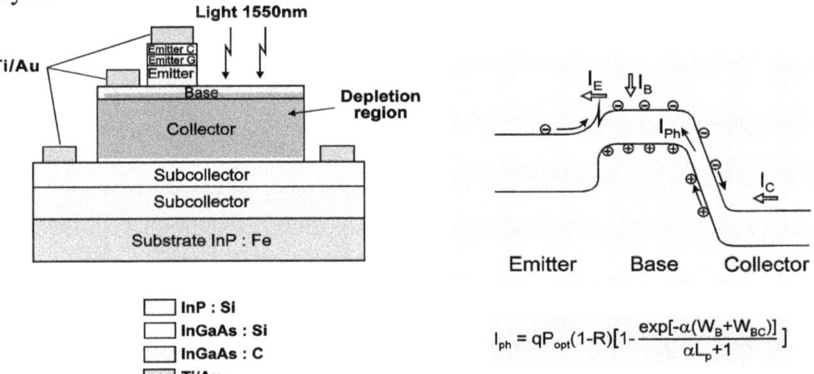

Figure 79 : Schematic diagram of the n-p-n phototransistor and its corresponding energy-band diagram.

On the other hand, the heterojunction phototransistor similar to the HBT, is a three-terminal device used in a common emitter-mode operation. As shown in figure 80, HPT with the optical window on the base region can be seen as an HBT with the base terminal connected to a PIN photodiode The incident optical signal generates a photocurrent similar to a current source applied to the base terminal. So, the phototransistor is an current-controlled current-source device.

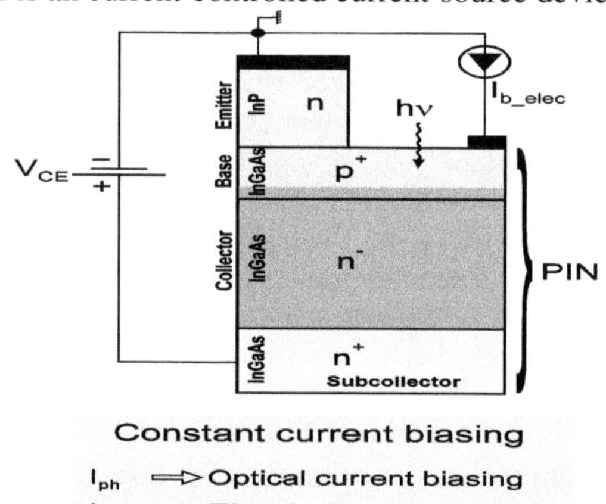

Figure 80 : Bias condition of the three-terminal phototransistor

5.3.3. HBT Phototransistor Technology

HPTs are fabricated using an in-house three-mesa technology, figure 81. All the samples were grown on Fe-doped semi-insulating InP substrates by chemical beam epitaxy (CBE) [205]. The layer structure is shown in table 4. Silicon and Carbon are the n-type and the p-type dopants, respectively. The three-mesa technology is used to make contacts with the emitter, base and collector layers. Emitter, base and collector mesa are delimited using successively dry and wet etching. Ti/Au and a rapid thermal annealing under nitrogen flux are used for emitter, base and collector contacts. Contact resistances of 5×10^{-7} and 10^{-5} Ohm-cm^2 are usually obtained for n- and p-type ohmic contacts, respectively. Finally, polyimide is used for planarization and isolation between the two levels of interconnection and no antireflection coating was used.

Device layer	Material	Doping (cm^{-3})	Thickness (nm)
Emitter Cap	n^+ : InGaAs	Si : $1x10^{19}$	100
	n^+ : InP	Si : $1x10^{19}$	50
Emitter	n : InP	Si : $2x10^{17}$	150
Base	p^+ : InGaAs	C : $2x10^{19}$	60
Collector	n^- : InGaAs	Si : $1x10^{16}$	500
Subcollector	n^+ : InGaAs	Si : $1x10^{19}$	200
Subcollector	n^+ : InP	Si : $1x10^{19}$	300
Substrate	Fe : InP	semi-insulating	450 µm

Table 4. Phototransistor epitaxial layer parameters.

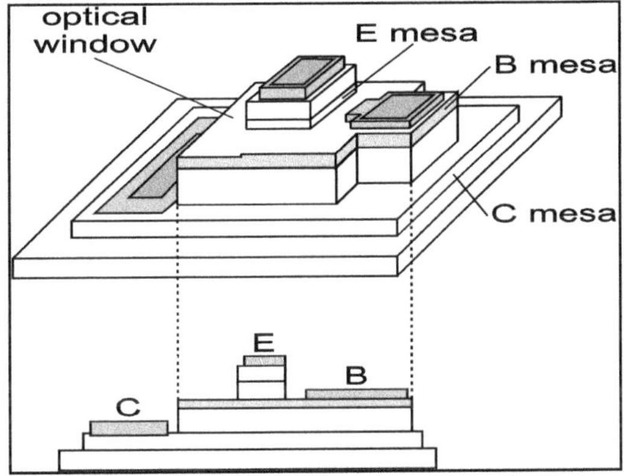

Figure 81 : Three-mesa technology used for HPTs.

5.3.4. HBT Phototransistor Performances

The HPT investigated in this work has the following geometric characteristics : emitter/base junction area, base/collector junction area and optical window area equal to 16 μm^2, 154 μm^2 and 48 μm^2, respectively. Both, electrical and optical measurements were obtained by using an on-wafer probe station, with and without illumination.

5.3.4.1. Electrical Characteristics

The electrical characteristics were obtained without illumination and S-parameters were measured with a network analyser ranging from 250 MHz to 65 GHz. The frequency dependence of the current gain H_{21} and the Mason's unilateral power gain G_{max}, shown in figure 82 a, were calculated from the S-parameters. Both, the unity current gain cut-off frequency f_t and the maximum frequency of oscillation f_{max} are dependent on the collector current Ic, and the best performances were f_t = 58 GHz, f_{max} = 20 GHz for Vce = 1.6 V, Ic = 9 mA and Ib = 500 μA, as is displayed in figure 82 b.

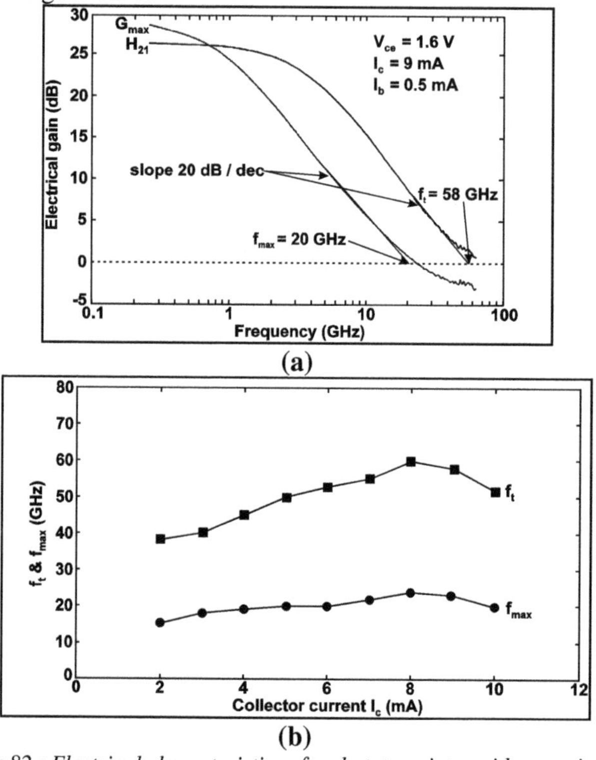

Figure 82 : Electrical characteristics of a phototransistor with an emitter area of 16μm² and a base/collector area of 154 μm². (a) Current and power gains as a function of frequency. (b) f_t and f_{max} frequencies as a function of the collector current Ic.

5.3.4.2. Optical Characteristics

5.3.4.2.1. HPT as a Direct Photodetector

Next, the phototransistor was investigated as a photodetector at a wavelength of 1.55 μm. The intensity of the light was modulated by a RF signal ranging from 130 MHz to 20 GHz. The light coupling was provided via a lightwave probe with a lensed single mode fibre with an illumination spot size equals to 5 μm. The experimental setup is shown in figure 83. Figure 84 displays the frequency photoresponse of the device under two bias conditions. One was the photodiode mode operation (PD-mode) with Vce = 1.6 V and Vbe = 0 V (without transistor effect) and the other one was the transistor mode operation (Tr-mode) with Vce = 1.6 V and Vbe > 0 V (with transistor gain). The best optical characteristics were obtained in the same bias conditions as that used for electrical characterisation, i.e., Vce = 1.6 V and Ic = 9 mA. However, in optical characterisation the base current is the addition of two terms : the base current due to the injected electrical current and the base current due the photogenerated carriers. The total base current was 500 μA. This value was obtained with an average optical power P_{avg} = 700 μW and an electrical current Ib = 325 μA. The modulation index of the laser beam was m = 50%, and the peak modulated component of the incident optical power was P_{mod} = 350 μW. The photoresponse R expressed in dB is equal to $20\mathrm{Log}_{10}[\Re]$, where $\Re$ is the responsivity in A/W. The optical gain G_{opt}, defined as the difference between the Tr-mode photoresponse and PD-mode photoresponse, was 24 dB at 130 MHz. Similar to the electrical cut-off frequency f_t, we defined the unity optical gain cut-off frequency f_c as the frequency for which the Tr-mode gain is equal to the low frequency PD-mode gain. f_c was estimated to 42 GHz. And the external DC responsivity $\Re_{DC}$, evaluated under the photodiode mode operation, was 0.32 A/W.

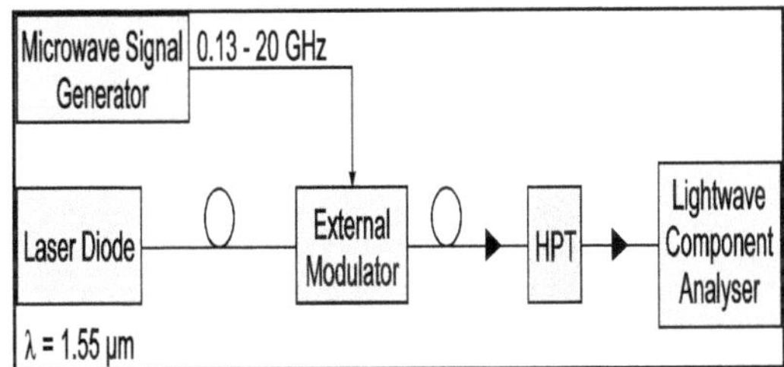

Figure 83 : Experimental setup used for photoresponse characterisation of HPTs.

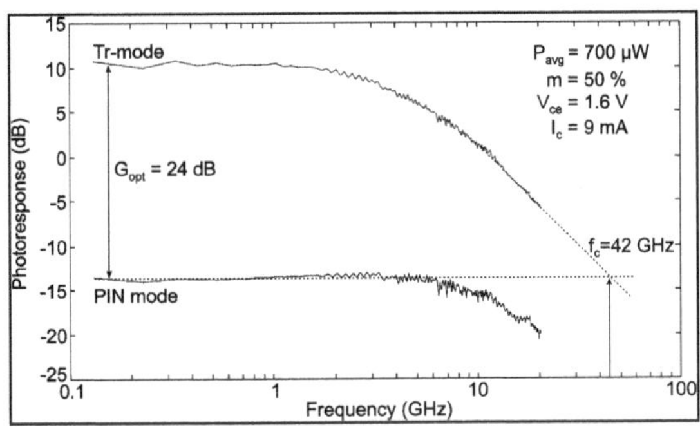

Figure 84 : Optical gain G_{opt}, and optical cut-off frequency f_c of the phototransistor.

5.3.4.2.2. HPT as an Optoelectronic Upconverter Mixer

We have taken advantage of the inherent non-linear properties of the phototransistor to achieve up-conversion of a modulated optical signal at lower frequency. For this mixing experiment, the intensity of the optical signal was modulated by an IF signal ranging from 200 MHz to 2.5 GHz. The average optical input power was 840 μW (-0.76 dBm) and m = 50%. The local oscillator was provided by a frequency synthesiser and injected into the base terminal. Two F_{LO} frequencies were used, 15 GHz and 30 GHz, both with an input power of -4.5 dBm. The experimental set-up for the mixing measurements is shown in figure 85. The IF signal component, F_{IF}, is up-converted to $F_{LO} \pm F_{IF}$ making use of the non-linearity of the transistor. Also, the IF signal is mixed with the second harmonic $2F_{LO}$ and so on, as is shown in the same figure.

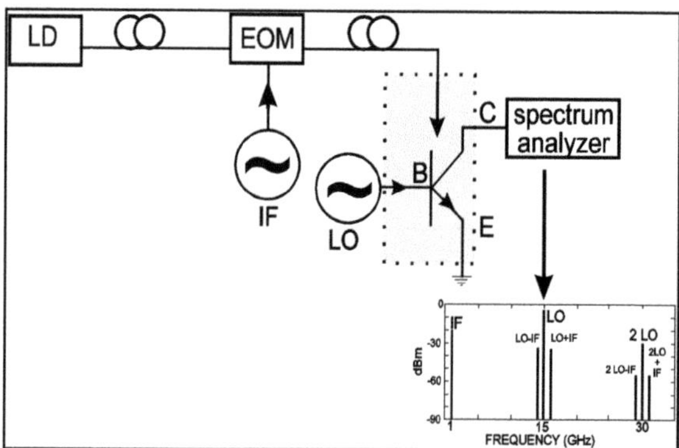

Figure 85 : Experimental setup for optoelectronic up-conversion measurements.

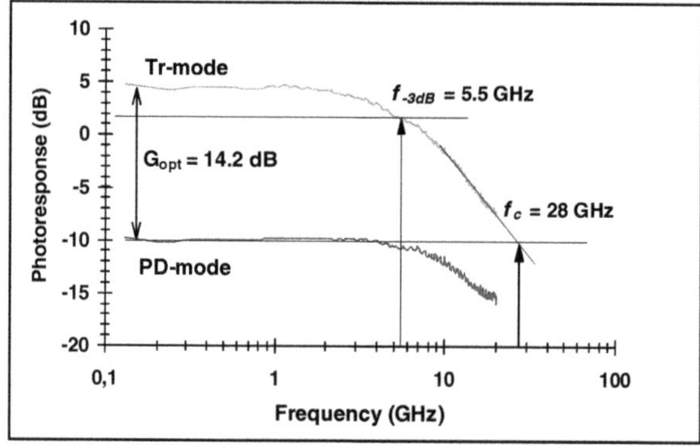

Figure 86 : Frequency photoresponse of the HPT under the same bias condition used for mixing experiment

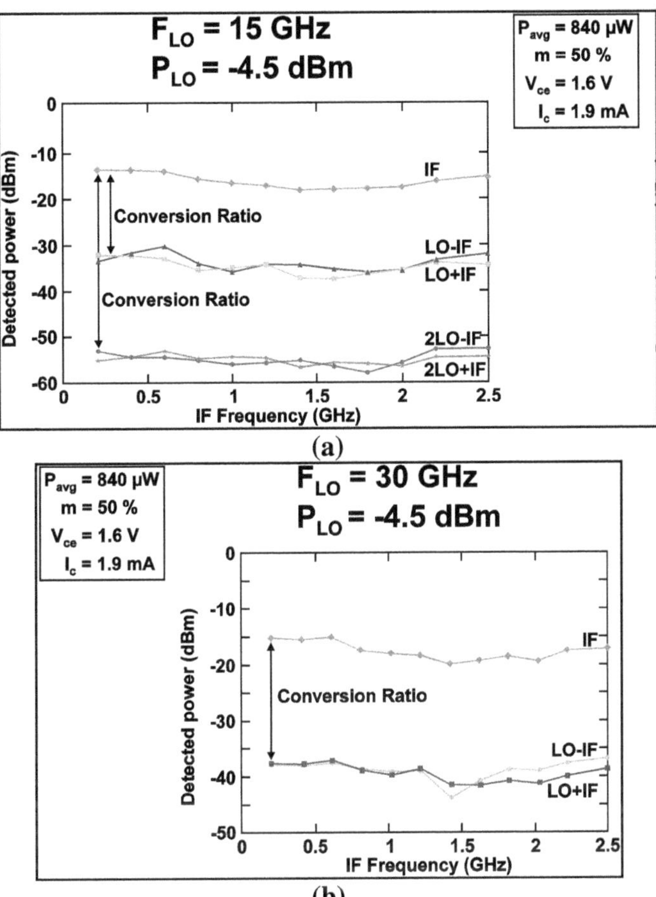

Figure 87 : Optoelectronic up-conversion to the 15 GHz band (a) and to the 30 GHz band (b), as a function of the IF frequency.

We show in figure 87a, the up-conversion of the IF signal when F_{LO} is equal to 15 GHz. This figure displays the output power of the IF signal and of the two up-converted signals, (15 $\pm$ F_{IF}) GHz, as a function of the IF frequency. Also, it is shown the up-converted signal levels with the second harmonic, (2x15 $\pm$ F_{IF}) GHz. The phototransistor was operated at V_{ce} = 1.6 V and I_c = 1.9 mA. These bias conditions were optimised in order to obtain the maximum up-converted level. At F_{IF} = 200 MHz, the detected level power of the IF signal and of the upper/(15+0.2) GHz and lower/(15-0.2) GHz side bands were -14.2 dBm, -34 dBm and -32.7 dBm, respectively. The conversion ratio, i.e., the radio of the up-converted signal power to the IF signal power, was around -19.2 dB at (15 $\pm$ 0.2) GHz. This value is nearly constant within the whole IF frequency interval.

This is a consequence of the 5.5 GHz f_{-3dB} bandwidth of the HPT photoresponse measured under the same mixing bias condition, as is shown in figure 86. From this curve we obtained an optical gain of about 14 dB within the whole IF frequency range. For the second harmonic, the detected power of the upper and lower side bands (30 $\pm$ 0.2) GHz were -53.3 dBm and -55.3 dBm, respectively. And the conversion ratio was around -40 dB. Since the detected IF signal includes the 14 dB of the optical gain, the conversion loss was estimated at -5 dB and -26 dB for (15 $\pm$ 0.2) GHz and (30 $\pm$ 0.2) GHz, respectively.

A second experience was developed using F_{LO} equals to 30 GHz. As shown in figure 87b, the output power of the up-converted components, (30 $\pm$ 0.2) GHz, is around -38.3 dBm with a conversion ratio of -24.2 dB. And the conversion loss was estimated to -11 dB. This result shows that a higher conversion efficiency is achieved using a fundamental oscillation frequency of 30 GHz with regard to the second harmonic of 15 GHz.

5.3.4.3. Noise Characteristics

To obtain the noise performances of HPTs, we measured the equivalent input noise current density $<I_{in}>$ and the signal-to-noise radio SNR as a function of the input optical power.

The equivalent input noise current density is referred as to the receiver input and the phototransistor was biased at the same bias conditions used for photoresponse experiments, but the device was not illuminated. The input optical power was substitute by the equivalent photocurrent I_{ph}. Figure 88 shows the equivalent input noise current density measured while varying the frequency from 1 GHz to 40 GHz. Within the whole frequency interval, $<I_{in}>$ was lower than 10 $pA/\sqrt{Hz}$.

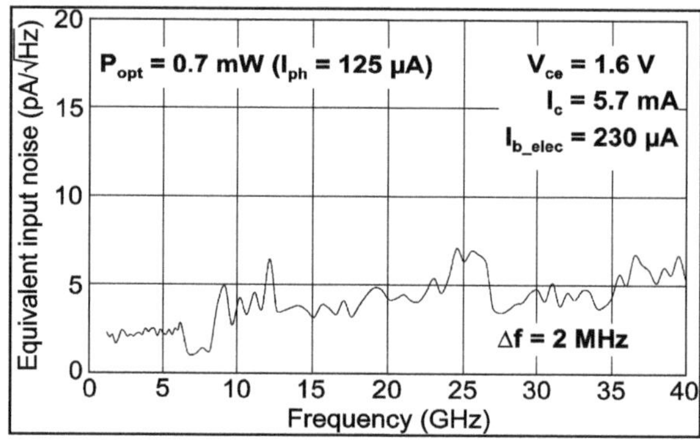

Figure 88 : HPT Equivalent input noise current density.

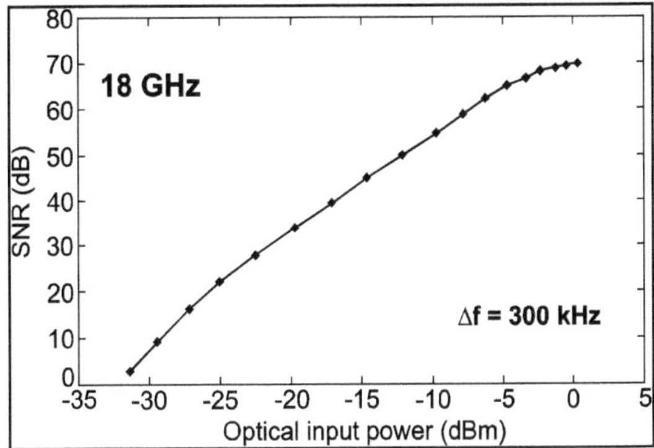

Figure 89 : Signal-to-noise ratio at 18 GHz.

The signal and noise power spectrums of HPT were measured at 18 GHz while varying the optical power from -35 dBm to 0 dBm. The noise bandwidth was 300 KHz. In this case the device is illuminated. Then the noise floor of the HPT is composed of three terms : the relative intensity noise (RIN) of the laser diode, the optical signal shot noise associated with and the receiver noise or equivalent input noise $<I_{in}>$, previously determinated.

The signal-to-noise ratio SNR, is shown in 89. As the optical input power decreases, the receiver noise gradually becomes dominant. So, at an optical power of -20 dBm, where the receiver noise $<I_{in}>$ is dominating, the SNR was 33 dB. This high value of SNR is caused essentially from the low noise characteristics of the electrical amplification of the HPT. In fact, the common-emitter HBT phototransistor operates as a transimpedance amplifier with a photodiode formed by the base-collector junction; therefore, it gives a lower noise floor than that of the PIN receiver in a 50-Ω system, as it was demonstrated by Suematsu et al.[197]. These authors found that at an optical power where the $<I_{in}>$ noise is dominating, the HPT has a higher SNR than the PIN receiver, because of lower noise floor.

5.3.5. Conclusion

We have reported the performances in the millimetre-wave band, of the InP/InGaAs HBT phototransistor as a direct photodetector and as an optoelectronic upconverting mixer. This device is able to replace the presently used device combination, that is the photodiode plus pre-amplifier, in high speed fibre radio communication. In particular, HBT phototransistor used as an optoelectronic upconverting mixer allows a microwave subcarrier of a few GHz to be converted to a mm-wave carrier, that is the high frequency carrier can be generated locally at the base station, instead of being generated at transmitter end and transmitted through fibres. Systems using local generation of high frequency subcarrier are potentially less expensive than systems which propagate such subcarriers on fibre. In this manner, the requirements for a high speed laser diode or an external optical modulator in the system can be avoided. Moreover, HBT phototransistor is compatible with the MMIC technology, therefore it can also realise a compact, simple and cost-effective MMIC photoreceiver in the millimetre-wave band.

5.4. HBT Phototransistor as an Optical Millimeter wave Converter – Part II : Simulation

C. Rumelhard, N. Chennafi, E. Namuroy
CNAM, 292 rue Saint Martin, 75141 Paris Cedex 03, France
E-mail : rumelhard@cnam.fr

Abstract

A first part describes the 2D numerical modeling of an heterojunction bipolar transistor illuminated by an optical beam. The numerical simulator is then used as a virtual measurement equipment to build an equivalent circuit non-linear model which can be used in non-linear simulators. In this model, the input signal(s) can be applied on the base or can be introduced into the transistor with a source of light used as a carrier. The characteristics of this model are compared to numerical and measurement results. This model is then used to simulate the phototransistor working as an up-converter where the local oscillator signal is applied on the base and the IF signal is applied through an optical carrier.

5.4.1. Introduction

The InP/GaInAs phototransistors which have been presented in the first part give interesting results. They show the possibility of these components to be used as optical millimeter wave amplifiers or mixers. But more complete circuits remain to be designed and these circuits will be realized in MMICs. Therefore, a complete electrical design of these circuits will have to be done before the technological realization and the success of this design needs that a non-linear model of the phototransistor be available to be used in circuit simulators working in time or in frequency domains. Numerous non-linear equivalent circuit models have been presented for the different types of HBTs [207,208]. But none of them allow the introduction of an input signal through an optical beam used as a carrier. The principle of this equivalent circuit was proposed earlier [209] but it remained to be realized and tested in different circuits. Moreover, the exact way of working of a phototransistor is always at least two dimensional and is not known in details. So, it is very useful to have at it's disposal a numerical or physical simulator being able to show the behavior of a phototransistor according to the way it is illuminated [210] or for different configurations.

So, the objective of this presentation is to describe the use of a numerical or physical simulator to get the static and dynamic responses of a phototransistor. Then, an equivalent circuit model of this phototransistor is constituted. The numerical simulator is used to have a better understanding of the different elements of the equivalent circuit. It is also used as a virtual measurement equipment to test the validity of the model for a large range of biasing voltages and currents. It can be used also to describe different configurations of phototransistors. The measurements of the phototransistor are then used to extract some parasitic elements of the model and to confirm the validity of the model. Finally, the model is used in a frequency domain circuit simulator to get the responses of the phototransistor when it is used as an up-converter.

5.4.2. 2D Numerical Simulations of the HPT

The numerical simulator is built with a set of modules which are found in a library of algorithms developed for the numerical modeling of semiconductors. These modules are available in a library called ATLAS [211].

The electrical part of the simulation is done in a classic way. Several basic equations have to be solved in the numerical simulation. The first one is Poisson's equation:

$$div(\varepsilon \nabla \Psi) = -\rho \tag{25}$$

where ε is the dielectric constant of the material, Ψ is the local voltage potential and ρ is the local charge density.

The electric field $\vec{E}$ is extracted from the voltage potential through the relationship:

$$\vec{E} = -\vec{\nabla}\Psi \tag{26}$$

Two other relationships come from the carrier continuity equations for the electrons and holes:

$$\frac{\partial n}{\partial t} = \frac{1}{q} div \vec{J}_n + G_n - R_n \tag{27}$$

$$\frac{\partial p}{\partial t} = \frac{1}{q} div \vec{J}_p + G_p - R_p \tag{28}$$

in which the electron and hole currents are $\vec{J}_n$ and $\vec{J}_p$ and the generation and recombination rates for the electrons and holes are respectively G_n, G_p, R_n and R_p while q is the electron charge.

The currents which are taken into account are the drift and diffusion currents for electrons and holes:

$$\vec{J}_n = nq\mu_n\vec{E} + qD_n\nabla n \tag{29}$$

$$\vec{J}_p = nq\mu_p\vec{E} + qD_p\nabla p \tag{30}$$

with D_n, D_p being the diffusion coefficients for electron and holes. The displacement currents are given by:

$$\vec{J}_{disp} = \varepsilon\frac{\partial\vec{E}}{\partial t} \tag{31}$$

In these relationships

- the mobility takes into account the electric field, the concentrations of charges and possibly, the temperature,

- the recombination can come from the Shokley-Read-Hall recombination, the Auger recombination, the radiation recombination and the surface recombination,

- for the electrical part, the carrier generation rate comes only from impact ionization.

Starting from these basic relationships, the different steps leading to a numerical simulations are given now.

The first step is the choice of the configuration of the component. If the component to simulate is described on figure 89, it is seen that except for the shape of the collector electrode, a 3D simulation is not necessary. A 2D simulation will be done with a vertical cross section of the transistor taken along the horizontal line.

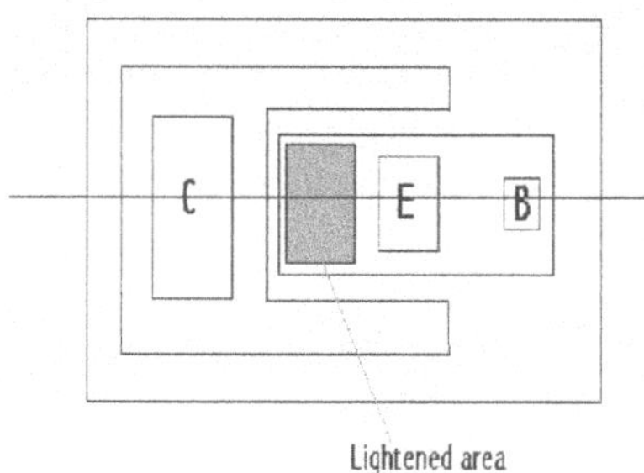

Figure 89 : Diagram of the phototransistor.

Then the different materials are defined with their parameters (dielectric constant, doping levels,...) and the corresponding physical models (mobility,...). Figure 90 gives a good example of these choices when starting from the transistor of figure 89.

Then, the meshing must be defined. The general method is a "finite boxes" method allowing local refinements in a finite difference scheme. The objective is the optimization of the meshes number to get a convergence in the computations and to optimize the computer time, but there's no general method to constitute the best meshing. The only indications are that the meshing must be refined in critical zones like in high doping gradients zones, in heterojunctions and in high recombination rates zones.

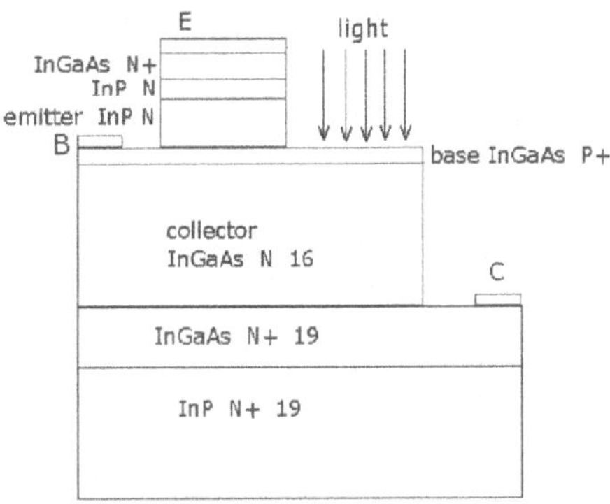

Figure 90 : Vertical cross section of the phototransistor.

The numerical method which is chosen is a block method which constitutes a mixed solution between the Gummel method consisting in computing separately each unknown while the others are kept constant and the Newton method in which all the unknown are computed in the same time.

With these first steps, a transistor is described and it is possible to get static and dynamic response of a HBT as a function of electrical biasing and of the amplitude of the electrical dynamic signal applied on the transistor.

It remains now to introduce an optical beam. In our case, it will be between base and emitter as indicated on figure 90. The position and direction of this beam are defined. In our case, the beam is perpendicular to the interface between air and semiconductor. In the semiconductor, the effect of the optical source is to add an optical generation term in the carrier continuity equation. This term is given hereafter:

$$G_{opt} = \eta_0 \frac{P\lambda}{hc} \alpha \exp(-\alpha x) \tag{32}$$

in which P is the power and λ the wavelength of the optical beam and η_0 is a transmission coefficient. This beam is absorbed in the material according to an attenuation coefficient α which is proportional to the imaginary part of the refractive index of the material:

$$\alpha = \frac{4\pi}{\lambda} \mathrm{Im}(n) \tag{33}$$

in which n is the complex refractive index of the material. In our case, this index is given in table 5. On this table, it is seen that for 1.77 μm, the material is almost transparent.

Figure 91 shows another way to consider the above relationship. For a constant optical power input, the number of electron-hole pairs which are generated and therefore the available current increases as a function of wavelength. This is shown on the continuous curve. But, in the same time, the absorption coefficient decreases and this effect is shown on the dotted curve.

λ (μm)	Re(n)	Im (n)
0.83	3.71	0.40
0.88	3.68	0.40
0.95	3.66	0.38
1.03	3.65	0.34
1.13	3.63	0.32
1.24	3.61	0.30
1.38	3.57	0.22
1.55	3.55	0.17
1.77	3.52	0.00564

Table 5 : Variation of the refractive index as a function of wavelength in GaInAs.

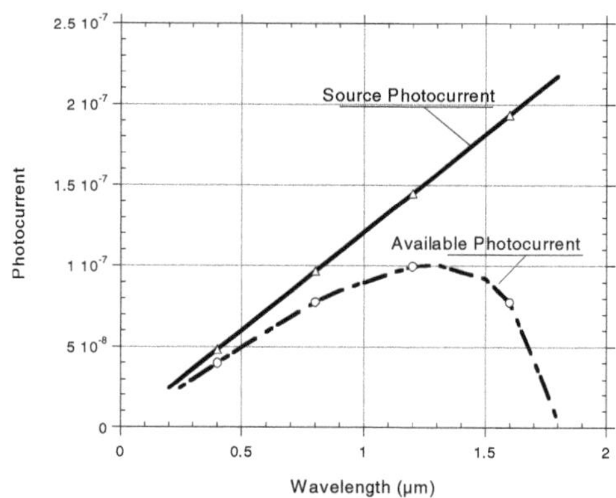

Figure 91 : Available current due to the absorption of light as a function of wavelength.

It is now possible to get static and dynamic responses of the phototransistor. The description of the transistor is given on figure 90 but all the results will be with a transistor having a depth of 1μm. The static responses will be shown hereafter on the curves showing the static response of the equivalent circuit model. Figure 92 shows the dynamic response when the transistor is illuminated with an optical power of 100 W/cm^2 with a modulation index of 50% and a wavelength of 1.55 μm. The output of the transistor is terminated with a short circuit.

The lower curve is obtained when the emitter and base are short-circuited, i.e. the transistor is working as a photodiode. This mounting is used to get the basic response of the photodiode. Then the base is biased and the response as a function of frequency is given on the upper curve. It is then possible to get a gain by a comparison of the two curves and also to determine a cut-off frequency of this transistor working with a signal carried by an optical input.

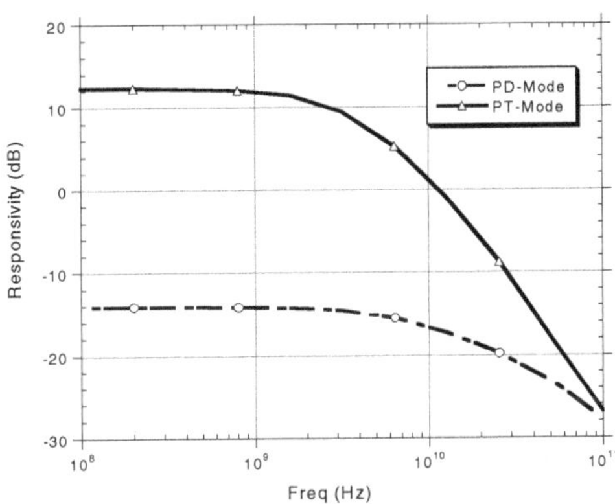

Figure 92 : Dynamic responses of the phototransistor in diode and transistor modes.

5.4.3. Large Signal Equivalent Circuit Model

We are now going to describe a second step in the modeling of the phototransistor. It consists in the building of a large signal equivalent circuit model which can be used in a circuit simulator. And the results of simulations with this equivalent circuit will be compared with numerical simulations and also with measurements.

The first step is to choose an equivalent circuit model. The choice is a classical modified Ebers-Moll model [208]. This model (figure 93) includes the base emitter and base collector capacitances. The relationships giving the currents are:

$$I_{C1} = C_1 I_{SO}\left[\exp\left(\frac{qV_{B'C'}}{n_{cl}kT}\right) - 1\right]$$
$$I_{C2} = C_2 I_{SO}\left[\exp\left(\frac{qV_{B'E'}}{n_{el}kT}\right) - 1\right] \quad (34)$$

$$I_{C2} = \frac{I_{EC}}{\beta_R}$$
$$I_{E2} = \frac{I_{CC}}{\beta_F} \quad (35)$$

Where *C1, C2* are constants; *ne, ncl* are ideality factors, *βR, βF* are reverse and forward β coefficients.

Then we have to represent the effect of the optical input. A solution proposed earlier [211] is to have a current source placed between the base and the collector like on the diagram of figure 93.

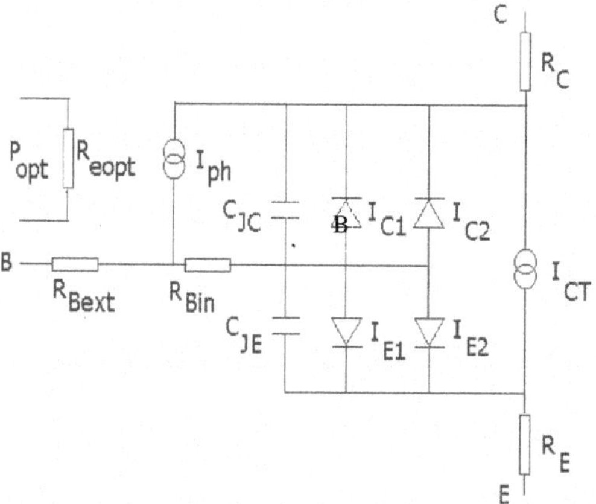

Figure 93 :Equivalent circuit non linear model for a phototransistor.

The 2D numerical simulations are a good help to check this choice. Figure 94 shows an example of this current as a function of the optical power input. This curve is plotted for collector-emitter voltages between 0 and 2 V and for base-emitter voltages between 0 and 1.2 V. The photonic current is always the same whatever the voltages are. So, it is a good demonstration that this current source is a very good representation of the optical power. In other simulations of microwave optical links, in which there are for instance lasers, the optical power is represented by a voltage. And the current source is linked to the input voltage by a responsitivity. Moreover, we consider that the optical power is introduced in the phototransistor by an optical fiber. To be able to characterize this input by S parameters, the matching of the input must be considered. It is the reason of the introduction of a matching resistance R_{eopt} at the input of this model.

Now, it remains to compare this equivalent circuit model to the numerical or physical simulations and to measurements.

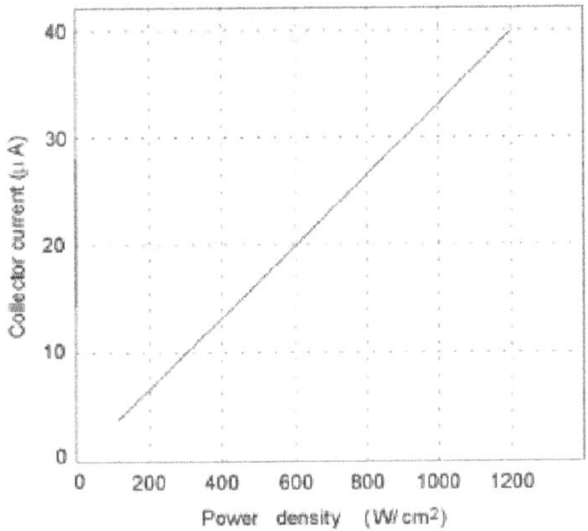

Figure 94 : Photocurrent as a function of Light power for V_{CE} varying between 0 and 2 V and V_{BE} between 0 and 1.2 V.

Figure 95 shows the static response simulation as a function of the optical power input for a 2D simulation and for the equivalent circuit model. There is a good agreement between the two.

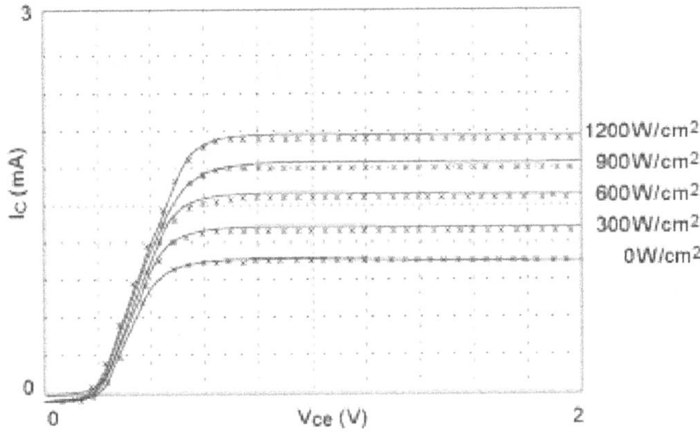

Figure 95 : Static responses of the numerical and equivalent circuit models of phototransistors.

R_{bext} = 100 ohms	C_{JC} = 15 fF
R_{bin} = 20 ohms	C_{JE} = 9.5 fF
R_C = 12 ohms	
R_E = 10 ohms	R_{eopt} = 50 ohms

Table 6 : Element values of the model.

The dynamic simulations would also show the same type of agreement. These different simulations give also the possibility to extract the different values of the equivalent circuit parameters which are shown on table 6.

5.4.4. Comparison with Measurements

Now, comparisons of this equivalent circuit model with measurements are going to be done for a static response and for a dynamic response.

Measurements give the possibility to extract some parasitic elements of the equivalent circuit like R_{B1} given on table 6.

Figure 96 shows a comparison between equivalent circuit and measurements for the static characteristic of the transistor.

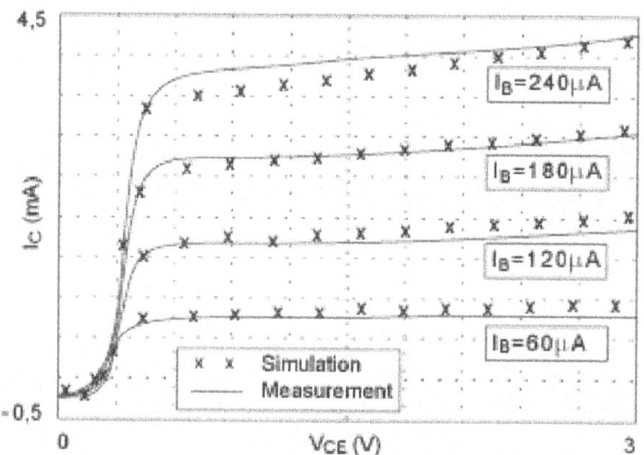

Figure 96 :Comparison of measurements and simula-tions for a phototransistor having a width of 4 μm.

Figure 97 shows the dynamic response of the phototransistor in photodiode mode, i.e. with emitter and base short circuited while figure 98 shows the phototransistor mode. With this model, the cut off frequency is around 10 GHz.

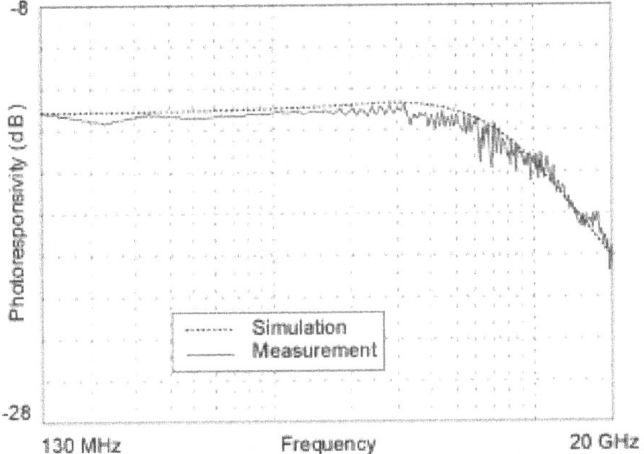

Figure 97 : Simulation and measurement of the phototransistor in photodiode mode.

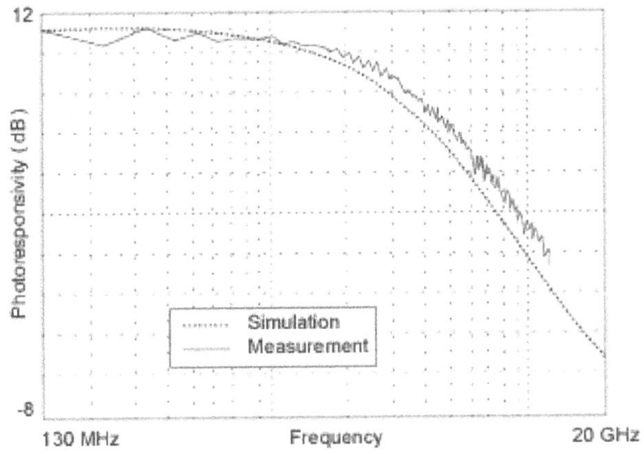

Figure 98 : Simulation and measurement of the phototransistor in transistor mode.

5.4.5. Simulation of an up Converter

Now, the equivalent circuit model is going to be checked to see the behavior of a phototransistor used as an up-converter. A local oscillator electrical signal, having a frequency of 15 GHz, is applied on the base of the transistor with a generator having a 50 ohms impedance. An IF signal is introduced into the transistor through a modulation of the optical input. And the signal corresponding to the sum of these two frequencies is extracted from the collector of the transistor. The results of this simulation for a LO signal of 15 GHz and a power of 0 dBm and an IF signal of 2.5 GHz introduced by an optical beam having an optical power of 0.84 mW modulated with an index of 50% are presented on figure 99. On this figure, the signals are taken on the collector of the transistor. The amplitudes of LO+IF, LO-IF and IF signals can be identified. The power of LO+IF is -39 dBm. But this value is obtained with no matching circuit on the base or on the collector. On these electrodes, the terminal impedances are only 50 ohms.

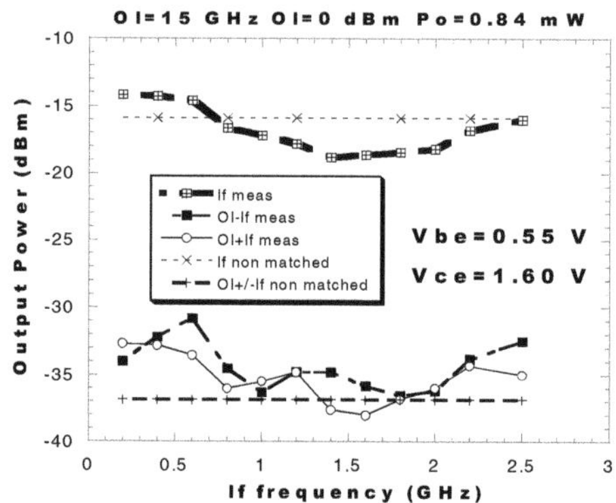

Figure 99 : Simulation and measurement of the signals taken on the collector for the phototransistor working as an up-converter.

It is quite evident that the amplitude of the up-converted signal could be improved by adding circuits around the transistor. But it is another subject.

5.4.6. Conclusion

Several items have been explored in this presentation.

A first one was the description of the numerical simulation of a phototransistor. This tool can be used to optimize the configuration or the distribution of the different layers of a phototransistor.

This tool can also be used to establish a large signal equivalent circuit model and then to find the values of the different elements.

Then the results of simulation with this equivalent circuit were compared with numerical simulations and also with measurements.

And finally, this model was used to describe the behavior of the transistor working as an up-converter.

Now, this model is ready to be used in the design of different circuits implying a phototransistor like a complete up-converter with its matching and filtering circuits.

The model itself can also be improved, for instance by the adding of noise sources and thermal effects.

6. REFERENCES

[1] J.Hong et al
Matrix-grating Strong Gain-Coupled DFB lasers with 34 nm CW tuning range
IEEE PTL Vol 11 N°5, 999

[2] F.Delorme et al
High reliability of high power and widely tunable 1.55μm DBR laser for WDM applications
IEEE Selected Topics, Vol3 N°2, April 1997

[3] P-J.RIgole et al
Improved tuning regularity over 100 nm in a vertical grating assisted codirectional coupler laser with a superstructure grating DBR
ECOC'95, Tu A 3.6

[4] H.Ishii et al
Wavelength stabilisation of a superstructure grating DBR laser for WDM networks
ECOC '97 N° 448

[5] T.Koch and al
ECOC'95

[6] H.Bissessur et al
ECOC'99

[7] M Shimizu, F Koyama and K Iga
Trans. IECE Japan n° E74, 3334 (1991)

[8] W Nakwaski and M Osinski
IEEE J. Quantum Electron. n° 27, 1391 (1991)

[9] W W Chow, S W Koch and M Sargent
Semiconductor-Laser Physics
Springer-Verlag, 1994

[10] A V Uskov, J G McInerney, J R Karin and J E Bowers
Opt.. Lett., Mar 1998

[11] D D Cook and F R Nash
J. Appl. Phys. n° 46, 1660 (1975)

[12] N K Dutta, L W Tu, G Hasnain, G Zydzik, Y H Wang and A Y Cho
Electron. Lett. n° 27, 208 (1991)

[13] Y-G Zhao, J G McInerney and R A Morgan
Opt. Eng. n° 33, 3917 (1994)

[14] Y-G Zhao and J G McInerney
IEEE J. Quantum Electron. n°31, 1668 (1996)

[15] Y-G Zhao and J G McInerney
IEEE J. Quantum Electron. n° 31, 1950 (1996)

[16] C J Chang-Hasnain, J P Harbison, G Hasnain, A C von Lehmen, L T Florez and N G Stoffel
IEEE J. Quantum Electron. n° 27, 1402 (1991)

[17] These lasers were provided by Dr R A Morgan, then at AT&T Solid State Technology Center, Breinigsville, PA

[18] R A Morgan, G D Guth, M W Focht, M T Asom, K Kojima, L E Rogers and L E Challis
IEEE Photonics Technol. Lett. n° 4, 364 (1993)

[19] H Li, T L Lucas, J G McInerney and R A Morgan
Chaos Solitons and Fractals n° 4, 1619 (1994)

[20] M Shimizu, F Koyama and K Iga

Japan. J. Appl. Phys. n° 27, 1774 (1988)
[21] Y A Wu, G S Li, R F Nabiev, K D Choquette, C Caneau and C J Chang-Hasnain
IEEE J. Sel. Top. Quantum Electron. n° 1, 629 (1995)
[22] F Koyama, K Morito and K Iga
IEEE J. Quantum Electron. n° 27, 1410 (1991)
[23] M Wu, L Buckman, G Li, K Y Lau, C Chang-Hasnain
Proc. 14th IEEE Semicond. Laser Conf p145 (1994)
[24] Y A Wu, G S Li, R F Nabiev, K D Choquette, C Caneau and C J Chang-Hasnain
IEEE J. Sel. Top. Quantum Electron. n° 1, 629 (1995)
[25] D V Kuksenkov, H Temkin and S Swirhun
Appl. Phys. Lett. n° 67, 2141 (1995)
[26] T Mukaihara, N Ohnoki, Y Hayashi, N Hatori, F Koyama and K Iga
IEEE Photonics Technol Lett n° 7, 1113 (1995)
[27] J Li, J-F Seurin, S L Chuang, K D Choquette, K M Geib and H Q Hou
Appl. Phys. Lett. n° 70, 1799 (1997)
[28] F Brown-deColstoun, G Khitrova, A V Fedorov, T R Nelson, C Lowry, B E Hammons and P D Maker
Chaos Solitons and Fractals n° 4, 1575 (1994)
[29] K Choquette, K Lear, R Schneider, R Leibenguth, J Figiel, S Kilcoyne, M Hagerott-Crawford, J Zolper
Proc SPIE n° 2382, 125 (1995)
N Badr, I H White, M Tan, Y Houng , S Y Wang
Electron. Lett. n° 30, 1227 (1994)
[30] Y J Yang, T G Dziura, S C Wang, G Du and S Wang
IEEE Photonics Technol Lett 3, 9 (1991)
[31] P D Floyd, M G Peters, L A Coldren and J L Merz
IEEE Photonics Technol Lett n° 7, 1388 (1995)
[32] R Michalzik and K J Ebeling
IEEE J. Quantm Electron. n° 31, 1371 (1995)
K L Lear, K D Choquette, R P Schneider Jr and S P Kilcoyne
Appl. Phys. Lett. n° 66, 2616 (1995)
[33] R A Morgan
Proc SPIE n° 3003, 14 (1997)
[34] T J Rogers, D L Huffaker, H Deng, Q Deng and D G Deppe
IEEE Photonics Technol Lett n° 7, 238 (1995)
[35] S P Hegarty, G Huyet and J G McInerney
Proc SPIE n° 3286, paper 26
[36] J Geddes, J Lega, J Moloney, R A Indik, E M Wright and W J Firth
Chaos Solitons and Fractals n° 4, 1261 (1994)
[37] T Mukaihara, F Koyama and K Iga
Electron. Lett. n° 28, 555 (1992)
[38] K D Choquette and R E Leibenguth
IEEE Photonics Technol Lett n° 6, 40 (1994)
K D Choquette, K L Lear, R P Schneider and R E Leibenguth
IEEE J Sel Top Quantum Electron n° 1, 661 (1995)
[39] A Chavez-Pirson, H Ando, H Saito and H Kanbe
Appl Phys Lett n° 62, 3082 (1993)
[40] K D Choquette, K L Lear, R E Leibenguth and M T Asom
Appl Phys Lett n° 64, 2767 (1994)
[41] T Yoshikawa, H Kosaka, K Kurihara, M Kajita, Y Sugimoto and K Kasahara

Appl. Phys. Lett. n° 66, 908 (1995)
[42] D V Kuksenkov, H Temkin and T Yoshikawa
IEEE Photonics Technol Lett n° 8, 977 (1996)
[43] T Mukaihara, F Koyama and K Iga
IEEE Photonics Technol. Lett. n° 5, 133 (1993)
[44] K D Choquette, D A Richie and R E Leibenguth
Appl Phys Lett n° 64, 2062 (1994)
[45] K D Choquette, R P Schneider, K L Lear and R E Leibenguth
IEEE J. Sel Top Quantum Electron. n° 1, 661 (1995)
[46] A K Jansen van Doorn, M P van Exter and J P Woerdman
Appl Phys Lett n° 69, 1041 (1996)
[47] A K Jansen van Doorn, M P van Exter and J P Woerdman
Appl Phys Lett n° 69, 3635 (1996)
[48] J H Ser, Y-G Ju, J-H Shin and Y-H Lee
Appl Phys Lett n° 66, 2769 (1995)
[49] D Sun, E Towe, P H Ostdiek, J W Grantham and G J Vansuch
IEEE J. Sel. Top. Quantum Electron n° 1, 674 (1995)
[50] T Numai, K Kurihara, K Kuhn, H Kosaka, I Ogura, M Kajita, H Saito and K. Kasahara
IEEE J Quantum Electron n° 31, 636 (1995)
[51] Y Kaneko, S Nakagawa, T Takeuchi, D E Mars, N Yamada and N Mikoshiba
Electron Lett n° 31, 805 (1995)
[52] M Takahashi, P Vaccaro, K Fujita, T Watanabe, T Mukaihara, F Koyama and K. Iga
IEEE Photonics Technol Lett n° 8, 737 (1996)
[53] Y-G Ju, Y-H Lee, H-K Shin and I Kim
Appl Phys Lett n° 71, 741 (1997)
[54] T H Russell and T D Milster
Appl Phys Lett n° 70, 2520 (1997)
[55] Z Pan, S Jiang, M Dagenais, R Morgan, K Kojima, M Asom and R Leibenguth
Appl Phys Lett n° 63, 2999 (1994)
[56] S P Hegarty, G Huyet, J G McInerney and H Li
Proc SPIE n° 2994, 190 (1997)
[57] S Jiang, Z Pan, M Dagenais, R A Morgan and K Kojima
Appl Phys Lett n° 63, 3545 (1994)
[58] M San Miguel, Q Feng and J V Moloney
Phys Rev n° A 52, 1728 (1995)
[59] J Martin-Regalado, M San Miguel, N B Abraham and F Prati
Opt Lett n° 21, 351 (1995)
[60] C Serrat, N B Abraham, M San Miguel, R Vilaseca and J Martin-Regalado
Phys Rev n° A 53 R3731 (1996)
[61] M Travagnin, M P van Exter, A K Jansen van Doorn and J P Woerdman
Phys Rev n° A 54, 1647 (1996)
[62] A K Jansen van Doorn, M P van Exter, M Travagnin and J P Woerdman
Opt Commun n° 133, 252 (1997)
[63] T. Mukai et al
Detuning characteristics and conversion efficiency of nearly degenerate four-wave mixing in a 1.5 μm traveling-wave semiconductor laser amplifier
IEEE J. Quantum Electron., 1990, vol. 26, N° 5, pp. 865-875
[64] A.M. Salehi
Non linear models of Traveling-Wave optical amplifiers
Electronics Letters, vol. 24, no 14, 1988, pp. 835-837

[65] Y. Yamamoto
Ed., Coherence, amplification and quantum effects in semiconductor lasers
Chapitre 8, 1991, J. Wiley and sons, ISBN 0-471-51249-4
[66] Y. Chung, et al
Intermodulation distorsion in a multiple quantum well semiconductor optical amplifier
IEEE Photon. Technol. Lett., vol. 3, N° 2, Feb. 1991, pp. 130-132
[67] G. P. Agrawal
Population pulsations and non degenerate four-wave mixing in Semiconductor Laser Amplifiers
J. Opt. Soc. Am. B, 1988, vol. 5, no 1, pp. 147-159
[68] T. E. Darcie et al
Intermodulation distorsion in optical amplifiers from carrier-density modulation
Electron. Lett., Dec. 1987, Vol. 23, , 1392-1394
[69] M. Eiselt et al
Decision gate for all-optical data retiming using a semiconductor laser amplifier in a loop mirror configuration
El. Lett., 1993, vol. 29, n° 1, pp. 107-109
[70] J.P. Solokof et al
A terahertz optical asymmetric demultiplexer (TOAD)
PTL, 1993, vol. 7, n° 5, pp. 787-790
[71] I. Glesk et al
demonstration of all-optical demultiplexing of TDM data at 250 Gbit/s
Electronics Lett., 1994, vol. 30, n° 4, pp. 339-341
[72] C. Joergensen et al
All-optical wavelength conversion at bit rates above 10 Gbit/s using semiconductor optical amplifiers
IEEE J[al] of sel. topics in QE, 1997, vol. 3, n° 5, pp. 1168-1179
[73] Joindot and M. Joindot
Les Télécommunications par Fibre Optique
Dunod et CNET-ENST ed., Paris, 1996. ISBN 2 10 002787 5
[74] Topical meetings on optical amplifiers and their applications: O.S.A. Ed
[75] Optical fiber communication conferences: O.S.A. Ed
[76] European conferences on optical communication
[77] W.E.Stephens and T.R.Joseph
J. of Lightwave Technology (LT-5) n° 3, March 1987, pg.380-387
[78] D.Kasemet et al
High frequency Analog Fiber Optic Systems
SPIE vol.1371 (1990), pg.104-114
[79] M.Nazarathy et al
J. of Lightwave Technology (11) n° 1, January 1993, pg.82-103
[80] N.Shaw and A.Carter
Microwave Journal, October 1993, pg.92-102
[81] M.M.Howerton et al
IEEE Photonics Technology Letters (8) n° 12, December 1996, pg.1692-1694
[82] D.Dolfi et al
Applied Optics (35) n° 26, September 1996, pg.5293-5300
[83] M.Varasi, M.Signorazzi, M.Massani and A.Vannucci
Optoelectronic Signal processing for Phased Array Antennas III
SPIE vol.1703 (1992), pg 566-572
[84] J.L.Jackel, C.E.Rice and J.Vaselska
Applied Phys. Lett. (41) 1982, pg.607

[85] M.De Michelli et al
Opt. Lett. (8) 1983, pg 114

[86] M.Varasi, A.Annulli and A.Vannucci
Materials and Technologies for Optical Communications
SPIE vol.866 (1987), pg.110-116

[87] S.T.Vhora, A.R.Mickelson and S.Asher
J.Applied Physics (66) n° 11, December 1989, pg.5161-5174

[88] X.F.Cao, R.V.Ramaswamy and R.Srivastava
J. Lightwave Technol.,(10) 1992, pg.1302

[89] H.Nishiharra, M.Haruna and T.Suhara
Optical Integrated Circuits, McGraw Hill, 3rd edition (1989)

[90] R.C.Alferness
IEEE Trans. Microwave Theory and Tech. (MTT-30) 1982, pg.342

[91] H.Chung, W.S.C.Chang and E.L.Adler
IEEE J. Quantum Electronics (27) 1991, pg.608

[92] M.Varasi, A.Vannucci and M.Signorazzi
Integrated Optical Circuits
SPIE vol.1583 (1991), pg.165-169

[93] R.D.Alferness, S.K.Koroty and E.A.J.Marcatili
IEEE J. of Quantum Electronics (DE-20) n° 3, March 1984, pg 301-309

[94] D.Erasme and M.G.F.Wilson
Integrated optical Circuit Eng. III
SPIE vol.651 (1986), pg.154-161

[95] D.W.Dolfi and T.R.Ranganath
Electronics Letters (28) n° 13, June 1992, pg.1197-1198

[96] K.Kawano
IEEE J. of Quantum Electronics (29) n° 9, September 1993, pg 2466-2475

[97] Z-Q.Lin and W.S.C.Chang
IEEE/LEOS Summer Topical on Broadband Analog Optoelectronics Devices and Systems
July 1990, Post-deadline paper n° 2

[98] R.B.Childs and V.A.O'Byrne
IEEE J. on Selected Areas in Communications (8) n° 7, September 1990

[99] L.M.Jhonson and H.V.Russell
Optics Letters , October 1988, pg.928-930

[100] J.D.Farina, B.R.Higgins and J.P.Farina
Optical Fiber Communication Conference, vol. 2, 1996 OSA Technical Digest Series, pg.283-285

[101] J.J.Pan
SPIE vol.1102, pg.20-29

[102] P-L.Liu, B.J.Li and Y.D.Trisno
IEEE Photonics Technology Letters (3) n° 2, February 1991, pg 144-146

[103] W.B.Bridges and J.H.Schaffner
IEEE Trans. on Microwave Theory and Techniques (43) n° 9, September 1995 pg. 2184-2197

[104] H.Skeie and R.V.Johnson
Integrated Optical Circuits
SPIE vol. 1583 (1991), pg.153-164

[105] M.K.Jackson, V.M.Smith, W.J.Hallan and J.C.Maycock
J.of Lightwave Technol. (15) n° 8, August 1997, pg.1538-1545

[106] E.Gartel and D.Sipes
Optoelectronic Signal Processing for Phased Array Antennas IV

SPIE vol.2155 (1994), pg.264-274
[107] G.H.Smith, D.Novak and A.Ahmed
Electronics Letters (33)1, January 1997, pg. 74-75
[108] E.Vergnol, F.Devaux, D.Jahan and A.Carenco
Electronics Letters (33)23, November 1997, pg. 1961-1962
[109] M.Izutsu, S.Shikama and T.Sueta
IEEE J. of Quantum Electronics (QE-17) 11, November 1981, pg.2225-2227
[110] G.A.Vawter et al
Optical Technology for Microwave Applications VI
Optoelectronic Signal Processing for Phased Array Antennas IV
SPIE vol.1703 (1992), pg.321-331
[111] B.Desormiere, C.Maerfeld and J.Desbois
J. of Lightwave Technol. (8) 4, April 1990, pg.506-513
[112] S. Meyer, J. Guéna, J.C. Leost, E. Penard, M. Goloubkoff
A new concept of LANs for microwave links hooked on a fibre optic backbone
IEEE-MTT Symposium, Atlanta, 1549 (1993)
[113] F. Devaux
Thesis, Université de Paris Sud, Orsay (1993)
[114] S. Chelles, R. Ferreira, P. Voisin, A. Ougazzaden, M. Allovon, A. Carenco
Efficient polarization insensitive electroabsorption modulator using strained InGaAsP-based quantum wells
Appl. Phys. Lett., n° 64, 3530 (1994)
[115] N. Souli, F. Devaux, A. Ramdane, P. Krauz, A. Ougazzaden, F. Huet, M. Carré, Y. Sorel, J.F. Kerdiles, M. Henry, G. Aubin, E. Jeanney, T. Montallant, J. Moulu, B. Nortier, J. B. Thomine
20 Gbits/s high-performance integrated MQW TANDEM modulators and amplifier for soliton generation and coding
IEEE Photonics Technology Letters, n° 7, 629 (1995)
[116] A. Ramdane, F. Devaux, N. Souli, D.Delprat, A. Ougazzaden
Monolithic integration of multiple-quantum-well lasers and modulator for high speed transmission
IEEE Journal of Selected Topics in Quantum Electronics, n° 2, 326 (1996)
[117] E. Vergnol, F. Devaux, D. Jahan, A. Carenco
Fully integrated millimetric single-sideband lightwave source
Electronics Letters, n° 33, 1961 (1997)
[118] C. Minot, H. Le Person, J.F. Palmier, D. Tanguy, E. Penard, J.C. Harmand, J.P. Medus, J.C. Esnault
Millimetre-wave superlattice oscillators for optically-fed wireless communications
Technical Digest of Microwave Photonics 96, Kyoto (1996)
[119] C. Minot, N. Sahri, H. Le Person, J.F. Palmier, J.C. Harmand, J.P. Medus, and J.C. Esnault
Millimetre-wave negative differential conductance In GaInAs/AlInAs semiconductor superlattices
Superlattices and Microstructures, n° 23, 1323 (1998)
[120] J.F. Palmier, C. Minot, H. Le Person, J.C. Harmand, N. Bouadma, J.C. Esnault, D. Arquey, F. Heliot, J.P. Medus
Reflection gain up to 6 dB at 65 GHz in GaInAs/AlInAs superlattice oscillators
Electron. Lett., n° 32, 1506 (1996)
[121] K. Kurokawa
Injection locking of microwave solid-state oscillators
Proceedings of the IEEE, n° 61, 1386 (1973)

[122] J.F. Cadiou, J. Guena, E. Penard, P. Legaud, C. Minot, J.F. Palmier, H. Le Person, J.C. Harmand
Direct optical injection locking of 20 GHz superlattice oscillators
Electron. Lett., n° 30, 1690 (1994)

[123] D. Tanguy, E. Penard, P. Legaud, C. Minot
Direct optical injection locking of superlattice millimetre-wave oscillators
Technical Digest of Microwave Photonics 97, Duisburg (1997)

[124] D.Jäger
Slow wave propagation along variable Schottky contact microstrip line
IEEE Trans. Microwave Theory and Techn., MTT-24, pp. 566-573, 1976

[125] von Wendroff, M.Welters and D.Jäger
Optoelectronic switching in travelling wave MSM photodetectors using photon energies below bandgap
Electron. Lett. Vol.26, 1874-1875 , 1990

[126] M.Block and D.Jäger
Optically controlled travelling wave MSM photodetector for phased array applications
Workshop Proceedings 21st EuMC, Stuttgart , pp. 22-27, 1991

[127] D.Jäger
Travelling wave optoelectronic devices for microwave and optical applications
Proc. PIERS, MITCambridge, p.327, 1991

[128] F.Buchali, I.Gyuro, F.Scheffer, W.Prost, M.Block, W. von Wendroff, G.Heymann, F.J.Tegude, P.Speier und D.Jäger
Interdigitated electrode and coplanar waveguide InAlAs/InGaAs MSM photodetectorsgrown by LP MOVPE
Proc. 4th IPRM Conference, Newport, USA, pp.596-599, 1992

[129] D.Jäger
Microwave photonics
ESPRIT/WOIT-conference, Edinburgh 1991
and in „Optical information technology", ed. S.D.Smith and R.F. Neale, Springer Verlag, pp.328-333, 1993

[130] R.Kremer, G.David, S.Redlich, M.Dragoman, F.Buchali, F.J.Tegude and D.Jäger
A high speed MSM travelling wave photodetector for InP-based MMICsProc.MIOP´93, Sindelfingen, pp.271-275, 1993

[131] D.Jäger, R.Hülsewede and R.Kremer
Travelling wave MMIC Schottky, MSM and pn diodes for nicrowave and optoelectronic applications
Proc. CAS´93, Bucharest, pp.19-29, 1993

[132] M.DragomanM.Block, R.Kremer, F.Buchali, F.J.Trgude and D.Jäger
Coplanar InAlAs/InGaAs/InP microwave delay line with optical control
Proc. IEEE LEOS Annual Meeting, Boston, pp.684-685, 1992

[133] G.A.Vawter, V.M.Hietala, S.H.Kravitz, M.G.Armendaritz
Unlimited bandwidth distributed optical phase modulators and detectors: design and fabrication issues
Proc. Optical Microwave Interactions, France, 1994

[134] Giboney, M.J.W.Rodwell, J.E.Bowers
Travelling wave photodetectors
IEEE Photon, Technol. Lett. Vol.4, No.12, pp.1363-1365, 1992

[135] D.Jäger
Characteristics of travelling waves along nonlinear transmission lines for monolithic integrated circuits. A review
Int.J.Electron, Vol.58, 649-669, 1985

[136] A.Stöhr, O.Humbach, S,Zumkley, G.Wingen, B.Bollig, E.C.Larkins, J.D.Ralston and D.Jäger
InGaAs/GaAs multiple quantum well modulators and switches
Optical and Quantum Electronics, Vol.25, pp.S865-S883, 1993

[137] H.Hayashi and K.Tada
GaAs Travelling wave directional coupler optical modulator /switch
Appl Phys.Lett, Vol. 57, pp.227-228, 1990

[138] J.D.Ralston, S.Weisser, I.Esquivas, E.C.Larkins, J.Rosenzweig, P.J.Tasker and J.Gleissner
Control of differential gain, nonlinear gain and damping factor for high speed application of GaAs-based MQW lasers
IEEE J.Quantum Electron., Vol.29, No.6, pp.1648-59, 1993

[139] K.Kato, S.Hata, K.Kawano, J.Yoshida and A Kozen
A high efficiency 50 GHz InGaAs multimode waveguide photodetector
IEEE J. of Quantum Electron., Vol.28, No.12, pp.828-830, 1994

[140] A.Stöhr, O.Humbach, R.Hülsewede, A.Wiersch and D.Jäger
An InGaAs/GaAs MQW Optical switch based on field-induced waveguides
IEEE Photon. Technol.Lett., Vol.6, No.7, pp.828-830, 1994

[141] S.D.Offsey, W.J.Schaff, P.J.Tasker and L.F.Eastman
Optical and microwave performance of GaAs-AlGaAs and strained layer InGaAs-GaAs-AlGaAsgraded index separate confinement heterostructure single quantum well lasers
IEEE Photon. Technol. Lett., Vol.2, pp. 9-11, 1990

[142] S.Weisser, J.D.Ralston, E.C.Larkins, I.Esquivias, P.J.Tasker, J,Fleissner and J.Rosenzweig
Efficient high-speed direct modulation in p-doped $In_{0.35}$ Ga $_{065}$ As/GaAs multiquantum well lasers
Electron.Lett., Vol.28, No.23, pp.2141-2143, 1992

[143] D.A.Tauber, R.Spickermann, R.Nagarajan, T.Reynolds, A.L.Holmes,Jr.,and J.E.Bowers
Inherent bandwidth limits in semiconductor lasers due to distributed microwave effects
Appl.Phys.Lett., Vol.64, No.13,1994

[144] Alex L. Ivanov, H. Haug, S. Knigge, and D. Jäger
Mesoscopic semiconductor switching element with giant electro-optical nonlinearities due to intrinsic photoconductivity
Proc. Int. Conf. Optical Properties of Nanostructures, Sendai, 1994, Jpn. J. Appl. Phys., Vol. 34 (1995), Suppl. 34-1, pp. 15-18

[145] David, W.Schroeder, D. Jäger, and I. Wolff
2D electro-optic probing combined with field theory based multimode wave amplitude extraction: A new approach to on-wafer measurement
1995 IEEE MTT-S International Symposium, May 15-19, Orlando, USA

[146] Kremer, S. Redlich, L. Brings, D. Jäger
Optically controlled coplanar transmission lines for microwave signal processing
Microwave and Millimeter-Wave Photonics, Special Issue of the IEEE Transactions of Microwave Theory and Techniques - vol. 43, no. 9, pp.2408 2413, 1995

[147] Jäger, R. Kremer, and A. Stöhr
High-speed travelling-wave photodetectors, modulators and switches
IEEE/LEOS Laser and Electro-Optics Society Summer Topical Meetings 1995 Conference Digest, ISBN 0-7803-2448-0, pp. 57 - 58, Keystone Resort Colorado, USA, August 1995 (Invited Paper)

[148] Kremer and D. Jäger
Optically controlled coplanar transmission lines for adaptive microwave signal processing applications
International Semiconductor Conference (CAS), 18th Edition, Oct. 11-14, 1995 Sinaia, Romania

[149] Jäger and R. Heinzelmann
Semiconductor devices for photonic integrated circuits
Int. Conf. Nonlinear Optical Physics & Appl. (ICNOPA '95), Harbin, China (invited paper)

[150] Massini, R. Heinzelmann, A. Stöhr, D. Jäger, D. Pedini, G. Bellanca, P. Bassi, and R. Sorrentino
Theoretical analysis of ultrafast electrooptical InGaAs/GaAs MQW modulators and comparison to experimental devices
25th EuMC '95, Conference Digest, pp. 170-174, Bologna, Italy, September 1995

[151] Jäger, R. Kremer, and A. Stöhr
Travelling-wave optoelectronic devices for microwave applications
1995 IEEE MTT-S Int. Microw. Symp. Digest, Vol. 1, pp. 163-166 (invited paper)

[152] Jäger and G. David
MMIC characterization using electro-optic field mapping
1995 Int. Symp. Signals, Systems and Electronics (ISSSE '95), San Francisco (invited paper)

[153] Jäger and G. David
Two dimensional imaging of fields in monolithic integrated circuits using optical techniques
Proc. Microwaves and RF '95, pp. 132-136, London (invited paper)

[154] David, P. Bussek, and D. Jäger
High resolution electro-optic measurements of 2D field distributions inside MMIC devices
Proceedings of CLEO 96, Anaheim, USA, 1996, pp. 450-451

[155] David, R. Tempel, I. Wolff, and D. Jäger
In-circuit electro-optic field mapping for function test and characterization of MMICs
1996 IEEE MTT-S Int. Microwave Symp., June 17-24, San Francisco, USA 1996, pp. 1533-1536

[156] Stöhr, R. Heinzelmann, R. Buß, and D. Jäger
Electroabsorption modulators for broadband fiber electro-optic field sensors
in Applications of Photonic Technology 2, eds. G.A. Lampropolous and R.A. Lessard, pp. 871-876, 1998

[157] Alles, Th. Braasch, R. Heinzelmann, A. Stöhr, and D. Jäger
Optoelectronic devices for microwave and millimeterwave optical links
11th Int. MIKON '96, Conference Proceedings, Workshop "Optoelectronics in Microwave Technology", pp. 68 - 76, Warsaw, 27-30 May 1996 (invited paper)

[158] Berini, A. Stöhr, K. Wu, D. Jäger
Normal Mode Analysis and Characterisation of an InGaAs/GaAs MQW Field Induced Optical Waveguide Includinig Electrode Effects
IEEE/OSA J. Lightwave Technol., Vol. 14, No. 10, pp. 2422 - 2435, October 1996

[159] Heinzelmann, A. Stöhr, Th. Alder, R. Buß, and D. Jäger
EMC measurements using electrooptic waveguide modulators

International Topical Meeting on Microwave Photonics MWP '96, Conference Proceedings, Technical Digest, December 3-5, 1996, Kyoto, Japan

[160] Jäger
Optically Controlled Microwave Devices
International Topical Meeting on Microwave Photonics MWP '96 technical digest, December 3-4, 1996, Kyoto, Japan

[161] Alles, U. Auer, F.-J. Tegude, and D. Jäger
High-speed Travelling-Wave Photodetectors for Wireless Optical Millimeter Wave Transmission
MWP '97, Duisburg/Essen

[162] Stöhr, R. Heinzelmann, T. Alder, D. Kalinowski, M. Schmidt, M. Groß, and D. Jäger
Integrated Optical *E*-Field Sensors using TW EA-Modulators
International Topical Workshop on Contemporary Photonic Technologies, CPT '98, Jan. 12-14, Tokyo, 1998

[163] Heinzelmann, A. Stöhr, M. Groß, D. Kalinowski, T. Alder, M. Schmidt, and D. Jäger
Optically Powered Remote Optical Field Sensor System Using an Electroabsorption-Modulator
1998 MTT-S International Microwave Symposium and Exhibition, Baltimore Maryland, June 7-12, 1998

[164] Alles, U. Auer, F.-J. Tegude, and D. Jäger
Distributed Velocity-matched 1.55 m InP Travelling Wave Photodetector for Generation of High Millimeterwave Signal Power
1998 MTT-S International Microwave Symposium and Exhibition, Baltimore Maryland, June 7-12, 1998 (accepted)

[165] Stöhr, K. Kitayama, and T. Kuri
Chirp Optimized 60 GHz Millimeter-Wave Fiber-Optic Transmission Incorporating EA Modulator
ECOC '98, Sept. 98, Madrid

[166] Jäger
Advanced Microwave Photonic Devices for Analog Optical Links
MWP '98, 12-14 Oct. 1998, Princeton, New Jersey (invited)

[167] Jäger
Fiber Optic Links for Microwave and Millimeterwave Systems
28th EuMC '98, 5-9 Oct. 1998, Amsterdam

[168] Stöhr, K. Kitayama, and D. Jäger
Error-Free Full-Duplex Optical WDM-FDM Transmission Using An EA Transceiver
MWP '98, 12.-14. Oct. '98, Princeton, New Jersey, USA

[169] F. Moncrief
LEDs Replace Varactors for Tuning GaAs FETs
Microwaves, vol. 18, No. 1, pp. 12-13, January 1979

[170] H. J. Sun, R. J. Gutmann, J. M. Borrego
Optical Tuning in GaAs MESFET Oscillators
1981 MTT-S International Microwave Symposium Digest, June 1981, Los Angeles, CA., pp. 40-42

[171] H. J. Sun, R. J. Gutmann, J. M. Borrego
Photoeffects in Common-Source and Common-Drain Microwave GaAs MESFET Oscillators
Solid State Electronics, vol. 24, No. 10, 1981, pp. 935-940

[172] A. De Salles, J. R. Forrest

Initial Observations of Optical Injection Locking of GaAs Metal Semiconductor Field Effect Transistor Oscillators
Applied Physics Letters, vol. 38, no. 5, pp. 392-394, March 1981

[173] D. C. Buck, M. A. Cross
Optical Injection Locking of FET Oscillators Using Fiber Optics
1986 MTT-S International Microwave Symposium Digest, June 1986 Baltimore, MD., pp. 611-614

[174] D. Warren, et al
Simulation of Optically Injection-Locked Microwave Oscillators Using a Novel SPICE Model
IEEE Transactions on Microwave Theory and Techniques, vol. MTT-36 November 1988, pp. 1535-1539

[175] P. R. Herczfeld, et al
Optical Phase and Gain Control of A GaAs MMIC Transmit-Receive Module
1989 International Microwave Symposium Digest, May 1989

[176] A. Paolella, P. R. Herczfeld
Optical Gain Control of a GaAs MMIC Distributed Amplifier
Microwave and Optical Technology Letters, vol.1, no.1, pp.13-16, March, 1988

[177] W. D. Jemison, T. Berceli, A. Paollela, P. R. Herczfeld, D. Kasemset, A. W. Jacomb-Hood
Optical Control of a Digital Phase Shifter
1990 MTT-S International Microwave Symposium Digest, Dallas, Texas, May 1990

[178] A. Paollela, A. Madjar, P.R. Herczfeld, D. Sturzebecher
Optically Controlled GaAs MMIC Switch Using A MESFET as an Optical Detector
1990 International Microwave Symposium Digest, Dallas TX, May 1990

[179] C. Rauscher, L. Goldberg, S. Yurek
GaAs FET Demodulator and Down Converter for Optical-Microwave Links
Electronic Letters, vol. 22, no. 13, pp. 705-706, 19th June, 1986

[180] L. Goldberg, C. Rauscher, J. F. Weller, H. F. Taylor
Optical injection Locking of X-Band FET Oscillator using Coherent Mixing of GaAlAs Lasers
Electronic Letters, vol. 19, no. 20, pp.848-850, September 1983

[181] H. R. Fetterman, D. C. Ni
Control of Millimeter Wave Devices by Optical Mixing
Microwave and Optical Technology Letters, vol. 1, no. 1, pp.34-39, March 1988

[182] D. C. Ni, H. Fetterman, W. Chew
Millimeter Wave Generation and Characterization of a GaAs FET by Optical Mixing
IEEE Transactions on Microwave Theory and Techniques, vol. MTT-38, no. 5 pp. 608-613, May 1990

[183] J. Graffeuil, P. Rossel, H. Martinot
Light Induced Effects in GaAs FETs
Electronics Letters, Vol. 15, No. 14, pp. 439-441, July 1979

[184] H. Mizuno
Microwave Characteristics of an Optically Controlled GaAs MESFET
IEEE Transactions on Microwave Theory and Techniques, vol. MTT-31, July 1983, pp.596-599

[185] J. L. Gautier, et al
Optical Effects on the Static and Dynamic Characteristics of a GaAs MESFET

IEEE Transactions on Microwave Theory and Techniques, vol. MTT-33 September 1985, pp. 819-822

[186] R. N. Simons, K. B. Bhasin
Analysis of Optically Controlled Microwave/ Millimeter-Wave Device Structures
IEEE Transactions on Microwave Theory and Techniques , vol. MTT-34, no. 12 pp. 1349-1355, December 1986

[187] R. N. Simons, K. B. Bhasin
Microwave Performance of an Optically Controlled AlGaAs/GaAs High Electron Mobility Transistor and GaAs MESFET
1987 International Microwave Symposium Digest, June 1990

[188] A. Madjar, A. Paollela, P.R. Herczfeld
Photo Avalanche Effects in A GaAs MESFET
Microwave and Optical Technology Letters, February 1990

[189] A. A. De Salles
Optical Control of GaAs MESFETs
IEEE Transactions on Microwave Theory and Techniques, vol. MTT-31, Oct. 1983, pp. 812-820

[190] R. B. Darling
Transit-Time Photoconductivity in High-Field FET Channels
IEEE Transactions on Electron Devices, vol. ED-34, 2, February 1987, pp. 433 444

[191] A. Madjar, A. Paollela, P. R. Herczfeld
Analytical Model for Optically Generated Currents in GaAs MESFETs
IEEE Transactions on Microwave Theory and Techniques, August 1992, pp. 1681-1691

[192] A. Madjar, A. Paollela, P. R. Herczfeld
Modelling The GaAs MESFET's Response to Modulated Light at RF and Microwave Frequencies
IEEE Transactions on Microwave Theory and Techniques

[193] A. Madjar, A. Paollela, P. R. Herczfeld
Modelling The Optical Switching of MESFETs Considering The External and Internal Photovoltaic Effects
IEEE Transactions on Microwave Theory and Techniques

[194] G. Papaionannou, J. Forrest
On the Photoresponse of GaAs MESFETs: Backgating and Deep Trap Effect
IEEE Transactions on Electron Devices, vol. ED-33, no. 3, pp. 373-378, March 1986

[195] P. C. Claspy, S. M. Hill, K. B. Bhasin
Microwave HEMT Photoconductive Detectors
Applied Microwave, November/December 1989, pp. 82-90

[196] J.C. Campbell and K. Ogaw
Heterojunction phototransistor for long-wavelength optical receivers
J. Appl. Phys.,1982, Vol. 53(2), pp. 1203-1208

[197] E. Suematsu and N. Imai
A fibre optic/millimetre-wave radio transmission link using HBT as direct and an optoelectronic up-converter
IEEE Trans. MTT, 1996, Vol. 44, pp. 133-143

[198] H. Ogawa, D. Polifko and S. Bamba
Millimetre-wave Fibre Optics Systems for personal Radio Communication
IEEE Trans. MTT, 1992, Vol. 40(12), pp.2285-2292

[199] E. Suematsu, H. Ogawa

Frequency response of HBTs as photodetectors
IEEE Micro. Guided Wave Lett., 1993, Vol. 3(7), pp. 217-218

[200] H. Fukano
High-speed InP-InGaAs heterojunction phototransistors employing a nonalloyed electrode metal as a reflector
IEEE J. Quantum Elect., 1994, vol.30, pp. 2889-2895

[201] C.Gonzalez, C.Palma, J.Thuret, J.L.Benchimol, M.Riet
InP/InGaAs HBT phototransistor as optoelectronic converter up to millimetre wave bands
MWP'97, Duisburg/Essen, 1997, Post-deadline Techn. Digest, pp. -12

[202] H. Sawada, N. Imai
Self-oscillating optoelectronic up-converter using an heterojunction bipolar transistor
OFC'96 Technical Digest, 1996, pp. 169-170

[203] H. Sawada, N. Imai
A self-oscillating optoelectronic up-converter using an heterojunction bipolar transistor up to millimetre-wave bands
MWP'96 Technical Digest, Tokyo, 1996, pp. 153-154

[204] C. Gonzalez, J. Thuret, J.L. Benchimol and M. Riet
Optoelectronic Up-converter to Millimetre-wave Band using an Heterojunction Bipolar Phototransistor
ECOC'98, Madrid, 1998,Vol.1, pp. 443-444

[205] Y. Betser, D. Ritter, C.P. Liu, A.J. Seeds and A. Madjar
A single-State Three-Terminal Heterojunction Bipolar transistor Optoelectronic Mixer
J. Light. Techn., 1998, Vol. 16(4), pp. 605-609

[206] J.L. Benchimol, J. MBA, A.M. Duchenois, B. Sermage, P.Launay, D. Caffin, M. Meghelli and M. Juhel
CBE growth of Carbon doped InGaAs/InP HBTs 25 Gb/s circuits
J. Crystal Growth, 1998, vol.188, pp.349

[207] J. Ph. Fraysse, D. Floriot, Ph. Auxemery, M. Campovecchio, R. Quéré, J. Obregon
A non-quasi-static model of GaInP/AlGaAs HBT for power applications
1997 IEEE MTT-S Digest, pp. 379-382

[208] A. Samelis, D. Pavlidis
Analysis of the Large-Signal characteristics of Power heterojunction Bipolar Transistors Exhibiting Self-Heating Effects
IEEE MTT, Vol. 45, n° 4, April 1997

[209] S. M . Sze
Physics of Semiconductor Devices
John Wiley and Sons, 1981, p.784

[210] N. Chennafi, C. Rumelhard, C. Gonzalas, J. Thuret
Modeling the photoresponse characteristics of InP / InGaAs Heterojunction Phototransistors with different incident directions of beam light
GAAS'98, Amsterdam, oct. 5-9, 1998

[211] ATLAS User's Manual, Version 1.5.0, April 1997

CHAPTER 2 : ELECTRONICS FOR OPTICS : INTEGRATED CIRCUITS

1. INTRODUCTION

Due to the advantage of the wide bandwidth of the optical fibre, Microwave-photonics can cover the RF, microwave and millimeterwave spectrum, between approximately 1-100 GHz.

New demands on performance are needed to better exploit the benefits afforded in such a wide spectrum :

1) reduced size and weight, compact devices
2) high speed devices and systems
3) cost-effective optoelectronic transceivers that consume little power.

These requirements demand a complete integration of sources, transmitters and photodetectors as well. Monolithic integration is enabled by the ongoing development of photonics technologies, and in the last decade, development of OEMMIC's (Opto-Electronic Microwave Monolithic Integrated Circuits) has considerably improved the circuits performance.

For example, in a microwave-photonics system used for transmission to radio base stations, one configuration is to use just a source, fibre transmission and a photodetector plus an antenna. A configuration that uses a photodetector plus an optical amplifier is attractive, leading to the development of an integrated microwave-millimeterwave photoreceiver.

In addition, the integration of the optical source is required for high performance systems, and process compatibility is indispensable.

In this chapter, the basis of OEMMIC's used to operate at very high frequency and data rates is presented in the first part. This leads to a good understanding of the specific nature of optoelectronics devices and circuits.

Then, high-speed integrated circuits used in optoelectronic systems are discussed, with front-end modules specifically.

Finally, since the specificity of OMMIC's is to profit from large bandwidth advantages offered by the optical transmission, the crucial problem of wide-band matching is pointed out. The optoelectronics devices must fit to microwave circuitry requirements, and the reference impedance is of 50 Ohms. The end of this chapter presents solutions dedicated to both wide-band laser source matching and photodiode matching.

2. ELECTRONICS FOR OPTICS; INTRODUCTION TO MMICS

I. Darwazeh
Department of Electrical Engineering and Electronics
University of Manchester Institute of Science and Technology (UMIST)
P O Box 88, Manchester M60 1QD, UK
Tel: +44-161-200 4747; Fax: +44-161-200 4770; E-mail: darwazeh@umist.ac.uk

2.1. Introduction

Recent years have seen a growing demand for reliable system components for the optical communications market. Most of today's optical communication systems are designed to operate at multi Gbit/s and to span unrepeated distances in the 10s of kilometres range. Such demanding requirements necessitate the provision of reliable high speed circuits, to perform functions such as laser (or optical modulator) driving, low jitter clock generation and timing extraction and signal reception, amplification, filtering and recovery. Most of such functions are now achieved using MMIC implementation.

Optical receivers are considered to be the weakest link in the optical system chain, as their noise and frequency performance set the limits of maximum transmission distance and rate. As in most communications receivers, the most critical element is the receiver front-end amplifier. In an optical communication system, the receiver's front end has the function of faithfully translating the photocurrent (generated by a photodetector) into an equivalent output voltage, suitable for driving the rest of the receiver circuitry.

This short chapter aims to introduce the reader to MMICs. It is hoped that it will offer an introductory insight into the design and implementation of MMIC broadband/ high frequency optical receiver preamplifiers, so that readers will be able to use it as a basis for further studies of advanced topics in this area.

The main features of MMICs are discussed below. The discussion will concentrate on aspects of interest to circuit designers, such as MMIC element models and design methodologies. The discussion is limited to MMICs using FET based devices, mainly MESFETs. Newer types of MMICs with Heterojunction Bipolar Transistor (HBT) active elements are not discussed here.

2.2. MMICs

Monolithic Microwave Integrated Circuits (MMICs) are ICs, containing active, passive and interconnect components and designed to operate at frequencies exceeding 1 GHz. Most of today's MMICs are fabricated on III-V substrates , such as GaAs and InP [1,2].

Since the mid 1970s, The growth in military and commercial demand for reliable high frequency circuits led to a large investment in GaAs foundries mainly aimed at developing MMIC processes. Today, MMICs are widely used in applications ranging from specialist phased arrays and multi-Gbit/s optical communications components to mobile phone and home satellite receivers. MMICs are widely used in place of their earlier *hybrid* predecessors the Microwave ICs (MICs).

2.2.1. MMIC Basics

MMICs are the components of choice for most of today's high frequency applications. They offer several advantages over their discrete or MIC counterparts. Such advantages are summarised below:

- Reduced size and weight, compact design
- High reproducibility and repeatable performance
- Low cost for high volume production (however, high cost for low volume)
- High yield (although, not as high as Si ICs)
- Low (or well controlled) Electro Magnetic coupling

Their main disadvantages are shared with all other ICs in that it is difficult (if not impossible) to alter the performance once the IC is fabricated. Most MMICs are fabricated using foundry processes. Such processes are widely available both commercially and as proprietary processes. The key to successful MMIC design is to have well characterised devices and foundry models. The following sections will describe the basic structure of typical GaAs MMICs and the associated active and passive device models. Details of more advanced processes, such as HEMT and HBT ones, can be found in a variety of references such as [1,2 and 5].

2.2.2. GaAs MMIC Technology

An MMIC is composed of several layers, all structured on a semi-insulating GaAs substrate. A typical process can include up to 8 layers of ion implanted GaAs, mesa, Dielectric insulator(s) and metallisation. A typical MMIC cross sectional view (three metallisation/two dielectric layers is shown below.

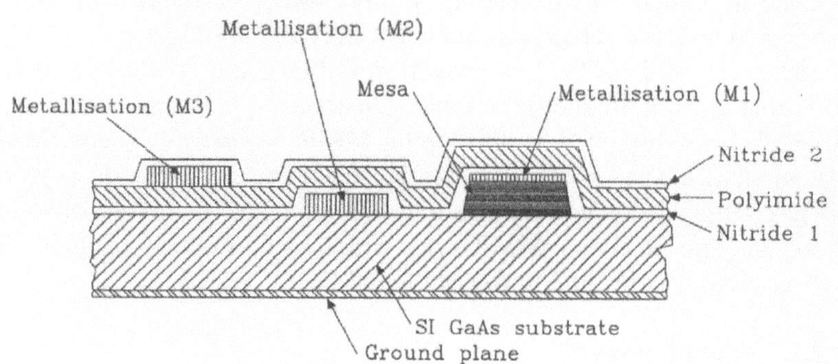

Figure 1 : MMIC cross-sectional view.

The GaAs substrate (2–4 in diameter) is usually "thinned" to a height ranging from 100 μm to 300 μm, depending on the process used. A typical GaAs process is comprises eight main layers, defining the passive and active elements used. These layers may be identified as:[3]

1. The active layer, usually two levels of doping (sub-layers); *n* active sub-layer (doping density $\cong 10^{17}/cm^3$) and *n*+ low resistance contact sub-layer (doping density $\cong 10^{18}/cm^3$). The layers are produced, in most cases, by ion implantation. For HEMT MMICs Molecular Beam Epitaxy (MBE) is normally used.
2. Isolation, carried out by mesa etching down to the SI GaAs. Defines the borders of the active devices and implanted resistors.
3. Ohmic contact, AuGe-Ni-Au is commonly used for contacts to the *n*+ layers of MESFETs drain and source terminals, diodes P terminals and to implanted resistors terminals.
4. Schottky (gate) metallisation, most critical layer (and manufacturing process) of the MMIC. This is the layer that defines the gates of the MESFETs. The gate metallisation is applied after the gate region is recessed to provide the appropriate pinch-off voltage. Metallisation is applied uses a three (or more) metal layers (Ti, Pt, Au). Two types of gates are commonly used; the **T** (or **mushroom**) type and the **Vertical** type. For commercially available foundries, gate widths range from 100 nm to 1 μm. The total metal thickness can be up to few microns.
5. First metal layer, Alloy metals are used to form contacts to the ohmic and to form the lowest layer of Metal-Insulator-Metal (MIM) capacitors.
6. Dielectric layer, usually Silicon Nitride or polymide, used for passivation of exposed semiconductor layers and as a dielectric layer for MIM capacitors. The dielectric constant of this layer can

be as high as 7, and with layer thickness ranging from 0.1-0.3μm, capacitance values from 100 – 700 pF/mm^2 can be manufactured, with break down voltages in excess of 50 Volts.

7. Second metal layer, few microns thick (low resistance) metal used to form top layers of MIM capacitors, interdigital capacitors, spiral inductors, transmission lines and other interconnect components such as air bridges.
8. Via holes, Chemical etching is used to "drill" holes through the GaAs substrate and connect metal layers to the metal plated back side of the substrate, which acts as a ground plane.

It is important to note that MMIC layer structure may differ from one process to another. Some foundries use three layers of metallisation and two layers of dielectric. Additional layers of high resistivity metal alloys, such as NiCr (Nichrome) can be used to construct resistors.

2.2.3. MMIC Elements

Active devices, passive components and interconnect components are needed to construct a working MMIC. Schottky gate MESFETs and diodes are the main active devices used in GaAs MMICs. Different types of resistors, capacitors and inductors are also used together with a variety of interconnect elements. The key to a good MMIC design is to have accurate DC and RF models of all the MMIC elements and to base the MMIC simulation on layout parameters and dimensions.

MMIC foundries provide users with layout based models, obtained from comprehensive sets of measurements of the different MMIC elements. In the following sections, the main elements used in GaAs MMICs are described and their models outlined [1–4].

2.2.3.1. MESFETs

The MESFET is built on the two active sub-layers of the GaAs substrate. The operational characteristics of a given MESFET are strongly dependent on its geometry and size. A cross section of a typical MESFET is shown below.

For a given MMIC process, the gain of the MESFET depends on its active area, which is defined by the total width (longer dimension) of the gate. Several gates can be "cascaded" to increase the gain. One of the most common geometries for doing that is known as the Π -geometry. This is shown in figure 3 below.

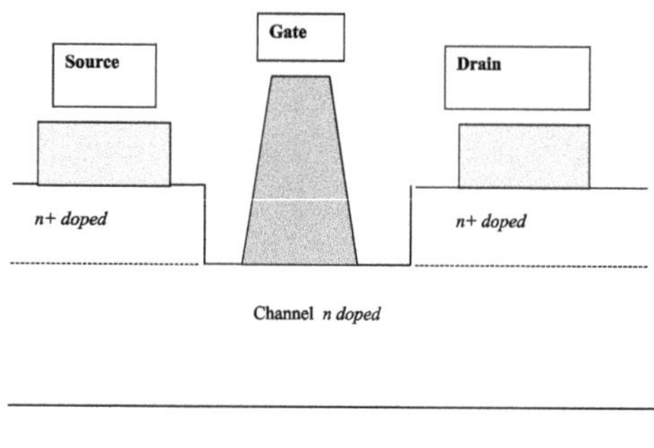

Figure 2 : Typical MESFET structure.

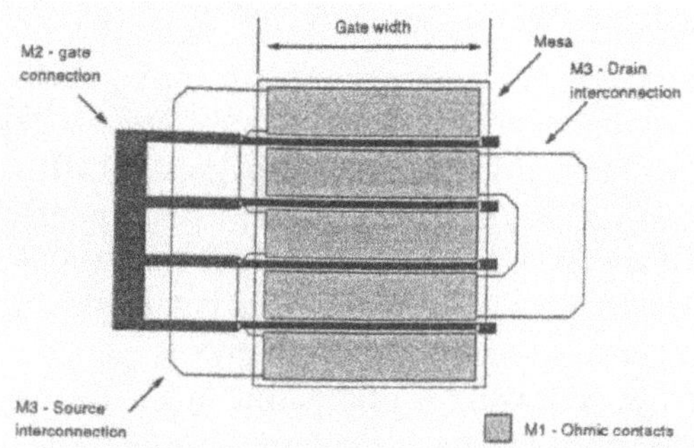

Figure 3 : Typical MESFET layout (4 finger Π - structure).

For this particular device, the MESFET active area is defined in the mesa layer and the drain and source ohmic contacts are on metal 1 (M1). Third level metallisation (M3) is used to interconnect the source/drain elements. This device can be viewed as a cascade of four identical MESFETs each having a single gate finger. The Π geometry is used to reduce the overall lateral size of the MESFET. The equivalent circuit of a MESFET (figure 4) is very similar to that of a JFET, except that for the higher frequency MESFET all parasitics must be accounted for in order to ensure correct modelling.

The MESFET can be divided into two parts, an intrinsic device, whose parameters are bias dependent and it models the active region of the device, and an extrinsic, bias independent, part that models the gate, source and drain contact metals. For high frequency modelling it is also important to note that the transconductance g_m is modelled as a complex quantity in order to account for the time τ_0 taken by the carriers to travel across the gate region.

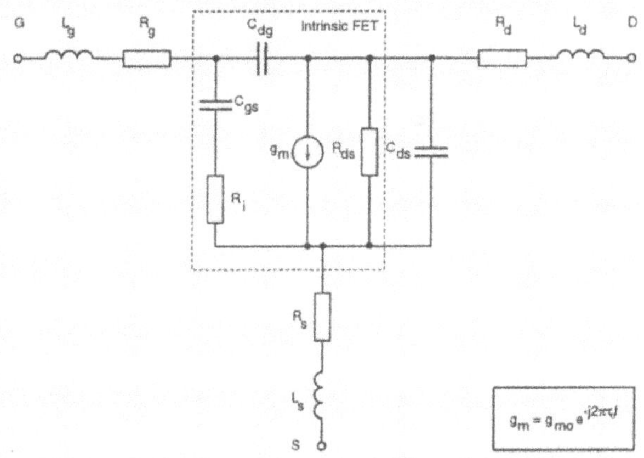

Figure 4 : MESFET equivalent circuit.

For frequencies in the tens of GHz region, more complex models than that of figure 4 may be required. Such models account for coupling capacitances across the device terminals (known as geometric capacitances) and for transmission line effects on the metal terminals.

The MESFET electrical behaviour is described by two models; a small signal model and a large signal one. The large signal model defines the relation between voltages and currents and can be used to extract the small signal parameters at a given bias point. MMIC foundries provide large signal model parameters based on one or more of the many models available in the literature (e.g. Curtice cubic, Materka and Tajima models) and implemented in different microwave CAD packages. Foundries also provide MESFET transfer characteristics (I_{ds} versus V_{ds} for different values of V_{gs}) and power transfer characteristics data. Designers are also provided with small signal model parameters (usually scalable with respect to device size/geometry). These parameters are provided at particular bias points (V_{ds} and I_{ds}) and if they were to be used the designer is restricted to such bias points.

For low noise designs, such as those of optical receiver amplifiers, noise models are of great importance. The intrinsic MESFET noise is modelled by two (gate and drain) correlated noise sources. Resistive elements generate thermal noise that adds to the intrinsic noise. Again, there are several models used to describe MESFET noise [5-8] and either one or more of such model parameters are provided by MMIC foundries.

The passive elements used in MMICs (resistors, capacitors, inductors, transmission lines and via holes), are normally described by equivalent circuit models that take into account all the parasitic elements associated with such devices and model their frequency behaviour. Due to limitation of space, such models are not discussed here, however, an excellent discussion can be found in [4].

2.3. MMIC Design Procedure

The MMIC design procedure comprises different steps involving circuit design, modelling and optimisation. For optical applications, it is always important to consider that the optoelectronic devices (photodetectors, Lasedr diodes, Modulators ..etc) are not "typical" microwave devices, that have to be 50 Ω matched and that innovative matching techniques may have to be considered for such applications [8-10]. The main difference between MMIC and lower frequency IC design is that for optimum MMIC operation (in terms of response, noise and stability) the non-ideal behaviour of all the MMIC elements need to be carefully considered. In addition, the MMIC lay out need to be considered with great care as electromagnetic coupling and transmission line effects acquire special importance at GHz frequencies. Apart from these considerations the circuit design techniques followed are identical to those of discrete microwave circuits.

A "well proven" MMIC design procedure can be summarised by the following steps:

1. Define MMIC target response
2. Design a basic circuit assuming that all the passive elements are ideal
3. Include full models of the least critical passive elements
4. Compare simulation results to target. Tune the active and passive elements to get as close to the target as possible
5. Include full models of all passive elements
6. Repeat step 4
7. Generate circuit layout. Minimise the overlapping of circuit components and crossing of transmission lines on different metallisation levels. If in doubt, try full EM modelling
8. Model all the layout components, account for all parasitic elements and the external connecting wires, decoupling capacitors. .etc, then re-simulate
9. Ensure that no design rules are violated
10. Resimulate and optimise

Well-established foundries provide designers with MMIC component models linked to layout parameters for use in an integrated design environment. In such cases, some of the steps mentioned above can be automated thereby simplifying the design process.

2.4. Summary

As most of today's optical communication systems operate at very high frequencies/bit rates, The use of MMICs for optical applications is increasing. MMICs are valuable for both narrow band and wide band and for low noise or high power applications. Good understanding of the MMIC process parameters and good appreciation of the specific nature of optoelectronic components is vital for a successful design. This chapter offers a glimpse into this interesting field.

References [1-8] below are of general nature looking at devices, circuits and specific system applications, while References [9-11] are concerned with specific MMIC designs for broadband optical receivers.

3. HIGH SPEED ICS FOR OPTOELECTRONIC MODULES

R. Lefèvre
OPTO+
Groupement d'Intérêt Economique
Route de Nozay, 91460 Marcoussis, France
France Telecom, CNET
E-mail : rene.lefevrer@cnet.francetelecom.fr

3.1. Abstract

Some basic principles about high bit rate digital transmissions are first reviewed ; high speed circuits for such system are analysed and front-end modules are more specifically addressed. Some devices, developed at OPTO+, are also presented.

3.2. Introduction

The first people to use digital optical communications was probably the US Indians but at a very low bit rate ! With the optical fibre as physical support, very high bit rate digital transmissions are now possible. Up to the seventieth years, analogue signal was used for information transmission and was practically voice channels. As mentioned by its name (analogue) the signal at the end of the link had to be analogous to the entering signal ; this meant a high linearity for the devices (optoelectronic and microelectronic devices) to keep a good signal quality. By sampling the analogue signal and coding it in a binary form, this drawback was overridden and time multiplexing became a reality. With the evolution of both optoelectronic and microelectronic components, some complementary multiplexing techniques such as Electrical Time Division Multiplex (ETDM), Optical Time Division Multiplex (OTDM) and Wavelength Division Multiplex (WDM) led, by mixing them to very high bit rate transmission systems (Some tenth of Terabit/s). Digital signals means relaxed performances in terms of linearity, noise immunity and information diversity ; video, audio and computer data signal are all represented by binary elements. For very high bit rate, the devices, mainly front-end devices, are made of III-V material (GaAs and InP) ; however, more recently SiGe represents a good challenge for some kind of circuits.

3.3. Basic Principles

To transmit analogue signals (ie modulated carrier) in a digital way, the signal is first sampled at a frequency which has to be twice the value of the highest spectrum frequency of the signal (Shannon theorem) (Figure 5). Then each sample is 8 bit coded ; the result is a series of 8 bit for each sample these bit can be equal to zero or one or a combination of zero and one depending on the value of samples. For example, a voice channel has 4kHz bandwidth, the sampling rate is then 8 kHz, 8 bit coding (256 levels) and the resulting bit rate is 64 kbit/s. The resulting pulse train is shown on figure 6.

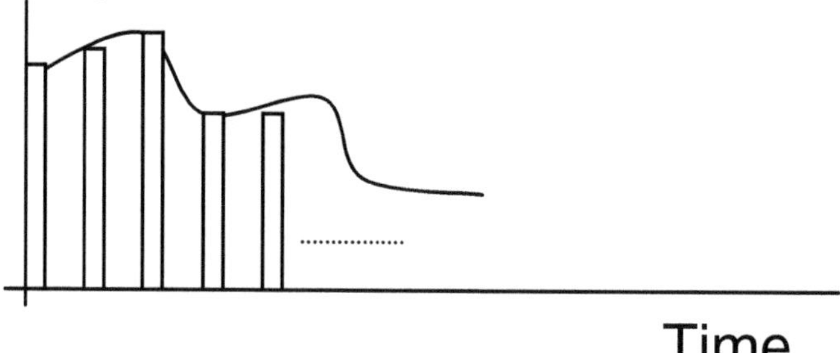

Figure 5 : Sampling principle of analogue signal.

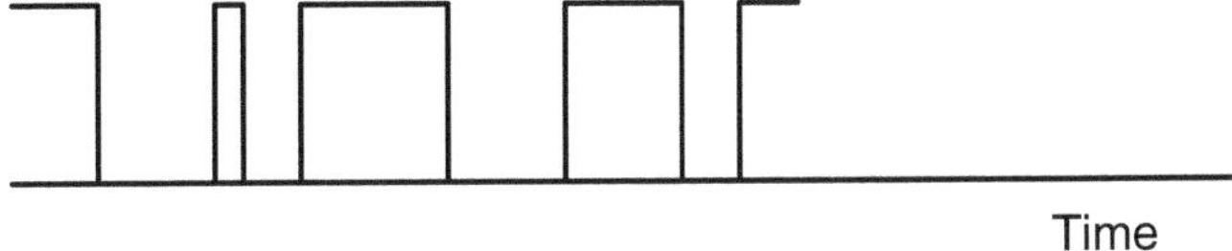

Figure 6 :Resulting signal after sampling and coding.

The multiplexing technique consists in narrowing the samples and interleave other samples from other signal (Figure 7).

1100101110011101	Channel 1
	11001011***00011110***10011101***10100111***
0001111010100111	***Channel 2***

Figure 7: coded Channels and their time multiplex.

The bit duration is given by the clock period, so a 64 kbit/s pulse stream, has a clock frequency of 64 kHz. Two main coding schemes are used for the high bit rate transmission: NRZ for Non Return to Zero (during the one bit duration) and RZ for Return to Zero (during the one bit duration). These code schemes have some impact on the bandwidth of the devices ; for NRZ, this bandwidth is from DC to 0.7 times the clock frequency while for RZ is from DC to clock frequency. If a given NRZ coded signal is observed on a scope synchronised by the clock frequency, the so-called eye-diagram is displayed on the scope (Figure 8).

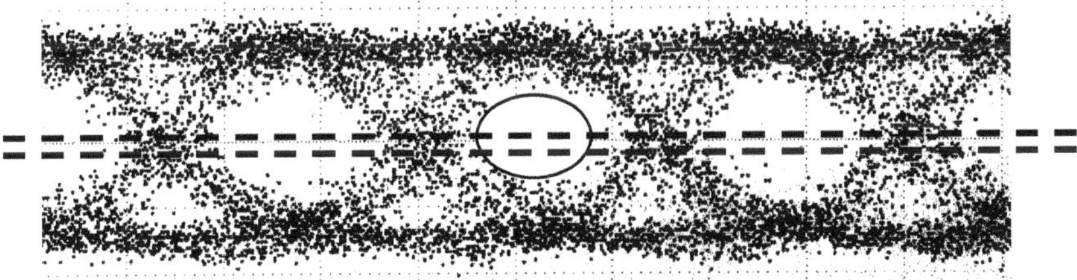

Figure 8 : Eye-diagram and decision window.

An error-free transmission shows a clear opened eye-diagram so that the decision window (represented by the two dotted lines on figure 4) may clearly separate the one level from the zero level. This is measured by the Bit Error Rate (BER) which gives the number of errors per second.

3.4. High Bit Rate Systems and Associated Front-End Modules

3.4.1. Time Division Multiplex (TDM)

This kind of transmission adresses two possibilities: Electrical Time Division Multiplex (ETDM) and Optical Time Division Multiplex (OTDM). In the former, data are multiplexed electronically; a high speed driver is needed to modulate a laser or an external modulator (above 10 Gbit/s) at the bit rate ; after propagation in the fibre, a very low noise photoreceiver is needed to convert the weak optical signal into an amplified electrical signal. In the case of OTDM, data are optically multiplexed and demultiplxed ; For example, at 40 Gbit/s Four 10 Gbit/s optical modulated pulse stream are multiplexed. This means lower speed for electronic but critical power budget for the link. OTDM can be an interesting way to multiplex N times 40 Gbit/s pulse stream each one being an ETDM pulse stream. For both systems, there is only one wavelength.

So the main critical modules are both the driver and the photoreceiver. These modules needs high speed electro-optic modulators and high speed electronic circuits and high speed photodetectors combined with a very low noise preamplifier. For the driver, the input impedance is 50 Ohms but the output load is about 5 Ohms in the case of a laser, or a capacitance in the case of a modulator; so the driver is a very specific circuit. For the photoreceiver, the input impedance of the preamplifier has to be high to get low noise and match to the high impedance of the photodetector while the output impedance is of 50 Ohms. The driver works with large signals while the photoreceiver works with small signals. Both circuits are very broadband circuits. Such modules are shown on figure 9.

Figure 9 : 20 Gbit/s driver and photoreceiver modules.

3.4.2. Wavelength Division Multiplex (WDM)

In this case 4 or 8 or more wavelengths are multiplexed and launched into a fibre ; each wavelength is itself an ETDM or OTDM pulse stream. Array of laser-modulator and drivers at the transmitter side and array of photoreceivers at the receiver side are needed; with monolithic integration, crosstalk between two adjacent channels has to be minimised.

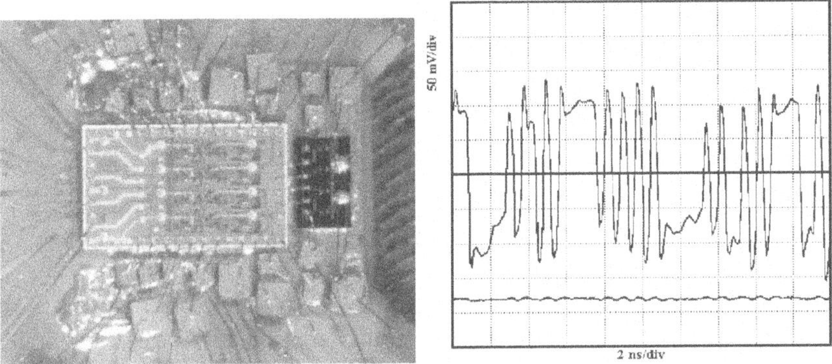

Figure 10 : Photoreceiver including an array of photodiodes and preamplifiers; crosstalk between two adjacent channels.

3.4.3. Radio on Fibre

A way for distribution of multimedia services to customers is to use millimetre wave; an antenna transmit a modulated RF carrier, frequency of which depending on the size of the covered area, up to a set of customers (set of buildings, home or building). The RF carrier can be transmitted to the antenna by mean of a fibre; this is the so-called radio on fibre distribution; in Europe three frequency band are allowed (28, 41 and 60 GHz); in such mode of distribution, the photoreceiver includes a narrow band preamplifier and a high speed photodiode. To get both high speed and high responsivity side illuminated photodetectors are required (Figure 11).

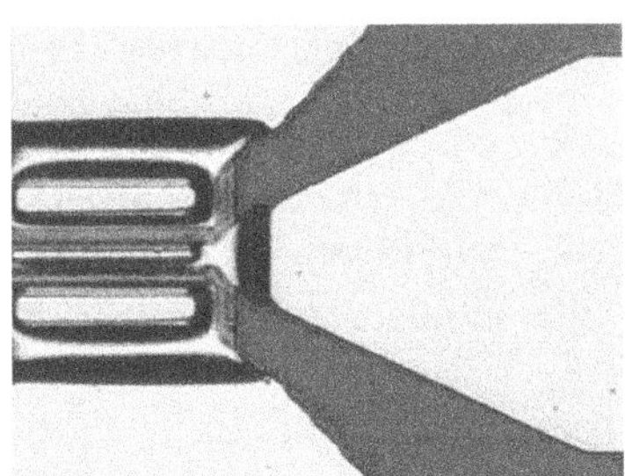

Figure 11 : Side illuminated photodiode and its 50 Ohms coplanar access line

3.4.4. III-V Microelectronic

To get such high speed, high mobility materials are needed; III-V materials as GaAs or InP offers these characteristics: Semi-insulating substrates allow the realisation of low loss passive components such as coplanar or microstrip lines. The main components used are the High Electron Mobility Transistor (HEMT) and Heterojunction Bipolar Transistor (HBT). The cut-off frequency of these components are about 105 GHz for GaAs and 340 GHz for InP and allow the realisation of very high speed digital and analogue circuits [12]; SiGe allow now high speed digital circuits [13]; InP microelectronic allows the monolithic integration of both optoelectronic and microelectronic components on a same substrate.

GaAs and InP microelectronic are then well suited for front-end modules realisation such as transmitter and low noise photoreceivers. A module can be represented schematically as on figure 12. Inside a package, there are an IC cascaded with an optoelectronic component (external modulator or photodiode) and mounted on an alumina substrate; coplanar lines are used to ensure the connection between the IC and the electrical connector while fibre has to be accurately positioned in front of the optoelectronic component (so-called pigtailing).

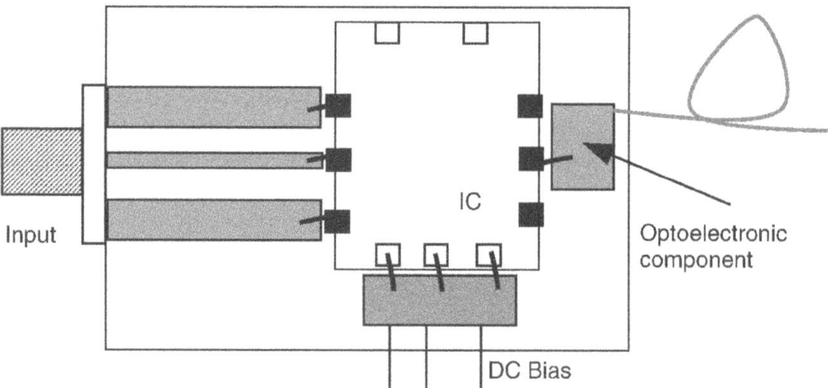

Figure 12 : Structure of an optoelectronic module.

For a transmitter, the IC is a driver while the optoelectronic component is an external modulator (Figure 13) which needs between 2 and 5 V of driving voltage over a DC to clock frequency bandwidth. IC architectures are based on differential pair, below 10 Gbit/s, and on distributed amplifiers for higher bit rates.

In the design of such a module, all parasitic elements du to the wire bonding, decoupling bias components and package itself must be modelled to get the wanted frequency response of the module. Large signal models of both transistors and optoelectronic component has to be used in the CAD softwares.

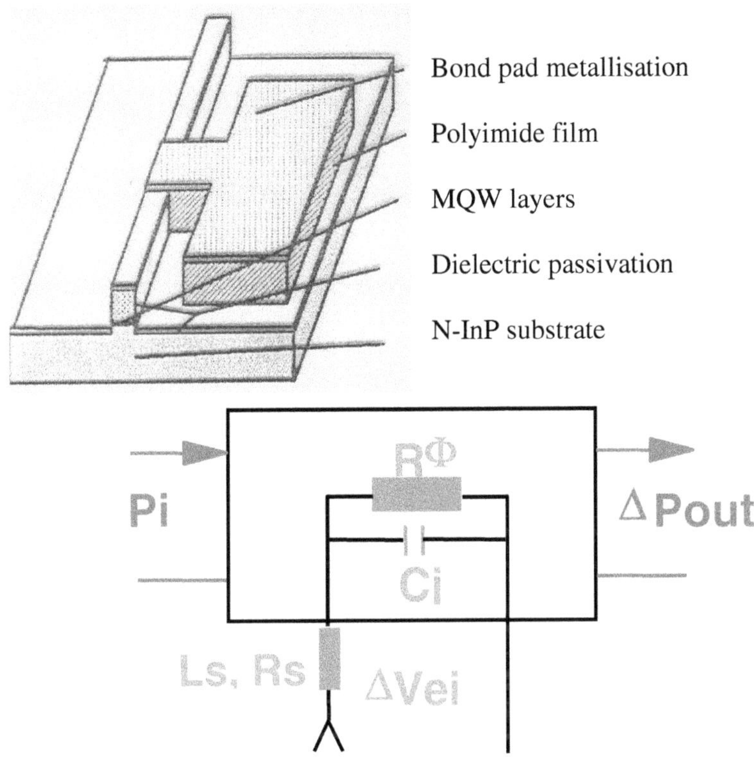

Figure 13 : Electroabsorption modulator and its equivalent Computer Aided Design circuit.

For a photoreceiver, the IC is a low noise transimpedance amplifier, which is operating in small signal; so both linear, and noise models of transistors and photodetector have to be used in the CAD software. The noise level mainly depends on both the dark current and capacitance of the photodiode as well as of transition frequency of transistor. The main architectures are the feedback amplifier (Figure 14) or a classical amplifier cascaded with an equaliser to levelled; the first architecture are based on differential pair while the second is based on distributed amplifiers.

Such modules operating up to 40 Gbit/s have been realised at OPTO+ (Figures 15 and 16) [14].

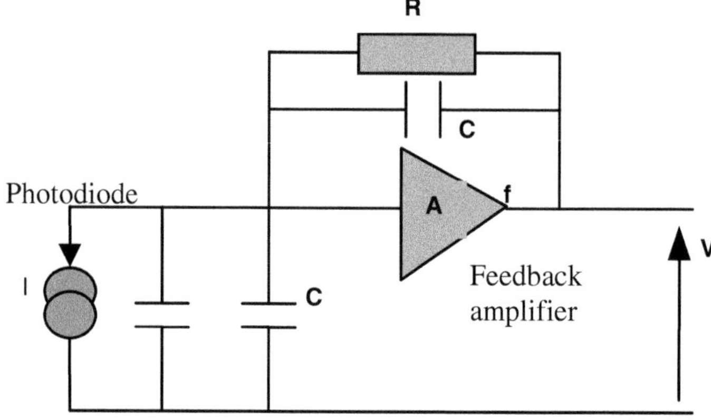

Figure 14 : basic scheme of photodiode cascaded with a feedback amplifier.

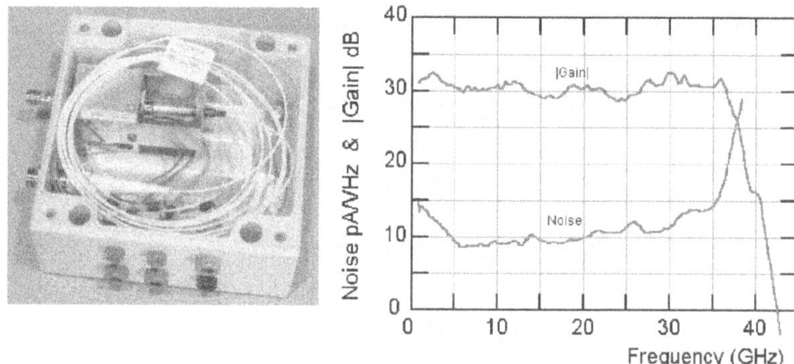

Figure 15 : 40 Gbit/s Photoreceiver module and its gain and noise performances.

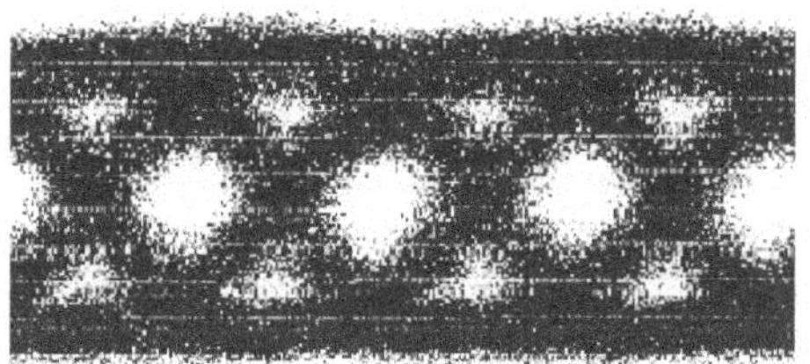

Figure 16 : 40 Gbit/s driver module and its output electrical eye-diagram.

3.5. Conclusion

High speed lasers, electroabsorption modulators and photodetectors with high responsivity are now existing. For microelectronic components, GaAs is a mature technology which is widely used by foundries in Europe and USA; InP microelectronic is under development and allows OEICs. SiGe microelectronic shows high speed digital circuit realisations. This is a chance for development of high quality services in optical communications.

4. HIGH EFFICIENCY OPTICAL TRANSMITTER AND RECEIVER MODULES USING INTEGRATED MMIC IMPEDANCE MATCHING AND LOW NOISE 50 Ω AMPLIFIER

M. Schaller, Ph. Duême, C. Fourdin, P. Nicole, J. Chazelas,
R. Blondeau, M. Crakowski, J.C. Renaud, P. Richin, F. Deborgies
Thomson-CSF Detexis Photonics and Microwaves
55 quai Marcel Dassault, 92 214 St Cloud Cedex – FRANCE
e-mail : Michel.Schaller@detexis.thomson-csf.com

4.1. Summary

Wideband photonic links consisting in combination of laser, photodiode and optical fiber present important losses. This is mainly due to passive impedance matching techniques implemented between microwave interfaces and optical transducers for preserving the system of spurious reflections.

An international fruitfully cooperation between 3 companies: Thomson, Miteq, and Diamond has leaded to overcome this drawback and develop a new generation of link with improved efficiency and innovative features.

4.2. Introduction

The increasing complexity of microwave systems in the field of Radar, Electronic Warfare and Telecommunication's applications implies the use of more and more microwave links. The intrinsic features of fibers (extremely low losses 0.2 dB/km) allow preserving propagating signals along kilometers. However conversion losses between electrical and optical information cancel the fiber advantages, especially when large bandwidth of frequencies are addressed. That currently leads to have more than 30 dB of losses on 10 Gbits/s links for instance and limit the application domains like analog microwave transportation.

Engineers have recently demonstrated it was possible to realize impedance transformers by Microwave Monolithic Integrated Circuits (MMIC), made on Gallium Arsenide (GaAs) substrates, represent key components regarding performance optimization (dynamic range, noise figure, consumption, losses....).

This paper describes both principle and implementation of 2 types of wide band Microwave impedance transformers using advanced and novel concepts:

- A distributed transimpedance amplifier dedicated to laser matching
- A specific transimpedance amplifier used for photodiode matching.

and show their implementation within photonic/ microwave modules.

4.3. Impedance Matching Problem

All the descriptions made hereafter concern direct modulation techniques. In this concept, well named, laser's current is directly modulated by microwave signal coming out from system.

Information is converted in optical modulated power and coupled into a fiber. At the other extremity of the fiber, Photodiode converts incoming optical power in electrical current. The modulated part corresponds to the main microwave signal.

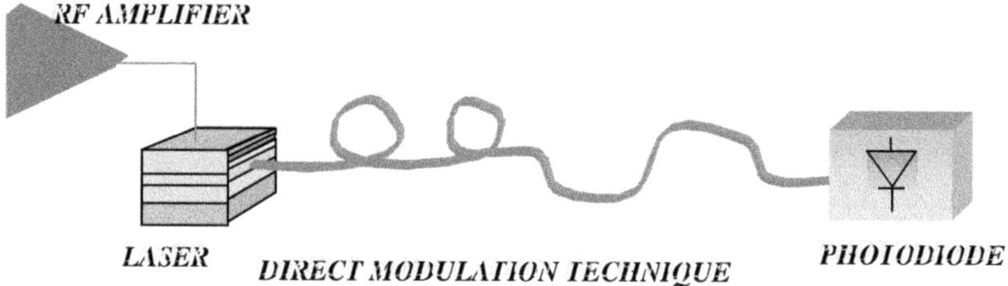

Figure 17 : Direct modulation principle.

Microwave systems use standard impedance of 50 ohms. Any component, which presents a different figure, induces return losses (or mismatching) and transmission losses regarding the propagating signal.

Using amplification could compensate transmission losses. However return losses cause dramatic degradation within system when it occur and must be avoided by designing.

Regarding wide-band photonic links, designers had not any choice excepted passive resistor integration between microwave and photonic devices. Implemented in line on laser side, in parallel on the photodiode side.

The following figures show both synoptic of conventional optical link and pictures of photonic transmitter and receiver.

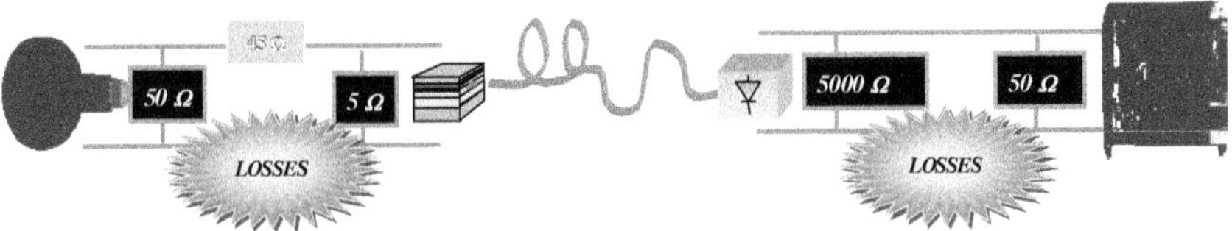

Figure 18 : Conventional matched direct modulation optical link.

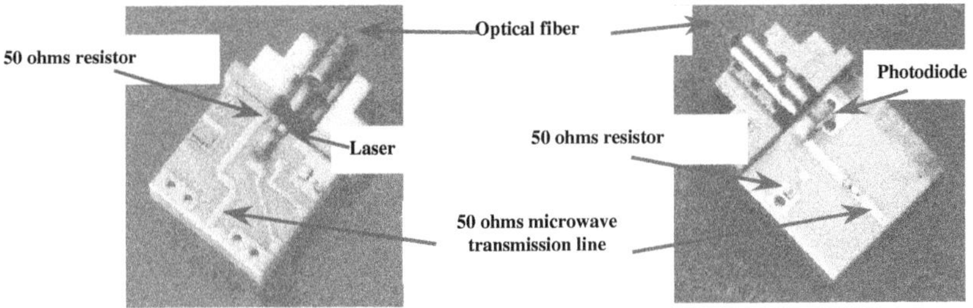

Figure 19 : Optical transmitter. *Figure 20 : Optical receiver.*

Resistor technique perfectly matches the interfaces of both worlds over wide frequency band but dramatically increases transmission losses of the links. In addition signal to noise ratio is degraded and dynamic range as well.

4.4. MMIC Impedance Transformer for Laser Diode

Distributed (or traveling wave) amplifiers have already been used as photodiode amplifiers, either with a conventional 50 Ω input impedance [15] or with a low input impedance (25 Ω) [16] to improve the input RC-bandwidth. Low output impedance distributed amplifiers have also been done for power purposes [17].

Here is reported a Distributed Amplifier with low output impedance designed for the direct modulation of a laser. The well known distributed configuration has been adapted to low output impedance by setting the drain line characteristic impedance near 5 Ω instead of classical 50 Ω impedance. This allows to avoid the previously described 45 Ω series resistance that matches the low laser diode input impedance (typically 5 Ω up to 20 GHz) but creates at least 10 dB losses at the very beginning of the transmission link. The laser bias current is supplied through an on-chip bias-T.

The circuit, manufactured with the VLN02 HEMT 0.25 μm gate process from THOMSON/TCS (France), is shown, connected to a Distributed Feedback laser diode, on the photograph of following figure.

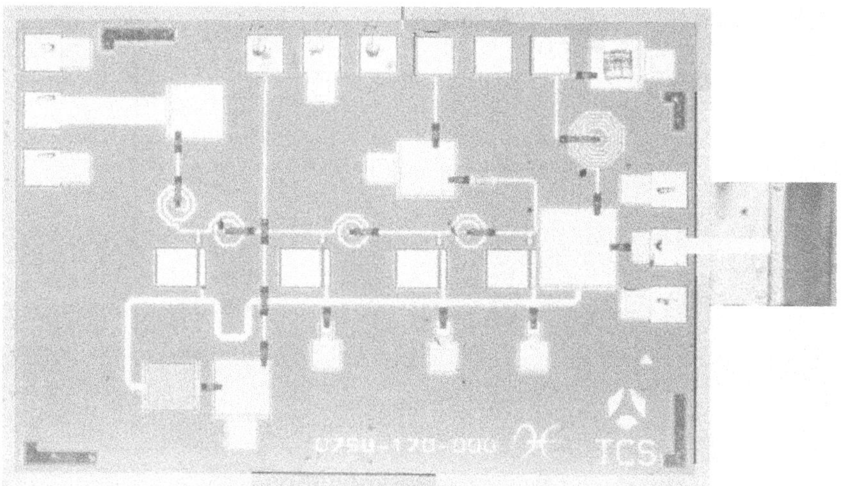

Figure 21 : Distributed Impedance Transformer Amplifier connected to a DFB laser . Input: 50Ω. Output: 10Ω. (1.8mm x 1.3mm).

A 7 dB intrinsic gain with less than 12 dB input and output return losses have been obtained over the 1-18 GHz range, with 50 Ω at the input and 10 Ω at the output as reference impedances.

On wafer measurements have shown a very good agreement to computed aided design values as presented on the following figure.

An improved design has just been completed by using UMS' design rules. It allows expanded bandwidth at lower frequency, going down 250 MHz. MMIC output power must be increased by 3 dB reaching figure close to 13 dBm.

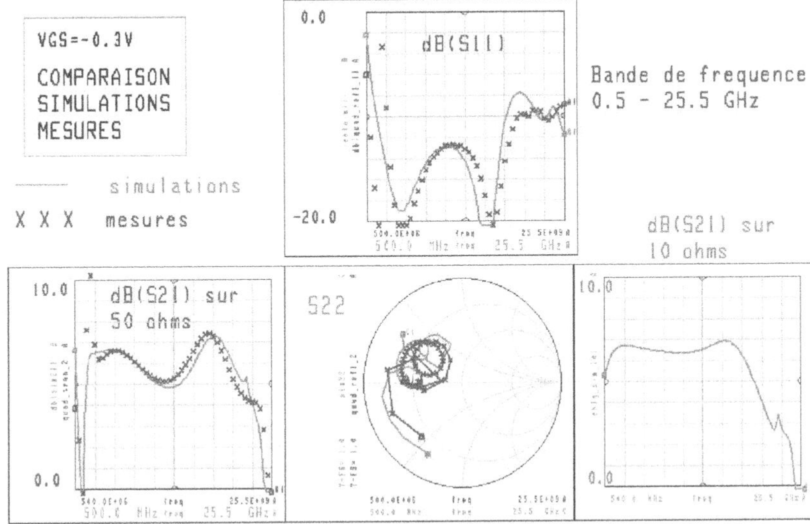

Figure 22 : MMIC impedance transformer for laser diode: on wafer measurements.

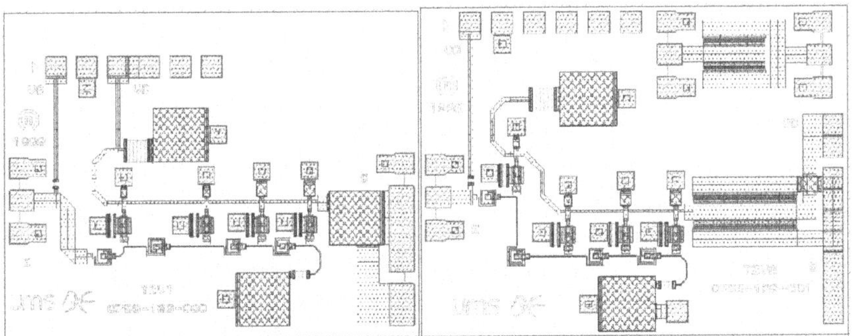

Figure 23 : Laser impedance transformers *Figure 24 : with shift diode.*

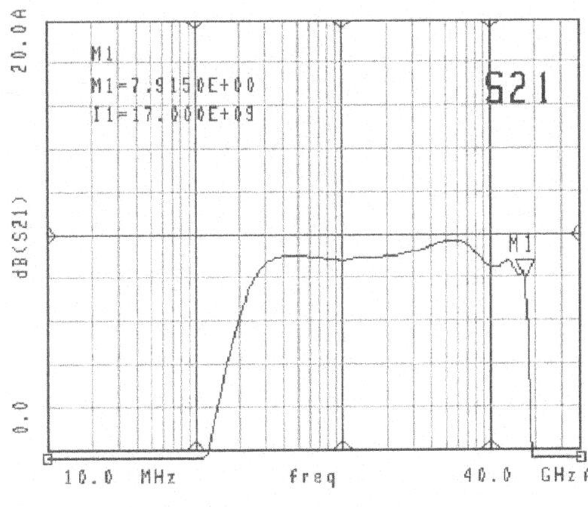

Figure 25 : S_{21} response without shift diode.

Thomson LCR has provided the laser diode, implemented for experimentation. This component presents a very linear behavior up to 10 GHz, well suited for analog applications. In addition, intrinsic noise figure is convenient and expressed as Relative Intensity Noise (RIN dB/Hz) with average value around -150 dB/Hz.

Figure 26 : DFB laser diode.

4.5. MMIC Impedance Transformer for Photodiode

The concept of this amplifier is based on the facts that, the parasitic R.C factor (resitor*capacitor) is a limitation in terms of frequency band and the combined resistor must be preserved as high possible for conversion efficiency and noise contribution.

For the first time in the microwave domain, the Bootstrap technique is used to create an active feedback loop aimed at canceling the voltage across the photodiode. This leads the current across the parasitic capacitance to be quite canceled and the photodiode to act ideally as a pretty pure current source. Basic theory and photograph are shown on the following pictures.

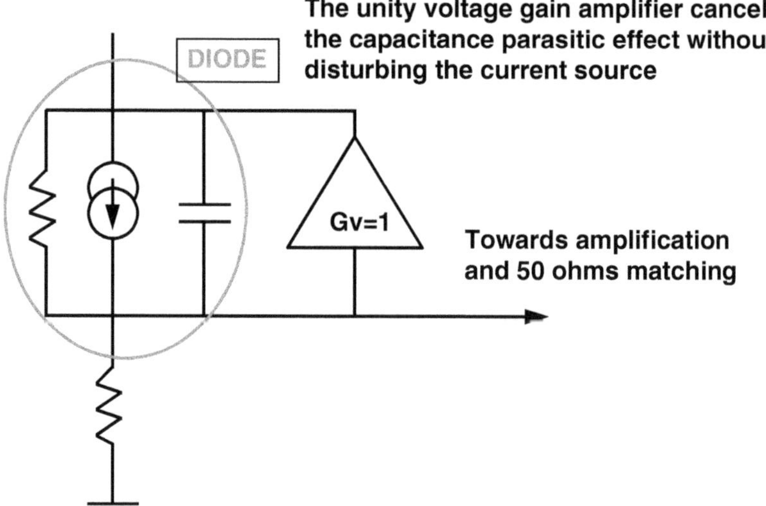

Figure 26 : Bootstrap basic diagram.

The ideal amplifier that cancels the voltage across the photodiode is fabricated in practice by using a FET in a common drain configuration, which has been the first generation.

A distributed amplifier 125 Ohms/ 50 Ohms, used in combination, improve the efficiency by maintaining a high value of associated resistor and the 50 ohms output matching. It represents the second generation recently design with UMS' design rules.

In addition this component includes innovative concept for transistor feeding. This principle called saturated loads allows lower frequency bandwidth expansion.

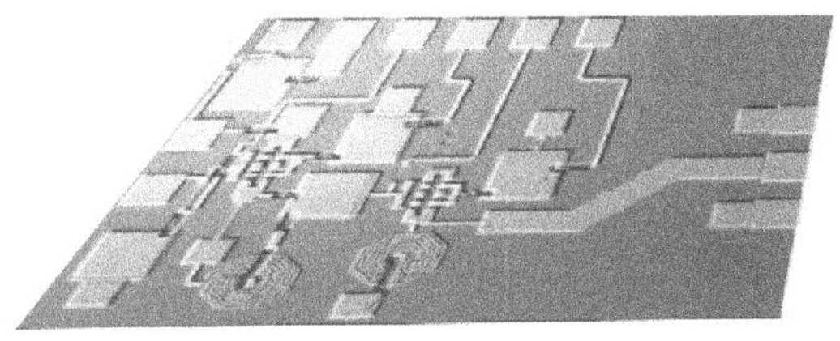

Figure 27 : Bootstrap stage (first generation) Amplifier chip (1.3mm x 0.9mm) - Input connected to photodiode. Output to 50 Ω.

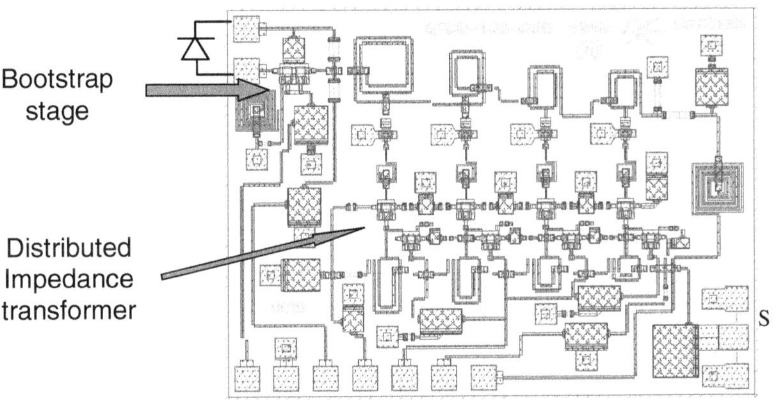

Figure 28 : Combined architecture: bootstrap and distributed amplifier.

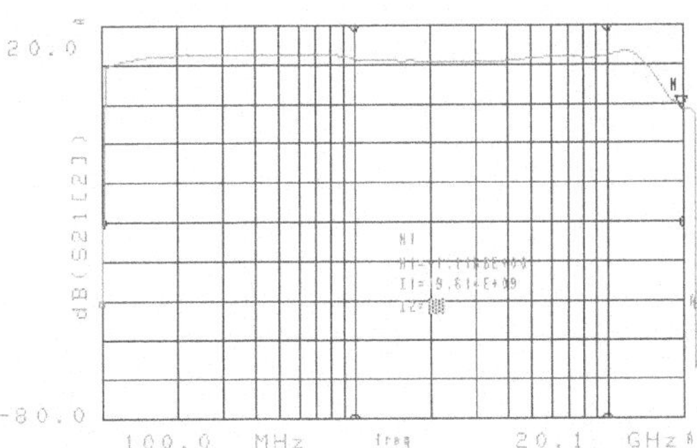

Figure 29 : S_{21} response.

4.6. Global Link Performances

A global link has been performed for measurements, combining the described components. In order to improved gain and noise figure of the link, additional stage of low noise amplification have been designed and added by Miteq company (USA) using hybrid technology.

The following figure shows an illustration of completed link.

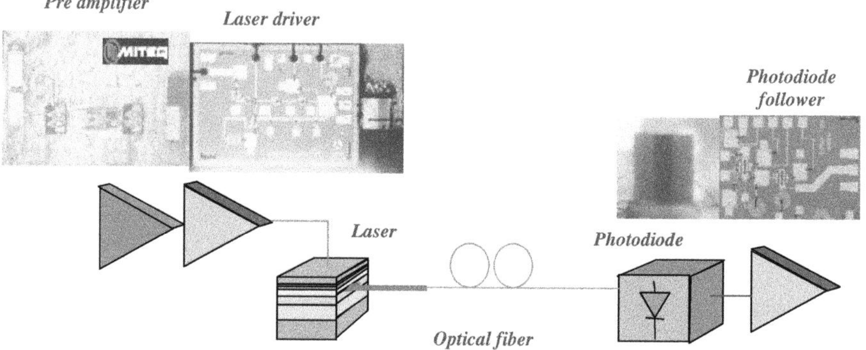

Figure 30 : Overall link with pre amplification.

Optical components from Thomson LCR are closely connected to MMICs in order to reduce parasitic bonding effects.

First generation link tests have been performed by using copper/ tungsten carriers for thermal dissipation problems.

Second generation will include Peltier cooler into transmitter housing and soon further multichip controller dedicated to thermal, optical power and bias control.

Active Impedance matching principle brings 15 to 20 dB of pure improvement compared to conventional technique. Beside this, 15 dB amplification shared between Miteq LNA and intrinsic MMICs gain bring the rest and allow to achieve 35 to 40 dB more efficiency than commercial links.

Regarding noise figure, commercial features are around 50 dB, leading to limit application field to long distance transportation or delay lines, chip resistor suppression is equivalent to noise reduction close to gain efficiency.

Gain is 10 dB on 200 MHz-12 GHz bandwidth limited on X band by DFB laser.

Curves on next page resume the state of art on mid 99.

Compared and summarized results for gain and noise figure of 3 types of optical links: commercially available, first generation of impedance transformers and second generation are presented on following graphs.

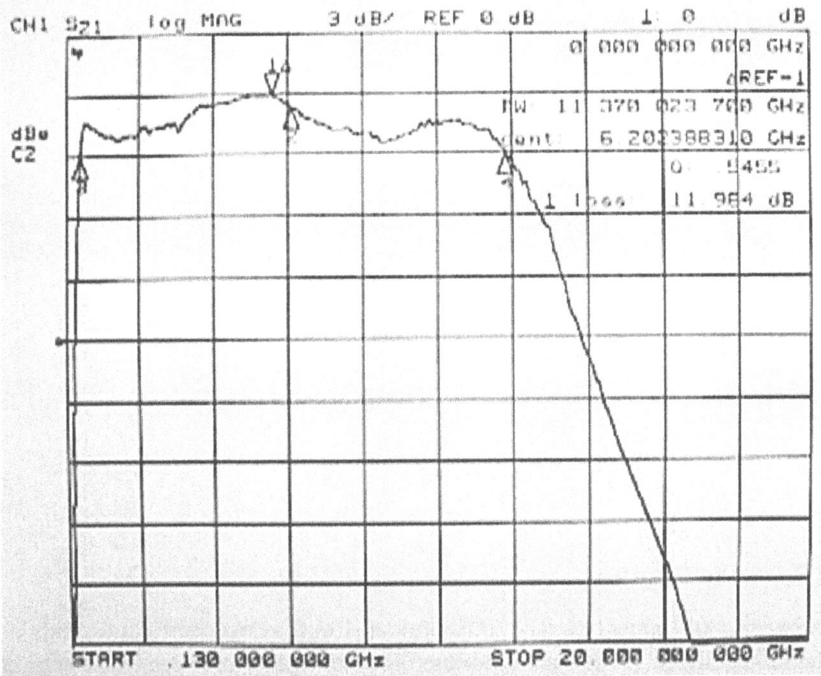

Figure 31 : This graph presents the best results obtained for S_{21} overall link using the second generation of MMICs just put under test end of June 99.

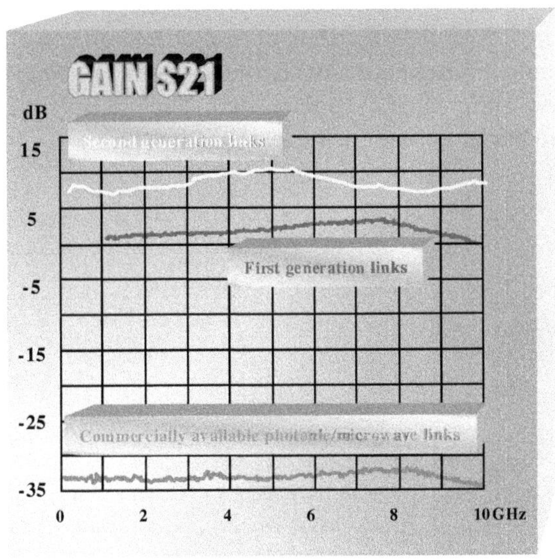

Figure 32 : Gain comparison.

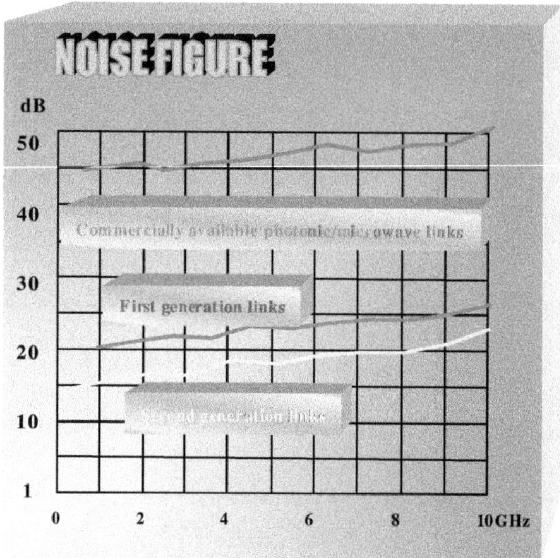

Figure 33 : Noise Figure comparison.

4.7. Packaging

A second objective for the optical links is to fabricate modules including all the requested functions like voltages and currents control or thermal regulation (only for laser). In this way, specific design of housing have been made with special features for receiving chip controllers, peltier cooler, microwave and optical connectors. Different kinds of housing are in progress for assessing the component installment. Some samples are presented on next figures.

One first step is to install chip regulators in charge to stabilize voltages and currents. Both negative and positive are requested. They deliver plus or minus 5 Volts, therefore bridge resistor achieve specific values.

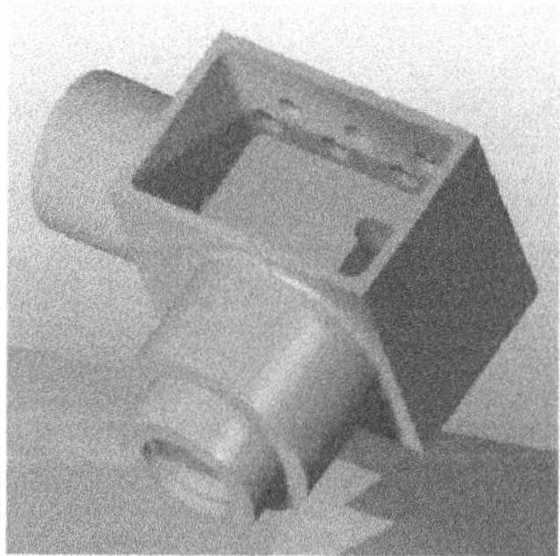

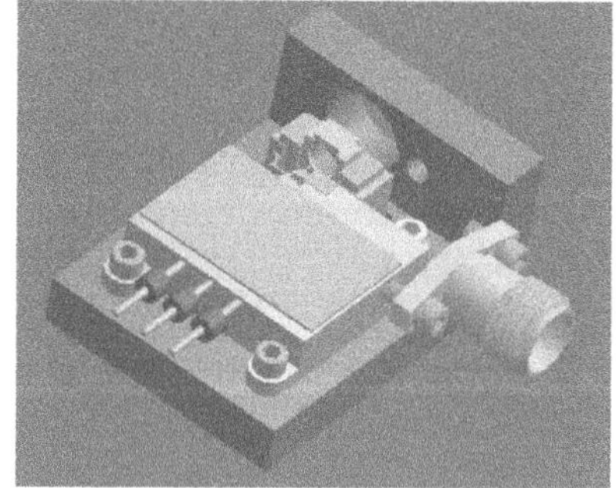

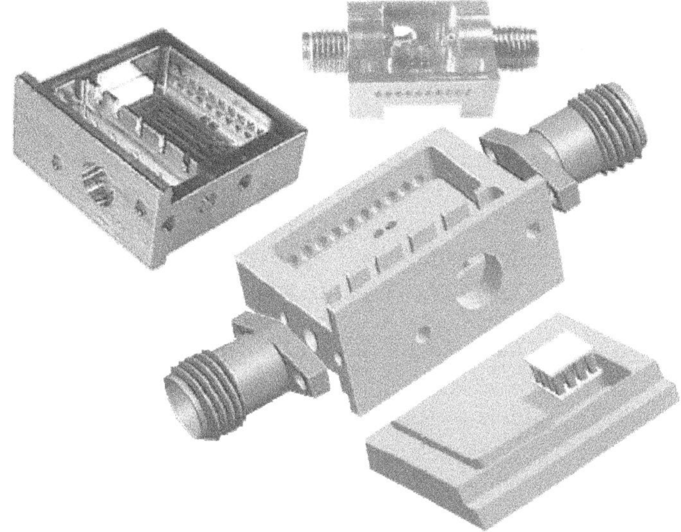

Figure 34 : Compared designs of housing.

Figure 35 : Component installment within package.

4.8. Conclusions

MMIC Impedance Transformers using advanced and novel concepts were reported. These very promising devices, obtained after a single foundry run, are key components for Digital and Analogue Optical Links. They have demonstrated the feasibility of high efficiency optical links in real integration situations with gain and reduced noise figure.

Conventional optical packaging excludes any current and voltage regulator obliging end user to implement himself requested external circuits. Using chip controllers and regulator we have demonstrated the opportunity to consequently reduce the volume of both functions transmit and receive modules.

The efficiency improvement, mainly regarding analogue applications, allows to obtain better noise figure and dynamic and start to convince users to prefer this means of microwave transportation for a lot of applications.

5. REFERENCES

[1] J. Golio
Microwave MESFETs and HEMTS
Artech House. ISBN 0-89006-426-1

[2] F. Ali and A. Gupta, *(Ed)*,
HEMTs and HBTs: Devices, Fabrication and Circuits
Artech House ISBN 0-89006-401-6

[3] R. Soares *(Ed.)*
GaAs MESFET Circuit Design
Artech House ISBN 0-89006-267-6

[4] I. Robertson *(Ed.)*
MMIC Design
IEE. ISBN 0-85296-816-7

[5] P. Ladbrooke
MMIC Design: GaAs FETs and HEMTS
Artech House. ISBN 0-89006-314-1

[6] R. Goyal *(Ed.)*
Monolithic Microwave Integrated Circuits: Technology and Design
Artech House ISBN 0-89006-309-5

[7] G. Gonzalez
Microwave Transistor Amplifiers
Prentice Hall, ISBN 0-13-254335-4

[8] B. Wilson, Z. Ghassemlooy and I. Darwazeh *(Ed)*
Analogue Optical Fibre Communications
IEE, ISBN 0-85296-832-9

[9] P. Monteiro, A. Borjak, F. da Rocha, J. O'Reilly and I. Darwazeh
10 Gbit/s Pulse Shaping Distributed Based Transversal Filter Front-End for Optical Soliton Receivers

IEEE Microwave and Guided Wave Letters, Vol. 8, No. 1, pp 4-6, Jan. 1998.

[10] A. Borjak, P. Monteiro, J. O'Reilly and I Darwazeh
High Speed Distributed Amplifier Based Transversal Filter Toplology for Optical Communication System
IEEE Transactions on Microwave Theory and Techniques, Vol. 45, No. 8, pp.1453-1458, August 1997.

[11] Iqbal and I. Darwazeh
23 GHz Baseband HBT Distributed Amplifier for Optical Communication Systems
Proceedings of 28th European Microwave Conference (EuMC-98), Amsterdam – Holland, Oct. 1998.

[12] T. Otsuji et al.
An 80-Gbit/s Multiplexer IC using InAlAs/InGaAs/InP HEMT's
IEE J. of Solid-state Circuits, Vol. 33 N° 9, Sept. 1998

[13] M. Neuhäuser, H.M. Rein
Low-noise, high gain Si-Bipolar preamplifiers for 10 Gbit/s optical fiber links – Design and realisation
IEEE Journal of Solid-State Circuits, Vol 31 N°1 Jan. 1996

[14] E. Legros et al.
High-sensitivity 40 Gbit/s photoreceiver using GaAs P-HEMT distributed amplifiers
Elect. Letters, Vol. 34 N°13, June 1998, pp 1351

[15] S. Kimura, Y. Imai, Y. Miyamoto
Development of a low-impedance travelling wave amplifier based on InAlAs/InGaAs/InP-HFET for 20 Gb/s optoelectronic receivers
1996 Conference on Indium Phosphide and Related Materials, pp 642-645

[16] S. Van Waasen, G. Janssen, R.M. Bertenburg, R. Reuter, F.J. Tegude
Novel Distributed Baseband Amplifying Techniques for 40-Gbit/s Optical Communication
IEEE GaAs IC Symposium, 1995, pp 193-196

[17] Ph. Dueme, G. Aperce, S. Lazar
Advanced design for wide-band MMIC power amplifiers
IEEE GaAs IC Symposium, 1990, pp 121-124

CHAPTER 3 : MODELING METHODS FOR OPTOELECTRONICS

1. INTRODUCTION

The increasing demand for processing and transmitting more and more information at a faster data rate leads, on the circuit level, to highly optoelectronic integrated circuits OEMMIC's. These packed circuits have been discussed in chapter 2.

The high density of devices in OEMMIC's may adversely affect the circuit performances, due to unwanted effects such as crosstalk, unintended radiation effects etc... This creates a need for modeling analysis and design tools, which is the object of this chapter.

The first part of this chapter is dedicated to some modeling methods for optoelectronics circuits. Modeling optoelectronics circuits is a difficult task since in microwave-photonics technology, integration techniques use different substrates of various properties. The modeling of guided-wave optics has become of increasing importance in optically interconnected modules and subsystems.

Then, careful circuit design based on advanced design tools is necessary. Tools for microwave and optic co-simulation are thus presented as a second part of this chapter.

Finally, full-wave analysis is often required for three dimensional (3-D) passive interconnections. Very few full-wave electromagnetic simulators based on solving Maxwell equations are able to work both in the microwave and optical range of frequency, because they require to mesh the analyzed structure for both wavelength ranges and the complexity becomes prohibitive. In the last part of this chapter, we discuss a global simulator treating at the same time problems of dispersion, crosstalk, packaging effects of 3-D passive structures. Examples of microwave and optical components are analyzed in this chapter with the same circuit simulator.

2. FOUNDATIONS FOR INTEGRATED OPTICS MODELING

I. Montrosset, G. Perrone
Department of Electronics - Politecnico di Torino
C.so duca degli Abruzzi 24, I-10129 Torino, Italy
Email : montrosset@polito.it

2.1. Introduction

In recent years the modeling activity in guided wave optics has expanded rapidly thanks to the availability of evermore powerful desk computers at lower cost. Modeling becomes increasingly important as a low cost alternative to repeatedly running device processing systems and characterization experiments to optimize the devices with respect to required output characteristics. Furthermore the simulation allows the evaluation of the device performance in a complex system.

The key factors in device modeling can be identified as follows:

- the representation of the physical mechanisms of the interactions involved;
- the waveguide and technology characterization;
- the evaluation of the mode and field evolution in the waveguide system;
- the formulation and solution of the equations describing the device behavior.

In this section we present the basic available techniques related with the last two parts, few examples of applications and some general considerations.

2.2. The Evaluation of the Fields in an Optical Circuit

Simpler devices are based mainly on propagation effects; e.g.: splitters, couplers, demultiplexers, etc. Typically, when a structure can be approximated as linear and longitudinally invariant, the knowledge of the modal fields and of the corresponding propagation constants is sufficient to describe the device behavior. When there is a longitudinal variation (range dependence) and the power is exchanged between the modes use can be made of numerical techniques such as Beam Propagation Method or of semi-analytical techniques as Coupled Mode Theory that allows the reduction of the numerical complexity of the full propagation problem.

2.2.1. Modal Analysis

The waveguides we will discuss are those typically used in Integrated Optics (IO) i.e. weakly guiding structures operating in the mono or quasi-monomode regime, as opposed to fiber devices. The most commonly used modal analysis techniques are either fully numerical or semi-analytical.

The fully numerical techniques such as Finite Element (FEM) [1,2,3], Finite Difference (FD) [4,5,6] methods and the Method of Lines (MoL) [7,8] can be used to solve the wave equation in vectorial or scalar form.

The scalar formulation follows from the so-called quasi-TE or quasi-TM approximation that assumes one transversal component of the electric or of the magnetic field to be dominant (Ψ) and the other negligible. Thus for quasi-TE modes $\Psi = E_y$ and $E_x \approx 0$; for quasi-TM modes H_y and $H_x \approx 0$. It then follows :

$$\left\{\nabla_t^2 + \left[k_0^2\left(n^2 - N_{eff}^2\right) + O\left(k_0^{-2}\right)\right]\right\}\Psi = 0$$

From Ψ it is then possible to compute the other four components of the electromagnetic field (H_x, E_z, H_z for quasi-TE modes and E_x, H_z, E_z for quasi-TM modes); being the longitudinal components smaller than the transversal ones in weakly guiding structures, they are frequently not relevant in many practical cases.

The numerical analysis proceeds by discretizing the structure in rectangular (FD, MoL) or triangular (FEM) elements. The wave equation is then reduced to a matrix problem through the discretization of the differential operator or by the projection of field expansion functions (moments method, Galerkin formulation), etc. From these procedures the problem becomes a normal or generalized eigenvalue problem.

The vectorial formulation is more accurate for waveguides with strong refractive index discontinuities while the scalar wave equation is more suitable for structures with lower Δn variations.

The most popular semi-analytical technique is the intuitively appealing Effective Index Method (EIM) [9]. It is obtained by assuming $\Psi(x,y) \approx X(x;y)Y(y)$ in the quasi-TE or quasi-TM wave equation and neglecting the derivative of X respect to y. This assumption allows a factorization of the wave equation and one obtains:

$$\begin{cases} \dfrac{\partial^2 X}{\partial x^2} + k_0^2\left(n^2 - n_{eff}^2(y)\right)X = 0 \\ \dfrac{\partial^2 Y}{\partial y^2} + k_0^2\left(n_{eff}^2(y) - N_{eff}^2\right)Y = 0 \end{cases}$$

where the solution of the first equation in depth gives the so called effective refractive index n_{eff} and X, and the solution of the second equation gives the modal effective index N_{eff} and the field Y. The first step is equivalent to reducing the dimensionality of the problem from two to one as schematically shown in the two examples in figure 1.

EIM is computationally simple and fast and is particularly efficient as a first step for the waveguide optimization process; around the parameter values found more accurate optimizations can be found with fully numerical techniques. A comparison between FEM and EIM shows, for example in the case of $Ti:LiNbO_3$ waveguides, a very good agreement between the normalized propagation constants and the field distribution in depth with an error of 20-30 % for the lateral spot size in the monomode regime [10].

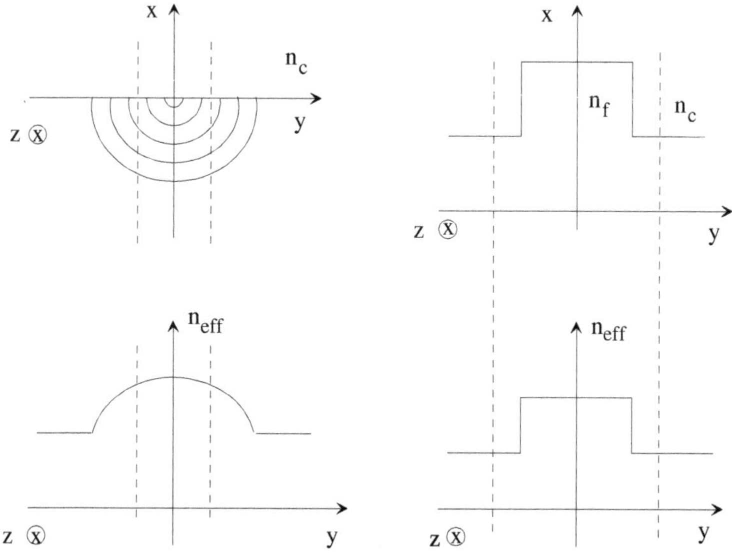

Figure 1 : Diffused and ridge waveguides and their equivalent one dimensional structures.

In many practical applications which involve particular interactions between photons and media, the wave equation should be solved self-consistently with the equations describing the interactions.

Active semiconductor waveguides provide an interesting example in which the field equation

$$\nabla_t^2\Psi + k_0^2\left(n^2(x,y,N,P) - N_{eff}^2\right)\Psi = 0$$

and the carrier diffusion and recombination equation

$$D\frac{\mathrm{d}^2N}{\mathrm{d}x^2} = -\frac{J}{ed} + R(N) + \frac{c}{n_g}\Gamma_x g(x)\Psi P$$

have to be solved self-consistently for each value of the modal power P. In this case we can assume

$$n(x,y,N,P) = n_0(x,y) + \Delta n(x,y,N,P)$$

where Δn is the perturbation due to the carrier injection and the stimulated recombination. Using the EIM and the usual linear relation for the variation of gain and refractive index with carrier concentration it is possible to write :

$$\tilde{n}_{eff}(y) \cong n_e(y) + \tilde{\Gamma}_x\left[bN(y) + ja\left(N(y) - N_0\right)/2k_0\right]$$

where $\tilde{\Gamma}_x$ is the depth confinement factor and N the solution of the diffusion equation. The imaginary part of $\tilde{n}_{eff}$ accounts for the saturated ($P \neq 0$) and unsaturated ($P = 0$) local gain.

Another interesting example is that of dielectric waveguides doped with active ions. In this case is typically necessary to account for the population inversion due to the pump, for the saturation effects due to pump and signal and for different physical mechanism of interaction between the atomic excited states.

In some cases the quasi two-level system approximation can be used as for the case of Er doping with 1480 nm pumping. In this case the equations for the populations of the upper metastable state (N_2) and of the ground state (N_1) to be solved self-consistently with the wave equations are :

$$N_2(\underline{r})\left(1+\frac{P_p\Psi_p(\underline{r_t})}{I_p^{ae}}+\frac{P_s\Psi_s(\underline{r_t})}{I_s^{ae}}\right)=\left(\frac{P_p\Psi_p(\underline{r_t})}{I_p^{a}}+\frac{P_s\Psi_s(\underline{r_t})}{I_s^{a}}\right)N_0$$

$$g_s=\sigma_{e,s}\iint N_2(\underline{r_t})\Psi_s(\underline{r_t})d\underline{r_t}-\sigma_{a,s}\iint N_1(\underline{r_t})\Psi_s(\underline{r_t})d\underline{r_t}$$

$$\alpha_p=-\sigma_{e,p}\iint N_2(\underline{r_t})\Psi_p(\underline{r_t})d\underline{r_t}-\sigma_{a,p}\iint N_1(\underline{r_t})\Psi_p(\underline{r_t})d\underline{r_t}$$

where N_0 is the Er doping distribution, P_i, Ψ_i, $\sigma_{e,a,i}$, I_i, are the total power, the emission and absorption cross-sections and the saturation intensity for the pump ($i = p$) and the signal ($i = s$); g_s and α_p are the local signal gain and pump absorption distributions [11].

In this case the very weak variation of the refractive index induced by the ion doping allows the computation of the fields $\Psi_{p,s}$ from the refractive index distribution of the undoped structure and the evaluation of the modal gain and absorption in a perturbative way.

While this procedure can always be used for rare earth doped waveguides, it is frequently not correct for semiconductor active waveguides when operating at high pumping and high field regime and in case of gain guiding structures.

2.2.2. Analysis of Range Dependent Structures

Here we will discuss two methods: the Beam Propagation Method (BPM) and the Coupled Mode Theory (CMT) which both have a wide range of applications in the analysis of linear and nonlinear devices.

2.2.2.1. The Beam Propagation Method

The BPM is a powerful numerical technique for the solution of the wave equation. The original formulation was based on the use of the FFT

[12,13]. However, more recent variations have employed Finite Difference [14] or Finite Element algorithms. We confine ourselves to a short presentation of the paraxial scalar FD class of BPMs.

For the sake of simplicity, we will discuss only the two-dimensional (2D) case. The starting point is the paraxial Helmholtz equation :

$$2jk_0n_0\frac{\partial\Psi}{\partial z}=\frac{\partial^2\Psi}{\partial x^2}+k_0^2\left(n^2(x,y,z)-n_0^2\right)\Psi$$

that is a parabolic partial differential equation and where n_0 is the bulk refractive index. The differential operators can be replace by difference operators by proper discretization of the domain. If we suppose a uniform mesh, the points on the grid (X,Z) are given by $X=m\Delta x$ and $Z=n\Delta z$, with integer *m* and *n*. The FD schemes used can be subdivided into explicit and implicit. In the explicit schemes [e.g. 14] a relation that links only one point in the unknown section $z+\Delta z$ to the previous sections is found. The most common example is the so called "three level scheme" :

$$\Psi_m^{n+1}=\Psi_m^{n-1}+\frac{\Delta z}{jk_0n_0\Delta x^2}\left[\Psi_{m+1}^n+\Psi_{m-1}^n\right]+\frac{\Delta z}{jk_0n_0}\left[-\frac{2}{\Delta x^2}+k_0^2n^2-k_0^2n_0^2\right]\Psi_m^n$$

that relates one point in section n+1 with those in the two preceding ones. Explicit methods are usually very simple to implement and fast but pose serious problems of stability.

On the contrary, in implicit schemes a relation linking several points in section n+1 is written. One of the most famous algorithm is the so called Crank-Nicolson scheme [15,16], a two level, unconditionally stable scheme that links three points in the unknown section with three points in the preceding one :

$$\Psi_{m+1}^{n+1}+a_m^{n+1}\Psi_m^{n+1}+\Psi_{m-1}^{n+1}=\Psi_{m+1}^n+a_m^n\Psi_m^n+\Psi_{m-1}^n$$

The resulting system of equations is of tridiagonal type and can be solved in an efficient way.

A key issue in practically implementing a numerical algorithm to analyze an integrated optical circuit is the problem of correct boundary conditions. Boundary conditions must guarantee that all the radiation modes are free to escape from the computational window without the introduction of unphysical reflections. Absorbing boundary conditions (i.e. a fictitious layer with high losses) were very popular in the past, while today are more common the so called "Transparent Boundary Conditions" (TBC). In the form proposed by Hadley [17,18], they are easy to implement into a 2D-FD scheme because they imply the modification of the computed field at boundary points by imposing a phase variation that satisfies the radiation conditions of outgoing energy flux.

Many forms of BPMs have been recently proposed; with proper modifications [19] these are capable of:

- analyzing range dependent structures in 2D and 3D problems;
- taking into account loss, gain due to current injection or pumping and nonlinear effects;
- computing waveguide modes;
- accounting for reflections and anisotropy;
- analyze pulse propagation and interactions in nonlinear media [19,20,21].

Above all, the BPM is straightforward to implement. The most problematic part of a BPM code is, for general structures, that part dedicated to the definition of the structure itself. Today, BPM is very used for the analysis of complex structures because various BPM packages including also a mask layout generator are on the market.

2.2.2.2. Coupled Mode Theory

A detailed treatment of this method has been given in many books [22,23]. Here we summarize its main features. Coupled Mode Theory represents the propagation in an actual waveguide $\varepsilon(x,y,z)$ in terms of the modes of a more simple "unperturbed" one $\varepsilon_0(x,y;z)$.

The choice of the unperturbed structure is equivalent to the definition of a complete set of orthogonal functions used to represent the electromagnetic field evolution. In the case when this set is independent of the longitudinal coordinate we have the so called normal modes; when the basis functions are longitudinally dependent we have the so called local normal modes. The equivalence theorem allows us to represent the difference between the two waveguides ($\Delta\varepsilon = \varepsilon\text{-}\varepsilon_0$) in terms of an equivalent dielectric polarization :

$$\vec{P} = \Delta\underline{\underline{\varepsilon}}\,\vec{E}$$

Expanding $\vec{E},\vec{H}$ in term of the forward (+ν) and backward (-ν) modes of the unperturbed structure we have :

$$\vec{E}=\sum_{\nu}(a_{\nu}+a_{-\nu})\vec{e}_{\nu t}+\sum_{\nu}(a_{\nu}-a_{-\nu})e_{\nu z}\hat{z}$$

and equivalently for the magnetic field; the evolution equation for each mode ν becomes:

$$\frac{\mathrm{d}a_{\nu}}{\mathrm{d}z}+j\beta_{\nu}a_{\nu}=-j\omega\iint\vec{P}\cdot\vec{e}_{\nu}^{\,*}\,\mathrm{d}x\mathrm{d}y$$

By extraction of the fast varying terms

$$a_{\nu}(z)=c_{\nu}(z)\exp\left[-j\beta_{\nu}z\right]$$

and expanding $\vec{P}$ using modal fields one obtains for each mode :

$$\frac{\mathrm{d}c_\nu}{\mathrm{d}z} = \sum_\nu K_{\mu\nu}\, c_\nu \exp[j(\beta_\mu - \beta_\nu)z]$$

where : $K_{\mu\nu} \cong -\dfrac{j\omega}{2P}\dfrac{\beta_\mu}{|\beta_\mu|}\iint \vec{e}_\mu^{\,*}\cdot \Delta\underline{\underline{\varepsilon}}\cdot \vec{e}_\nu\, \mathrm{d}x\mathrm{d}y$

is the coupling coefficient between the modes μ and ν, the summation extends to all guided modes and to the continuous spectrum and P is the power normalizing factor.

From this system of differential equations the relevant ones are only those that approximately satisfy the *phase* matching condition :

$$\beta_\mu - \beta_\nu - m\frac{2\pi}{\Lambda} = 2\delta_{\mu\nu} \approx 0$$

where Λ is the period of the refractive index variation. After some manipulations, the typical form of the equations to be solved when only two modes (μ and ν) are coupled is $\dfrac{\mathrm{d}c_\mu}{\mathrm{d}z} = (-j\delta_{\mu\nu} + K_{\mu\mu})c_\mu + K_{\mu\nu}\, c_\nu$

A similar expression holds for mode ν.

2.3. Practical Applications

In the simulation and design of IO components, the choice of the analysis technique to be used depends very much on the device structure and on the relevant effects we are looking for. In many cases all the previously described techniques can be alternatively used and their choice depends on the availability of the codes for the analysis and on the knowledge background of the researcher.

2.3.1. Examples of Use of the Modal Technique

The range of application of the modal analysis is limited to structures that are or can be approximated as linear structures and in which there are no modal coupling effects. The evaluation of the modal gain and effective index in active waveguides is one of the possible examples; the total gain can then be obtained by integration of the local gain function that can be computed considering also saturation effects and pump depletion.

Another interesting example is the study and analysis of coupled waveguide systems where the modes of the complete structure in every longitudinal section can be computed. The exchange of power between adjacent waveguides is represented in this case as an interference phenomenon among the field distributions of the various modes of the whole structure propagating with different propagation constants.

Finally, we have to remember that modal analysis is also the basis of coupled mode theory (§2.2.2.2.).

2.3.2. Examples of Use of Coupled Mode Theory

As a first example we treat the case of a perturbation (Δn) of the complex dielectric constant such as that due to gain in the active layer of a dielectric waveguide.

The relevant coupling coefficient in this case is the self-coupling coefficient :

$$K_{\mu\mu} \cong -\frac{j\omega}{2P}\frac{\beta_\mu}{|\beta_\mu|}\iint \Delta n^2 |e_\mu|^2 \,\mathrm{d}x\mathrm{d}y$$

and the only relevant equation, if the perturbation does not change significantly the modal distribution is:
the solution of which is straightforward. If we suppose that we have a constant gain (g) in the active region, in the TE case we obtain :

$$\frac{\mathrm{d}c_\mu}{\mathrm{d}z} = K_{\mu\mu} c_\mu$$

$$K_{\mu\mu} \cong \frac{g}{2}\frac{\iint_a |e_\mu|^2 \,\mathrm{d}x\mathrm{d}y}{\iint_{\Re} |e_\mu|^2 \,\mathrm{d}x\mathrm{d}y} = \Gamma\frac{g}{2}$$

corresponding to a modal field gain and where Γ is the so called modal field confinement factor.

Similar procedure can be followed to evaluate the variation of refractive index due to the electro-optic effect; in this case:

$$\Delta n(x,y) = -\frac{1}{2}n^3 r\frac{V}{G}e_m(x,y)$$

where r is the relevant electo-optic coefficient, V and G the applied voltage and the electrode gap and e_m the normalized electric field distribution due to the electrodes. One obtains:

$$\Delta n_{eff} = -\frac{n^3}{2}r\frac{V}{G}\Gamma_0$$

where Γ_0 is the overlapping integral:

$$\Gamma_0 = \iint |e_\mu|^2 e_m(x,y)\,\mathrm{d}x\mathrm{d}y$$

and e_μ is normalized to have $\iint |e_\mu|^2 \, \mathrm{d}x\mathrm{d}y = 1$

As a second class of examples, we consider the case of a two mode interaction due to a periodic structure. In the grating case, we can assume $n(x,y,z+\Lambda) = n(x,y,z)$, so it is convenient to Fourier expand the refractive index with respect to the longitudinal coordinate

$$n^2(x,y,z) = \sum_q A_q(x,y)\exp\left[jq\frac{2\pi}{\Lambda}z \right]$$

and assume $n_0^2(x,y) = A_0(x,y)$ that guarantees that the self coupling coefficient $K_{\mu\mu}$ is zero. The coupling coefficient between mode μ and ν with the same polarization through the m-th order grating harmonic that satisfies the phase matching condition will be :

$$K_{\mu\nu} = -\frac{j\omega\varepsilon_0}{2} \iint A_m \vec{e}^{\,*}_\mu \cdot \vec{e}_\nu \, \mathrm{d}x\mathrm{d}y$$

The coupled mode equation for mode μ becomes :

$$\frac{\mathrm{d}c_\mu}{\mathrm{d}z} = -j\delta_{\mu\nu}c_\mu + K_{\mu\nu}c_\nu$$

where $\delta_{\mu\nu}$ is the detuning with respect to the phase matching (Bragg) condition; a similar relation holds for the other mode ν.

In the case ν = -ν the contro-directional coupling between the forward and backward propagating components of the same mode is obtained. This is the basis for the realization of a frequency selective mirror.

The same equation can be obtained for the analysis of coupled waveguides. This approach differs from what suggested in section 2.3.1 because the modes used to represent the interaction are those of each waveguide considered as isolated. For this problem there are then two alternative approaches and this corresponds also to two different interpretations of the propagation phenomenon: interference between the modes of the complete structure for the modal approach and exchange of power between the modes of the isolated waveguides for CMT.

Acousto-optic interaction can be also reduced to this case. The interacting modes μ and ν should be of different polarization and slightly frequency shifted to satisfy "phase matching conditions in time" that is equivalent to energy conservation in the photon-phonon interaction. One obtains :

$$K_{\mu\nu} = -j\frac{\pi}{\lambda}\sqrt{\frac{(n_{TE}n_{TM})^3}{\rho \mathrm{v}_a^3}}\, p\sqrt{I_a}\,\Gamma_0$$

where $n_{TE,TM}$ are the effective indexes of the coupled modes, v_a is the acoustic wave velocity, ρ is the material density, p is the relevant photo-elastic coefficient, I_a is the intensity of the acoustic wave (SAW in integrated optics) and Γ_0 is the overlapping integral between the electric field distribution of the optical field and the acoustic wave one.

In many cases we have to consider at the same time self coupling ($K_{\mu\mu}$), coupling effects ($K_{\mu\nu}$) and a detuning ($\delta_{\mu\nu}$) between the interacting modes. For example, in the case of a DFB laser the propagation equations become :

$$\frac{dc_\mu}{dz} = -j\tilde{\delta}\, c_\mu + K_{\mu\nu} c_\nu$$

$$\frac{dc_\nu}{dz} = +j\tilde{\delta}\, c_\nu + K_{\nu\mu} c_\mu$$

having called μ and ν are the forward and backward propagating components of the same mode, $K_{\mu\nu}$ the coupling coefficient due to the grating and $\tilde{\delta} = \delta_{\mu\nu} + jg_\mu = \left(\frac{2\pi}{\lambda} n_\mu - \frac{\pi}{\Lambda}\right) + \frac{j}{2}\left(\Gamma_{\mu\mu} g - \alpha\right)$ accounts for the detuning and for the modal gain due to current injection and waveguide modal losses due to scattering and absorption.

As a conclusion we can say that CMT can be applied to study a large variety of IO devices and the simple form of final equations (a system of two coupled differential equations with constant or z-variable coefficients) allows one to obtain simple analytical or numerical solutions. Furthermore CMT allows a direct physical interpretation of the device behavior that is also important to analyze the results of the alternative analysis based on pure numerical techniques (e.g. BPM).

2.3.3. Examples of Use of BPM

A special feature of beam propagation techniques is that they include automatically the information of the evolution and of the interaction of all the guided and radiation modes. While guided modes are discrete and finite in number and well representable with CMT, radiation effects can only be represented with a continuos spectrum of modes that is much less easy to represent using other techniques.

For this reason BPM and similar techniques are primarily used to study and design low loss branch waveguides splitters, transitions (S-bands, tapers, etc.), longitudinally varying devices, radiation effects in devices designed with modal or CMT, spurious radiation and interference effects in a cascade of devices integrated on the same chip, etc.

From our personal experience we find BPM very useful in combination with modal analysis or CMT for the final refinement of previously designed devices neglecting unwanted radiation effects.

BPM techniques have been also combined with carrier diffusion equations in semiconductors and population inversion equation to study complex spatial interaction effects in optical amplifiers and lasers [24].

2.3.4. A Final Example

In the previous paragraphs we discussed and highlighted the specific features of the various techniques and their complementarity and the possibility of alternative approaches to the same problem. As a final example we present a case of combined use of CMT and BPM for the analysis of an integrated acousto-optic device in which both the interacting fields are guided [25]. This element is the building block of tunable filters, switches, add-drop elements, etc. The device structure is presented in figure 2 where the large guides represent the acoustic waveguides made by titanium diffusion in $LiNbO_3$ in the cladding region. At the center of the straight acoustic waveguide is indicated the optical waveguide.

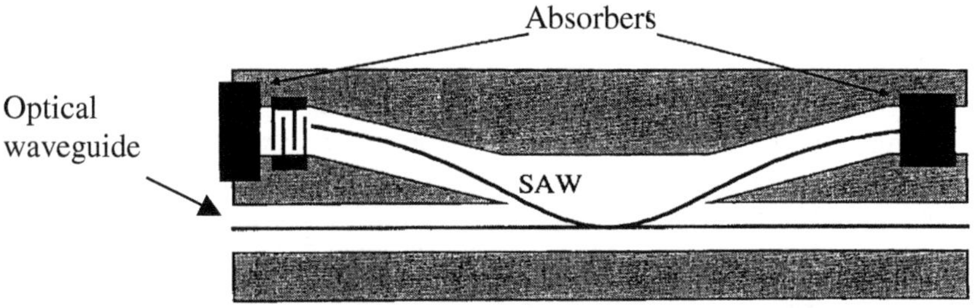

Figure 2 : Layout of the acousto-optic device.

The problem of evaluating the acousto-optic interaction can solved using CMT as indicated in section 2.3.2 while the evaluation of the acoustic field can be done using a scalar approximation by BPM.

Figure 3 presents a map of the computed acoustic field. Figure 4 shows the comparison between measured results for the acoustic field along the optical waveguide and the computed ones obtained using different acoustic velocities in the titanium diffused region. Finally, in figure 5 are compared simulation and measured results for the modal conversion efficiency as a function of the optical frequency.

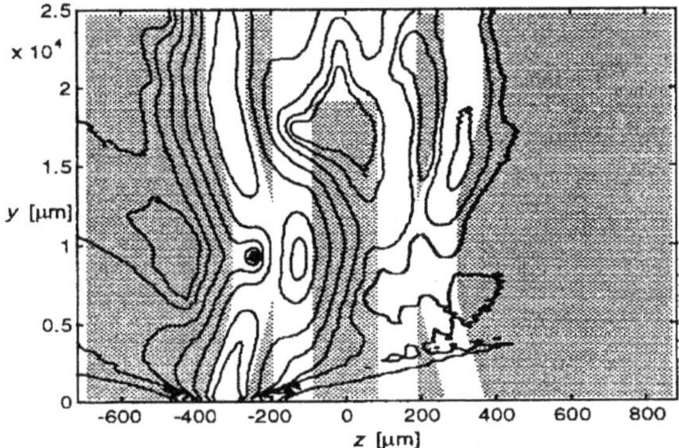

Figure 3 : Map of the computed acoustic field; contours are at 0dB,-3dB, -5dB, -10dB, ...,-30dB.

2.4. Conclusions

In this section we presented an overview of the basic simulation techniques available for the analysis and design of integrated optical components and circuits. We emphasized the overlap and complentarity of the various techniques for the solution of specific problems. This is important both for their combined use and also for validation purposes.

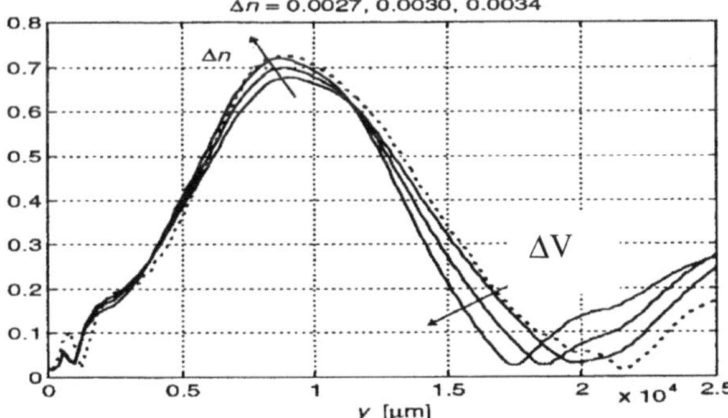

Figure 4 : Acoustic field distribution in the optical waveguide region for different acoustic velocities in the cladding region and (...) experimental results from Pirelli.

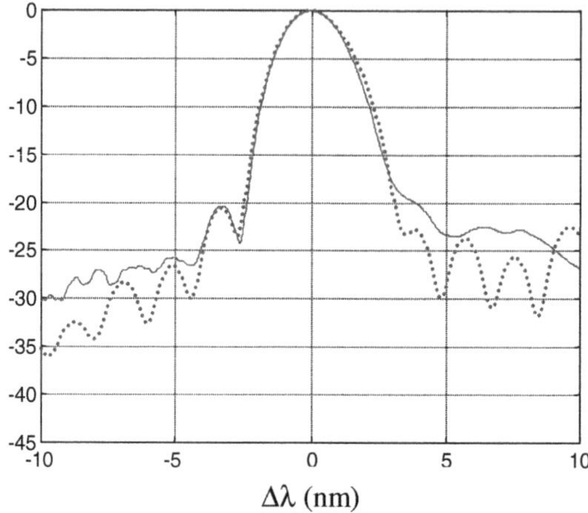

Figure 5 : Measured from Pirelli and simulated (...) conversion efficiencies vs. the wavelength variation.

3. TOOLS FOR MICROWAVE-OPTIC CO-SIMULATION

D. Breuer, D. Hewitt, I. Koltchanov, A.J. Lowery, R. Moosburger
Virtual Photonics Inc., Helmholtzstr. 2-9, D-10587 Berlin, GERMANY
e-mail: d.breuer@virtualphotonics.com

3.1. Abstract

This paper discusses issues of microwave-photonic co-simulation. After explaining the basic building blocks like direct or external modulated transmitters, transmission fiber, optical amplifiers and receiver results for different applications like CATV and millimeter-wave fiber-radio communication systems are presented. These numerical results are based using a simulator called "Photonic Transmission Design Suite".

3.2. Introduction

The field of microwave-photonics may be defined as the study of photonic devices where lightwave signals are modulated by microwave or millimeter frequencies and their application in microwave systems.;Commercial applications include the remoting of antennas for cellular micro-cellular radio using analog fiber links, the distribution of cable-television signals, and signal processing using optical techniques for phased-array antenna beam forming.

The electrical excitation of microwave–photonic systems is quite different to the traditional base-band digital lightwave communication systems. Information is usually analog or digitally modulated on to a number of RF, microwave, or millimeter-wave sub-carriers and non-linearity of the photonic system is often of critical importance to keep cross talk between channels to a minimum.

Many of the tools necessary for successful modeling of microwave-photonic systems are the same as base-band digital lightwave systems, namely:

- A full range of photonic device and optical network element models at different levels of abstraction be available.
- The diverse physical processes describing components such as semiconductor lasers, EDFA optical amplifiers, and non-linear fibers require detailed numerical models encapsulating the full behavior of the component.
- Different optical signal representations are provided that suit a particular aspect of system behavior or design.;Optical signal data exchange can be organized in blocks or by transmitting individual samples. The Block mode is more suitable for system simulation where signals flow unidirectionally along fiber from transmitter to receiver and is the most efficient form of simulation. Passing data bidirectionally between optical modules is necessary where signal reflections will modify device behavior or where bidirectional signal passing is an essential part of system operation.

Modeling of microwave-photonic systems often involve dealing with multiple sub-carriers at frequencies in the MHz range with small frequency separations in addition to optical carriers in the THz range.;Special modeling techniques are required to meet the measurement and system requirements of high dynamic range and narrow frequency resolution.

In this paper we describe general modeling techniques for two examples of microwave-photonic systems.;A schematic of a subcarrier multiplexed (SCM) broadcast lightwave system which may form part of a general cable television (CATV) network is shown in figure 6. In the simplest system the different subcarriers are modulated by standard AM-VSB video signals and the composite multiplexed signal directly drive a semiconductor laser. No electrical format conversion takes place in this AM CATV system but the cost is a stringent noise and distortion specification on the laser. Typical CATV systems;may use around 110 RF subcarriers ranging from 50 MHz to 860 MHz.;The carrier spacing is 6 MHz for the NTSC frequency plan and 8 MHz for a PAL system.

A directly modulated laser will suffer frequency chirp in the optical domain and cause distortion due to dispersion in the optical fiber. An alternative is to externally modulate the laser with a Mach-Zehnder modulator using the electro-optic effect. However this modulator is non-linear and pre-distortion circuits are required to meet system specifications.

Multichannel analog AM-VSB and digital M-Quadrature amplitude modulation (QAM) subcarrier multiplexed;video lightwave transmission systems are currently being installed by telecom and CATV companies and allow simultaneous delivery of both broadcast analog video and interactive digital video/data channels. Around 30 channels of 64/256 QAM operating at a bit rate around 30 Mb/s offer a high bandwidth efficiency (5-7 b/s/Hz) and robust transmission with respect to nonlinear distortion and noise.

Millimeter-wave fiber-radio communication systems are attractive possibilities for the efficient transmission of signals to low cost cellular and micro-cellular millimeter-wave transmitters. Simulation results are shown for a 10 microwave subcarrier system where a Mach-Zehnder modulator is used to efficiently generate two optical carriers from a single laser source.

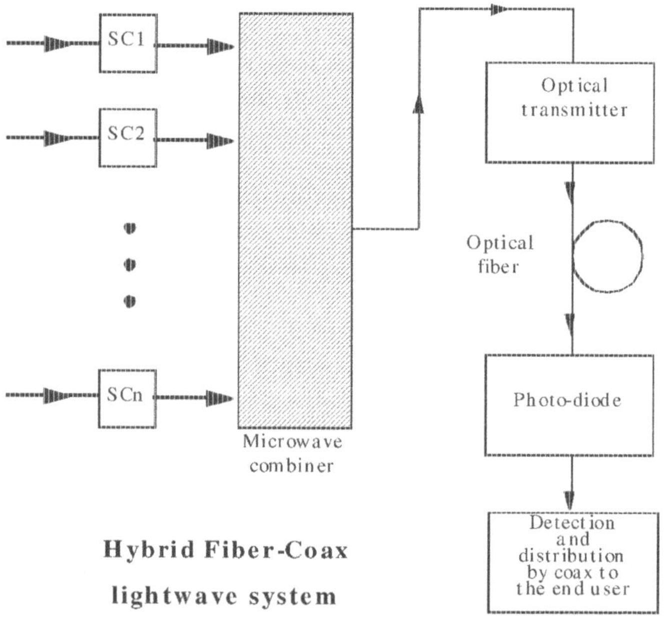

Figure 6 : Block diagram of a subcarrier multiplexed (SCM) lightwave system.

3.3. Building Blocks for Simulation

In the following sections the basic modules like transmitters consisting of direct or external modulated lasers, transmission fibers, optical amplifiers and receivers for building a microwave optic co-simulation will be reviewed. Particularly the requirements and impairments of the different physical devices with respect to co-simulation are explained.

3.3.1. Transmitter

One of the most critical components for an analog lightwave system is the transmitter, since it should provide high average output power with low noise and an extremely linear optic-electric transfer characteristic. The different available transmitters are divided into two categories: direct and external modulation. Each of them may be used at an operating wavelength of 1300 nm or 1550 nm., whereby the latter one corresponds to the low loss window of;the fiber, where Er-doped fiber amplifiers are easily available.

In the direct modulated scheme the analog current which like in SCM systems consists of the different subcarriers is used to modulate the intensity of the laser. This scheme takes advantage of the intrinsically linear current-intensity characteristic of laser diodes above the laser threshold. The principle of analog modulation for semiconductor lasers is shown in figure 7, where a LI-(light power-current) characteristic is shown together with input and output signals.

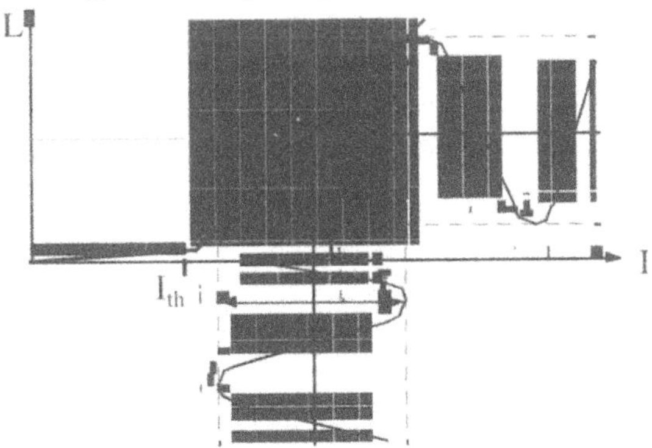

Figure 7 : Analog modulation scheme for semiconductor laser. The laser is biased at the laser current Ib and modulated by adding the modulation current to Ib.

The optical output signal is a replica of the electrical analog input signal. However, due to the nonlinear gain suppression and spontaneous emission the laser LI-characteristic is not perfectly linear. Furthermore, the nonlinear gain suppression and a finite response time of the laser (given by the inverse relaxation frequency) lead to a variation of the carrier density and gain at modulation. Because of the gain - refractive index coupling (alpha-factor) the latter results to the chirp and additional signal distortions. But even for a laser transmitter with a perfect linear electric-optic transfer function the broadband distortions are generated as soon as the electrical drive signal of the laser falls below the laser threshold current (laser clipping). From these considerations it is rather obvious, that the appropriate simulation laser model must include all the mentioned above effects. This restricts the choice of the laser modules to a transmission-line laser model (TLLM) or a rate equation model [26]. The latter is simpler being single-mode, unidirectional and considering the laser as a;lump device. It gives a higher computation efficiency; however if the mentioned approximations are not acceptable (e.g. for investigation of back-reflections into the laser) the TLLM must be used.

Using external modulation the electric analog signal is applied to modulate a continuos optical wave (CW) outside the laser cavity. To minimize distortions the LV-transfer characteristic of the external modulator must be as linear as possible. Typically Mach-Zehnder (MZ) or electro-absorption modulators are used; however both do not have a linear transfer characteristic. For example the MZ-modulator is known to have a sinusoidal transfer function [27]. A variety of techniques like predistortion linearization of the electrical drive signal have been developed to overcome the different limitations [28].

3.3.2. Fiber

Neglecting effects of polarization the forward propagation (+z-direction) of optical pulses in fibers is described by the scalar nonlinear Schrödinger Equation (NLSE) for the complex pulse envelope A [29].

$$\frac{\partial A}{\partial z} = \left(-\frac{i}{2}\beta_2 \frac{\partial^2}{\partial T^2} + \frac{1}{6}\beta_3 \frac{\partial^3}{\partial T^3} - \frac{\alpha}{2} + i\gamma |A|^2 \right) A \qquad (1)$$

with $\gamma = \frac{n_2 \omega_0}{c A_{eff}}$ (2)

and T == t - z/ v_g denotes the transformation to a frame of reference moving with the group velocity v_g. For simplicity stimulated Raman scattering is not included in equation (1). Stimulated Brillouin scattering occurs due to interaction of the optical wave with acoustical phonons in the fiber and is a back-scattering process. This means that part of the input power is back-scattered to the transmitter and the input launch power is limited [30].

The first two terms on the right hand side of equation (1) describes chromatic dispersion. The dispersion parameters β_2 and β_3 result from expansion of propagation constant $\beta(\omega)$ around the center frequency ω_0 and describe dispersion effects up to third order. Fiber dispersion is usually given by the dispersion D and the dispersion slope $S=dD/d\lambda$. For high bit rate transmission chromatic dispersion is one of the main limiting factors because dispersion induced pulse broadening leads to intersymbol interference. Moreover severe signal distortions occur if a highly chirped transmitter is used like in direct modulation scheme since the incident frequency modulation is converted to an intensity modulation. Mixing of the induced intensity modulation with the original intensity modulation leads to intermodulation distortions.

The fiber loss α is described by the third term on the right hand side of equation (1) and is given in dB/km. Figure 8 shows how dispersion and loss in principle affect the signal quality.

The last term on the right hand side of equation (1) describes fiber nonlinearity. It is proportional to the pulse intensity $|A|^2$. γ as defined in equation (2) is the nonlinear coefficient related to the nonlinear refractive index n_2, the effective fiber core area A_{eff}, and the velocity of light c. The nonlinear effects included in the NLSE (equation (1)) are self-phase modulation (SPM), cross-phase modulation (XPM) and four-wave mixing (FWM). Single channel transmission is effected by SPM only whereas in multi-channel transmission the combined effects of SPM, XPM and FWM lead to signal degradation [10]. Figure 9 shows how in principle SPM effects pulse transmission. Due to the power dependence of the nonlinear index of refraction a phase change is induced on the pulse itself. The leading edge is red shifted and the trailing edge blue shifted. This nonlinear phase change leads to a chirp and may broaden the corresponding spectra. In interaction with the chromatic dispersion this leads to signal distortions.

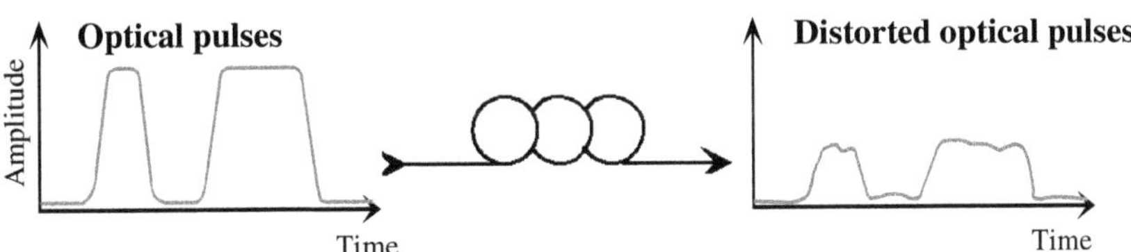

Figure 8 : After transmission through a dispersive, lossy fiber the optical output pulses are broadened and the amplitude is decreased.

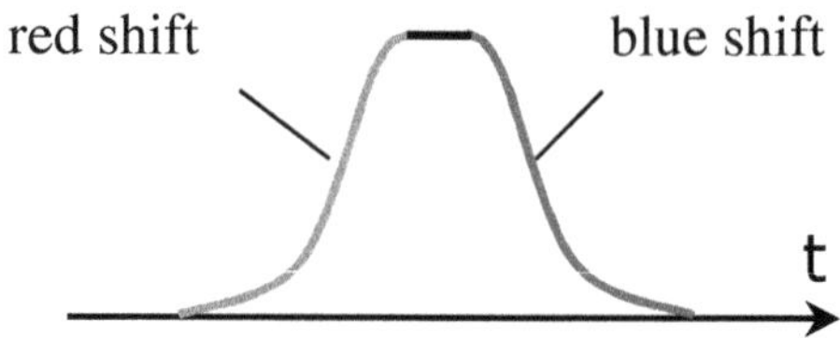

Figure 9 : SPM induced chirp leads to a red shift on the leading edge of the pulse and a blue shift on the trailing edge.

Since SPM affects the optical wave via its interaction with chromatic dispersion cancellation of the nonlinear fiber degradation can be achieved under certain circumstances. The resulting pulses are called Solitons and play a major role in long-haul undersea systems.

Due to the complicated interaction of dispersion and nonlinearity within the fiber analytical solutions are rarely obtainable. Therefore the NLSE has to be solved numerically. In most cases the well known split-step fast Fourier method is used [29]. This is a semi-spectral method where dispersion due to its time derivatives is calculated in the frequency domain and nonlinearity in the time domain.

3.3.3. Amplifier

Due to the fiber loss α the signal power decreases along the fiber. Therefore to increase the loss budget of a link optical amplifiers are used. If the amplifiers provide enough optical power, it is possible to split the signal among multiple receivers, as required to reach many end users. Two kinds of optical amplifiers are available: semiconductor optical amplifiers and rare-earth doped fiber amplifiers.

Most practical systems today work at transmission wavelength around 1550 nm, corresponding to the low loss window of the fiber. In this wavelength regime erbium-doped fiber amplifiers (EDFAs) are used. The rare-earth doped fiber amplifiers are characterized by a broadband gain spectrum which is typically in the range of 35 nm, corresponding to a frequency range of about 4 THz. But even more than 80 nm has been experimentally demonstrated for multistage C- &;L-band amplifiers [31]. This allows for simultaneous amplification of numerous wavelength channels. EDFAs are commercially available with gains of 20–30 dB and saturation output powers in the order of 25 dBm. Typical noise figures are in the range of 4-5 dB. The spectral shape can be slightly modified by co-doping (e.g. Al-codoping leads to a more flat gain curves of EDFA). Very important is, that the active ion lifetime (approximately 10 ms) is very long in comparison with the bit duration and therefore provides practically time-independent gain. In EDFAs no cross-talk appears due to amplifier saturation.

The semiconductor optical amplifiers (SOA) have the advantage of being very compact compared to EDFAs, and moreover allow for a simple choice of arbitrary operating wavelength by the "bandgap design". They are also quite broadband (~ 20 nm) but have a somewhat higher noise figure than an EDFA. However, the most important disadvantage if SOAs are used for amplification is the small carrier lifetime which leads to strong nonlinearities (for example, time-dependant saturation). This results in undesired cross-talk between different WDM channels. On the other hand exactly this disadvantage makes it possible construct a lot of nonlinear devices such as frequency converters, OTDM demultiplexer, optical-phase conjugators etc., which is the main application area of the SOA's.

Besides signal amplification noise due to spontaneous emission is added to the output signal (see figure 10). This noise contribution has to be considered by looking at the overall system performance.

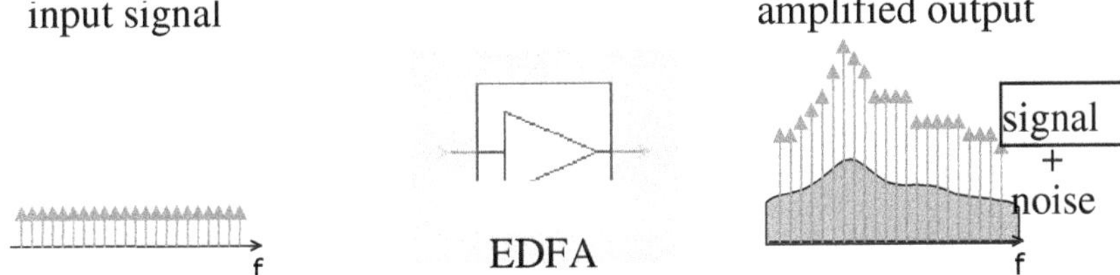

Figure 10 : Schematic showing the amplification of the input signal by an erbium-doped fiber amplifier. Additionally to the signal amplification noise occurs due to spontaneous emission.

In the limit of high gain the noise power from the amplified spontaneous emission (ASE) can be described by

$$P_n = 2n_{sp}\, h\nu\,(G-1)\,B_0\,,$$

where G is the gain of the amplifier, B_0 is the optical bandwidth and n_{sp} specifies the population inversion of the amplifier. The factor 2 corresponds to the fact that standard single-mode fibers support two orthogonal polarizations. For high gain the noise figure NF of the amplifier is given by NF=$2n_{sp}$. In a system using a cascade of optical amplifiers the total noise accumulates from amplifier to amplifier stage. The signal to noise ratio (SNR) can then be estimated by

$$SNR = \frac{P_{launch}}{NF\big((G-1)h\nu B_0 N\big)}\,,$$

where P_{launch} specifies the average power launched into the fiber and N determines the number of amplifiers in the transmission line, which are all supposed to provide the same gain.

A schematic of a rare-earth doped fiber amplifier which comprises a length of amplifying fiber (erbium doped), an optical pump to provide the energy necessary for population inversion and a WDM-coupler to combine the input signal and the pump is shown in figure 11.

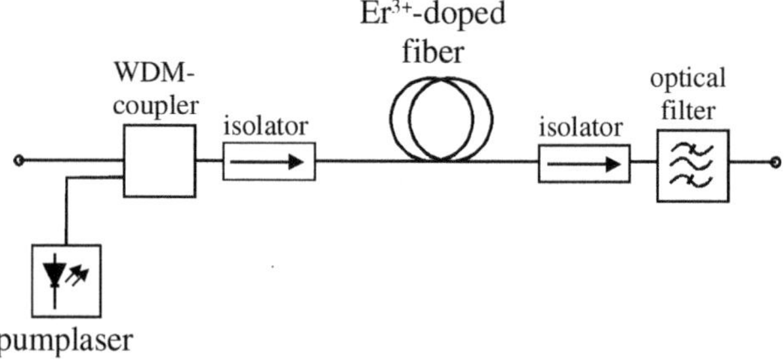

Figure 11 : Schematic of a rare-earth doped fiber amplifier.

The isolators are inserted to avoid back-scattering due to reflections at splices etc. The optical filter is used to limit the output ASE noise bandwidth.

3.3.4. Receiver

At the receiver the optical power is back converted to an electrical signal using a photodiode. Although commercially available PIN photodiodes are inherently linear at the considered optical power levels, the coupling of the detector to a preamplifier may impose noise and signal distortions. The generated photocurrent is related to the received optical power by

$$I = \eta \frac{e}{h\nu} P ,$$

where $h\nu$ is the photon energy and η the quantum efficiency of the photodiode. Taking into account the electrical noise sources like shot noise n_{sh}, thermal noise n_{th} and dark current i_d, the electrical current can be written as

$$i(t) = i_s(t) + n_{sh}(t) + n_{th}(t) + i_d .$$

In amplified optical transmission systems the ASE noise manifests as an additional intensity noise at the receiver. Two beat terms occur: ASE-ASE beating and signal-ASE-beating. The ASE-ASE beating can be significantly reduced by inserting an optical filter in front of the receiver, which limits the total ASE noise [32]. The more stringent limit therefore occurs due to the signal-ASE beating. For well designed receivers, laser RIN and signal-ASE beating represents the dominant receiver noise sources [32].

3.4. Application Examples

As application examples a few results for CATV and millimeter-wave fiber-radio communication systems are presented below. More results will be shown within the presentation.

3.4.1. Analog Video CATV Systems

In analog CATV systems multiple electrical subcarriers are multiplexed together and the composite signal is the used to drive a directly modulated laser or an external MZ-modulator. Often the distortion characteristics of a laser may be measured by using two carrier test set to estimate the performance of a multicarrier system. Figure 12 shows a schematic setup to measure two tone intermodulation distortions of a directly modulated laser.

The composite drive signal is built of two RF tones at f_1=500 MHz and f_2=525 MHz. The electrical drive signal together with the laser output spectrum are shown in figure 13 and figure 14. The output laser spectrum is significantly broadened due to the laser chirp which leads to additional signal distortions when transmitted over a dispersive fiber.

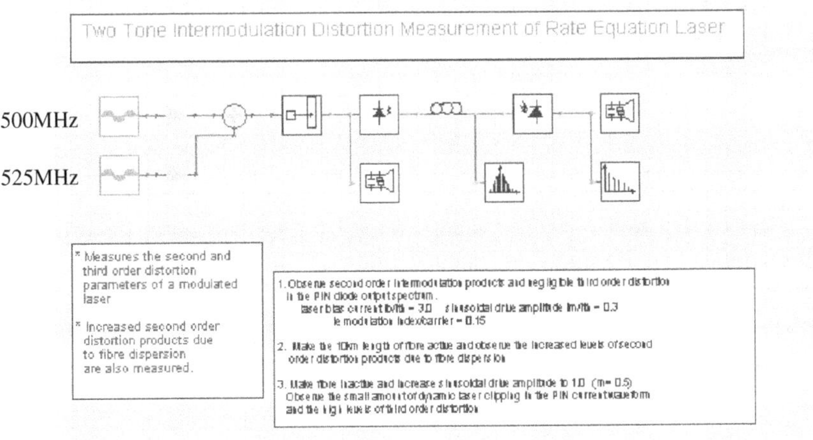

Figure 12 : Schematic setup to measure two tone intermodulation distortions of a directly modulated laser.

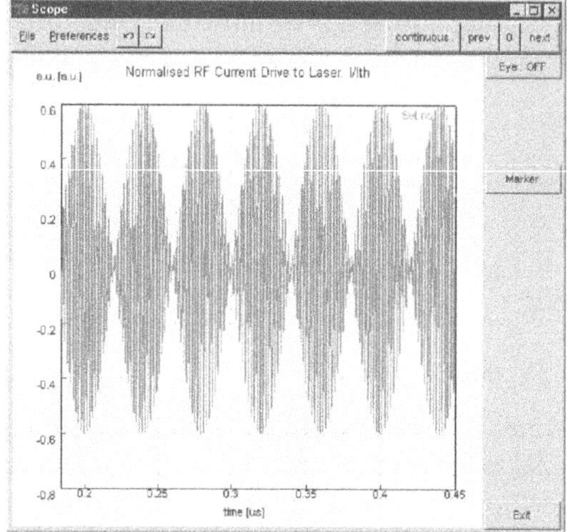

Figure 13 : Composite two tone drive signal.

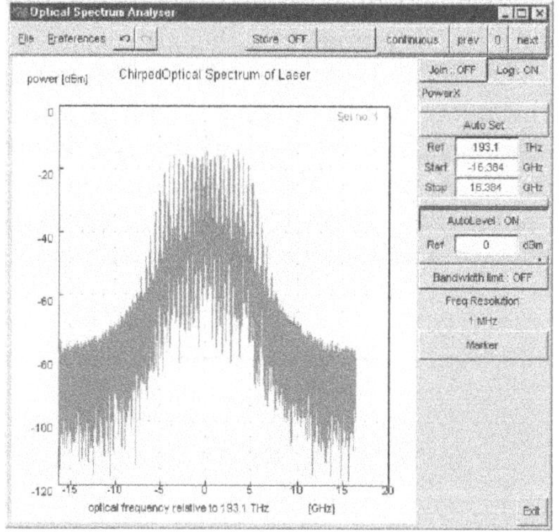

Figure 14 : Output spectrum of directly modulated laser

Due to the non-ideal laser characteristic intermodulation distortions at frequencies $2f_1$, $2f_2$ and f_1+f_2 occur (figure 15) in the RF-spectrum. These intermodulation distortions increase significantly if the signal is transmitted over 10 km of a high dispersive fiber (D=16 ps/km/nm) (see figure 16).

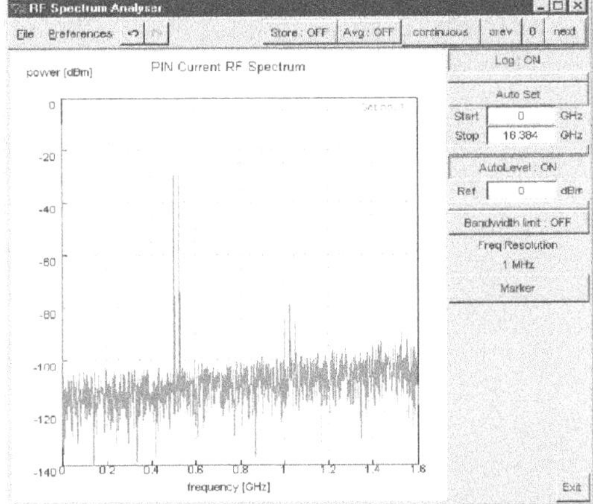

Figure 15 : Electrical RF-spectrum without fiber (modulation index m=0.15).

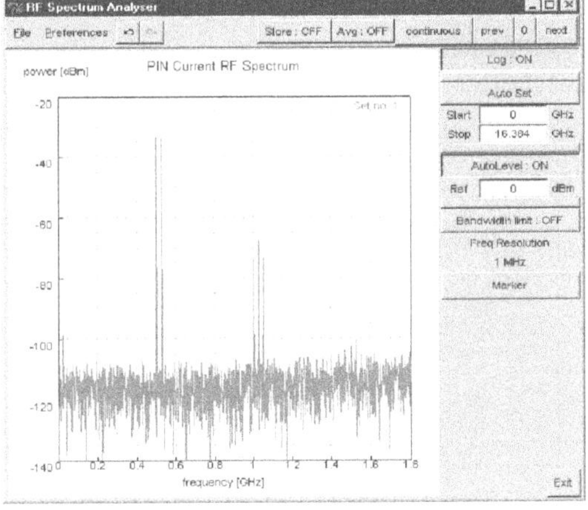

Figure 16 : Electrical RF-spectrum after 10 km fiber (modulation index m=0.15).

Besides these distortions due to fiber dispersion and nonlinear transfer characteristic additional degradations occur when the composite electrical drive signal falls below the laser threshold. In this case clipping occurs, which leads to broadband signal distortions and an increased number of;intermodulation products. Usually a amount of clipping is tolerated in order to maximize the capacity of a single laser. Figure 17 shows the typical RF-pin diode current for laser clipping. A non-symmetrical current wave form with strong nonlinear distortions is visible. The corresponding RF-spectrum with a large number of intermodulation products is shown in figure 18.

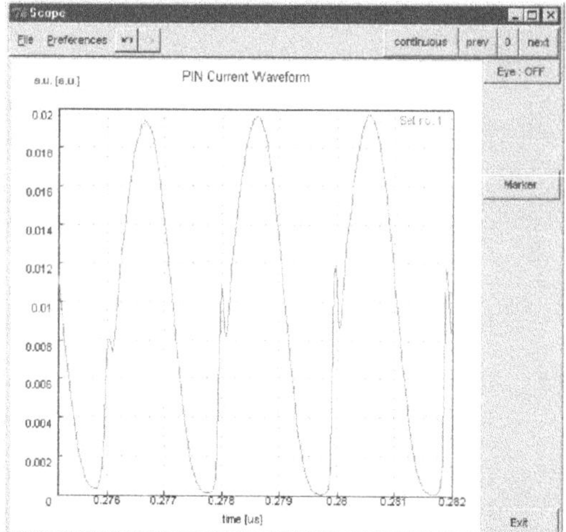

Figure 17 : Pin-diode RF-current showing clipping distortions.

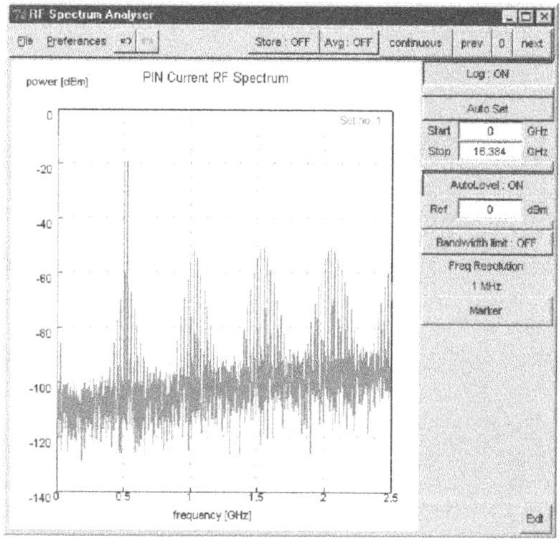

Figure 18 : RF-spectrum with increased number of intermodulation products due to laser clipping.

An 80 channel CATV system with a NTSC carrier frequency spacing of 6 MHz is shown in figure 19. It uses an external Mach-Zehnder modulator with a predistortion driver circuit to compensate for the inherent distortion of the modulator.

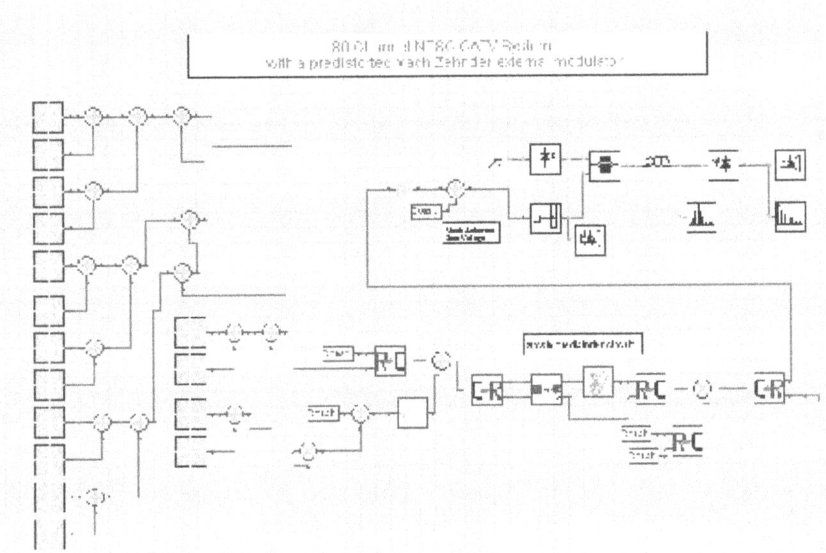

Figure 19 : Schematic for an 80 channel NTSC CATV system with a pre-distorted Mach-Zehnder modulator

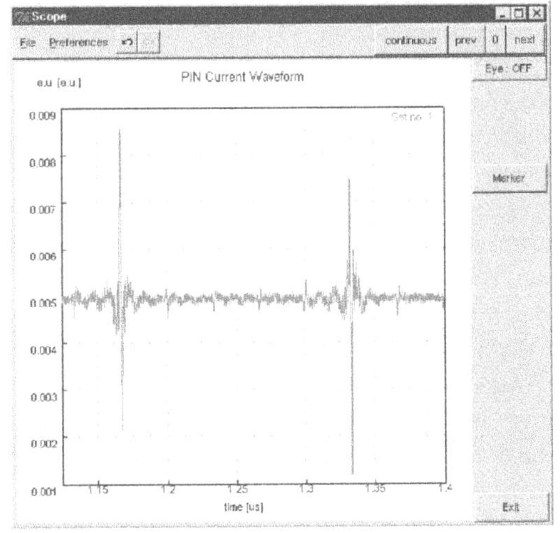

Figure 20 : Pin-diode composite RF-current.

The receiver pin-diode current waveform shown in figure 20 shows the very sharp pulse waveform that 80 sinusoidal carriers will generate as a composite signal. No inband or outband distortion products are generated with this pre-distorted external modulator and dynamic range or signal to noise is limited by the spontaneous noise of the laser and receiver noise as shown in figure 21. There is no laser chirp generated so the fiber dispersion does not contribute to second order distortion products.

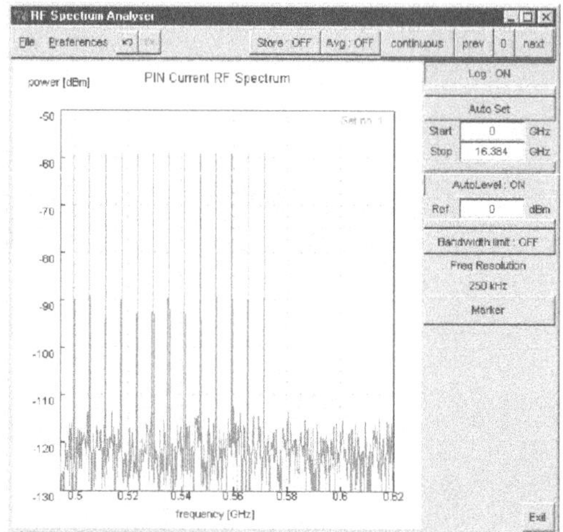

Figure 21 : RF-spectrum showing no intermodulation.

3.4.2. Millimeter Wave Fiber Radio Systems

The use of millimeter-wave radio for future broadband service provision is expected to be in great demand and radio over fiber is an attractive technology for the transmission of signals to cellular millimeter-wave transmitters. Direct current modulation of laser diodes is not practical at millimeter frequencies and an optical self-heterodyne system provides a practical alternative [33]. Two optical carriers are generated from a single semiconductor laser using a Mach-Zehnder modulator.;The frequencies of the two carriers are displaced by the required millimeter-wave frequency and the two carriers are separated by an optical filter following their generation. One of the carriers is modulated by ten microwave sub-carriers as shown in the simulation schematic (figure 22).

Both carriers are then sent via a fiber splitting and distribution network to the receiver. Each remote receiver needs only a filter, a microwave amplifier, and an antenna to transmit modulated microwave signals to each customer in the cell. There is high coherence between the two optical carriers at the receiver with beat frequency carrier linewidths less than 1kHz [34].

Figure 23 shows the optical spectrum after generating the two sidebands with the Mach Zehnder modulator. Higher level sidebands with better carrier suppression is achieved by changing the Mach- Zehnder bias level and drive voltage to the approximate double sideband suppressed carrier mode of operation rather than the usual AM modulator mode. The composite drive to the second MZ modulator is shown in figure 24 where the level is close to 100% modulation.

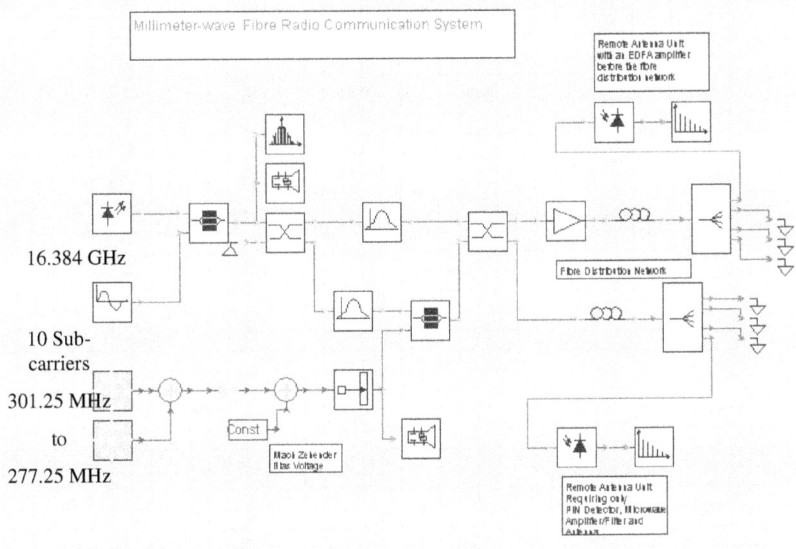

Figure 22 : Schematic setup of millimeter wave fiber radio system.

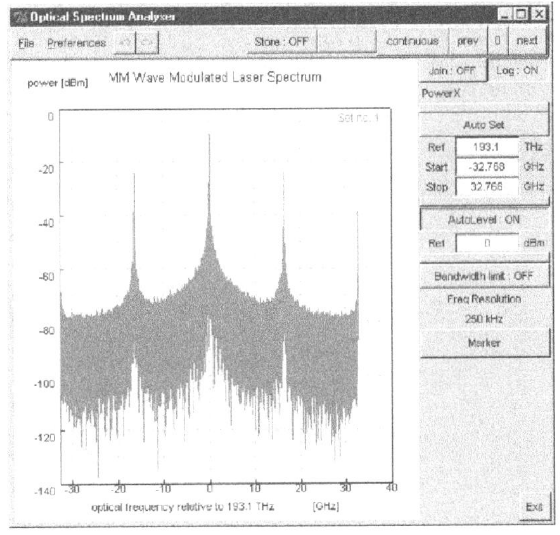

Figure 23 : Optical spectrum after generating the two sidebands with the MZ-modulator.

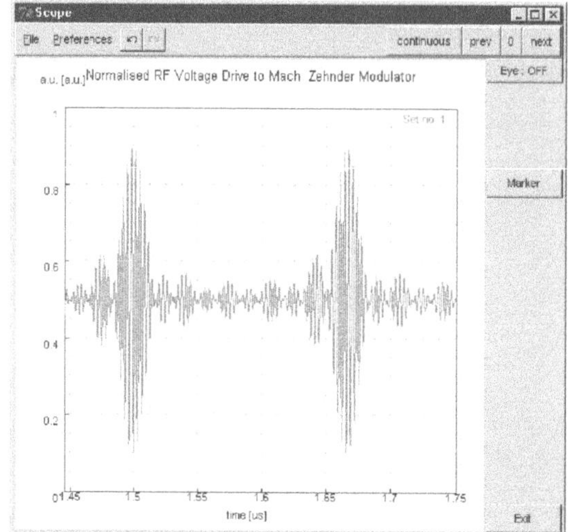

Figure 24 : Composite drive signal to the second MZ-modulator.

The receiver spectrum with no optical amplification in the fiber distribution network after an eight way split and transmission through 20 km of fiber is shown in figure 25 and would be suitable for digital modulation of the carriers. A higher level received signal is observed in figure 26 after being amplified by the EDFA. Note that third order intermodulation products arising from the Mach-Zehnder nonlinearity now appear above the receiver noise.

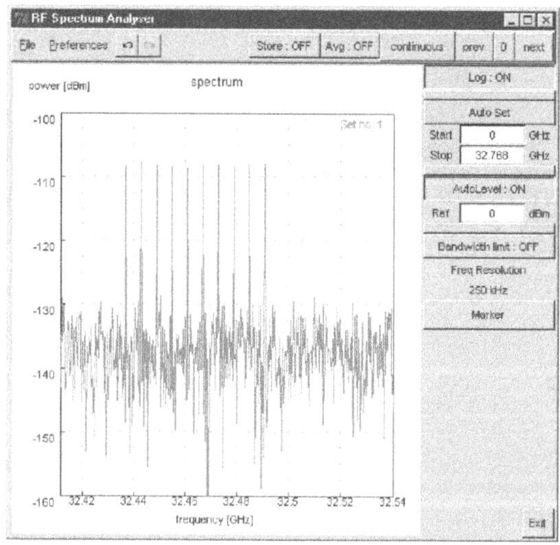

Figure 25 : Receiver spectrum without optical amplification after an eight way split and 20 km fiber transmission.

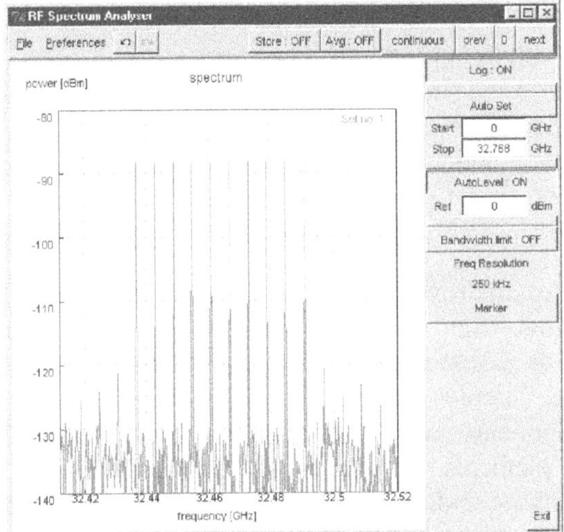

Figure 26 : Receiver spectrum with optical amplification after eight way split and 20 km fiber transmission

3.5. Summary

The principle building blocks for microwave-optical co-simulation have been explained and application examples for CATV and millimeter wave fiber radio systems have been presented. With a combined co-simulation tool available in the Photonic Transmission Design Suite [26] the whole transmission path from microwave to optic and back can be modeled and an investigation and optimization of the overall system performance is possible.

4. THE TLM METHOD – APPLICATION TO MICROWAVES AND OPTICS

F. Ndagijimana, P. Saguet, C. Golovanov, O. Jacquin
LEMO / ENSERG, UMR 5530 INPG-UJF-CNRS, B.P.257, 38016 Grenoble Cedex 1, FRANCE
e-mail : fabien@enserg.fr

4.1. Abstract

The TLM (Transmission Line Matrix) method is a full-wave technique used to simulate the propagation of the electromagnetic waves in complex media whatever the number of dielectrics and metal objects. Usually developed in time domain, the TLM method enables the calculation of device responses to a given excitation in terms of electromagnetic field distribution, power flow, voltages, currents, etc.

In this presentation we address following topics :

- an introduction to EM simulators and the TLM theory,
- the application to microwave circuits,
- the simulation of lumped passive and active non linear components,
- new application to optical integrated wave-guides.

4.2. Introduction

The simulation of complex microwave circuits including passive and active components requires the use of circuit simulators or electromagnetic simulator to account for propagation, radiation and electromagnetic interference. Circuits simulators are based on the use of Kirchhoff rules and components are described by their electrical equivalent network whose elements are known by compact empirical formulas. When coupling effect, radiation and general electromagnetic interference are to be taken into account in complex circuits, Electromagnetic (EM) simulators are to be used ; at least for most critical parts of the circuit. EM simulators are based on Maxwell's equations and can be developed in time domain or in frequency domain. Furthermore, depending on the symmetry of the problem, EM simulations can be performed in 2 dimensions, or in 3 dimensions. A two dimensional implementation enables the computation of propagation characteristics : propagation constant, characteristics impedance and the field distribution in the transversal cross-section. When discontinuities have to be simulated for complex configurations, 3D simulators are used. Recently 2.5D simulators have been implemented for the simulation of planar integrated circuits including vias.

Both circuit simulators and full-wave electromagnetic simulators are used to provide this kind of analysis but both are limited when the frequency of the signals increases. The circuit simulators are handling reduced RLC networks, which are able to translate the electromagnetic behaviour into electrical one. However, problems occur for large frequency domain because the provided RLC network and its reduction are strongly affected by the frequency. Full-wave electromagnetic simulators require the meshing of the analysed domain. When the dimensions of the lumped elements are small compared to the interconnection lengths and to the wavelength the computational effort becomes quickly extremely expensive even using a variable mesh. In this case, only a global simulator allows treating in the same time problems such as dispersion, crosstalk, package effects and problems due to the non-linear behaviour of the lumped elements.

We present a simulator based on the 3D TLM and its implementation is developed in section 4.3. Section 4.4 and section 4.5 deal with the application of the 3D TLM technique to microwaves and optical components respectively.

4.3. The TLM Method : 2D and 3D Implementation

The TLM technique is usually developed in time domain for the analysis of 3D electromagnetic structures of arbitrary shape extended to analyse three-dimensional hybrid problems consisting of distributed and lumped components. It computes the time domain response of such structures to arbitrary excitation in 3D space, and extracts their frequency characteristics, such as S parameters and return loss, via discrete or fast Fourier transform [36]. It also visualises the time evolution of the field distribution in a generated-solution mode for field propagation in time.

The TLM technique is based on a set of regularly or irregularly interconnects nodes in a cartesian mesh. The excitation of a voltage pulse located on a given node and propagates from a node to another on connection arms. Pulsed arriving (incident pulses) simultaneously on different arms of a node are scattered according to the scattering matrix of the node. This matrix is related to the mesh size, the permittivity and the permeability of the simulated media. The new set of pulses (scattered pulses) propagates again on connecting arms. This is referred to as the iterative process. The time domain response of a point will be given by the record of pulses in the corresponding node during the iterative process. Applying a Fourier transform to the time domain response, we obtain a frequency domain response.

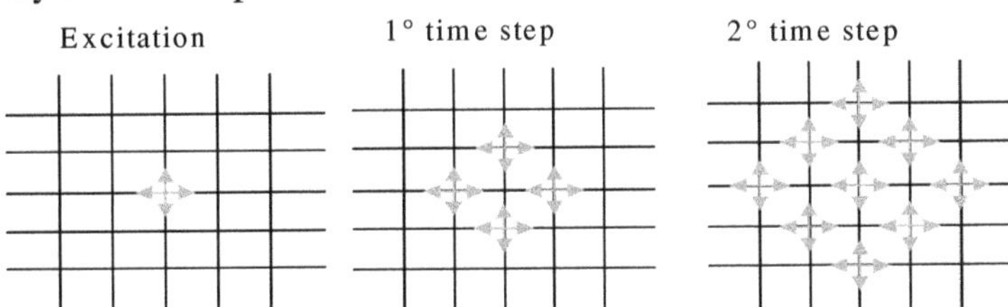

Figure 27 : Iterative process in a 2D TLM mesh : Scattering and propagation.

The accuracy of the simulation and the equivalence between simulated pulses and the electromagnetic fields are mainly related to the type of node : 2D or 3D.

4.3.1. The 2D TLM Node

The 2D node consists in the connection of 2 transmission lines in a parallel or serie's configuration and can be represented by an electrical equivalent network. The scattering matrix of a node is determined considering the reflection and transmission of voltages arriving on different arms at the same time. The permittivity is simulated by the use of an open stub and the conductivity losses are accounted for using a matched stub.

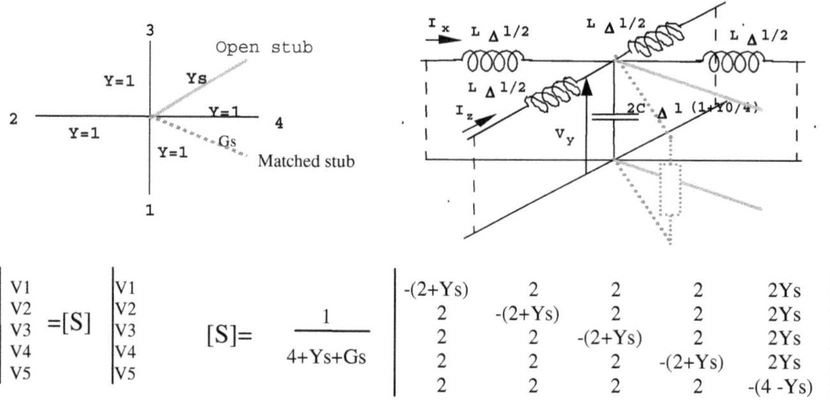

$$\begin{vmatrix} V1 \\ V2 \\ V3 \\ V4 \\ V5 \end{vmatrix} = [S] \begin{vmatrix} V1 \\ V2 \\ V3 \\ V4 \\ V5 \end{vmatrix} \qquad [S] = \frac{1}{4+Ys+Gs} \begin{vmatrix} -(2+Ys) & 2 & 2 & 2 & 2Ys \\ 2 & -(2+Ys) & 2 & 2 & 2Ys \\ 2 & 2 & -(2+Ys) & 2 & 2Ys \\ 2 & 2 & 2 & -(2+Ys) & 2Ys \\ 2 & 2 & 2 & 2 & -(4-Ys) \end{vmatrix}$$

Stub characteristics : $Ys = 4(\varepsilon_r-1)$ and $Gs=\sigma\ Z0\ \Delta l$

Figure 28 : Parallel 2D node with permittivity and loss stubs and its scattering matrix.

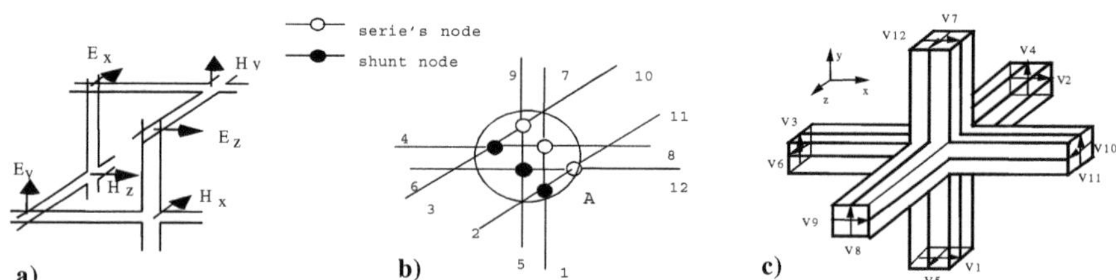

Figure 29 : Asymmetrical expanded, asymmetrical condensed and symmetrical condensed 3D nodes.

4.3.2. The 3D TLM

The first 3D node was derived directly from the discretisation of Maxwell's equations and was an asymmetrical expanded node. To avoid the calculation of EM field components on different edges of a parallelepiped, an asymmetrical condensed node has been proposed, and later, to avoid the asymmetry, the condensed symmetrical node has been developed [37].

The 3D node is characterised by 12 connecting arms and 6 stubs simulating the graded mesh, the permittivity, the permeability and losses.

As in the 2D TLM mesh, the characteristics of the media (ε_r, μ_r), and the graded mesh are simulated by means of stubs. For example for the symmetrical condensed node SCN the scattering matrix is developed as shown in figure 30.

For an incident pulse V1 :

$$V1 = \frac{-Y_x}{2(4+Y_x)} + \frac{Z_z}{2(4+Z_z)} \qquad V9 = V2$$

$$V2 = \frac{4}{2(4+Y_x)} \qquad V11 = -V3$$

$$V3 = \frac{4}{2(4+Z_z)} \qquad V12 = \frac{-Y_x}{2(4+Y_x)} - \frac{-Z_z}{2(4+Z_z)}$$

$$V13 = V2$$

$$V18 = Z_x V3$$

With stub admittance :

$$Y_x = 2\left[\frac{vw}{u}\frac{\varepsilon_r}{c\Delta t} - 2\right]; Y_y = 2\left[\frac{uw}{v}\frac{\varepsilon_r}{c\Delta t} - 2\right]; Y_x = 2\left[\frac{uv}{w}\frac{\varepsilon_r}{c\Delta t} - 2\right]$$

$$Z_x = 2\left[\frac{vw}{u}\frac{\mu_r}{c\Delta t} - 2\right]; Z_y = 2\left[\frac{uw}{v}\frac{\mu_r}{c\Delta t} - 2\right]; Z_x = 2\left[\frac{uv}{w}\frac{\mu_r}{c\Delta t} - 2\right]$$

Figure 30 : Example of scattered pulses and stub Impedance and admittance.

The implementation of the 3D TLM technique implies the following simulation issues :

- Reflection on conductors,
- Open media and absorbing boundaries,
- Dispersive and lossy dielectrics.

The reflections on boundaries are simulated by the application of an appropriate reflection coefficient to the incident voltage pulses. To ensure the synchronism of pulses arriving at a node before scattering, each boundary is located half distance from nodes. For a perfectly conducting wall, we apply R=-1. In the case of a dispersive wall, the frequency dependent reflection coefficient is converted to time domain and can be implemented in the iterative process. In the case of absorbing boundaries simulating the open space, different schemes have been investigated. Figure 31 gives a comparison of different techniques used for absorbing boundary implementation. An example of implementation of a TLM software based on the SCN node is presented in figure 32. Usually a graphical user interface facilitates the geometrical description and the mesh generation, as well as the signal processing.

ABC Technique	**Achieved reflection coefficient**	**Limitations**
Line matching	-15 to -20 dB	Normal incidence
Space-Time Extrapolation	-30 to -40 dB	Instabilities
Higdon's formulation	-40 to -50 dB	Instabilities, prediction of incidence angle
Béranger's perfect matching layers (PML)	-40 to - 60 dB	Implementation
Diakoptics	Perfect matching	Memory - CPU

Figure 31 : Comparison of absorbing boundary conditions (ABC).

Module	Operations
DIS3D (user graphical interface)	Geometrical description of the structure Computation of the TLM mesh Excitation and outputs
TLM code (Fortran or C)	File control and Data processing and Iterative process : Excitation Scattering Propagation Reflection of wall Output storage: EM fields, Voltages, currents
TLMcad (user graphical interface)	Data processing (time domain – frequency domain) Calculation of S-parameters, characteristic impedance (Zc), propagation constant (γ)

Figure 32 : An example of 3D TLM Software implementation.

4.4. Application to Microwave Components

Suppose we are analysing a microstrip line with a bend as shown in figure 33.

The aim of the TLM simulation is to provide :

- EM fields distribution
- propagation characteristics (Zc, β)
- S-parameters of the discontinuity
- Electrical equivalent network

Prior to the simulation, an appropriate excitation must be applied. With a Dirac pulse is characterised by an extremely wide frequency spectrum, the bandwidth of the simulation is only limited by the TLM mesh. The gaussian pulse provides a limited spectrum that must be lower than the TLM mesh bandwidth in order to avoid distortion during propagation. The quasi-step excitation is useful when a comparison of simulated voltages to Time Domain Reflectometry (TDR) measurements is necessary. Figure 35 shows an example of reflected and transmitted responses of a discontinuity.

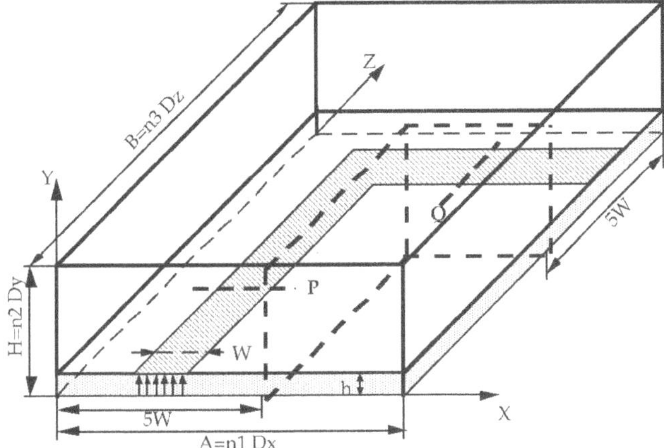

Figure 33 : An example of microwave discontinuity : microstrip bend.

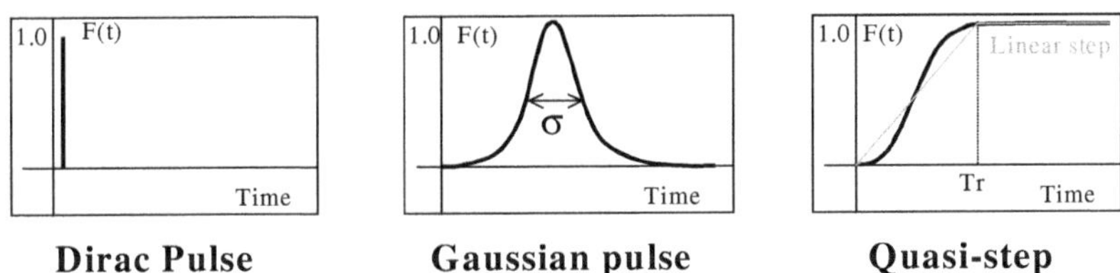

Figure 34 : Different forms of excitation.

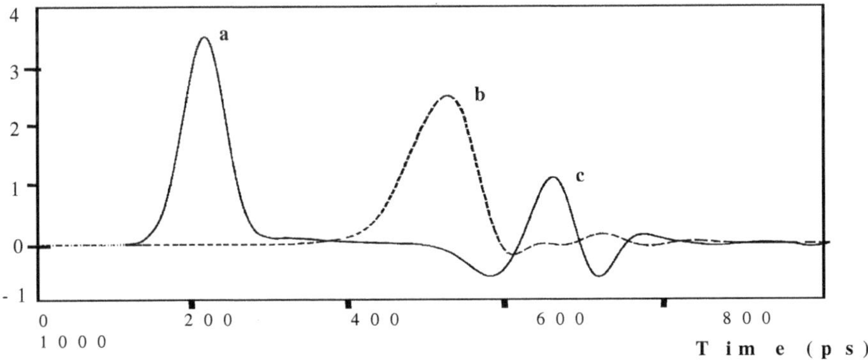

Figure 35 : Reflected (c) and transmitted (b) signals in a discontinuity with a gaussian excitation (a).

4.4.1. Calculation of Propagation Characteristics

Frequency domain behaviour of a microwave device are obtained using a Fourier transform. Prior to this operation, an appropriate "time windowing " is essential to separate incident, reflected and parasitic signals [38]. The propagation constant is calculated from the phase difference in two points on the strip, with a known distance :

$$E_y(\omega, z=0) = \int_{-\infty}^{+\infty} E_y(t, z=0) e^{-j\omega t} dt$$

$$E_y(\omega, z=L) = \int_{-\infty}^{+\infty} E_y(t, z=L) e^{-j\omega t} dt$$

$$e^{-j\beta(\omega)L} = \frac{E_y(\omega, z=L)}{E_y(\omega, z=0)} \; ; \; \beta(\omega) = \frac{\omega}{c}\sqrt{\varepsilon_{reff}(\omega)}$$

When a quasi-TEM mode is assumed, the characteristic impedance can be calculated from the knowledge of the voltage and the current. From the EM fields, different definitions of Zc can be used :

$$Zpi = \frac{2P}{II^*} \qquad V = \int_l E dl \qquad I = \oint H dl$$

$$Zpv = \frac{VV^*}{2P} \qquad \text{with} \qquad \mathrm{P} = \frac{1}{2}\mathrm{Re}\left(\oint_S \mathrm{E x H^* ds}\right)$$

$$Zvi = \sqrt{Zpi \; x \, Zpv}$$

It appears that for a TEM mode Zpi and Zpv converge to Zvi. In other cases, the actual (measured) characteristic impedance is accurately approximated by Zvi.

4.4.2. Simulation of Discontinuities : S-Parameters and Electrical Network Extraction

The S-parameters are calculated from the power flow in defined access ports, and for a quasi TEM mode, following formulas give a good accuracy when access ports are located far away from the discontinuity, from the excitation and from the boundaries.

$$|S_{11}| = \sqrt{\frac{V_{refl} I_{refl}^*}{V_{inc} I_{inc}^*}} \qquad |S_{E1}| = \sqrt{\frac{V_{trans} I_{trans}^*}{V_{inc} I_{inc}^*}}$$

The extraction of an equivalent network of the discontinuity is performed by the comparison of the time domain response from TLM and the response of the equivalent network from SPICE. In frequency domain the comparison is made between the S-parameters and a microwave circuit simulator like MDS (Hp-Eesof).

The simulation can also provide the radiation of the discontinuity (as for an antenna) when the EM near-fields have been recorded on a closed surface containing the discontinuity.

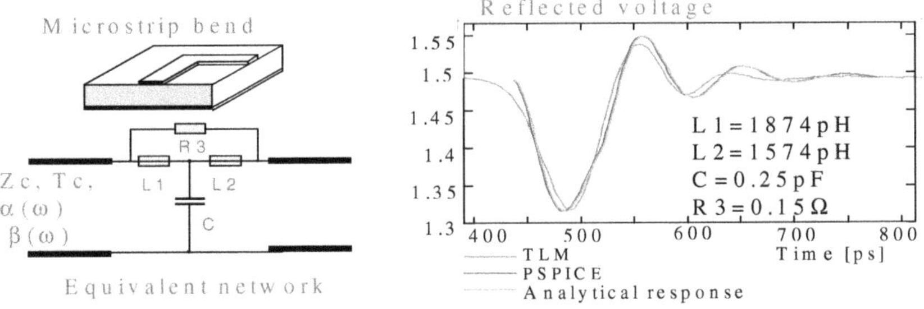

Figure 36 : Extraction of electrical network from TLM simulation.

4.4.3. Simulation of Lumped Elements and Non Linear Devices

An EM solver generally considers the Maxwell's equations in each point of the simulated domain. When the interaction between a lumped component and the environment is to be simulated, the mesh of the lumped device would result in a prohibitive memory consuming.

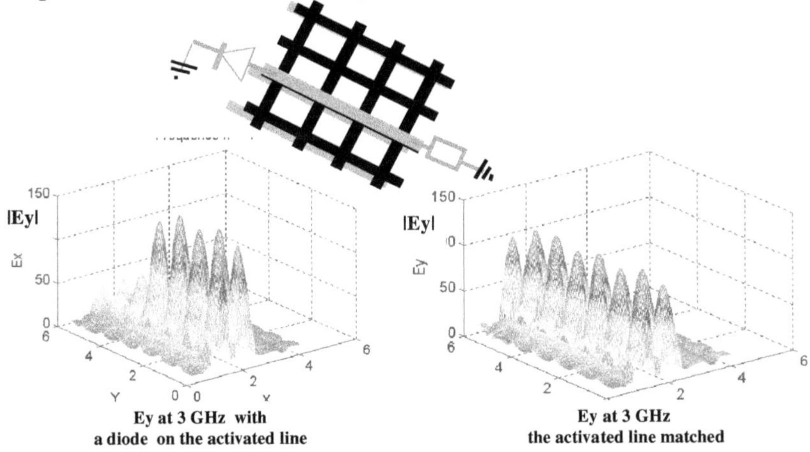

Figure 37 : Simulation of lumped diode in meshed ground plane device : Electromagnetic fields [39].

The I (V) equation of the component is implemented in connecting nodes in the TLM mesh and solved at each time step. The difficulty is to rely the I (V) equation corresponding to global quantities to incident and scattered voltages pulses corresponding to local quantities at each time step, in a given region of simulated domain

In the example above and for microwave component simulation in general, the TLM simulation enables the calculation of :

- propagation characteristics of a trip over a meshed ground,
- the coupling effect to other strips through the meshed ground plane

- the EM field distribution and the radiation of the device

4.5. Application to Optics

Integrated optic structures are usually modelled using 2D mode solvers (field distribution in the cross section) or field solvers (field propagation along optical directional coupler, Y junctions, taper, etc.). The most common among fields solvers is the Beam Propagation Method (BPM) and now a great number of versions of BPM [36] has been developed. These Methods are based on the paraxial approximation or on the slowly varying field approximation and are well suited to structures with weak guiding (small Δn), and smooth index variation along the propagation direction. For components characterised by small dimensions, high optical field confinement, abrupt index discontinuities along of the direction propagation, multiple reflections leading to resonance and important diffraction phenomena, the use of full wave electromagnetic solvers is essential.

Here the TLM solver is applied "Photonic band gap device " consisting in a resonant structure based on a set of air gaps of different dimensions along the propagation axis (Figure 38). The goal of the simulation is to control the efficiency of the propagating mode excitation, the transmitted and the reflected waves.

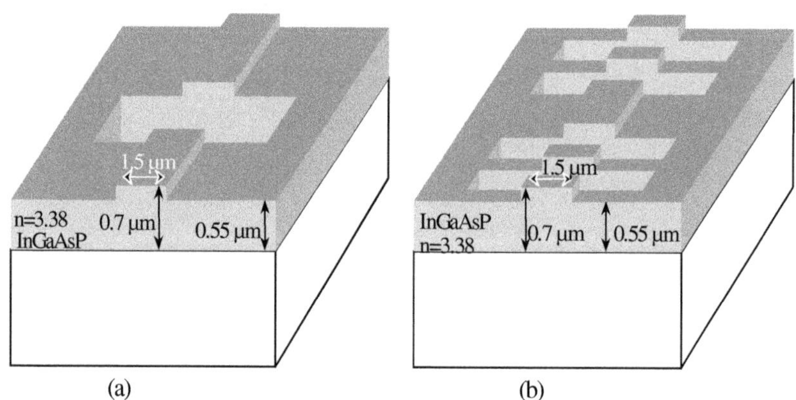

Figure 38 : Elementary photonic band gap devices : a) air gap, b) resonant structure.

In order to limit the memory storage required by the TLM method to a reasonable value, simulated devices are limited to a few wavelengths. Thus, the spatial field distribution of the propagating mode has to be correctly injected to avoid propagation of either radiation modes or evanescent modes. This requires to couple the TLM technique with a mode solver in order to control the excitation efficiency [40].

An example of EM field distribution the device is shown in figure 39. TLM results show that air gaps are good reflectors and that the light remain confined inside the resonator. The simulation has permitted to analyse the evolution of field distribution along the propagation in the structure. Additional simulations showed that the influence of etching depth in the propagation losses. However, the accuracy of the results is very sensitive to the absorbing boundary conditions implemented in the TLM software.

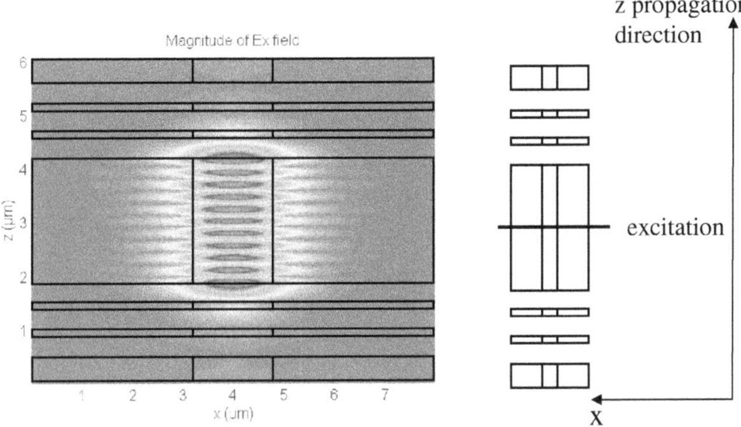

Figure 39 : EM field distribution in an optical integrated component (from TLM simulation).

4.6. Summary

The TLM technique is based on the propagation of voltage pulses in a transmission line network. Equivalence equations rely the voltage pulses on the interconnected transmission lines and the EM field components in the simulated media. For this, the simulated domain is meshed in an orthogonal grid. The primary result of a TLM simulation is the EM field distribution in Time Domain. Absorbing boundaries are essential for the simulation of open media.

The application to microwave devices permits the calculation propagation characteristics, S-parameters, radiation and electromagnetic interference. The simulation of a discontinuity can lead to the extraction of the electrical equivalent network from time domain of frequency domain results. New developments have extended the simulation of lumped linear and non linear devices.

The application to optical integrated components permits to account for reflection and diffraction phenomena, leading to relevant results in photonic band gap devices where the EM field distribution is calculated.

5. REFERENCES

[1] C. Yeh, K. Ha, S.B. Dong, W.P. Brown
Appl. Opt., vol. 18, pp. 1490-1504, 1979.

[2] N. Mabaya, P.E. Lagasse and P. Vandenbulcke
IEEE Trans. MTT, vol. 29, pp.600-605, 1981.

[3] M. Zoboli and P. Bassi
The Finite Element Method for Anisotropic Optical Waveguides
in G. Stegeman and G.C. Someda "Anisotropic and Nonlinear Optical Waveguides"', Elsevier, 1991.

[4] M. Koshiba, K. Hayata and M. Suzuki
IEEE Trans. Microwave Theory Tech., vol. 33, pp. 900-905, 1985.

[5] B.M.A. Rahman and J.B. Davies
IEEE Trans. Microwave Theory Tech., vol. 32, pp. 922-928, 1984.

[6] K. Hayata, M. Koshiba, M. Eguchi and M. Suzuki
IEEE Trans. Microwave Theory Tech., vol. 34, pp. 1120-1124, 1986.

[7] R. Pregla, W. Pascher
The Method of Lines
in T. Itoh ed. "Numerical Techniques for Microwave and Millimeter wave Passive Structures", J. Wiley Publ., New York, 1989.

[8] V. Rogge and R. Pregla
J. Opt. Soc. Am. B, vol. 8, pp. 459-463, 1991.

[9] G.B. Hocker and W.K. Burns
Appl. Opt., vol. 16, pp. 113-118, 1977.

[10] E. Strake, G. P. Bava, I. Montrosset
J. Lightwave Tech., vol. LT-6, pp. 1126-1135, 1988.

[11] M. Dinand, W. Sohler
J. Quantum Electron., vol. QE-30, pp. 1267-1276, 1994.

[12] M.D. Feit, J.A. Fleck
Appl. Opt., vol. 17, pp. 3990-3998, 1978.

[13] J. Van Roey, J. Van der Donk, P.E. Lagasse
J. Opt. Soc. Am., vol. 71, pp. 803--810, 1981.

[14] Y. Chung, N. Dagli
IEEE J. Quantum Electron., vol. 26, pp. 1335-1339, 1990.

[15] A.R. Mitchell
Computational methods in partial differential equations
John Wiley, New York, 1969.

[16] R. Accornero, M. Articlia, G. Coppa, et al.
Electronics Letters, vol. 26, pp. 1959-1960, 1990.

[17] G.R. Hadley
Opics Letters, vol. 16, pp. 624--626, 1991.

[18] G.R. Hadley
IEEE J. Quantum Electron., vol. 28, pp. 363-370, 1992.

[19] Integrated Photonics Research
1998 OSA Technical Digest Series, n. 4.

[20] B. D'Agens, S. Balsamo, I. Montrosset

J. Sel. Topics in Quantum Electron., vol. 3, pp. 233-244, 1997

[21] G. Perrone, F. Sartori, I. Montrosset
Physics and Simulation of Optoelectronic Devices VI, SPIE vol. 3283, pp. 983 989, 1998.

[22] H. Kogelnik
Theory of Optical Waveguides
in T. Tamir ed. "Guided Wave Optoelectronics", Springer-Verlag Berlin, 1988.

[23] D. Marcuse
Theory of Dielectric Optical Waveguides, Academic Press, New York, 1974.

[24] G.P. Agrawal
J. Appl. Phys., vol. 56, pp. 3100-3109, 1984.

[25] A.Bove
Graduation Thesis , Politecnico di Torino, May 1998

[26] BroadNeD and PTDS are a product of Virtual Photonics Incorporated, Berlin, Germany

[27] A. H. Gnauck, T. E. Darcie, G. E. Bodeep
"Comparison of direct and external modulation for CATV lightwave transmission at 1.55 μm wavelength"
Electronics Letters, 1992, 28(20), pp. 1875-1876

[28] M. J. Nazarathy, J. Berger, A. J. Ley, I. M. Levi, Y. Kagan
"Progress in externally modulated AM CATV transmission systems
IEEE Journal of Lightwave Technolgy, 1993, 11, pp. 82-105

[29] G. P. Agrawal
"Nonlinear fiber optics
second edition, Academic Press Inc.(1995)

[30] D. A. Fishman, J. A. Nagel
Degradations due to Stimulated Brillouin Scattering in Multigigabit Intensity Modulated Fiber-Optic Systems
IEEE Journal of Lightwave Technology, 11(11), pp.1721-1728, 1993

[31] Y. Sun, J. W. Suhlhoff, A. K. Srivastasa, A. Abramov, T. A. Strasser
A gain-flattened ultra wide band EDFA for high capacity WDM optical communications system
European Conference on Optical Communications, pp. 53-54, 1998

[32] R. C. Steele, G. R. Walker, N. G. Walker
Sensitivity of Optically Preamplified Receivers with Optical Filtering
IEEE Photonics Technology Letters, pp. 545-547, 1991

[33] R. Hofstetter,;H.Schmuck, R. Heidemann
Dispersion Effects in Optical Millimeter-Wave Systems using Self-Heterodyne Method for Transport and Generation
IEEE Trans. Microwave Theory Tech., Vol. 43, No. 9, pp 2263-2369 Sept 1995

[34] J.J.O'Reilly, P.M.Lane, R.Heidemann, R. Hofstetter
Optical Generation of very narrow linewidth wave signals
Electronics Letters, vol. 28 pp. 2309-2311, Dec. 1992

[35] D. Marcuse, A. R. Chraplyvy, and R. W. Tach
Effect of fiber Nonlinearity on Long-Distance Transmission
IEEE Journal of Lightwave Technology, vol. 9, pp 121-128, 1991

[36] W.J.R Hoefer
The transmission Line Matrix Method. Theory and Application
IEEE. Trans.MTT-33, n°10, pp 882-893, 0ct. 1995.

[37] P.B. Johns
A symmetrical Condensed Node for the TLM method
IEEE Trans. MTT-35 n°4 April 1987.

[38] C. Boussetta, F. Ndagijimana, J. Chilo, P. Saguet
Electrical Modelling of Packaging Discontinuities : A General Methodology Based on the Three-Dimensional TLM concep
Intern. Journ. of Microwave and Millimeter-Wave Computer Aided Engineering, Vol.5 no.2 1995 John Wiley & Sons.

[39] C. Golovanov, F. Ndagijimana, P. Saguet
Global Simulation of a Multilayer Interconnection with Lumped Non-linear Elements
Electrosoft 99 (Software Applications in Electrical Engineering, Computational Mechanics Publications), Seville 17-19 May 1999.

[40] O. Jacquin, F. Ndagijimana, P. Benech
Application of the TLM technique to integrated optic components modelling
Third International Workshop on Transmission Line Matrix (TLM) Modeling – Theory and Applications, Oct-1999 Nice – France.

CHAPTER 4: MICROWAVE-PHOTONICS SYSTEMS

1. INTRODUCTION

Today, the optical transmission of microwave signals offers in conjunction with their low loss propagation over very wide frequency bandwidth, a high immunity to electromagnetic perturbations, which opens new avenues for the insertion of new concepts and photonic architectures in microwave systems.

Due to a great improvement in the performances of optoelectronic components over the last ten years, photonics becomes one of the major technology for advanced telecommunication, wireless and radar systems. Further progress in the near future will have a very significant impact on the design of new microwave system architectures. As examples, architectures for optically controlled phased array antennas were demonstrated.

Such antennas will be use in a large number of applications such as radar, communication and electronic warfare. In order to satisfy this multifunctional aspects, it will be necessary to distribute these antennas on ground based areas as well as the aircraft surface. Multistatic systems will impose multiple remoting of antennas with respect to their processing units.

In all cases, it appears a need for low loss link able to remote the control of the antennas as well as distribution and processing of very wideband microwave signals (typ. 1-20 GHz).

This chapter is divided in 5 parts covering the microwave photonics domain from the component side to the antenna and satellite system applications:

The first part is dedicated to the microwave optical link as a basic building block for the system applications,

The second and third parts are related to telecommunication applications, wirelss and broadband access networks ,

The forth part deals with optical beamforming approaches for antenna applications

The last part is covering the satellite communications applications of fiber optic link

2. MICROWAVE OPTICAL LINKS

2.1. Analog Optical Links : Models, Measures and Limits of Performances

C.H. Cox, III
Research Laboratory of Electronics, MIT, 77 Massachussets Avenue, Cambridge, MA, 02139,USA, e-mail : ccox@mit.edu

2.1.1. Abstract

We present the small signal models for direct and external intensity-modulation analog links. We then discuss three of the most common measures of performance for analog links: gain, noise figure and intermodulation-free dynamic range, IMFDR. The limits of noise figure with passive matching and IMFDR under linearization are also presented.

2.1.2. Introduction

To convey analog signals over an optical fiber generally requires that the analog signal is impressed on the optical carrier via any one of a number of optical modulation devices and recovered at the destination end of the fiber via some form of a photodetection device. As we will see in the discussion below, the RF performance of such a combination of electro-optic devices is usually insufficient to interface directly with the RF system. Thus it is common to augment the performance of the intrinsic link with pre- or post-amplifiers. However, when we wish to study the design of the intrinsic link, amplifiers tend to obscure the tradeoffs involved. Therefore in the discussion below we will focus exclusively on the intrinsic link; i.e. the link without any amplifiers.

In principal we could modulate any of the parameters of the optical carrier, just as is done with modulation of an RF carrier. In practice only intensity modulation of the optical carrier is used at present, although optical FM is under investigation at several research institutions. Thus we will further limit our discussion below to intensity modulation.

There are many measures by which we could characterize the performance of an analog optical link. We will focus on three of the most common and basic ones in the discussion below, all of which are borrowed from the RF community. One of them is the gain of the intrinsic link. Of the gain definitions that have been developed by the RF community, the one that has proven most useful for analog links is the available power gain. Consequently we define the intrinsic link gain (i.e. the link gain without any amplifiers) to be available power gain between the input to the modulation device and the output of the photodetection device. We will use gain here in the general sense of the term where negative gain denotes loss. Gain vs. frequency, or bandwidth, of the link will included in the gain discussion.

The other two measures of link performance, noise figure and intermodulation-free dynamic range, will be defined at the beginning of the sections devoted to those topics.

Optical fibers are nearly an ideal transmission medium. The optical loss can be extremely low, about 0.2 dB/km and they do not limit bandwidth directly, however dispersion can distort a broadband spectrum over long lengths of fiber. Thus for the purposes of this discussion we will assume that the fiber length is sufficiently short that the deleterious effects of fibers can be neglected.

The choice of operating wavelength for a fiber link involves many practical and economic aspects, but does not directly enter into the modeling to be presented below. Thus although most of the experimental results we discuss were done at 1.3 μm, they could have been obtained at either of the other principal wavelengths for optical links – 1.55 or 0.85 μm – with appropriate scaling for the particular wavelength.

As we will see from the discussion below, there are two main methods for imposing the intensity modulation onto the optical carrier. However the photodetection method is always a photodiode. For bandwidths up to about 10 GHz, photodiodes have been demonstrated with nearly ideal optical-to-electronic conversion efficiency, as measured by the slope efficiency, A/W. Further, photodiodes with bandwidths up to 500 GHz have been fabricated, albeit with a decrease in slope efficiency as the bandwidth increases. Distortion generated in the photodiode is also generally negligible except when very high optical powers or high linearity modulation devices (i.e. linearized) are used. As we will see, all these measures of performance are well in excess of those obtainable from either of the modulation methods, so photodiodes generally do not limit the performance of an analog link.

All the topics discussed above, as well as additional topics, are covered in more detail in the forthcoming book by Cox, *Analog Optical Links: Theory and Practice* which was scheduled to be published by Cambridge University Press in 2000.

2.1.3. Direct Modulation

A straightforward way to achieve intensity modulation is to have the analog signal modulate the intensity of a laser. The only laser at present with sufficient bandwidth to be of practical interest is the diode laser. We are now in a position to assemble a directly modulated link by combining a diode laser with a photodiode detector.

It can be shown that the intrinsic gain of such a link is proportional to – and when impedance matched equal to the square of the slope efficiencies of the laser and photodiode. Each of these terms enters as the square because these devices are responsive to the RF current, which in turn must be squared to get the RF power.

As shown by the representative transfer curves in figure 1, there is a range of optical powers over which both devices are linear, i.e. there derivatives – which are the slope efficiencies – are constant. Over this range of optical powers we would expect that the intrinsic gain would be constant. Indeed this is what we observe experimentally, as seen by the data shown in figure 1 for a Fabry-Perot and distributed feedback (DFB) type of diode lasers.

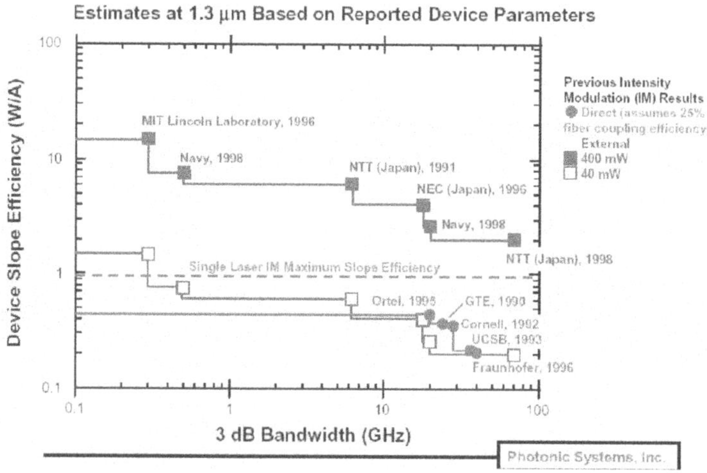

Figure 1 : Direct and external modulation slope efficiency.

Intrinsic gain and bandwidth for some of the directly modulated links that have been reported in the literature and plotted on figure 2 see also [5]. From these data we can conclude two facts. One is that virtually all directly modulated links have RF loss, which in many cases can be substantial. The loss can be partially – and in one case completely – overcome by trading excess bandwidth to improve the gain. The other fact is that the majority of maximum link bandwidths are limited to about 10 GHz. This reflects a limitation imposed by the relaxation resonance of commercial diode lasers, which lags behind the best laboratory devices whose maximum modulation frequency is about 30 GHz.

The link loss is primarily a consequence of the low slope efficiency of diode lasers, which is typically about 0.1 W/A with the best around 0.3 W/A. The slope efficiency of a single diode laser is limited by energy conservation. However, recently the cascade laser has been applied to links resulting in the first demonstration of a broad bandwidth, directly modulated link with intrinsic gain.

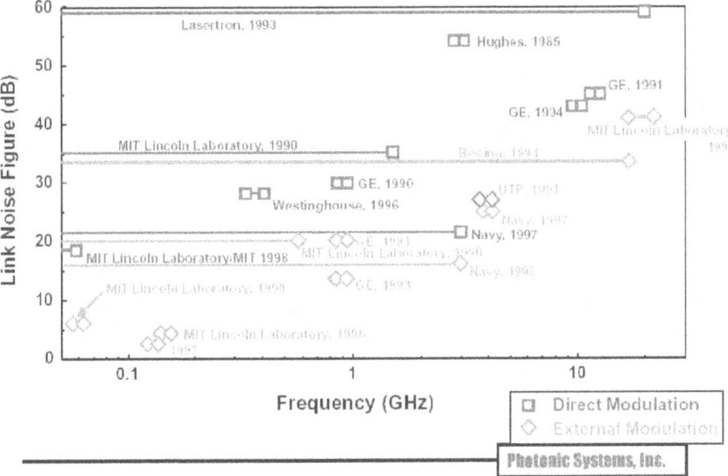

Figure 2 : Performance of analog link : direct and external modulation.

2.1.4. External Modulation

To implement external modulation, the laser operates CW and the intensity modulation is imposed via a device external to the laser. By far the most common external modulator in use today is based on a Mach-Zehnder interferometer fabricated in the electro-optic material lithium niobate. We can now assemble an externally modulated link by combining a CW laser, a Mach Zehnder modulator and a photodiode.

It is possible to arrange the variables in the expression for externally modulated link gain such that there is a term with the units of W/A. In other words, we can assign a slope efficiency to the combination of a CW laser and an external modulator. This formalism permits us to use the same expression for link gain as before – i.e. the product of the square of slope efficiencies. This formalism also exposes the optical power dependency of the external modulation slope efficiency, which has a significant impact on both the gain and noise figure of externally modulated links. We discuss the impact on gain here and defer the impact on noise figure to the noise figure section.

As indicated at the bottom of figure 1, the square of the average optical power through the modulator appears in the expression for the small signal slope efficiency. Thus an external modulation link does not have a unique slope efficiency, but rather a family of slope efficiencies, depending on the average optical power. The impact of this fact on link gain is shown by the external modulation gain data shown in figure 3, where we see the intrinsic link gain increases as the square of the average optical power.

It is interesting to note that for sufficiently high average optical power, an externally modulated link can have positive intrinsic gain. One way to appreciate the basis for this gain is to consider the RF powers at the link input and output. The input RF power drawn by the modulator depends on its impedance but is independent of the average optical power flowing through the modulator. Conversely, the RF power produced by the photodiode clearly depends on this same power; in the extreme case of no optical power on the photodiode, there is no RF power from the photodiode. Thus as the optical power is increased from zero, the link RF output power increases, but the link RF input power remains fixed. Viewed from this perspective, the gain from an externally modulated link is no more unusual than gain from an active electronic device such as an FET.

The intrinsic gains vs. frequency, which have been reported for a variety of external modulation links, are presented in figure 2, see also [5]. In contrast to the analogous direct modulation data, these data show that positive intrinsic gain, both narrow and broad bandwidth, have been demonstrated in external modulation.

Diode pumped, solid state lasers are the most common CW source for externally modulated links because of their high average optical power and low relatively intensity noise, RIN. Diode lasers are under development for such applications.

2.1.5. Noise Figure

Noise figure, NF, is another important link parameter, especially for applications where low-level signals are involved, such as antenna remoting. We use the same definition of noise figure that has been developed by the RF community: the ratio of the signal-to-noise ratio at the link input to the signal-to-noise ratio at the link output. By definition, the input noise is taken as thermal noise at 290 degrees Kelvin. This form of the noise figure definition makes it clear that NF is a measure of the degradation in the SNR as the signal passes through the link.

For the purposes of analyzing the noise figure of direct and external modulation links, it is more useful to use the following alternate, but equivalent, expression for noise figure:

$$NF(dB) = 10\log\left(\frac{(S/N)_{in}}{(S/N)_{out}}\right) = 10\log\left(\frac{N_{out}}{kTg_i}\right) \quad (1)$$

where g_i is the intrinsic link gain and $N_{out} = g_i N_{in} + N_{link}$. In general N_{link} will consist of the sum of laser RIN and thermal noise of the modulation and photodetection devices. We assume for this discussion that the laser RIN is negligible, i.e. it is at the shot noise limit.

We now wish to investigate the scaling of link noise figure with average optical power. For RIN at the shot noise limit, the dominant contribution to N_{link} is shot noise, which increases as the first power of optical power. Recall from the preceding that the directly modulated link gain is independent of optical power. Substituting these facts into equation 1 we obtain:

$$NF = 10\log\left(\frac{\text{constant} + \text{shot noise } (\propto \text{P})}{kTg_i(\text{independent of P})}\right) = \text{constant} + NF(\propto \text{P}) \qquad (2)$$

Thus for a directly modulated link the noise figure should increase with average optical power.

For the externally modulated link we make the corresponding substitutions into equation (1) to obtain:

$$NF = 10\log\left(\frac{\text{constant} + \text{shot noise } (\propto \text{P})}{kTg_i(\propto \text{P}^2)}\right) = \text{constant} + NF(\propto \text{P}^{-1}) \qquad (3)$$

In other words, the noise figure of an externally modulated link decreases as the optical power increases. The reason for this apparently paradoxical result is that although the noise power at the photodiode is increasing linearly with optical power, the link gain is increasing quadratically with optical power. Thus when the effect of this noise at the link output is translated back to the link input – which is what noise figure represents – the result is as given above.

At higher optical powers, we see that the Fabry-Perot noise figure increase with optical power. However at lower optical powers for the Fabry-Perot and for virtually all powers for the DFB, the laser RIN is greater than the shot noise limit, which violates the assumption made at the start of the derivation. The external modulation data shows the expected decrease in noise figure with increasing optical power, at least initially. At higher optical powers the noise figure appears to be approaching a limit, which we discuss next.

In figures 2 and 3, the noise figures for the links reported are plotted against the corresponding bandwidth for these links. It appears that the only low noise figures have been achieved are in links with positive intrinsic gain.

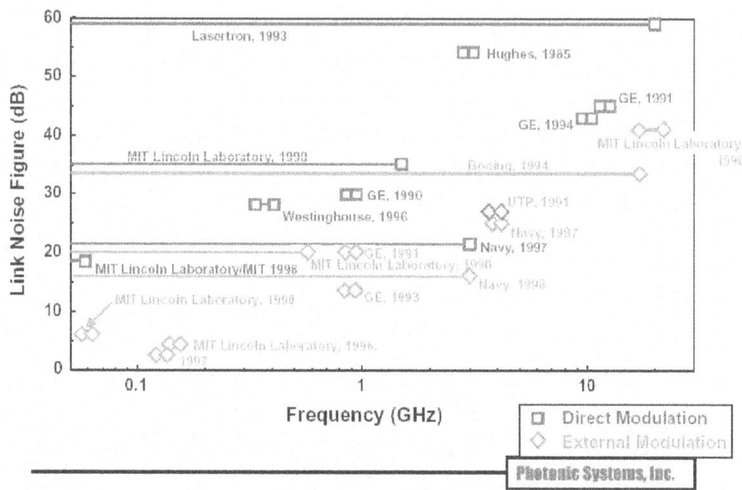

Figure 3 : Performance of analog link : direct and external modulation.

To appreciate the basis for this effect, consider again equation 1, but this time we substitute in the expression for N_{out}, then write the two terms separately, canceling the common gain terms; the result is :

$$NF(dB) = 10\log\left(\frac{N_{out}}{kTg_i}\right) = 10\log\left(\frac{N_{in}}{kT} + \frac{N_{link}}{kTg_i}\right) \qquad (4)$$

Consider now two limiting cases for equation (4). For g_i>>1, the second term in equation 4 is negligible compared to the first. This condition gives a noise figure which is independent of link gain. Alternatively for g_i<<1, the second term dominates over the first, which yields a noise figure that is proportional to $1/\ g_i$ It is encouraging that none of the reported data lie below these limiting curves.

To examine the detailed shape of the noise figure curve, [1] constructed a high-gain, external modulation link with variable gain. He then measured the noise figure at various values of link gain. The resulting data are plotted in figure 4, together with the theoretically predicted curve. The predicted curve, which is the above limits with a correction for the loss of the actual matching circuit, matches the experimental data quite well. Note in particular that from a noise figure view point, a link with 1 dB of loss has a much higher noise figure than a 1 dB attenuator.

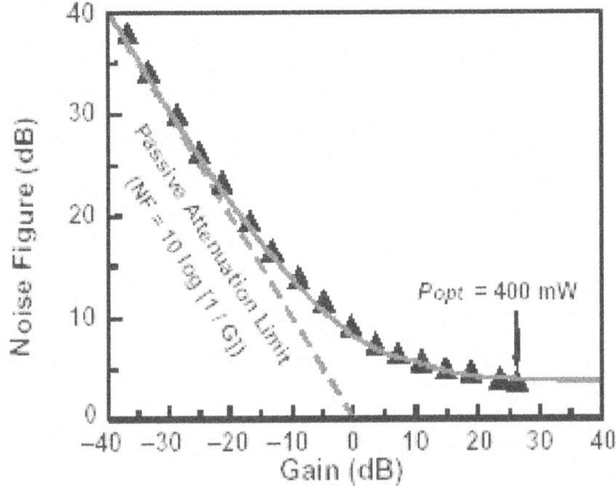

Figure 4 : Record performance of analog link : external modulation – noise figure.

2.1.6. Intermodulation-free Dynamic Range

The third principal link parameter we will discuss is the intermodulation-free dynamic range, IMFDR. Any practical device has some – albeit small if it is to be useful for analog modulation – non-linear component to its other wise linear transfer function. This non-linearity will generate distortion products at the link output. The IMFDR is defined as the SNR for which the distortion terms just equal the noise floor. Consequently the IMFDR is less than the SNR dynamic range. Thus it is important to distinguish these two measures of dynamic range. The two most common IMFDRs are the second- and third-order ones.

Figure 2 presents data on the reported third-order, IMFDRs for both direct and external modulation links [3]. The IMFDR of the basic modulation device can be improved by 10 to 15 dB by using additional means to linearize the device transfer function. Diode predistortion is perhaps the most common technique used with direct modulation, whereas concatenation of two modulators is the most common technique for external.

It has been proposed [4] and demonstrated [2] that linearization of both second- and third-order terms, i.e. broad bandwidth linearization, increases the noise figure by at least 10 dB, whereas linearization of the third-order only, i.e. narrow bandwidth linearization, does not incur such a penalty. The basis for this noise figure penalty with broad bandwidth linearization is not presently understood.

Although the IMFDRs achieved to date with present linearization methods are sufficient for many applications (see figure 5), it is instructive to ask: what is the maximum improvement that is possible?

All the present linearization techniques achieve the increased IMFDR by actually generating more intermodulation terms but with opposite phase so that the net result is a reduced intermodulation power. This category of techniques tends to leave the intercept point approximately unchanged, which means that the increased IMFDR comes from a rotation of the distortion line; i.e. after the third-order is cancelled, the fifth-order dominates, etc.

However a linearization technique that also increased the intercept point would actually translate the distortion curve. This translation could continue until the IMFDR equaled the SNR dynamic range. Some typical numbers for a Mach Zehnder modulator link will put this discussion in context. A typical link using a standard – i.e. unlinearized – Mach Zehnder modulator has an IMFDR of ~110 dB in a 1 Hz bandwidth. The same link with a linearized Mach Zehnder modulator has an IMFDR ~130 dB; the SNR dynamic range of this link would be ~160 dB. Therefore for an externally modulated link, there is potentially about 30 dB further improvement that should be possible with future linearization techniques.

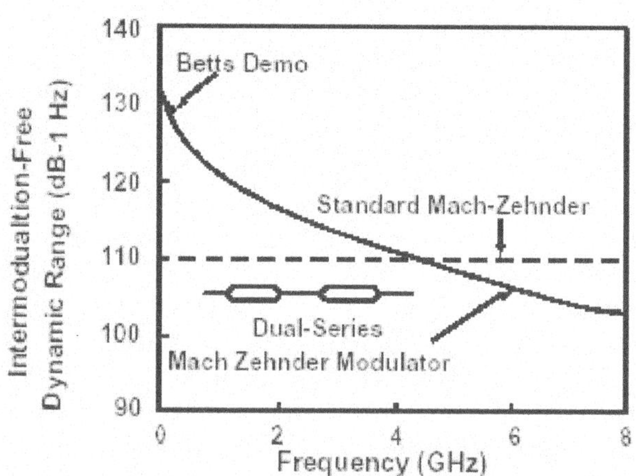

Figure 5 : Record performance of analog link : external modulation Intermodulation Free Dynamic Range (IMFDR).

2.1.7. Summary

The table below summarizes the performance of present analog optical links.

Parameter	Direct Modulation	External Modulation
Wavelength	0.85, 1.3, 1.55	1.3, 1.55
Maximum modulation frequency (GHz)	30	150
Intrinsic link gain (dB)	-5 to -35	-30 to +30
Noise figure (dB)	20 to 60	4 to 30
IMFDR Standard (dB-Hz$^{2/3}$)	100 to 114	112
Linearized (dB-Hz$^{4/5}$)	120	130

Consequently for high performance applications, external is preferred. Unfortunately external modulation is also the more expensive of the two techniques.

We have also investigated the limits to noise figure with passive matching and IMFDR under linearization.

All the topics discussed above, as well as additional topics, are covered in more detail in the forthcoming book by Cox, *Analog Optical Links: Theory and Practice* which was scheduled to be published by Cambridge University Press in 2000.

2.1.8. Acknowledgement

The author would like to thank the members of his former research group at MIT Lincoln Laboratory for assistance in collecting many of the results discussed here: Ed Ackerman, Gary Betts, Mike Corcoran, Roger Helkey, Scott Henion, Robert Knowlton, Fred O'Donnell, Joelle Prince, Kevin Ray, Gil Rezendes, Harold Roussell, Mike Taylor, Rob Taylor, John Vivilecchia and Allen Yee.

2.2. Optoelectronic and Optical Devices for Applications to Microwave Systems

P. Richin, D. Mongardien
Thales Research & Technology, 91404 Orsay Cedex
Now in Alcatel Optronics Nozay

2.2.1. Optoelectronic Interfaces

Until now, most of the transmissions are based on the amplitude modulation of the optical carrier and direct detection. Coherent systems were investigated previously for telecommunication needs but the use of EDFA appear to be a more realistic answer to the requirements. However, in the field of microwave applications, some laboratory demonstration make use of the coherent aspect for the microwave signal processing and control.

The transmitter design must be selected from two kinds of principles: direct or external modulation.

A simple (and cheaper) way is the direct modulation of the semiconductor laser diode with a modulated injected current. It is single transverse mode in order to have an efficient coupling with the fibre and either multi (Fabry Perot) or single longitudinal mode (DFB – Distributed FeedbBack – or DBR – Distributed Bragg Reflector -).

The alternative is the external modulation of a continuous wave source such as a power semiconductor laser or a diode pumped solid state laser. These modulators might be fabricated on semiconductor (Mach Zehnder or electroabsorption principle, with possible monolithic integration with a DFB laser), lithium niobate (Mach Zehnder principle) or polymer (Mach Zehnder principle) substrates.

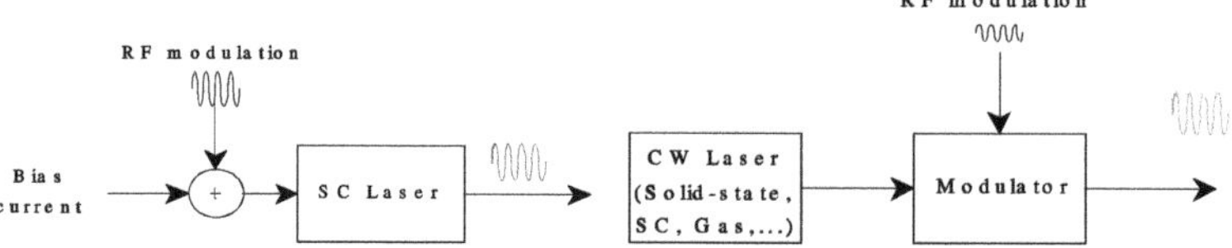

For the receiver, even if both MSM and PIN had been investigated in laboratories, most of the development involve either surface illuminated PIN photodiodes (for application up to Ku band) or side illuminated or waveguide PIN photodiodes (for frequency requirement up to 60 GHz and above or for high optical input power).

All the characteristics and performances of these above components are detailed elsewhere. A wide range of commercially available products is available.

2.2.2. Fibres

The following figure is a summary of the fibre history.

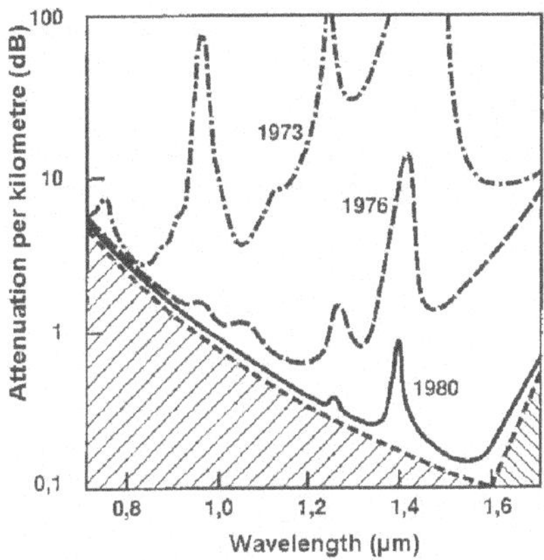

The first window to be considered in the early 70's, is centred at 0.8µm. The corresponding links made use of GaAs emitters (Light Emitting Diode or Laser Diode, depending on the bit rate or bandwidth), multimode fibres with step or graded index profile and typical attenuation of 3dB/km, Si PIN photodiodes. Typical applications are in the field of short range transmissions, bit rate in the range of tens to hundreds of Mb/s, and limited S/N ratio due to the modal noise. Several years later, specific single mode fibre (polarization preserving fibre) allow to build up different sensors (gyrometer, temperature, strain,...).

The second window offers acceptable attenuation (0.5dB/km) and minimum material dispersion. The total dispersion at λ ($1200\text{nm} < \lambda < 1600\text{nm}$) of these standard fibres, can be calculated:

$$D(\lambda) = \frac{S_0}{4} \times \left[\lambda - \frac{\lambda_0^{\ 4}}{\lambda^3} \right] [\text{ps/(nm.km)}]$$

λ_0 the zero dispersion wavelength $1300 < \lambda_0 < 1320\text{nm}$

S_0 zero dispersion slope – example: 0.092ps/nm².km)

The telecommunications required much lower attenuation and high bandwidth. Most of the current long distance systems consist of 1.5µm single longitudinal mode emitter associated with a single mode fibre: within this 3rd window, the attenuation is minimum (0.25dB/km) and total dispersion (material + waveguide) can be minimized (dispersion shifted fibre). It is calculated at λ ($1500\text{nm} < \lambda < 1600\text{nm}$)

$$D(\lambda) = \frac{S_0}{4} \times \left[\lambda - \frac{\lambda_0^{\ 4}}{\lambda^3} \right] [\text{ps/(nm.km)}]$$

λ_0 the zero dispersion wavelength $1535 < \lambda_0 < 1565$nm

S_0 zero dispersion slope – example: 0.092ps/nm².km)

Dispersion flattened fibres had been also optimized for both 2nd and 3rd window.

The different typical diameters of a standard fibre are 9μm/50μm/125μm/250μm for respectively the mode, the optical cladding, the mechanical cladding (silica) and the primary coating.

As mentioned above, the polarisation fibres are of interest for sensor applications or as a pigtail between a laser and an external modulator. Their sizes are similar to that of telecom fibres but their structures are quite different as depicted below.

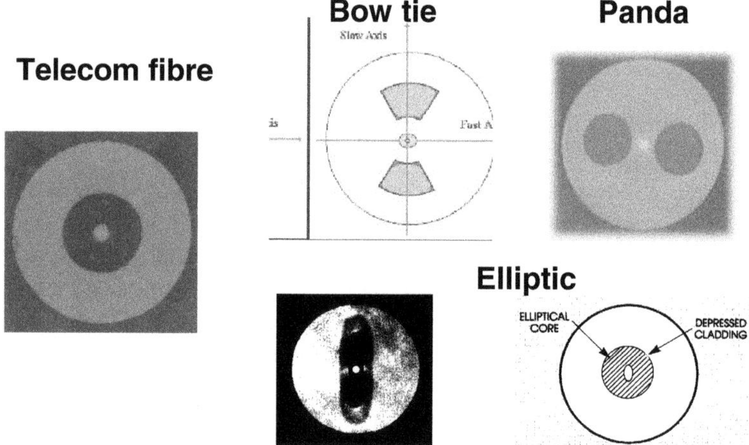

One must mention that for very short range and inexpensive links, the multimode plastic optical fibres (step index profile, core diameter ~1mm) and optoelectronic E/R working at visible wavelength (typically 0.6μm) are under pre-development.

For practical use in a real environment, to protect the fibre, number of manufacturers propose various cables with either tight or loose cabling, cylindrical or ribbon structures, with a diameter smaller than 1 mm and up to a few centimeters, depending on the number of fibres (1-2-4-8 or more) and of the required protection.

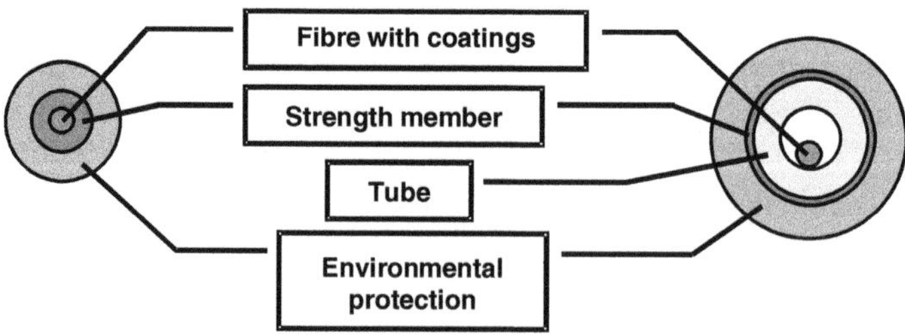

2.2.3. Connectors

One must not forget that an efficient and reliable fibre to fibre coupling might be the key point of a system. The fusion splicing of fibre is the preferred solution for low loss (down to 0dB) and permanent connexions. Otherwise, one must select a connectors among the market offer, with typical insertion and return losses respectively below 0.5dB and higher than 40dB (PC - Physical Contact -, ...) or 60dB (APC - Angled Physical Contact -, Optoball, EC, E2000, ...).

ST – Standard Telecom

SC/PC or SC/APC

Standard Connector

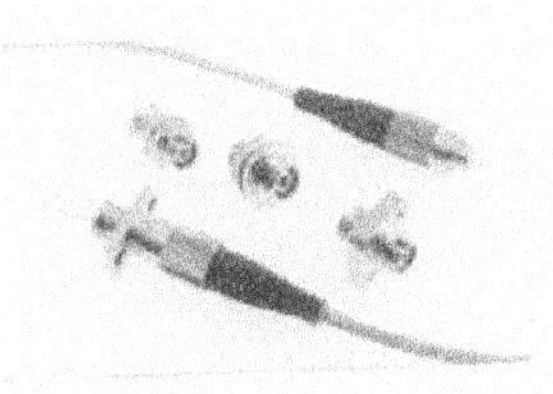

FC/PC or FC/APC

Fibre Connector

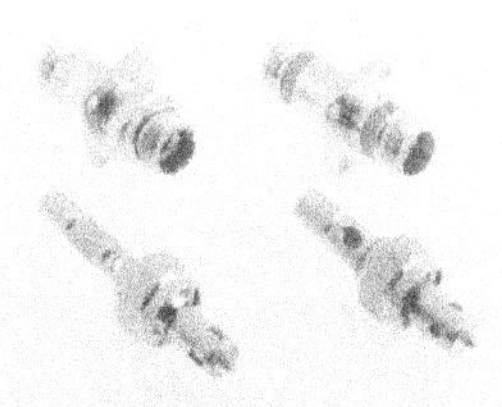

Optoball

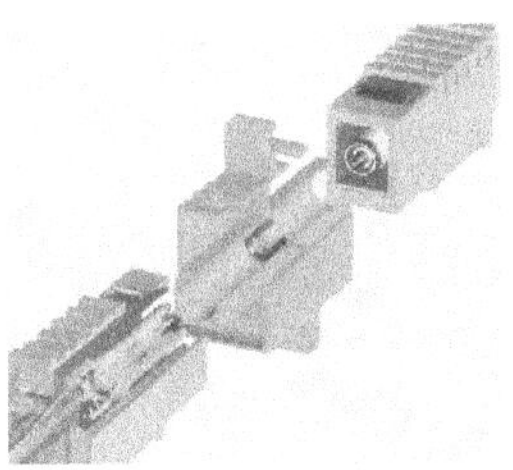

EC – European Connector

E 2000 – European 2000

The answer to any question about the maturity of such device, is that, today, several millions connectors are already installed and working well in communication equipments.

2.2.4. Couplers

The device has N output ports and N output ports. The general characteristics are (for instance, with a 2 port coupler as shown, with P_i for the optical power on port "i"):

Excess Loss: EL = 10 Log $(P_1/(P_3 + P_4))$
Insertion Loss: IL = 10 Log (P_1/P_4)
Directivity: D = 10 Log (P_1/P_2)
Return Loss: RL = 10 Log (P_1/P_5)

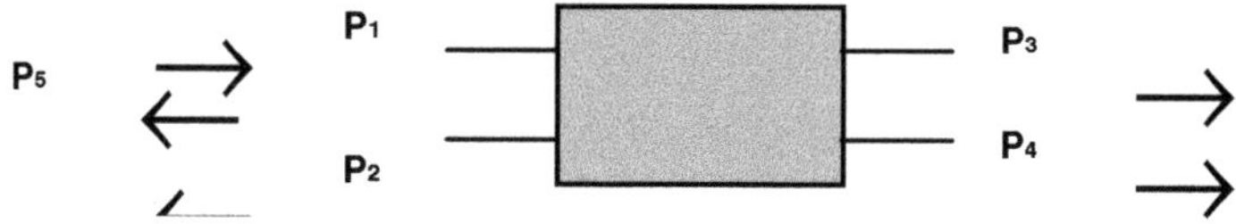

The maximum number of ports for commercially available products, is generally 8 or 16. Within systems, this kind of couplers are use as splitter, mixer or add-drop devices. Most of them are manufactured either with fibres or with an integrated optic technology.

2.2.4.1. Fibre Couplers

This basic component is composed of 2 fibres fused together and stretched in order to get an evanescent field coupling between both singlemode waveguides.

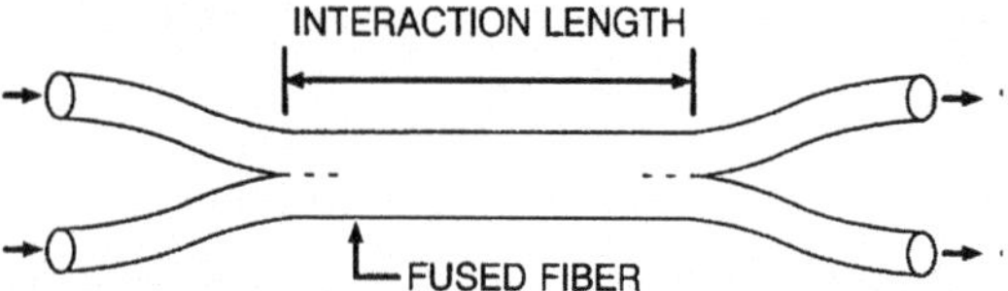

The standard coupling ration are 50/50 or 90/10. With such element, tree couplers (1->N, N = 4-8-16 ...) or star couplers (NxM) can be assembled. For a 2 ports device, the standard characteristics are:

Insertion loss (dB)	Directivity (dB)	Return loss (dB)
< 3.5	> 50	> 50

2.2.4.2. Integrated Optics

The single mode waveguides are fabricated on different substrates: glass (with ion exchange techniques), Silicon (doped silica deposition by CVD or Sol gel process, for instance), semi conductor (GaAs or InP). This kind of technical approach is mainly interesting for devices with a great number of port (it is a collective manufacturing process attractive for reducing the costs), or when one require the hybridization with other semiconductors components (on SC substrates). The performances are slightly worst compared to those of the fibre couplers due to the propagation and to the fibre to PIC coupling loss.

Doped waveguides are under investigation world-wide in order to get lossless components.

2.2.5. Switches

Several technical approaches are possible.

2.2.5.1. Mechanical Switches

This is a “field proven” technology. They offer very low insertion losses (≈0.5dB) and a cross-talk below –60dB. However, the switching time of these “optical relays” is in the millisecond range.

2.2.5.2. Acousto Optic Switches

By using beam deflexion by a grating generated an acousto optic effect in a cell (GaP, LiNbO3, KTP block with an acoustic transducer), devices had been realized with switching time in the microseconds range. Insertion losses and cross-talk are worst (respectively 3-4dB and 20dB) compared to the previous solution, due to the fibre to fibre coupling through distant micro optic elements.

By using an acousto optic effect too, integrated optic devices had been demonstrated and some product are in pre-development. They lead to a reduction of the driving power of the device (10dBm Cf 30dBm typically).

2.2.5.3. Electro Optic Switches

The single mode waveguides with a design similar to that of the above “Fibre couplers” are realized on LiNbO3 substrate with integrated optic technology. Their switching time is in the nanosecond range, but most of the products available on the market are sensitive to the state of polarization of the light at the input of the device. Their insertion loss, cross-talk and driving voltage (for instance, respectively less than 5dB, 15dB min and 10V) limit their field of application.

2.2.5.4. *Other Technical Approaches*

Among the other solution which were investigated and now available, one can mention an integrated optic devices (polymer on silicon) driven by a thermo optic effect with the following characteristics:

Insertion loss	Isolation	Return loss	Switching time	Driving voltage
<2.5dB	>25dB	>50dB	<2ms	~8V

The use of SOA - Semiconductor Optical Amplifier – as an optical gate is also attractive: it can be used alone or integrated with a switch on a SC substrate in order to improved the performances of the component (to compensate fibre coupling and propagation losses). However, one must take care of the impact of this active components on the transmission performances (linearity, noise, ...).

2.2.6. Isolators

Using the Faraday effect, they allow the isolation between the upstream and downstream side of the device: typical applications are within modules to protect the laser from link reflexions and within EDFA to prevent them from lasing. They optimized for a given telecom window (mainly, for 1.3 or 1.5µm) and to be either polarization dependent or independent.

	Insertion loss	isolation	Return loss
One stage	0.5dB	30-40dB	>55dB
Two stage	0.7dB	45-50dB	>55dB

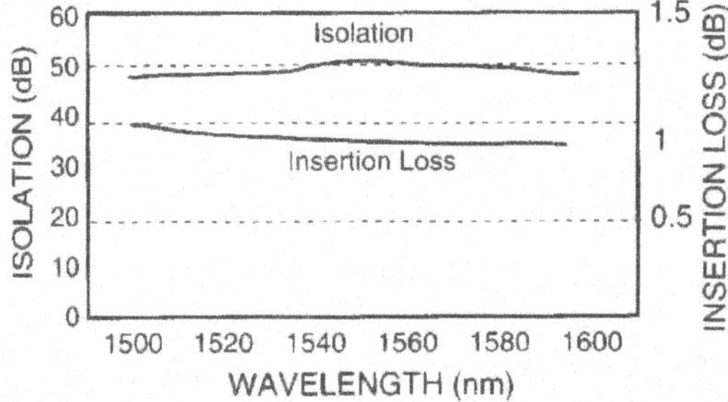

2.2.7. Optical Multiplexing

Depending of the channel spacing $\Delta\lambda$, one can consider WDM (Wavelength multiplexing, with DL in the nanometer range) and for smaller values of $\Delta\lambda$ (Tens of GHz), ODFM (Optical frequency multiplexing) or HDWDM (High density wavelength multiplexing). An ITU standard already exist in the 1.55µm, with $\Delta\lambda$ equal to 200GHz or smaller (100GHz, 50GHz, ...).

2.2.7.1. Gratings

As described below, the input (N wavelength) optical power is spatially diffracted in a given direction depending on the wavelength, and coupled in N corresponding output fibres.

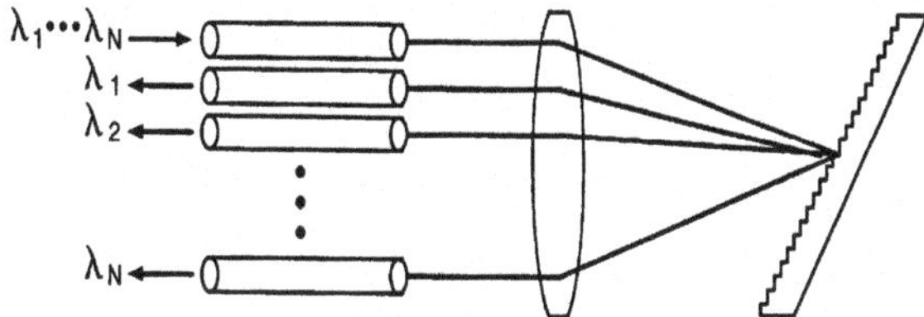

Field proven products, made with optical bulk elements, already exist on the market.

Number of channels	Insertion loss	OPTICAL isolation between channels
2 to tens	3dB	30-40dB

2.2.7.2. Optical Filters

The multi-dielectric coating technology allows the realization of band pass (see below) or dichroic filter (for instance, 1.3/1.5µm device with transmission of short wavelength and reflexion of the long wavelength).

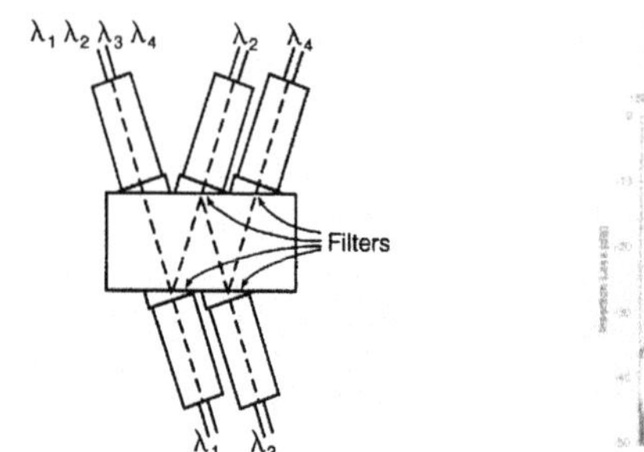

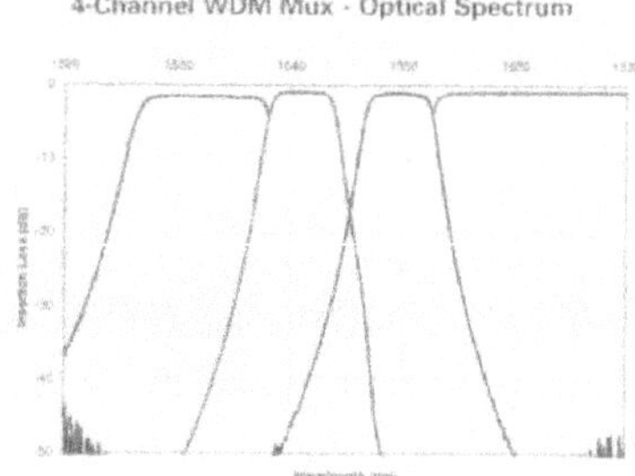

2.2.7.3. *Integrated Optics*

Polarization insensitive devices had been realized with silica/doped silica waveguides on silicon as represented below. The thermo optic effect allow to slowly control or tune the centre wavelength of the filter.

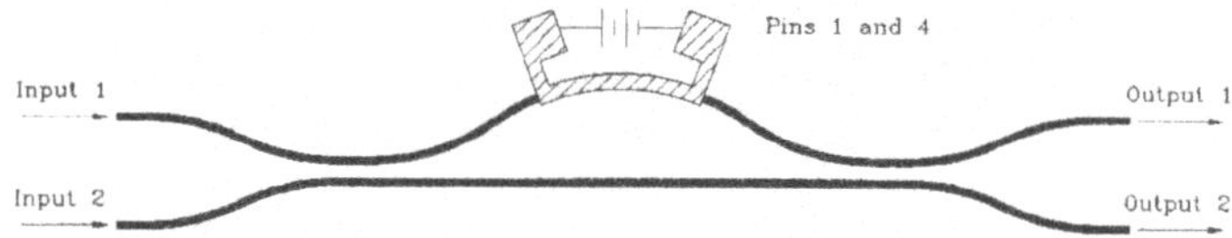

Basically, the channel spacing of such Mach Zehnder structure is a few nanometer but it can be far less by using cascaded devices.

The design such as a PHAsed ARray - PHASAR – lead to 1xN or NxM compact configurations with reduced channel spacing (100-200GHz).

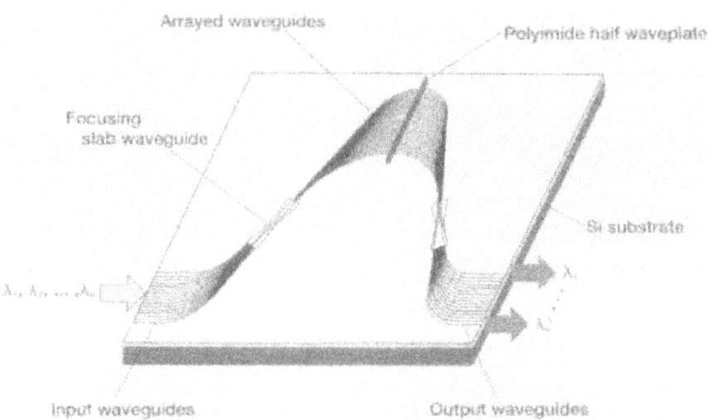

2.2.7.4. Fibre Bragg Grating

By using such fibre components, channel spacing down to 200-100GHz can be obtained with low loss (1-2dB range) and 20-30dB optical isolation.

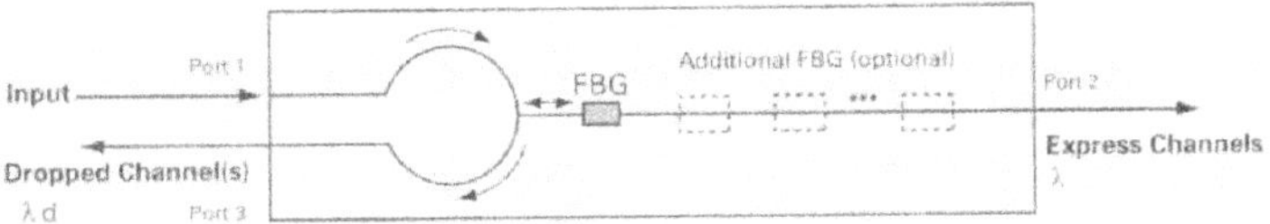

2.2.7.5. Fabry Perot Filter

Beside the multiplexing devices, the tunable filters are useful to reject unwanted channels or spontaneous wideband spectrum. A common product is based upon the Fabry Perot interferometer. The center wavelength can be adjusted either by modifying the thickness L of the cavity with piezoelectric translators or for other benchtop filter by manually tilting the cavity.

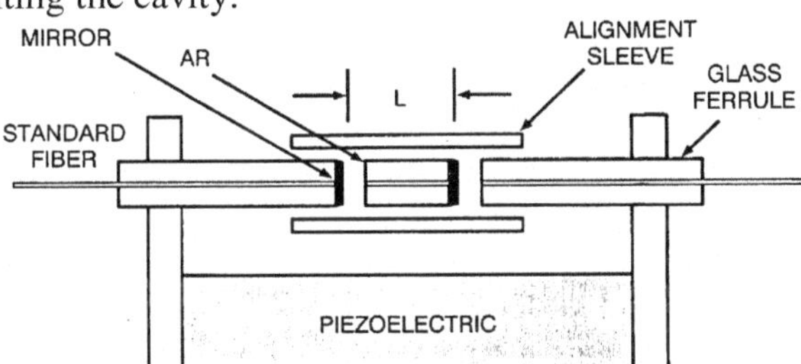

Today, similar components are under investigation by using the MOEMS (Micro Opto Electronic Mechanical Systems) technology on SC substrates (Silicon, GaAs, InP): it should lead to very compact and cheap elements.

2.2.8. Optical Amplifiers

2.2.8.1. Introduction

The aim of an optical amplifier is to provide gain for the input optical signal, without optical to electrical and electrical to optical conversions. This type of amplifier is transparent for the modulation signal (analogue or digital, frequency or bit rate, ...). Today, mainly two types of optical amplifiers are available: the erbium doped optical fibre amplifier (EDFA) and the semiconductor optical amplifier (SOA). They are working with the same basic rules: energy absorption and stimulated emission but they differ by the active material (rare earth ions and semiconductor) and by the way they are pumped (optically and electrically), so their properties and drawbacks are quite different. A recent interest has grown again for the Raman fibre amplifier (RFA) due to the need of more and more optical bandwidth, this type of amplifier uses non-linear scattering mechanism to provide gain.

2.2.8.2. Erbium Doped Fibre Amplifier (EDFA)

In the doped optical fibre amplifier, the gain medium is a length of optical fibre doped with a small amount of rare earth ions. The optical bandwidth of the amplifier is given by the energy levels of the rare earth ions used as dopant. Today, the most promising dopant is erbium which has a radiative transition around 1.55 μm, silica fibres with erbium dopants can be drawn to realise erbium doped fibre amplifiers (EDFA). Praseodymium ions, for instance, are currently investigated to obtain 1.3 μm amplifiers (PDFA).

2.2.8.2.1. Operating Principle

A schematic of an EDFA is shown in figure 6. It has a section of rare earth doped fibre. Pump light is generally produced by semiconductor laser diodes and is coupled in the active fibre with a wavelength division multiplexer (WDM). Optical isolators are used to prevent optical reflections and laser oscillation. Figure 6 shows co-propagative configuration of the amplifier, when the pump light propagates in the opposite direction to the signal, the amplifier is called counter-propagative.

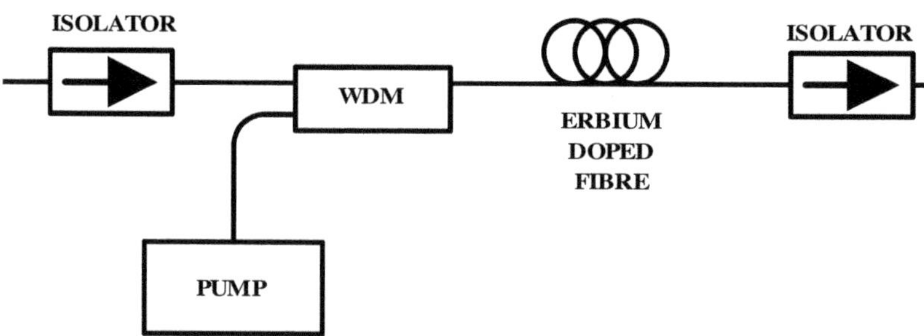

Figure 6 : Schematic of an EDFA.

The physical phenomenon in EDFA is the stimulated emission occurring in a population inversion medium. Population inversion is achieved through continuous optical pumping, which excites erbium ions from the ground level to the pumped level (Figure 7). The ground state ion is transferred rapidly to the pump level. Then, it drops, with fast non-radiative decay to the metastable level. The energy difference between the pump and metastable levels is lost. The lifetime of the metastable level is long (compared to the other levels one), so it acts as a reservoir of excited ions. Input signal photons can use the stored energy to produce other identical photons by stimulated emission.

The appropriate wavelengths of pump are given by the different energy levels of rare earth ions. For erbium, two of them are of practical interest (good efficiency and semiconductor pumping): 1480 nm, which is the same level as the emission one and 980 nm which is the upper one. When pumping at 1480 nm, stimulated emission at the pump wavelength, due to the erbium fluorescence, increases noise of the amplifier (by reducing population inversion) and decreases quantum efficiency (excited ions are used to produce pump photons). Using 980 nm pump wavelength provides ideal population inversion and best noise performance.

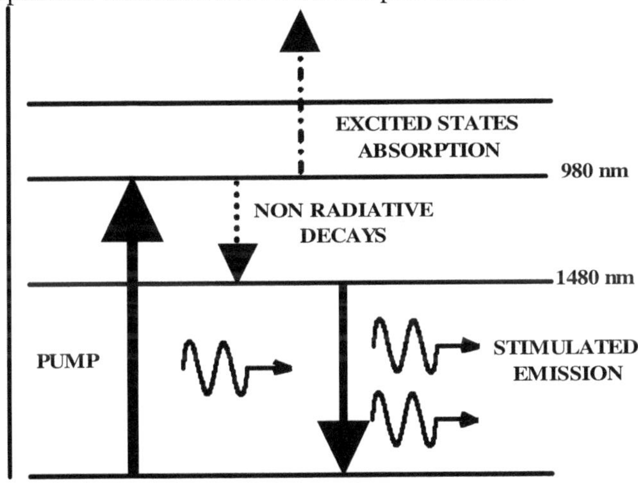

Figure 7: Energy levels of Erbium.

2.2.8.2.2. Basic Features

The EDFA is inherently compatible with optical fibre transmission and the gain is polarisation insensitive.

In term of optical gain, the EDFA works differently, depending on the input signal level. For small input optical power (less than a few μW) the optical gain is high, as the input power increases, the gain decreases, it saturates. The saturation optical power, defined as the output optical power obtained when the optical gain is divided by 2 (-3 dB), depends on the available optical pump power. Typical figures of optical output power are +14 dBm to +18 dBm, with standard single mode pump diodes, and 30 dB to 40 dB for the small signal gain.

One interesting EDFA characteristic is the highly linearity of the amplifier even in the gain compression regime. This is due to the long time constants (around 10 ms for erbium) for excitation and relaxation of the rare-earth ions laser level. This results in amplifier gain, which is slow to respond to changes in the level of the pump or the input signal. The instantaneous amplifier gain is independent of signal format (assuming a frequency modulation greater than a few tens of kHz for erbium) even when the amplifier works deep into saturation, at the maximum output power.

The optical bandwidth of EDFAs is related mainly to the spectral fluorescence of the rare earth ion and to the matrix (silica, fluoride glass) and co-dopants (Al, Ge, ...) used in the core of the fibre. For EDFA, co-doping with alumina allows 50 nm bandwidth to be achieved.

As an active system, EDFAs produce noise. It is related to the spontaneous drop of an excited ion from the metastable level to the fundamental one, producing a photon. Spontaneous emission is then amplified as it propagates in the doped fibre, its bandwidth is identical to the amplifier's one. The total noise is due to

- the quantum noise produced by the optical power of the amplified spontaneous emission (ASE)
- the beat noise between each spectral components of the ASE with itself
- the beat noise between the spectral components of the ASE and the signal.

Usually, the predominant noise source is the third one because, in small input signal regime, optical filtering allows reduction of the spontaneous-spontaneous beat noise and in saturation regime, the level of signal-spontaneous beat noise dominates.

The optical noise figure F_{opt} has been defined as the ratio of input signal-to-noise ratio and output signal-to-noise ratio, the input noise is reduced to the shot noise. For high gain amplifier (G>10), the optical noise figure is given by $F_{opt} = 2.n_{sp}$, where n_{sp} is the population inversion factor. In the ideal case, $n_{sp}=1$, and the optical noise figure is equal to 3 dB. Using the 980 nm pump wavelength, in EDFA, allows to reach the theoretical value of 1 for the population inversion (n_{sp}) and noise figure (= $2.n_{sp}$) of 3 dB have been reached. The practical noise figure is increased by input optical losses, thus giving 5 dB noise figure as typical value.

It must be pointed out that the assumptions made to define F_{opt} are:

- signal-to-noise ratios are defined electrically at the output of a square law detector, consequently, it is not only optical amplifier properties dependent but also electrical receiver,

- input noise is restricted to the shot noise. In general, the optical source noise (laser RIN) is usually the dominating source of noise.

So, the laser RIN has to be taken into account when evaluating the signal-to-noise ratio degradation of the optical link with the optical amplifier.

2.2.8.2.3. Commercially Available EDFAs

Today, EDFAs are commercially available products. The first generation was devoted to digital signals amplification and pumped at 1480 nm. The 980 nm pumped EDFAs came, more recently, to the market. For these amplifiers, semiconductor diodes, emitting at 980 nm are used as pump sources. They are proposed in single and double pumps configuration. They show lower noise figure than their 1480 nm pumped counterparts and provide high output power. They are now widely available (Alcatel, Nortel, Lucent, Pirelli, ...).

In order to get more output optical power, the erbium doped fibre can also be co-doped with ytterbium to increase the optical bandwidth absorption in the pump wavelength region. So, it is possible to use pump sources emitting at 1.06 μm where high power solid-state lasers (Nd:YAG lasers for instance) are available (ATX, Pritel, ...).

To increase the output optical power of EDFAs, the higher optical power (a few watts) emitted by multimode 980nm laser diodes can be used to pump double clad fibres. These fibres have a core, single mode at 1.5μm, doped with Erbium, (where the signal propagates), surrounded by a second core, multimode around 980nm, (where the pump power is coupled), and finally the optical clad. The second core plays also the role of optical clad for the signal wavelength. With this technique even, optical output power from 20 to 40 dBm are obtained within commercially available products.

2.2.8.3. Semiconductor Optical Amplifier (SOA)

2.2.8.3.1. Operating Principle

The semiconductor amplifier is derived from a conventional laser diode: the waveguide is formed in the pn junction on the substrate. Current injection into the waveguide allows excitation of the charge-carrier in the active region. By this way, the population inversion needed to get optical gain is obtained. Facet reflectivity is reduced (less than 10^{-4} reflectivity can be achieved by thin layers deposition) in order to suppress laser action. Then, the chip can be used as an amplifier. Two optical fibres couple light to and from the amplifier chip. They are tapered and lensed in order to improve the coupling efficiency of light into the amplifier waveguide (Figure 8). Provided that the input signal wavelength matches that of the gain medium, it experiences optical gain.

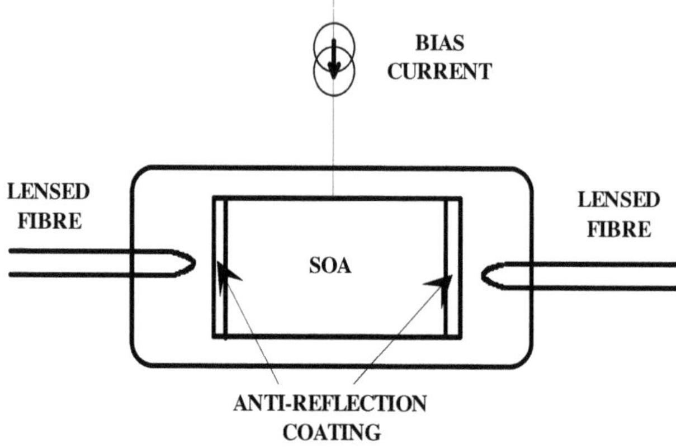

Figure 8 : SOA.

2.2.8.3.2. Basic Features

The SOA can be designed to work at different wavelength, for instance 1.3 or 1.55 μm which are the second and third telecommunication windows, respectively. It has typically a few tens of nanometres optical bandwidth. Because of the wavelength dependent properties of the facets anti-reflection coating, wavelength-dependent gain ripple is experienced in the SOA. Angling the laser waveguide with respects to the facets and multilayer facet coating are proposed to reduced the gain ripple, leading to value less than 0.2 dB.

Optical fibre to fibre gain of more than 30 dB has been obtained. The basic SOA shows light polarisation sensitivity due to the non symmetrical waveguide of the amplifier. It is difficult to make a completely optical equality between the thickness and the width of the active region while maintaining single-mode waveguide conditions. The proposed method is to use strained multiple quantum well (MQW) structure to have the gain coefficient as an extra design parameter. By this way, less than 0.5 dB of polarisation sensitivity between TE and TM modes are obtained.

As for the EDFA, the gain of the SOA decreases when the input optical power increases, and saturates. Due to the short time recovery of the gain of the SOA (minority charge-carrier conduction-band lifetimes of around 1ns), working in the saturation region induces signal distortion for modulation frequency less than a few gigahertz. So, high output saturation power is necessary to increase signal power dynamic range. Saturation output powers greater than +10 dBm are typical values. The gain clamping is also used to improve the linearity range of the SOA.

The noise produced by the SOA is due to amplified spontaneous emission (ASE), via spontaneous-spontaneous beat noise, signal-spontaneous beat noise and shot noise. The noise figure of the amplifier, due to signal-spontaneous beat noise, can be reduced but the input coupling losses which are typically of 3 to 5 dB increase the practical value. Noise figures of SOA are typically 6-8 dB. Reducing input losses is the major challenge in order to lower the noise figure. In practical systems, optical filtering of ASE optical power is required.

Recently, satisfactory SOA devices become commercially available, they are destined to amplify small signals ($P_{in} < -20$ dBm) but are difficult to chained (accumulation of ASE,...). It seems that SOA are much more dedicated to the realisation of photonic integrated circuits due to their compatibility of integration with other active (laser, photodetector, ...) and passive (multiplexer, coupler, ...) optical components. They also have major applications in signal processing with switching capacity, wavelength conversion, ...

2.2.8.4. Raman Fibre Amplifier (RFA)

The physical mechanism providing gain in Raman fibre amplifier (RFA) is a weak non-linear mechanism. A small fraction of the pump light is scattered by the molecular vibrational modes of the silica glass matrix of the fibre. Some of this scattering light is frequency shifted (from the pump wavelength) by an amount equal to the vibrational frequencies of the molecules. This frequency shift peaks around 440cm^{-1}, which gives 100nm for a pump wavelength equal to 1450nm, the gain spectrum will lie around 1550nm with 30nm bandwidth.

Due to the fact that the Raman effect is quite small in silica fibre, lengths of fibre in the order of tens of kilometres and pump powers as high as tens of mW to several W are required. Optical gain from 20 to 30 dB are obtained.

Noise in RFA may have different origins:

- transfer of the amplitude noise of the pumps to the signal, which can be avoided by using counter propagative configuration of pump power and signal,
- double Rayleigh scattering, which increases multi-path interference. It is characterised by measuring the optical noise figure with the electrical measurement. It can be reduced by dividing the amplifier into isolated sections.

- signal-spontaneous beating. The ASE level due to the Raman effect is low and noise figure less than 4-5 dB are achievable.

The scattering process has a response time in the order of femtoseconds. So, the RFA will behave as non-linear device when used in the saturation regime.

RFAs sill require further improvements to compete with EDFAs.

2.2.9. Conclusion

The investigations needed to fulfil the specifications of the telecom market led to the development of a very useful technology basis. However, compared to the fibre transmission of digital signals, one must take care of two aspects:

The microwave systems have specific and much more limiting requirements such as linearity, low noise, high dynamic range, spectral purity after transmission,..., which are not taken into account by the digital world.

For the proper design of an optic and microwave system, one must forget either the digital (see the above paragraph) or microwave rules (for instance, do not confuse the noise figure of microwave amplifier and of EDFA).

3. TELECOMMUNICATION SYSTEMS

3.1. Microwave and Millimetre-Wave Photonics for Telecommunications

D. Wake
BT Laboratories, Martlesham Heath, Ipswich, IP5 3RE, UK
E-mail: dave.wake@bt.com

Abstract

Telecommunications is a major application area for both microwave radio and optical fibre systems. This paper looks at how these very different technologies can be combined to produce a hybrid of the two known as fibre-radio. Fibre-radio brings together the complementary advantages of both types of system – the low loss/high capacity of optical fibre and the wireless capability of microwave radio. This synergy also provides additional benefits relating to the simplification of remote antenna sites for radio access or mobile networks, which will ensure that it has an important role to play in the future.

3.1.1. Introduction

Telecommunications is a huge global industry. Fundamental to this industry are the physical networks and systems that transport the required information (e.g. voice, data) from one location to another. Microwave and photonic systems are both used extensively; photonics mainly for core networks and microwaves mainly for mobile access. Both types of system have their own individual features that explain why they predominate in their respective areas. By combining the strengths of each of these basic technologies we create a hybrid - fibre-radio - where we gain additional advantages resulting from the synergy and interaction between optics and microwaves.

This paper is structured as follows. Section 3.1.2 looks at how, why and where microwave radio systems are used in telecoms networks and section 3.1.3 performs a similar role for optical fibre systems. Section 3.1.4 deals with fibre-radio; what it is, what its benefits are, the components it needs, and how it is used. Section 3.1.5 looks at some examples of recent research activities in fibre-radio and concluding remarks are given in section 6.

3.1.2. Microwave / Millimetre-Wave Systems in Telecoms

Microwave radio systems are used extensively in telecommunications. The most obvious example perhaps is cellular mobile telephony, but microwave radio is used for all types of telecoms network in varying degrees. The main examples are:

- core networks. Radio is used here because it gives a complementary approach to cable, which is especially useful for route diversity protection (i.e. the network is not broken if the cable is damaged). Another important advantage is that no continuous right of way is required to prepare the route. These links are characterised by large masts with high gain, highly directional antennas, with each section typically having a length of 30 km. Systems are mainly digital, typically 155Mbps, and use spectrally-efficient modulation schemes such as 64-QAM. There is a wide range of carrier frequencies used, for example in the UK, BT uses bands at 2, 4, 6, 7, 11, 18, 28 and 38 GHz.
- access networks. Radio has many advantages for access networks, especially where a useable copper network is not already in place, since cabling is extremely expensive. In this situation radio is not only economically attractive but also can be deployed rapidly and can be re-used if the link is cabled at a later stage. An example of radio

access in the UK is the Ionica system, which uses proprietary technology developed by Nortel working in the 3.4 – 3.5 GHz band.
- cordless communications. Applications include residential cordless telephones, office systems (cordless PBX), cordless terminal mobility (an outdoor system with some degree of mobility management) and fixed radio access. The old analogue systems are now being replaced by digital ones, which give reduced levels of interference and noise. An example is DECT (digitally enhanced cordless telecommunications), which is a European standard working between 1880 and 1900MHz. This band is spilt into 10 frequency carriers, each of which can carry 12 simultaneous calls using TDMA (time division multiple access).
- mobile communications. The growth of mobile telephony has been outstanding over the last two decades. Again, the old analogue systems have mostly been superseded by digital systems. A good example is GSM (global system for mobile communications), another European standard which now has over 100 million customers worldwide. GSM operates between 860-960MHz, although variants use bands at 1800 MHz (DCS1800) or 1900 MHz (PCS1900). Like DECT, the bands are divided into a number of carriers, each of which has eight timeslots.
- satellite communications. Satellites are used for mobile communications, where the wide area coverage they afford would be too expensive with a conventional cellular network. They are also used for international backbone links as an alternative to transoceanic cables. Current systems include INTELSAT, EUTELSAT and INMARSAT.

To summarise, microwave radio plays an extremely important role in all manner of telecommunications networks. Although signal propagation using radio is fraught with interference problems, the advantages of having a cordless or mobile connection to the network are worth the huge effort that has gone towards reducing the interference to acceptable limits. Indeed, mobile communications is one of the big growth areas in this industry, both in terms of number of users and capacity per user, as multimedia applications grow in demand. There is currently an intensive international effort aimed at specifying the next generation mobile networks, which will have the capacity to support many of these new multimedia applications.

3.1.3. Optics in Telecommunications

Transmission of information using light pulses along glass fibres has revolutionised telecom networks over the last two decades due to the low loss, low cost, high capacity, interference immune, small size and light weight properties of this medium. Current research is pushing the capacity limits of silica fibre (approximately 20 THz of bandwidth in two low-loss windows at 1.3 μm and 1.55 μm) over spans of hundreds of km. Optical transmission systems are used for all kinds of cabled network, the main types being:

- core networks. Optical fibre is used extensively for core networks. For example, in the UK alone there is over 20 million km of installed fibre in core networks, using transmission systems working typically at bit-rates of 2.5 Gbps with unrepeatered spans of 70 km. Increases in capacity for this installed infrastructure can be realised using multiplexing techniques such as wavelength division multiplex (WDM), in which each channel uses a separate optical wavelength, or optical time division multiplex (OTDM), where the optical pulses are shortened in duration and interleaved in time with others. Experimental OTDM systems have demonstrated bit-rates of 40 Gbps over a distance of 400 km [6] using techniques to overcome dispersion effects in the fibre. WDM has been used to demonstrate a total capacity of one Tbps over a distance of 55km using 50 separate wavelength channels [7].
- access networks. Optical fibre is also being introduced into the access network, firstly for businesses and ultimately to the residential customer. The research effort in optoelectronics for access networks has not been concerned with the high capacity, long span work that has been the dominant focus for core networks, but instead has concentrated on reducing cost. Most of the cost of a fibre-pigtailed laser, for example, is related to packaging the chip. Fibre pigtailing is a time-consuming process requiring skilled operators due to the sub-micron alignment required between fibre and chip. Mode expander technology has been developed to reduce the alignment tolerance so that fibre pigtailing can be done by machine. This technology is part of the chip; the output facet is designed so that the spot size of the light emerging is similar to that required for good coupling into optical fibre. This technique is used to produce optoelectronic modules on silicon motherboards with great ease, and has been demonstrated using lasers [8] and semiconductor optical amplifiers [9]. This approach overturns conventional wisdom that dictates that fibre-pigtailed optoelectronic components are expensive, and means that the deployment of optical access networks is not limited by the cost or performance of the optoelectronic interfaces.

3.1.4. Fibre-Radio Systems

Fibre-radio systems combine the wireless advantages of microwave radio with the low loss, high capacity benefits of optical fibre. Figure 9 shows the arrangement of such a system, which comprises a relatively long optical fibre link (typically 1 – 10km) from the central hub to the radio access point (RAP) and a relatively short microwave radio link (typically 10 – 100m) from the RAP to the customer. A technique known as 'radio over fibre' is used in the optical part to transport signals directly at the microwave carrier frequency.

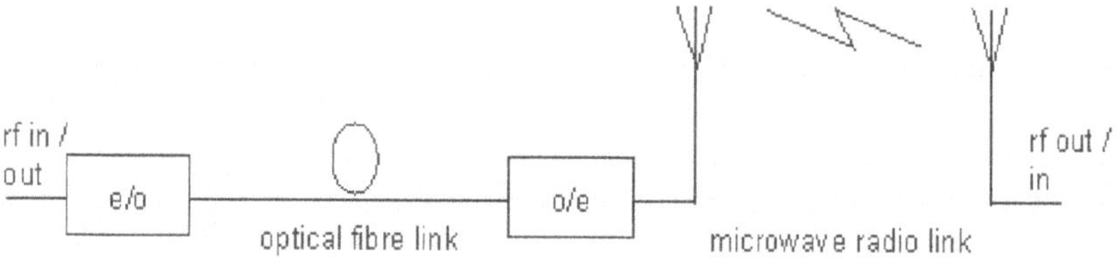

Figure 9. Arrangement of fibre-radio system.

The optical sub-system is therefore analogue in nature, although digital signals are usually being carried. Several radio carriers can be transported simultaneously using a technique known as subcarrier multiplex, which is analogous to frequency division multiplex for radio systems and wavelength division multiplex for optical systems.

Since the signals are transported over the optical link at the radio carrier frequency, functions such as upconversion and signal multiplexing can be done at the central hub rather than the radio access point. The benefits of fibre-radio stem therefore from the simplification of the RAP and the complementary concentration of system complexity at the central hub. This means that the RAP can be low cost, small size, lightweight and reliable, which means it should be easy to install and maintain. Centralisation means that expensive or sensitive equipment can be housed in a benign environment, where costs can be shared, network management can be simplified and resources allocated dynamically.

Applications of fibre-radio include fixed radio access (for example in a fibre to the radio distribution point architecture), cellular networks (to cover dark spots in coverage, microcells and picocells) and in-building networks for mobile/cordless telephony and wireless computing.

The benefits of fibre-radio come at a cost however. Since the optical link is analogue, it is especially prone to interference, noise and distortion. The optoelectronic components (lasers, modulators and photodiodes) must have low noise, be highly linear and have a frequency response sufficient for the radio carrier signals. Chromatic dispersion is a major problem for high frequencies and long spans, and the optical source must have an appropriate spectrum to avoid a power penalty. Standard optical fibre has a dispersion coefficient of 17 ps/nm/km at a wavelength of 1550nm, which causes a time lag between the optical carrier and the modulation sidebands. When the optical spectrum is detected at the end of the link, the resulting beat signals interfere. This causes a 3 dB rf signal degradation for a fibre span of 6 km at 20 GHz or 0.7 km at 60 GHz. For mm-wave systems therefore, the optical spectrum at the end of the link must only give rise to a single beat signal, which means it should only have two main components. Many types of optical source for mm-wave systems have been proposed over the last few years, each having pros and cons in terms of performance and practicality. A few examples are given in section 5.

Much recent effort has been made to produce optoelectronic components with the exacting requirements needed for analogue links. Some good examples are given below:

- low noise, high linearity DFB laser. A 1.3 μm strained MQW device, developed by Ortel Corp. [10], with a RIN (relative intensity noise) of -155 dB/Hz, a spurious-free dynamic range of 108 dBHz $^{2/3}$ and a slope efficiency of 0.65 mW/mA.
- high frequency DBR laser. A 1.55 μm strained MQW device, developed by Royal Institute of Technology, Sweden, [11], with a bandwidth of 30 GHz.
- high frequency PIN photodiode. An InGaAs waveguide photodiode developed by NTT [12], with a bandwidth of 110 GHz and a quantum efficiency of 50%.
- high power PIN photodiode. An InGaAs device developed by Ortel Corp. [13], with a bandwidth of 16 GHz and an optical power limit of 20 mW.
- high frequency Mach-Zehnder modulator. A GaAs/AlGaAs device, developed by GEC-Marconi [14], with a bandwidth of over 50 GHz.
- high frequency electroabsorption modulator. An InGaAsP MQW device, developed by FranceTelecom [15], with a bandwidth of 40 GHz, a drive voltage of 1.9 V and a fibre-to-fibre loss of 11 dB.

Although the devices outlined above are mostly in the research phase of development, there are commercial microwave photonic systems available, designed for picocell or microcell applications for coverage and capacity enhancements where demand is high, such as city streets, airports, railway stations etc. Companies producing these products include Ortel [16], Anacom [17] and Mikom [18].

3.1.5. Recent Research

Current applications of microwave photonics for telecoms are centred around the cellular radio bands at 900, 1800 and 1900 MHz. There is considerable research interest, however, in producing links and components for future broadband wireless systems, which may operate at frequencies up to 70 GHz. Some of these activities are highlighted here:

- FRANS. This was a 4th framework European collaborative project, with the objective of specifying and developing mm-wave fibre-radio field trials for broadband interactive services [19]. It started in 1995, and is due to finish this year. Two different system approaches have been taken for the demonstrators, but each has a downlink operating around 30 GHz. The mm-wave generation scheme used in each case involves driving a Mach-Zehnder modulator in such a way that the optical carrier is suppressed [20]. The frequency separation between the modulation sidebands (the beat frequency generated by the remote photodiode) is at twice the drive frequency, so the component and drive oscillator requirements are considerably relaxed. Achievements to date include a 25 Mbps QPSK transmission experiment at 28 GHz over 12 km standard single mode fibre with a 1:8 split and a 60 GHz field experiment at a data rate of 140 Mbps over 46 km of installed standard single mode fibre.
- electroabsorption modulator (EAM) transceiver. This device consists of a semiconductor optical waveguide inside a pn junction, where the waveguide core is electroabsorptive, i.e. absorption of light in the waveguide can be controlled by a dc bias voltage. Because photocarriers are generated in this process, the device can be used as a photodiode as well as its conventional use as an optical modulator [21]. In other words the EAM can act as a transceiver in a configuration as shown in figure 10. Here the EAM acts as a photodiode for the downlink and as a modulator (of the downstream light) for the uplink. The advantage of this approach is that no light source is needed at the RAP (and therefore no control circuitry), which makes the RAP much simpler. Full duplex bidirectional links can be achieved by using different frequency carriers for each direction , i.e. frequency division duplex. This has been demonstrated in an experiment where 140Mbps QPSK signals were transmitted over 13km using a carrier frequency of 1.3 GHz for the downlink and 140MHz for the uplink without significant degradation [22].

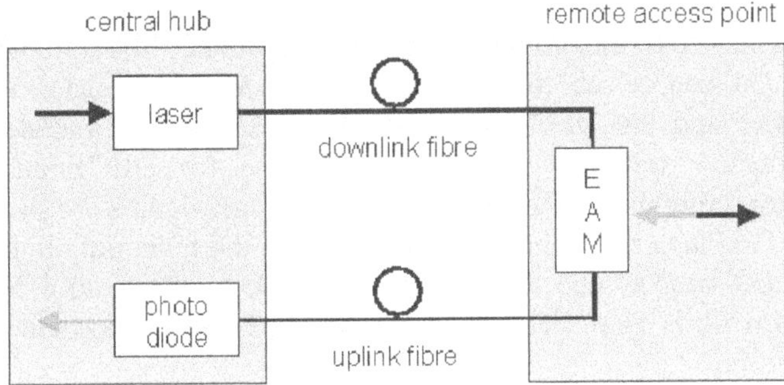

Figure 10. EAM transceiver configuration.

- 'passive picocell'. This concept also uses an EAM as a remote transceiver, but in this case it is not biased [23]. Furthermore there is no remote amplification or processing which means that the RAP needs no power supply. It consists only of a single EAM device and an antenna. This concept takes simplification of the RAP to the limit. Since we are relying solely on the rf signal power generated by the EAM from the downstream light, the range of the radio link is confined to around 10 - 100m depending on the propagation environment, the antenna type and the radio system. This radius is consistent with requirements for picocells – hence the name of the concept. A demonstration system has been set up using radio LAN (2.4 GHz) and DECT (1.9 GHz) in a layout shown in figure 11. The inset photographs show the EAM module (labelled 'passive optical transceiver) and the RAP (labelled 'passive base station').

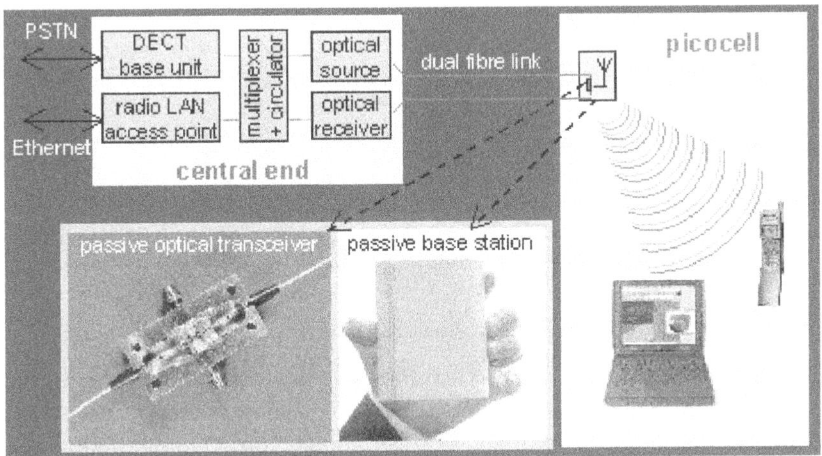

Figure 11 : 'Passive picocell' demonstration system.

At the central hub, DECT and radio LAN signals are multiplexed together and the composite signal modulates the intensity of the optical source. In the picocell, the EAM demodulates the optical carrier and the DECT and radio LAN signals are radiated to their respective terminals (cordless telephone for DECT and wireless laptop computer for radio LAN). The return signals are picked up by the RAP antenna, where they remodulate the remaining light from the optical source, and are transported back to the central hub by the return fibre. Here they are demultiplexed and fed to their respective base units.

- single sideband (SSB) optical source. This is a modulation technique designed to overcome fibre dispersion in long, high frequency links. The original experiment used a dual-electrode MZ modulator driven in such a way as to produce the optical carrier and only one of the modulation sidebands [24]. Since this produces only a single beat component in the photodiode, very little dispersion-induced power penalty is observed. This was demonstrated using an experimental layout shown in figure 12. The MZ modulator was biased at quadrature and the same rf signal was applied to both electrodes, one phase-shifted by /2. Less than 2dB penalty was observed for 2-20 GHz signals over a fibre span of 80 km compared with the conventional double sideband case where deep fades were observed at frequencies of 6.6, 11.8, 15.2 and 17.9 GHz.

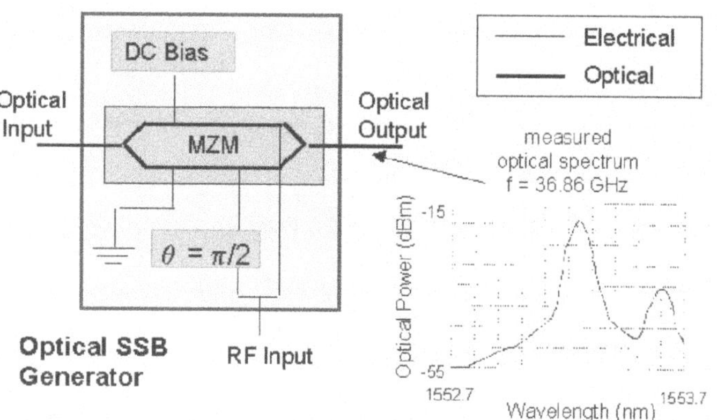

Figure 12 : Experimental arrangement for single sideband optical source.

- fibre grating laser (FGL) heterodyne. Optical heterodyne is an alternative dispersion-tolerant technique for generating mm-wave signals by beating two cw optical signals on a photodiode. The required mm-wave signal is the difference frequency. It can either use two single-mode lasers or one two-moded laser [25]. The usual problem with the two laser approach is that the purity of the beat signal is inadequate unless sophisticated control loops are employed to reduce phase noise. These loops are difficult to engineer if the phase noise, or linewidth, of the lasers is high (which is the case for semiconductor lasers for example). FGLs are single-mode lasers which use a semiconductor gain region and a frequency-selective feedback grating made from optical fibre, as shown in figure 13.

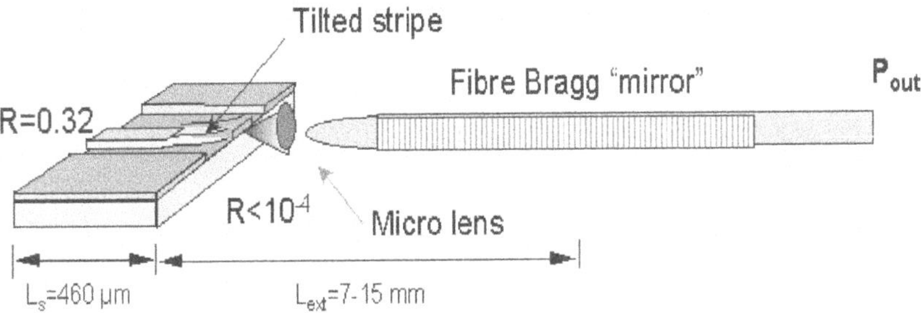

Figure 13 : Fibre grating laser.

The output of this device is ideally suited to optical heterodyne, i.e. stable and and narrow linewidth (< 50 kHz), which means that a simple feedback loop is all that is required to give added stability if required. FGL heterodyne has been demonstrated and shown to be a practical means of producing pure, tuneable and stable mm-wave signals [26].

- 60GHz fibre-radio transmission experiment. A variety of techniques were developed to facilitate this experiment, which was configured as in figure 14:
- master / slave DFB laser technique gave a simple, practical and flexible means of generating 60 GHz signals [27]. Two DFB lasers were arranged in series, each contributing a single mode to the resulting two-moded output, in an optical heterodyne scheme. The second (slave) laser was driven by an electrical oscillator at a subharmonic of the difference frequency, which ensured that the beat signal was extremely pure.
- remote upconversion scheme allowed full transparency to radio signal format without dispersion penalty. Here, rf carrier and data were mixed at the RAP rather than conventionally at the central hub. Both signals were generated using separate laser systems; the master/slave technique for the carrier and a conventional low frequency laser for the data. A single fibre was used to transmit both signals using WDM. Although this seems more complicated, it has the advantages of transparency and flexibility.

- EAM transceiver (described above) removed complexity from the RAP and gave an elegant solution to the return path.

Using these techniques, 120 Mbps QPSK data was transmitted over 13 km of standard optical fibre and several metres of free-space after the RAP without significant penalty [28].

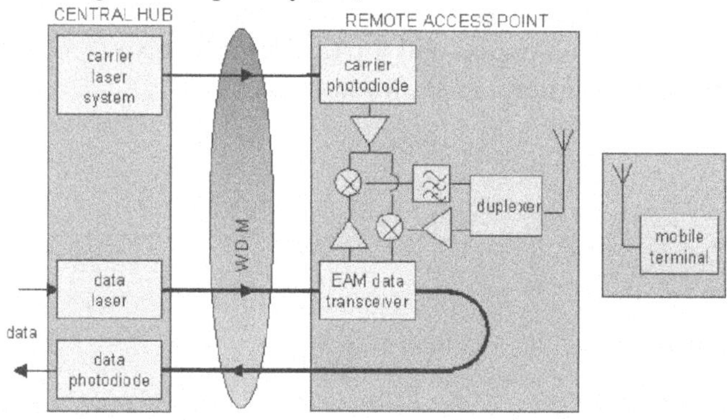

Figure 14 : 60 GHz fibre-radio transmission experiment.

3.1.6. Conclusion

We have seen that the telecommunications industry is a major user of microwave radio systems for a wide variety of applications, especially mobile communications. It is also a major user of optical systems, witnessed by the millions of km of installed fibre around the world. Combining the two, in a hybrid known as fibre-radio, not only gives the advantages of each transmission medium (low loss and high capacity from optical fibre and cable-free connectivity from microwave radio) but also allows considerable simplification of remote antenna sites. It is this synergy that will ensure that microwave photonics will make a big impact in future telecommunications networks.

3.1.7. Acknowledgements

The author would like to acknowledge the contributions made to this work by many colleagues at BT Laboratories, especially Dave Moodie, Laurent Noel, Derek Nesset and Dominique Marcenac. Thanks are also due to Graham Smith of the University of Melbourne and Steve Bennett of University College London for contributing figures 12 and 13 respectively.

3.2. Fibre Supported MM-Wave Systems

P. Lane
Department of Electronic and Electrical Engineering, University College London, Torrington Place, London, WC1E 7JE, UNITED KINGDOM
Phone : +44 171 419 3945 – Fax : +44 171 388 9325
e-mail : plane@eleceng.ucl.ac.uk

Abstract

This paper outlines the role and diversity of optical methods that can be used to generate and distribute mm-wave signals. How these methods can be used in a systems context is then described and finally a discussion of the way forwards in this rapidly evolving field is given.

Introduction

There is much current interest in the use of fibre optic technology to support the generation and distribution of mm-wave signals. Most of the applications envisaged for these systems are communications oriented, although there are a few other application areas of interest. Following this introductory section, the potential applications for fibre supported mm-wave systems (FSMS) will be outlined, and the rational behind the choice to work at demanding mm-wave frequencies will be explained. In section 3.2.3 a reference architecture for a FSMS will be given in order to clarify the notation and nomenclature that will be used in the remainder of the paper and section 3.2.4 will review optical mm-wave generation methods and discuss the performance issues associated with these different approaches. Section 3.2.5 will outline some of the issues that arise when these optical generation methods are taken forwards to real system deployments and future avenues of research will be explored. The paper will finish with a conclusion that looks ahead to the possible future of these systems.

3.2.1. Applications for Fibre Supported Mm-Wave Systems

Most of the potential application areas for FSMS are in the communications area. Many system designs and concepts are proposing to use mm-wave radio technology. These systems include wireless local loops (WLLs), mobile broadband systems (MBS), traffic management systems, and wireless local area networks (WLANs). The common themes running through all of these systems is that they use a radio interface, and they are broadband. Irrespective of the efficiency of the modulation scheme adopted in terms of what data rate it can place in a given spectral allocation, broadband systems will need large spectral allocations in order to accommodate the elevated data rates being considered. Due to the pressure that is on radio spectrum, very large allocations of bandwidth are only available at mm-wave frequencies. This is the factor that has fuelled the interest in mm-wave radio systems.

LMDS operating at 28 GHz provides an example of a deployed mm-wave WLL system. The system provides a broadband forward link to the customer, together with a lower rate return channel. LMDS targets internet access and interactive multimedia (including TV) as its main markets.

MBS is seen as the next step after UMTS/IMT-2000 and it is sometimes referred to as 4th generation mobile. The aim is to provide fully mobile access at rates approaching B-ISDN, i.e. 155 Mbit/s. Research and development has already started on these systems, and spectral allocations in the region of 40 and 60 GHz are being made.

As road traffic volumes continue to grow, there is interest in the use of communications and radar technology to reduce congestion. Proposed systems include radar to identify traffic conditions coupled with high-speed data links to vehicles so that alternative route information and maps can be downloaded to vehicle mounted display units.

Wireld LANs curently offer speeds of up to 1 Gbit/s while radio alternatives, such as HIPERLAN or IEEE802.11, can only offer a few Mbit/s. Work is underway to develop WLANS operating at mm-wave frequencies that will offer rates comparable with wired LANs.

This very wide range of applications clearly shows why there is considerable research activity in mm-wave systems, and the low-cost and flexibility that fibre based systems *could* offer is driving a major international effort in the area of FSMS.

3.2.2. A Reference Architecture

The topology of a typical fibre supported radio system is shown in figure 15. The data/optical interface needs to generate an optical signal that can be detected and processed to yield the required mm-wave signal. The optical network distributes this optical signal to a number of antenna units (AU) where the optical to RF conversion occurs. The AU radiates the RF signal and the mobile equipment (ME) receives the signal. The optical network may make use of optical amplifiers to support a high split ratio, and may also use wavelength division multiplexing (WDM) in conjunction with wavelength routing devices to allow different signals to be radiated from different antenna sites.

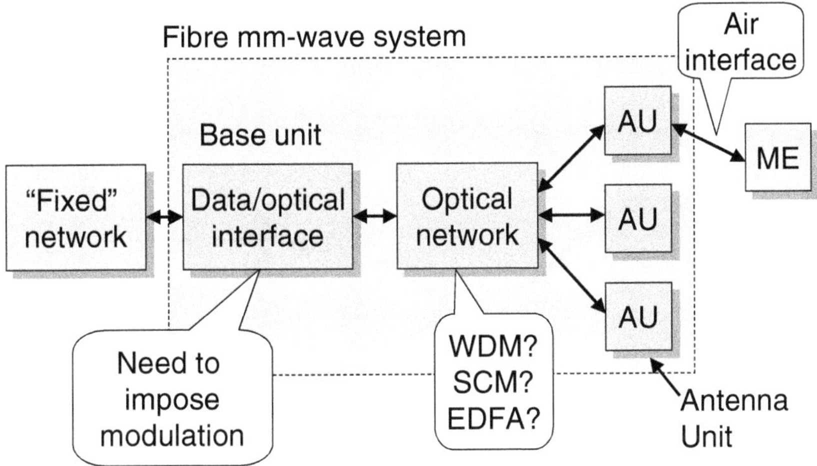

Figure 15 : Reference architecture for a fibre supported mm-wave radio system.

Based on the applications identified and discussed above, it is possible to identify a set of target requirements for a FSMS:

- The system must be able to deliver mm-wave signal to a remote antenna unit.
- The system must be able to modulate the mm-wave signal.
- The system should be transparent to the modulation format.
- The system should be able to operate over an optical network.
- The system must be able to provide capacity into area on an as needed basis.

From these requirements, it can be seen that a number of generation methods are appropriate. These will be discussed in the next section.

3.2.3. Generation Methods and Performance

Optical mm-wave generation methods can be divided into 2 broad classes :

- 3-term techniques corresponding to conventional amplitude modulation of the optical signal at the mm-wave frequency required
- 2-term techniques where the two optical components mix (heterodyne) on the photodetector to generate an electrical signal at a frequency equal to the separation of the two optical components
- The relationship between these two broad classes and how the methods that will be described in this section fit into these classes can be seen in the roadmap of generation methods shown in figure 16.

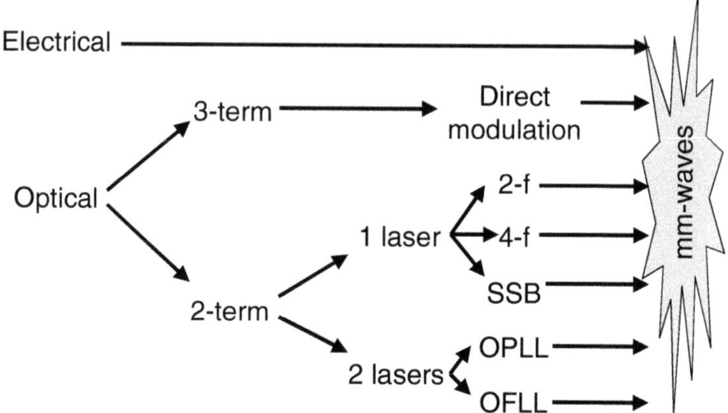

Figure 16 : Roadmap of mm-optical wave generation methods.

3.2.3.1. 3-Term Techniques

The 3-term approaches are very simple in concept since all that is needed is an amplitude modulator that can operate at the mm-wave frequency. Commercial devices are available now that will operate at around 40 GHz so the technique is suited to the low-end of the mm-wave spectrum. The method is transparent to modulation format; whatever modulation is present on the drive signal applied to the modulator will be present on the electrical signal generated when the optical signal is detected on a photodiode (within the constraints imposed by the linearity of the modulator). This technique is also attractive due the high electrical power that can be generated if a high modulation index is used. However, this method also has some serious limitations. One is that component development is needed for the higher end of the mm-wave spectrum of interest. Electroabsorption modulators (EAMs) are available with a good response to around 60 GHz[1]. Operation in higher bands around 70 and 90 GHz would require these devices to be developed further. The major limitation of this approach though is the impact that fibre dispersion has on the generated signal. The phase change experienced by the 3 different components due to fibre dispersion can be viewed as a rotation of the three phasors representing the signals. This rotation leads to a cyclic variation of generated power with fibre distance or frequency. At the frequencies of interest here, this effect limits the usefulness of 3-term techniques to fibre reaches of only a few km.

[1] Note that these are narrow band devices with a bandwidth of around 2 GHz at 60 GHz.

3.2.3.2. 2-Term Techniques

The 2-term techniques rely on the E-field non-linearity of the photodetector. The detector acts as a mixer that generates an electrical signal at a frequency equal to the separation of the two optical components. The advantage of this method is that the dispersion induced rotation of the phasors representing the two optical components only leads to a phase (not amplitude as in the 3-term case) change in the generated mm-wave signal. Since the phase origin can be arbitrarily defined, this is not an issue of any concern. This technique also allows the use of components that do not have to operate at the mm-wave frequency. This can lead to considerable cost savings. On the other hand, it is not obvious how to impose modulation onto the signal.

In summary, 3-term techniques are not appropriate for mm-wave signal generation mainly due to the severe impact of dispersion. 2-term techniques are therefore the favoured methods. 2-term techniques can be subdivided into those that use two lasers to generate the two optical components and those that use a single laser to generate the two optical components.

3.2.3.3. Specific 2-Term Generation Methods

3.2.3.3.1. Two Laser Techniques

The simplest 2-laser method is the optical frequency locked loop (OFLL) shown in figure 17.

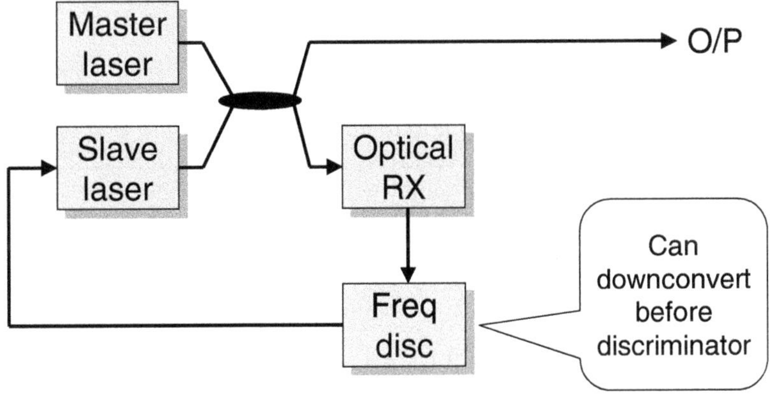

Figure 17 : Optical frequency locked loop.

Here, a frequency discriminator is used to generate an error signal if the frequency separation of the lasers is not as required. This error signal is used to tune the slave laser to restore the required frequency difference. The OFLL only maintains the correct average frequency separation. Short-term variations due to the phase noise of the lasers are ignored. The electrical signal generated therefore has a linewidth of twice the linewidth of the lasers. Generation of a high spectral purity mm-wave signal therefore requires very narrow linewidth lasers.

A variant on this method is the optical phase locked loop (OPLL) shown in figure 18.

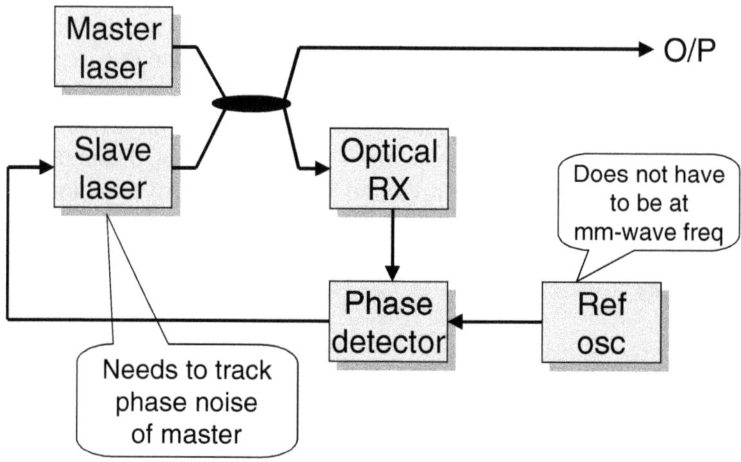

Figure 18 : Optical phase locked loop.

Here, instead of a frequency discriminator, a phase detector is used to generate an error signal depending on the phase error between the lasers. This method can generate very narrow electrical linewidths since the slave tracks the phase noise of the master laser. Complex lasers are required though due to the high-speed phase tuning.

In summary, OFLLs are relatively simple to implement but need narrow linewidth lasers for good performance, while OPLLs offer excellent performance at the cost of complex lasers.

3.2.3.3.2. Single Laser Techniques

Single laser methods rely on the generation of two optical components from the single laser through modulation. One technique, the 2-f method, uses a Mach-Zehnder modulator biased at minimum transmission. Driving the modulator with a sinusoid around this point generates a DSB-SC form signal where two optical components separated by twice the drive frequency are generated. This method is shown in figure 19.

A variant of this method is the 4-f method where the modulator is biased at maximum transmission and the drive level is adjusted to suppress the first order terms. This approach yields two optical components separated by 4 times the drive frequency.

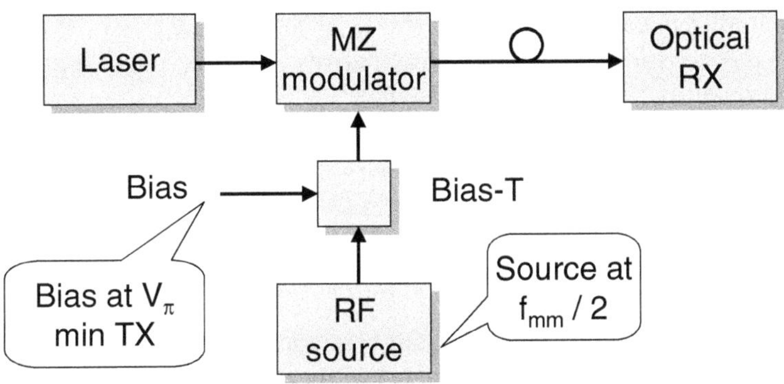

Figure 19 : 2-f generation method.

Another single laser technique is optical SSB where the original spectral line from the laser is retained and a single modulation sideband is generated on one side of this component. This can be achieved as shown in figure 20. A dual drive Mach-Zehnder modulator is biased at quadrature and each arm is driven by two quadrature signals at the required mm-wave frequency.

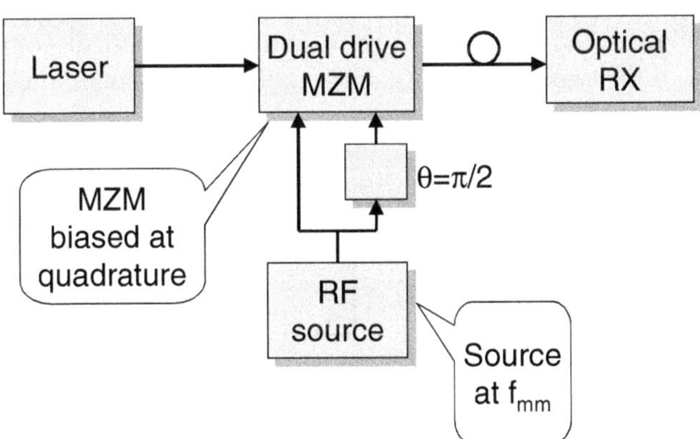

Figure 20 : Optical single sideband generation method.

Each of these techniques has advantages and disadvantages.
2-f method allows
- many components to operate at half the required frequency;

- the electrical linewidth is very narrow since the phase noise on each component is completely correlated
- the electrical power generated is low
- imposition of modulation is an issue.

4-f method allows many components to operate at a quarter of the required frequency

- the electrical linewidth is very narrow since the phase noise on each component is completely correlated
- the electrical power generated is very low
- imposition of modulation is an issue

optical SSB

-components are more complex and need to operate at the mm-wave frequency

- the electrical linewidth is again very narrow
- conversion efficiency is better than the 2-f and 4-f methods

The final choice of method is not straightforward. The 2-f method is simpler to realise, but the optical SSB offers more mm-wave power for a given drive level.

3.2.4. System Issues

There are two main issues associated with the use of 2-term generation issues :

- how to impose modulation
- how to provide a return link

Modulation can be imposed on to either both optical components or onto one of the two components. These options are shown in figure 21.

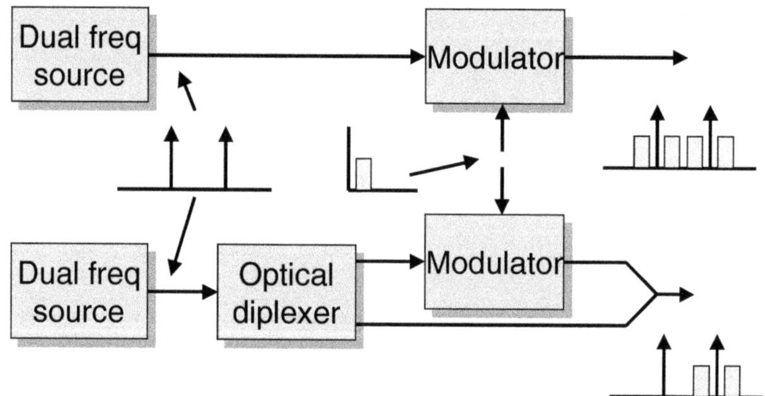

Figure 21 : Imposition of modulation onto a dual frequency optical source.

The first of these is simplest to implement, since no optical filter is required, but the second does yield better tolerance to dispersion. A problem with the second approach is the need to have the optical diplexer track the frequencies of the lasers.

A return link can be implemented by using the mm-wave forward channel to generate a mm-wave local oscillator at the antenna unit. This local oscillator can be used to downconvert the received mm-wave signal to a microwave IF that can be used to directly modulate a laser. This is shown in figure 22.

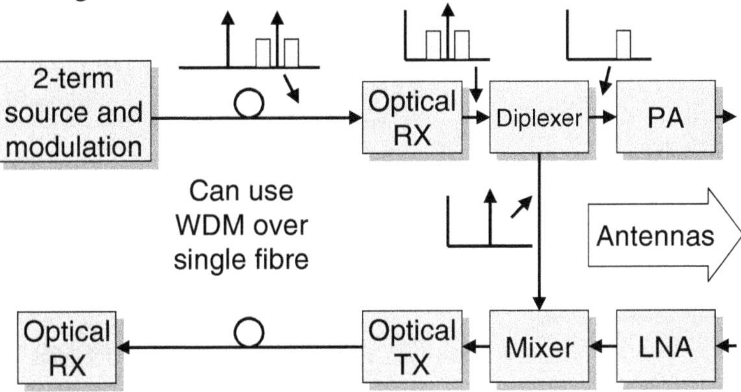

Figure 22 : Provision of a return channel.

Alternatively, an unmodulated optical signal can be sent from the base unit to the antenna unit, and this can be modulated by the IF signal using a low-cost microwave bandwidth modulator. These methods rely on the observation that much data traffic is asymmetric – the return channel data rates are much lower than the forward channel and therefore a microwave frequency is suited to the transport of the return channel signal.

3.2.5. Deployment of Fibre Supported Mm-Wave Radio Systems

There is much work ongoing in research laboratories in the fibre supported mm-wave area. Most of this work seems to be directed at further developments of generation methods. There is some work that is looking at systems deployment. Examples include the RACE II and ACTS funded projects MODAL and FRANS; the work of the Photonics Research Laboratory in Melbourne, Australia; work at BT Laboratories, UK; and work at the Communications Research laboratory, Japan. Research is also being carried out in the devices area. An attempt can be made to rank the importance of the research that is ongoing in this area by first identifying the issues that are impeding the deployment of FSMS. These include:

- the poor economic case for FSMS – mm-wave optoelectronic devices are expensive whereas MMIC based electrical technology is rapidly falling in cost and gaining in maturity, driven largely by the requirement for collision avoidance radar and autonomous cruise control in vehicles
- lack of a clear deployment scenario that would allow the flexibility offered by optical distribution to be realised.

Given the above, the areas that need to be addressed if FSMS are to be widely deployed are shown in figure 23.

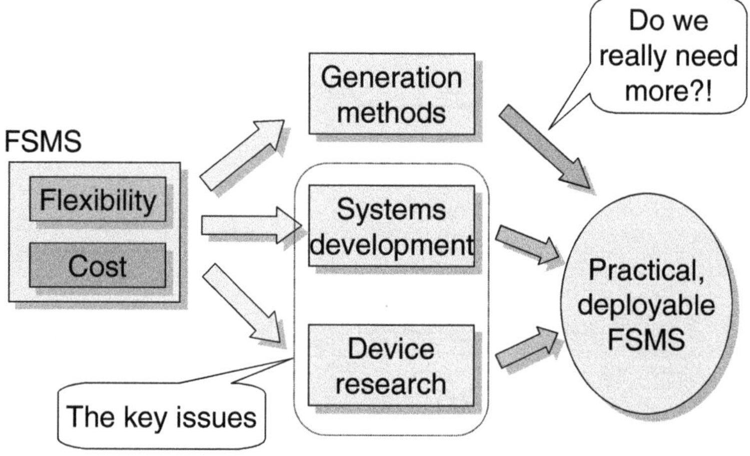

Figure 23 : FSMS research areas and priorities.

FSMS have the potential to offer huge flexibility in terms of providing capacity on an as needed basis. On the other hand, their high cost at the moment makes their deployment uneconomical. Two areas can therefore be identified where research efforts should be concentrated:

device research, especially the development of low-cost packaging methods, to drive down the cost of systems

systems deployment to identify architectures and topologies best suited to achieving the potential flexibility.

Since there are already many generation methods that all achieve roughly the same ends, it can be argued that more development in this direction is not a priority at the moment.

3.2.6. Summary

This paper has outlined a range of options for the remote delivery and generation of modulated mm-wave signals through the use of optical techniques. The optical generation of mm-waves is very attractive due to the flexibility that can be achieved by adopting this approach.

A simple classification of generation methods was given and how the various techniques being developed today fit into this classification was outlined.

Methods for the imposition of modulation onto the mm-wave signal, together with techniques to allow the provision of a return link were described.

Finally, suggestions were made as to where future research should focus were made. It is suggested that research should focus on device development to improve performance, but more importantly reduce costs, and on developing deployment concepts that will allow the flexibility potentially offered by these systems to be realised.

3.3. OPTICS AND MICROWAVES IN TELECOMMUNICATIONS NETWORKS TODAY AND IN THE FUTURE

M. Joindot
France Telecom R&D, Technopole Anticipa, 2 Avenue Pierre Marzin, F 22307 Lannion, FRANCE
michel.joindot@francetelecom.fr

3.3.1. A Brief History of Transmission Techniques in Telecommunications Networks

3.3.1.1. Before the Second World War : Neither Microwaves nor Optics

Up to the Second World War, long distance terrestrial telephone networks mainly used copper cables and baseband transmission (by multiplexing some 4 kHz wide telephone channels). Radio in UHF band was the only communication technique for intercontinental links and provided a very limited capacity (as an example some tens of voice channels between France and the United States in the thirties). Transatlantic cables existed (the first had been laid in the second half of the nineteenth century), but only for telegraph transmission : the available technology did not allow undersea amplification, which is needed for voice channels. On the contrary very low bandwidth 50 Bauds telegraphic data streams could be transmitted coast to coast without any in line reamplification.

Microwave research began before the War, and the first radio relay transmission was carried out over the Channel in 1936 : it must be noted it was digital transmission, the modulating signal being a telegraphic data stream.

Microwave technology was pushed by military needs (for instance radar) between 1940 and 1945 and the results found rapidly applications in civilian telecommunication networks after the end of the conflict.

3.3.1.2. The After War Analogue Era

New transmission systems appeared then. Coaxial cables were installed to connect the main cities, radio relay systems using FM modulation and carrier frequencies of some GHz were developed. Capacities up to 1800 voice channels on one radio system could be attained.

As far as international networks are considered, the first undersea coaxial cable TAT1 (having a capacity of 60 voice channels) could be laid in 1955 : amplifiers reliability was high enough to allow them to be immerged each two km on 6000 km distance. Other systems followed, TAT2 in 1960 and TAT 3 (138 channels) in 1963. The last analogue system reached a capacity of around 4000 voice channels, certainly very small compared with the optical systems of today, but representing also a complete revolution if compared with the some tens of HF circuits of the pre war era.

Capacity between Europe and North America was then dramatically increased. Another increase occurred in 1962 with the introduction of the first telecommunication satellite, whose bandwidth allowed not only voice but also TV transmission (the first one occurred between Pleumeur Bodou France and Andover Maine US, through Telstar satellite, in the night of July 10th 1962). A few years later, geostationary satellites provided permanent and constant quality transmission between earth stations. They used carrier frequency of some GHz, FM modulation and Frequency Division Multiple Access, large diameter earth antennas (to achieve a sufficiently high gain) and on board re -amplification in travelling wave tubes.

To illustrate the technical evolution, let us just survey the list of satellites of the international organisation Intelsat. Intelsat 1 launched in 1965 offered a capacity of 240 voice channels and one TV channel. Intelsat III, in 1968, 1500 voice and 2 TV channels over 450 MHz bandwidth. With Intelsat VA, launched in 1984, 15000 voice and 2 TV channels were available. The following satellites moved to digital technology.

3.3.1.3. Digital Technology Arrives

A very important event in telecommunications history is the digital revolution, based on Shannon's results. Any band limited signal (and then practically any telecommunication signal) can be recovered from its samples, provided the sampling frequency exceeds twice the maximum frequency of the spectrum. By sampling followed by coding, any signal can be transmitted as a data stream. It is periodically regenerated in repeaters along the transmission line in order to compensate for distortions and attenuation. These repeaters detect the signal and send on the next section a data stream only corrupted by the errors in the detection process.

The advantages of digital transmission compared to the analogue one are :

- Transmission quality becomes independent of the distance : it is determined by the bit error rate (BER) and it can be very easily shown that the BER at the end of the line is the sum of the BER of the different regeneration spans. As there is a sort of threshold in the relation between quality and BER (for instance for voice, quality is perfect if BER does not exceed 10^{-5}), quality does not depend on distance as far as the resulting BER is under the threshold.
- A digital transmission system can accept any type of signal (voice, TV, data, images...), because they are all converted into similar data streams (only the bitrate is different).

The first digital systems were introduced into the local networks, on copper wires, with a capacity of 2 Mbit/s (30 telephone channels) and 8 Mbit/s (120 telephone channels). The voice signal is sampled at 8 kHz and coded with 8 bits per sample, which results into a data stream at 64 kbit/s.

3.3.1.4. Digital Radio Relay Systems and Satellites

The first investigations about high capacity digital radio relay systems to be used in national trunk networks started around 1975 : between 1975 and 1990 research was very active in this domain on the following topics :

- High level modulation schemes in order to achieve a larger spectral efficiency (Binary and Quaternary Phase Shift Keying, 16 QAM, 32 QAM, 64 QAM...) ;
- Equalisation and more generally signal processing techniques : due to the bandwidth which is much larger than for analogue FM radio systems, selective fadings have to be considered, and must be compensated for in order to cope with the outage time requirements. Baseband transversal equalisers, which had been extensively studied by the data modems manufacturers in the sixties, provided the useful tool. Application to digital radio, at much higher data rates and on a

significantly different channel (in terms of stationarity), was a very active research domain from 1975 up to 1990.

Carrier frequencies were comprised between 2 and 15 GHz and total capacity could reach 8×140 Mbit/s (eight bi-directional radio channels with a bitrate of 140 Mbit/s). Regenerators were located in the radiotowers which could be seen in the country, and the maximum distance between them was roughly 50 km in the French network.

Figure 24 :The radio tower in Rennes.

The satellite systems became also digital, using digital modulation schemes with Time Division Multiple Access (TDMA). Several generations of intercontinental INTELSAT satellites with were launched to provide an always increasing capacity between Europe and United States. The eastern block had also its own organisation.

National satellites Telecom 1 and then 2 were developed in France in the eighties in order to provide interconnections with overseas territories. Progress in low noise receiver made earth stations with small diameter antenna (some meters) possible. The satellite did not only interconnect earth stations with 30 m diameter antennas : it could also be directly received by business users. This possibility was largely used to provide digital connection between points for which digital transmission through the (non completely digitalized) terrestrial network was not possible.

As an example Telecom 1, launched in 1984, used the bandwidths 6/4 GHz for fixed communications with overseas, and 14/12 GHz for business communications.

3.3.1.5. The Millimetre Waveguide

Propagation of electromagnetic waves through a guide had been theoretically demonstrated at the end of the 19th century, but due to the lack of microwave circuits and components at that time, no practical application was possible. Research about millimetre waveguide was initiated at Bell Laboratories just before the War, when the first sources and detectors became available.

After the War, an important research effort was conducted in many countries (Australia, France, Germany, Italy, Japan, United Kingdom, United States), in order to achieve high capacity and low loss transmission. The guide diameter was 50 or 60 mm, two techniques (helicoidal and dielectric coated waveguide) were investigated according to the countries, the propagation mode used was TE_{01}, and the usable bandwidth extended in the best case from 30 GHz to around 110 GHz. The used bandwidth in the French system extended from 31 to 60 GHz (21 channels separated by 550 MHz) and each carrier was PSK modulated by a data stream at 560 Mbit/s. Repeaters were located every 17 km along the line. The total capacity of the fully equipped system was approximately 100000 voice channels.

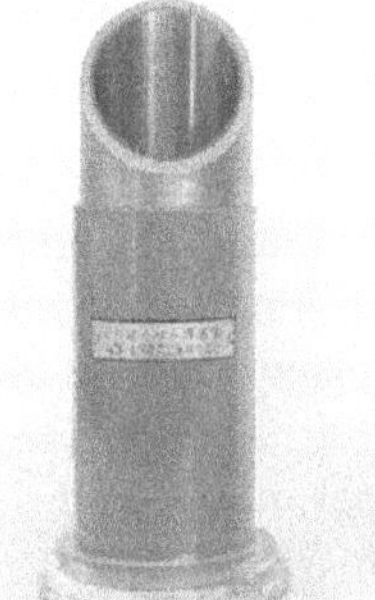

Figure 25 : An example of millimetre waveguide.

Millimetre waveguide should allow the increase of traffic by picture phone, whose development was expected. In fact it did not occur, and it appeared that optical fibbers could probably be in the future a more efficient solution.

Research on millimetre was then stopped in all the countries around 1975, although everything was ready for an industrial development of the systems. Nevertheless, a lot of results in microwave technology had been acquired, and they could be usefully reused for the aforementioned radio relay systems.

3.3.1.6. In Conclusion

In summary, at the end of the eighties, terrestrial long distance networks used coaxial and radio relay systems and the objective was to share the traffic equally between these two transmission media and secure one by the other. If a cable was cut, traffic could be re-routed onto radio during the time needed to repair the damaged link, and vice versa if a problem occurred on a radio system.

The same situation was encountered on international links. Over the oceans, satellite and undersea coaxial cables offered comparable capacities and the possibility of mutual securisation.

3.3.2. Optical Transmission

Using light to transmit information was an old idea, and very high capacities could be expected, due to the very high carrier frequency. But more than thirty years of research were needed before the first practically usable system appeared.

3.3.2.1. Basic Research Paved the Way

The history of modern lightwave communications systems began at the end of the fifties, with lasers, which were the first optical source, with high power and directivity, able to be used in a communication system. Gas lasers, needing high supply voltages and relatively large sized, were followed in 1965 by semiconductor lasers, which appeared much more promising, because of their small dimensions and the possibility of modulating them with low currents. More than ten years of research were necessary to achieve reliable enough semiconductor lasers working continuously at ambient temperature.

Free space optical transmission did not appear as promising because, contrary to what happens in radio, attenuation due to hydrometeors leads to outage times incompatible with the requirements of a telecommunication system. Nevertheless, some equipments are proposed today by manufacturers, for short-range transmission (for interconnection of buildings in cities) . Another possible application of free space optical transmission is the interconnection between satellites, for instance between a moving and a geostationary satellite which allows the first one to remain linked with the earth station, wherever it is around the earth. Nevertheless, telecommunication networks use practically only guided optical transmission.

The first research about optical fibres began around 1966, and at the beginning of the seventies, many laboratories were very active in this field. Constant progress was realised, allowing to achieve attenuations of 20 dB/km in 1975 and 0,2 dB/km in 1984.

3.3.2.2. The First Optical Transmission Systems

This last value opened the way to the implementation of practical systems competing with the other already existing, especially those using coaxial cables, in terms of repeater spacing as well as capacity. The first fibres to be used were multimode fibres, having a relatively large diameter and supporting then the propagation of many electromagnetic modes. Single mode fibres, i.e dielectric waveguides supporting only one mode, replaced later multimode fibres, allowing much higher bandwidths. Three so-called "transmission windows" are used in telecommunications The first one, around 0.8 μm, was historically the first to be used, because the first semiconductor lasers operated at this wavelength, but is no more employed, at least for high capacity transmission systems. Two other windows are yet in use now, the first around 1.3 μm, where the chromatic dispersion is minimum, the second around 1.55 μm, where the attenuation is minimum.

The first generation of lightwave systems, using the 0.8 μm window, began to be operated at the end of the seventies : they used multimode fibres and could transmit typically 50 Mbit/s, with a repeater spacing of roughly 10 km.

Between 1980 and 1990, lightwave communication systems at 1.3 and 1.55μm over singlemode fibres carrying hundreds of Mbit/s were introduced into the trunk network, with a repeater spacing up to 50 km, compared to 2 km for coaxial cables with the same bitrate. It became then evident that optical fibres could compete successfully not only with metallic cables, but also with radio relay systems, which could not provide such a high capacity. At the beginning of the nineties, digital optical systems at 1,55 μm, with a capacity of 2,5 Gbit/s and a repeater spacing of 100 km were available : it became then evident that optics outperformed radio in terms of capacity, repeating span, and transmission quality (because of the absence of outage due to propagation phenomena). These high capacity systems were then widely installed by the operators with the objective of completely eliminating the other transmission techniques in their backbone networks.

As far as undersea communication systems are concerned, the first optical one (TAT 8) was deployed in 1988, with a bitrate of 280 Mbit/s per fibre pair : repeater spacing reached 42 km. It was followed in 1991 by TAT 9, (560 Mbit/s per fibre pair) and later TAT11, with the same capacity.

3.3.2.3. Coherent Technique

Any history of optical transmission must mention coherent techniques. As it will be explained later, the sensitivity of optical receivers is determined by the thermal noise of electronic devices located behind the photodetector ; following exactly the same way as in radio at the beginning of the century, the idea was then to replace direct detection by heterodyne reception : received signal beats in the photodiode with a local oscillator, and an intermediate frequency (IF) signal carrying the modulation of the optical carrier is obtained at the IF output of the photodetector, acting as a mixer. This technique allows effectively to increase the receiver sensitivity and any modulation scheme can be used, for instance phase modulation which is not compatible with direct detection. A lot of research was devoted to coherent techniques between 1980 and 1990, especially to solve difficult problems like phase recovery or polarisation maintaining receivers. The activity in this domain decreased very rapidly around 1990, when the optical fibre amplifiers allowed to achieve the same sensitivity with less complexity.

Coherent receivers could nevertheless remain good candidates for free space intersatellites communications.

3.3.2.4. Optical Amplification

Amplification is a key function for telecommunication and it is well known that the invention of triode at the beginning of the 20th century brought very important changes, because it allowed amplification of signals which had not been possible before. During years and years, researchers worked on the crucial question of light amplification.

The first optical amplifiers to be studied were semiconductor amplifiers (SOA), which use the same physical phenomena as lasers : pumping in a material through electrical carriers injection (i.e through an electrical current providing the external energy) causes a population inversion. Electrons on the upper overoccupied level fall down onto the fundamental energy level again and emit correlatively photons at a wavelength corresponding to the energy difference between the levels. Most of them add in phase with those of the incident light and contribute to its amplification : this is the stimulated emission process. But other photons are emitted incoherently with the incident light : this is the spontaneous emission. These photons travel themselves through the amplifier, are amplified and constitute at the amplifier output the amplified spontaneous emission (ASE) noise, i.e the noise generated in the amplification process.

The first theoretical works were published at the beginning of the sixties and application to optical communication systems were proposed in the seventies. This is only in the beginning of the eighties that progress about semiconductor lasers allowed to consider SOA's as practically implementable devices and a lot of investigations was devoted to them.

In optical fibre amplifiers the active medium is a piece of rare earth (commonly erbium) doped fibre pumped by one (or eventually two) laser(s) diode emitting at a wavelength of 980 or 1480 nm. The amplification bandwidth of the first devices was around 12 nm, i.e 1500 GHz (and even around 24 nm for fluoride doped amplifiers). The output power of these erbium doped fibre amplifiers (EDFA) can be high (up to 20 dBm), which allows to increase the transmission length, with nevertheless limits due to the counterpart of non linear. Sometimes, remote pumping amplifiers are used : this is done in some undersea lightwave systems, where the active fiber itself is immersed, while pump, which is the most critical component in terms of reliability, is placed at the end of the link and feeds the amplifier itself through the fiber.

The first publications about EDFA appeared in 1987 and practically usable devices were available less than four years later. Compared to semiconductor amplifiers, fiber amplifiers are easier to implement in practical systems, and present the advantage to be polarisation insensitive : they can be used as power emitting amplifiers (boosters), preamplifiers at the receiving end, or in line amplifiers. In this last case they can replace electronic regenerators, as for instance in the last generation of transoceanic undersea lightwave systems. Nevertheless, the counterpart is noise, linear and non-linear distortions.

It is important to remark that, although they do not appear as the most promising candidate for amplification in optical communication systems, SOA's exhibit very interesting non linear properties, which make them key devices for optical signal processing (reshaping, sampling...) which will be more and more used in future optical networks.

The first amplified undersea system, TAT 12/13 was laid in 1995 : it has a capacity of 5 Gbit/s per fibre pair, with an amplification span of 45 km.

3.3.2.5. Wavelength Division Multiplexing (WDM)

EDFA's opened the way to Wavelength Division Multiplexing (WDM) : in order to utilise efficiently the large amplification bandwidth of amplifiers (35 nm, i.e 4000 GHz in the first amplifiers, and more today) simultaneous transmission of several optical carriers on one fibre appears very efficient because the cost of the amplifiers is shared between all the carriers. Inversely optical amplification allows to process all the carriers simultaneously, while optoelectronic regeneration needs one regenerating unit per carrier. But amplified WDM systems are subject to noise and propagation distortions : this is a come back to analogue transmission, with its drawback of distortion accumulation.

The first WDM systems available in 1995 had a capacity of 10 Gbit/s (four 2,5 Gbit/s channels with a channel spacing of 8 nm). As it will be explained later, the offered capacity increased rapidly and a total bit rate of 1 Tbit/s over one fibre will be soon available.

WDM provides then a potential enormous capacity onto one single fibre : the operators can increase the capacity of their existing networks without laying new cables. As an example, figure 26 depicts the measured spectrum of a 40 channels multiplex. The upper curve represents the quality factor Q, related with the bit error rate.

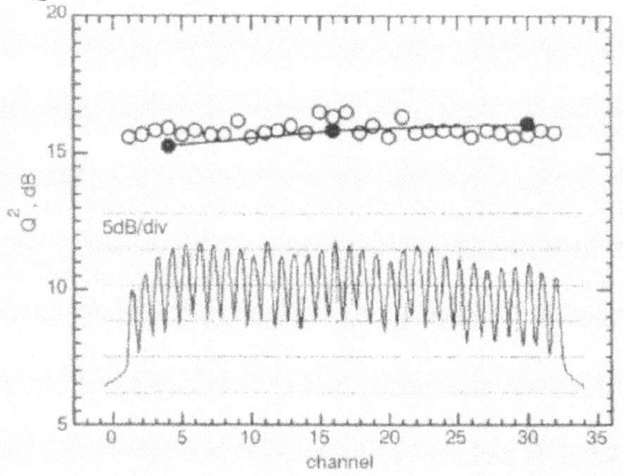

Figure 26 : Example of a WDM multiplex.

3.3.3. General Comparison Between Microwave and Optical Systems

3.3.3.1. Propagation Medium and Modulation

We already mentioned that microwave transmission uses generally free space propagation, while optical systems use generally guided propagation in fibres. An important consequence is that the received power varies as the squared inverse distance in the first case, and decays exponentially versus distance in the second. 36000 km of the best fiber available would have an attenuation of 7200 dB (!), while the losses on the same distance between a satellite and an earth station are only around 220 dB at 10 GHz.

No non-linearity occurs in the propagation medium in microwaves and non-linearities are located at the emitting end. For instance, satellite systems use travelling wave tube amplifiers, which affect the emitted signals through the well-known AM/AM and AM/PM conversions. Compensation of these distortions has been extensively investigated.

In optics, the transmission medium can be strongly non-linear, especially through the Kerr effect, which is the dependence between the refraction index and electromagnetic field intensity. The importance of these non linear effects for injected optical powers of some tens of dBm, should appear as surprising ; in fact, the critical parameter is the power density and the area over which this power is spread is very small, typically 50 μm^2 on an usual fibre.

Optical systems use binary on-off keying (intensity modulation) with direct detection, while microwave use generally heterodyning, and various modulation schemes, in particular multilevel schemes on order to achieve a better spectral efficiency. However some low cost microwave systems can use direct detection, in order to avoid a local oscillator.

3.3.3.2. Noise

Important differences occur also as far as noise is considered. The main noise source in microwaves is thermal noise, which is additive, white and gaussian. Noise in optical communication systems originates from three different contributions :

3.3.3.2.1. Quantum Noise

Optical power fluctuations carrying the information to be transmitted are converted into electrical current in the photodetector, which can be a PIN photodiode or an avalanche photodiode (APD) : incident photons illuminating the photoconductor junction create photoelectrons generated according to a random Poisson process. In the case of the PIN photodiode, the mean value of the photocurrent I is given by :

$$I = \frac{e\eta}{h\nu_s} P = SP , \tag{5}$$

where e is the electron charge, h the Planck's constant, ν_s the optical frequency and η the quantum efficiency, which can be viewed as the percentage of incident photons creating effectively a photoelectron, and P the incident optical power. Relation (5) shows that the detection process is quadratic, i.e photocurrent is proportional to the square of the incident electromagnetic field. Typical values of η and S at 1.55 µm are around 0.7

and 0.8 A/W. Shot noise inherently related with the corpuscular structure of light is also generated in the detection process. Its two-sided power spectral density is given by the classical Schottky's formula :

$$\gamma_g(f) = eI = \frac{e^2\eta}{h\nu_s}P \qquad (6)$$

3.3.3.2.2. Thermal Noise

The electrical signal at the photodiode output must be amplified in an electronic amplifier, which adds thermal noise (additive and gaussian). This noise is the largely dominant contribution in the receivers without optical preamplifier.

3.3.3.2.3. Amplified Spontaneous Emission Noise

An optical amplifier adds then its own noise to the amplified signal, as any electrical amplifier also does. Optical electromagnetic field associated to ASE noise can be modelled as a white gaussian process and the optical noise power spectral density (p.s.d) per mode at the output of the amplifier is written as :

$$\gamma_{esa}(\nu) = n_{sp}(G-1)h\nu_s \qquad (7)$$

where G is the power gain and n_{sp} the spontaneous emission factor characterising the amplifier. In the case of a semiconductor amplifier, (7) gives the noise p.s.d while, in the case of a fiber amplifier, the total power in a band B is $2\gamma_{esa}B$ because each mode is degenerated according to the two possible orthogonal states of polarisation.

As in the case of an electronic amplifier, the amount of noise due to the amplifier can be characterised by the noise figure. ASE noise can be considered as due to a noise source with a p.s.d γ_{ase}/G at the input of a noiseless ideal amplifier. Noise at the amplifier output can be viewed as the sum of shot noise and ASE noise and the signal to noise ratio can be written as :

$$\rho_{out} = \frac{GP_s}{\left(h\nu_s + n_{sp}(G-1)h\nu_s\right)B} = \frac{P_s}{F_a h\nu_s B} = \frac{\rho_{in}}{F_a} \qquad (8)$$

where ρ_{in} is the "intrinsic signal to noise ratio" at the amplifier input (cf. (16)) and F_a the amplifier noise figure expressed as :

$$F_a = n_{sp}\frac{G-1}{G} + \frac{1}{G} \qquad (9)$$

If G is large, noise figure equals practically n_{sp}.

The equivalent noise figure of a cascade of optical quadripoles (amplifiers, attenuators...) can be computed using exactly the same rules as in microwaves.

Although the ASE noise is gaussian, the non-linear detection process in the received photodiode results into two terms :

- The signal spontaneous noise (ssp) due to the beating of the noise with the signal, which is gaussian.
- The spsp noise, due to the quadratic detection of the ASE noise itself. As the square of a gaussian process, it is not gaussian, but is approximated as gaussian to obtain analytical expressions of the bit error rate.

3.3.4. Basic Propagation Phenomena in Microwave and Optical Propagation

3.3.4.1. Propagation Impairments in Microwave

In perfect propagation conditions, the medium can be considered as non-selective. Even if attenuation does naturally depends on frequency, the bandwidth of the modulated carriers is small enough to consider that there is no dependence within the spectrum of the transmitted signals.

Radio relay links or satellite channels exhibit usually this behaviour. During a small proportion of the time, they are affected by two sorts of fadings :

- Non selective fadings, due to rain, bring an additional attenuation which can lead to outage. The only countermeasure is an increase of emitted power.
- Selective fadings are due to the interference between the main path and a second one coming to the receiving antenna. The resulting signal distortion can be compensated for by baseband equalisation, decreasing the outage time.

It must be noted that mobile channels exhibit a different behaviour. The generally moving receiver receives the combination of a lot of rays with different delays, changing with time. Moreover, received signal is evidently affected by Doppler effect.

3.3.4.2. Chromatic Dispersion of Optical Fibres

For propagating signals, the optical fiber acts first as a linear medium the transfer function of which is characterised by chromatic dispersion, a basic physical parameter of the fiber.

Inside the bandwidth occupied by the transmitted signals around an angular frequency ω_s, phase can be expanded to the second order according to the formula :

$$\Phi(\omega) = \Phi(\omega_s) + \tau(\omega_s)L.(\omega - \omega_s) + \frac{\beta_2}{2} L(\omega - \omega_s)^2. \tag{10}$$

$\tau(\omega_s)$ is the group delay per unit length at angular frequency ω_s and β_2 the second derivative of the phase per unit length characterises group delay variations versus ω. Instead of β_2, one uses the chromatic dispersion parameter $D(\lambda_s)$, defined as the group delay variation for a length of 1 km and within a bandwidth corresponding to a wavelength variation of 1 nm around the wavelength λ_s.

$$D(\lambda_s) = -\left(6\pi \times 10^5\right) \frac{\beta_2(\lambda_s)}{\lambda_s^2}. \tag{11}$$

where $\beta_2(\lambda_s)$ and λ_s are respectively expressed in ps^2/km and nm. For the so called standard singlemode fibres (SSF) at 1.55 µm, $\beta_2(\lambda_s)$ and $D(\lambda_s)$ equal respectively -20 ps^2/km and 17 ps/(nm.km). An important parameter of the fiber is the zero dispersion wavelength λ_0, where $\beta_2(\lambda_0)$ cancels. For singlemode fibres, λ_0 is around 1.3 µm and dispersion is then much smaller around 1.3 µm, ensuring that propagating signals suffer less distortion, which explains why the 1.3 µm window has been used. But the counterpart is a larger attenuation than in the 1.55 µm window. Dispersion Shifted Fibres (DSF) exhibit around 1.55 µm a dispersion much smaller than standard fibres, typically less than 3.5 ps/(nm.km).

Impulses propagating on the fiber are broadened by chromatic dispersion : in the case' of gaussian shaped pulses, analytical formulae can be obtained.

Distortion suffered by the signal depends also strongly on the initial chirp. When the modulating signal is directly applied to the laser diode through the injection current (direct modulation), the optical output signal is not only intensity but also frequency modulated (chirped). This chirp depends on the line enhancement factor or Henry's factor α of the laser diode. It can be shown that when the product $\alpha\beta_2$ is positive, the temporal width of the impulse begins to decrease, reaches a minimum and increases again, while it continuously increases when $\alpha\beta_2$ is negative. This explains why direct modulation is non applicable on standard fiber controlling the chirp of an external modulator allows to increase the transmission distance.

3.3.4.3. Non Linear Distortions in Optical Fiber Communications

As already mentioned, the field intensity induces changes in the refractive index, which modifies the relative velocity of the spectral components of the signal and then induces a phase modulation onto the signal itself.. Two cases can be distinguished, according to the fact that the index is modified by the signal itself (self phase modulation) or by other signals travelling onto the fiber (Cross Phase Modulation).

3.3.4.3.1. Self Phase Modulation (SPM)

The phase modulation suffered by a pulse can be analytically computed in the ideal case of a dispersionless fiber. If $u(t,0)$ is the complex envelope of the signal at the input of an ideal dispersionless fiber of length L, the complex envelope of the output signal can be written as :

$$u(t,L) = u(t,0)\exp\left(-\frac{\alpha L}{2} + \mathrm{i}\Phi(t,L)\right),$$
$$\text{with } \Phi(t,L) = \gamma|u(t,0)|^2 \frac{1-\mathrm{e}^{-\alpha L}}{\alpha} = \gamma|u(t,0)|^2 L_{eff}, \qquad (12)$$

where α is the attenuation per unit length, L_{eff} is the effective length (equal to the length in a lossless fiber) and γ a coefficient characterising the non linearity, the value of which is around 3 W^{-1}/km. It is the be noted that γ depends of the effective mode area, i.e the area over which the energy is concentrated, which is typically between 50 and 80 μm^2 at 1.55 µm : the energy surface density is then a crucial parameter.

The maximum phase shift Φ_{max} occurs at the impulse centre (t=0) and is given by :

$$\Phi_{max}(L) = |u(0,0)|^2 \gamma \frac{1-\mathrm{e}^{-\alpha L}}{\alpha} = \gamma P_0 \frac{1-\mathrm{e}^{-\alpha L}}{\alpha} = \frac{L_{eff}}{L_{NL}} \text{ with } L_{NL} = \frac{1}{\gamma|u(0,0)|^2} = \frac{1}{\gamma P_0}, \qquad (13)$$

where P_0 is the peak power and L_{NL} the characteristic length, corresponding to a phase shift of one radian.

Comparing the length L with L_D and L_{NL} shows whether the system performance is limited by dispersion or non-linear effects. For instance, in the case of terrestrial communication systems with a typical launched power of 0 dBm (1 mW) L_{NL} equals 330 km, and assuming an attenuation of 0.2 dB/km (α=0,046 km^{-1}) the asymptotic phase shift after 330 km is only 0.06 radian. These orders of magnitude prove that self-phase modulation can be neglected in the case of these systems. On the contrary, non-linear distortions play an essential role in the case of amplified, due to the very important length (several thousands of kilometres), or even on shorter distances when a booster amplifier, allowing to launch 15 dBm or more into the fiber is employed.

Combined effects of chromatic dispersion and self-phase modulation will modify the impulse shape, and the induced effect will depend strongly on the sign of β_2.

3.3.4.3.2. Cross Phase Modulation (XPM)

When several optical carriers propagate simultaneously in a fiber, each of them undergoes not only the previously described SPM effect, but also another phase modulation, called cross phase modulation (XPM), due to the fact that index perturbation is induced by the total power propagating in the fiber. Description of this phenomenon is very complex : analytical formulas can be found assuming pure unmodulated carriers or a dispersionless fiber, and show that XPM depends on the total power of all the carriers, which means that this effect is dominant compared to SPM. But, in the actually interesting case of modulated carriers and chromatic dispersion, results can only be obtained using complex computer softwares.

3.3.4.3.3. Four Wave Mixing (FWM)

This effect occurring also when several carriers propagate simultaneously in a fibre, causes intermodulation and is through its effects very similar to what happens in multichannel radio systems. For instance, third order non-linearity creates beats between three carriers at angular frequencies ω_1 ,$\omega_2 = \omega_1 + \Delta\omega$ and $\omega_3 = \omega_1 + 2\Delta\omega$ and then intermodulation products at angular frequencies $p\omega_1 + q\omega_2 + r\omega_3$ where p, q, r are integers such that $|p|+|q|+|r|$ equals 3. Obviously, the intermodulation product at angular frequency $2\omega_2 - \omega_1 = \omega_3$ will perturb carrier at angular frequency ω_3. Analytical formulas exist in the case of pure carriers. But, as for XPM, phenomena are much more complex in the practical case of modulated channels. FWM influence depends on channel spacing and fiber dispersion : highly dispersive fiber will "mismatch" two neighbouring channels because the corresponding signals propagate with different group delay and dispersion and then reduce intermodulation. This is why low dispersion fibres are less favourable as far as FWM is concerned, especially for channels near the zero dispersion wavelength.

3.3.4.3.4. Stimulated Brillouin and Raman Scattering

The basic physical reason is the energy transfer from the optical field to the non linear medium, i.e the fiber, exciting vibrational modes of silica, the basic difference between the two effects being that acoustic (resp. optical) phonons are involved in SBS (resp. SRS). They are governed by the set of differential equations representative of a parametric amplification process :

$$\frac{dP_b}{dz} = gP_bP_p - \alpha P_s \frac{dP_b}{dz} = -gP_bP_p - \alpha P_b \ , \tag{14}$$

where P_b and P_p are respectively the probe and pump power, α the transmission factor of the fiber and g a gain characteristic of the phenomenon under consideration. The useful signal, acting as the pump, transfers its power to an interferer, denoted in (11) as the probe. We will just very briefly review the degradations due to both effects in optical communication systems, beginning by SBS.

When the pump power exceeds the Brillouin threshold (typically 1 to 3 mW in usual fibres), a backward propagating wave (probe), down shifted in frequency by 11 GHz in silica fibres and called Stokes wave, is generated at the expense of the signal acting as a pump. This will then cause an additional attenuation of the useful signal and also harmful effects due to this counterpropagating wave coming back to the emitter. Maximum value of gain g_B is around 5×10^{-11} m/W. The aforementioned threshold power was obtained for an unmodulated pump and increases when the pump bandwidth $\Delta\nu_p$ increases beyond the Brillouin gain bandwidth, typically 100 MHz, which means that frequency spreading due to modulation is a favourable factor. If a very high power is launched into the fiber, SBS can be prevented by a very low frequency modulation (dithering) of the laser, which broadens the linewidth without affecting the information signal.

In SRS, the frequency shift between pump and (here copropagating) probe is much larger (13 THz), amplification bandwidth is around 8 THz, but the maximum gain g_R is much smaller than for SBS, typically 10^{-13} m/W. The SRS power threshold is then much higher, around 300 to 600 mW for silica fiber at 1.55 μm. SRS is then not to consider for most of the optical communication systems.

3.3.4.4. Combined Effects of Linear and non Linear Effects

The quadratic detector is naturally not sensitive to the phase of the incoming optical signal and one then could believe that SPM or XPM do not have any consequence. But in fact, phase modulation is converted into amplitude modulation by chromatic dispersion : interaction between linear and non-linear effects is then essential.

Assuming a typical impulse shape, increasing for negative values of t, reaching its maximum value for t=0 and decreasing for t positive, relation shows that the frequency modulation induced by non linear effects causes a frequency increase for t positive (blue shift of the rear edge) and a frequency decrease for t negative (red shift of the front edge). If β_2 is positive, the group delay increases with frequency, the front edge (resp. the rear edge) propagates more (rep. less) rapidly, and then the pulse broadens monotonically. When β_2 is negative, linear and non-linear distortions induce opposite effects and the impulse narrows before, and then broadens monotonically again.

Exact resolution of the non linear Schrödinger equation which governs the propagation through the fiber shows that for specific conditions, linear and non linear effects can mutually compensate along the fiber, leading to a particular impulse shape which remains undistorted while propagating in a lossless fibre: this is the optical soliton.

3.3.4.5. Channel Stationarity

Microwaves is essential non stationary : in radio relay systems or geostationary satellites, it is perfect during practically all the time, and exhibits fadings during a short proportion of time, which determines outage time and then transmission quality. Mobile channel on the contrary is much more variable, and its transfer function is continuously changing as the receiver moves.

The physical parameters of an optical fiber, like chromatic dispersion, non-linear characteristics... are much more stable. Nevertheless, there exists a non-stationary phenomenon, Polarisation Mode Dispersion (PMD).

In a perfect fiber, any mode is degenerated : when birefringence occurs, because of an imperfect circularity, mechanical constraints, two modes with different propagation characteristics are present. The resulting effect on the photodetected signal is the presence of an echo, with a random delay and a random repartition of power between the two impulses. PMD is a very important effect in optical WDM systems. While manufacturing processes of fibres have been continuously improved in order to reduce PMD, compensating devices are studied in several laboratories in the world.

3.3.5. Noise Limitations in Optical WDM Systems

Accumulation of ASE noise generated by the in line optical amplifiers is a basic limitation of amplified optical systems. At the receiver side, a minimum optical signal to noise ratio (*OSNR*), defined as the ratio of the optical power to the ASE noise power in a given bandwidth (usually 0.1 nm) is required in order to achieve a minimum BER. The minimum power to be launched into the fibre can then be derived.

Assuming each amplifier exactly compensates for the attenuation of the fibre span between two amplifiers, it is easy to see that the noise power due to N amplifiers are simply added.

Using relation (3), the noise power in a bandwidth B is then :

$$P_{noise} = Nn_{sp}(G-1)h\nu_s B = \frac{L}{Z_a}\left[\exp(\alpha Z_a)-1\right]n_{sp}h\nu_s B \tag{15}$$

where Z_a is the distance between two amplifiers, L the total length, and α the fiber loss parameter.

This formula shows explicitely the exponential dependence of the noise power versus distance. As an example, less us assume two situations for a L=1000 km link, corresponding respectively to Z_a=50 km (20 spans) and 100 km (10 spans). In the second case, the amplifier gain is doubled compared to the first, and goes from 10 to 20 dB, causing an increase of 10 dB of the noise generated by each amplifier, while the amplifier number is divided by 2, which brings a gain of 3 dB. The resulting noise increase is then 7 dB ; this simple example shows how *OSNR* is a limiting factor.

Increasing the amplification span and (or) the number of channels or the bitrate requires to increase the total output power of the amplifiers.

3.3.6. Limitations due to Propagation on Fibres

3.3.6.1. Optical Fibre is not an Ideal Media

In the early days of optical communications, and before the advent of WDM amplified systems, fiber could appear as the ideal transmission medium, acting only as a perfect attenuator without introducing any distortion on the transmitted signals.

For example, in electronically regenerated systems at 2.5 Gbit/s, the distance between repeaters (100 km) was so small that chromatic dispersion did not play any detrimental role : it is negligible at 2.5 Gbit/s up to 500 km. In the absence of amplifiers, and due to the short range, the emitted power remained low and did not induce any non-linear effect.

The situation is completely different with amplified systems. Power and transmission length are much higher, which means that chromatic dispersion and non-linear effects play a significant role. With the increase of the bitrate per channel and the number of channels, propagation effects become more and more important, because narrower pulses are more affected by dispersion and launched power is higher, leading to more severe non linear impairments.

3.3.6.2. Optical Solitons

As already mentioned, linear and non linear effects compensate mutually in anomalous propagation regime : there exists theoretically a situation when compensation can be perfect, which results into an impulse propagating along the fibre without any deformation, the optical soliton. In fact this particular solution of the non linear Schrödinger equation exists only on an ideal lossless fiber, but this assumption is non essential and a quasi soliton can be defined on an actual lossy fiber, under some practically verified conditions.

Solitons has been considered some years ago as a very promising technique for long haul high bitrate transmission systems, especially undersea cables. Due to the fact that the linear effects must remain limited in order to keep the linear effects (and then the power) reasonably high, solitons used dispersion shifted fibre, and not standard fiber. This low dispersion was then not compatible with wavelength multiplexing, because of four wave mixing.

Two ways seemed then opened to reach high capacities, WDM and solitons on the long distances.

But, new propagation regimes, Dispersion Managed Solitons, were extensively studied theoretically and experimentally in the last years : only the average dispersion is kept small, but ot the local one, and this is obtained by properly designing the dispersion map. The basic stabilisation of pulses due to the mutual compensation of linear and non-linear effects is maintained, but the high local dispersion allows WDM.

As a consequence, WDM is today the universal technique to increase the capacity of the transmission systems, terrestrial as well as undersea.

3.3.7. State of the Art of WDM Technique

As already recalled, the first WDM optical systems were available in 1995, offering a capacity of 4×2,5 Gbit/s. Since that time, performance have been continuously improved, and the technical evolution in this domain, due to the very big demand to build high capacity transport networks. Today WDM technology is a dramatically active business, and a lot of start-ups have been and are presently created. Figure 27 shows what the state of the art looks like.

At 2.5 Gbit/s per channel, technology is completely mature and a lot of commercial systems are available, with up to more than 100 channels.

Transmission at 10 Gbit/s is more difficult, because the signal is much more sensitive to chromatic dispersion, while non-linear effects are higher, because of the increase of power required to achieve the necessary *OSNR* . Nevertheless, N×10 Gbit/s WDM systems are today available, with up to 64 channels.

40 Gbit/s per channel is a much more difficult issue and not any system is available today : but research has been very active for some years, demonstrations are presented in the international conferences by the major laboratories in the world, and the first systems should be available within two years. The objective was to demonstrate the possibility of transmitting more than one Tbit/s, a symbolic value, on one single fibre. Very recently, transmission of 7 Tbit/s on one fibre has been shown. As explained earlier, not only the total bitrate, but also the distance and the amplification span are of primary importance.

Above, at bitrates of 80 or 160 Gbit/s per channel, all the problems are infinitely more difficult, and no electronic components are available, with requires signals to be processed optically : some single channel demonstrations at 100 or 160 Gbit/s have already been presented, an active research is carried out in this domain, but the advent of such systems belongs to a far future.

WDM is then the key technology which allows the implementation of very high capacity backbone networks which are widely deployed in the world to cope with the expected traffic explosion in relation with the development of data caused by Internet. The offered capacity and the transmission quality are absolutely non-comparable with what could be offered by radio or satellites systems.

As an example, the proportion of transatlantic traffic on undersea cables and satellites was respectively 40 and 60 % : the tendency is that lightwave undersea systems will practically carry 100 % of traffic.

As it appears on figure 27 capacity can be increased by increasing either the number of channels or the bitrate per channel. Some key technologies, like amplification and error correcting coding are essential to design high capacity systems.

- The first optical amplifiers used the so called C (conventional) band extending from 1535 to 1560 nm. A continuous research effort allowed to use new bands, above and under the C band, the respectively S and L bands, and the maximum bandwidth usable today is around 120 nm, i.e 14 THz, naturally at the expense of simplicity, because larger the bandwidth is, more complex is the amplifier.

- Spectral efficiency η, defined as the ratio of the total capacity by the whole occupied bandwidth, is the parameter associated with an efficient utilisation of the available spectrum : the maximum value obtained today is around 0. 4 bits/s/Hz . A better η requires a closer channel spacing, whose minimum attainable value is related with sources and filters stability, and also fibre characteristics which determine interchannel effects. In radio, spectral efficiency has been increased by using multilevel (instead of binary) modulation schemes, at the expense of the power budget. This way has not been explored in optical systems at the moment.
- FEC (*Forward Error Coding*) has been extensively employed in satellites and data modems on voice channels : progress in microelectronics make them available today for high capacity systems and the BCH (239,255) code has even be normalised. The introduced redudancy allows error correction and then a lower BER at the receiver input. Typically, 10^{-4} at the decoder input results into a BER of 10^{-12} at the output : the associated gain of around 4 or 5 dB on received power allows to increase the transmission distance. The penalty due to the increase of in line bitrate (here around 0,3 dB) must be taken into account. Most of the 10 Gbit/s WDM systems include today error correcting coding, and this will be a general rule above. Correction capability can be increased with more complex coding techniques, like concatenated codes, but the redundancy is larger.
- Raman amplification is another method to increase the performance of optical transmission systems : it relies on the energy transfer, through Raman effect, between the signal and a pump signal injected into the fiber. Although it had been envisaged as promising before the advent of doped fiber amplifiers, Raman amplification was eliminated by them. Today, it is recognised as a possible solution for WDM systems : injecting the pump at the receiving end provides an amplification of the signal, compensating for the fibre loss and increasing the received level at the receiver input. This gain allows to increase the transmission distance or, for a given distance, to decrease the emitted power and then work in a more linear regime.
- Fibre is clearly a key element in the lightwave communication system. As already mentioned, two types have been manufactured and used for many years , standard single mode fibre G.652 and dispersion shifted fibre G. 653. The second one is not well matched to WDM transmission, because of its small dispersion and the resulting high level of intermodulation due to Four Wave Mixing. Since some years, all the fibre manufacturers have proposed new types of fibres, belonging to the *NZDSF (Non Zero Dispersion Shifted Fibres*) family.

The objective is to have a lower dispersion than G. 652, reducing then the cost of compensation, while keeping it high enough to prevent non-linear detrimental effects. In fact, the amount of dispersion changes from one manufacturer to the other within this family, and G.652 remains the best candidate for transmission up to 10 Gbit/s. Above it has also excellent potentiality and the question of the « best fiber » remains completely opened.

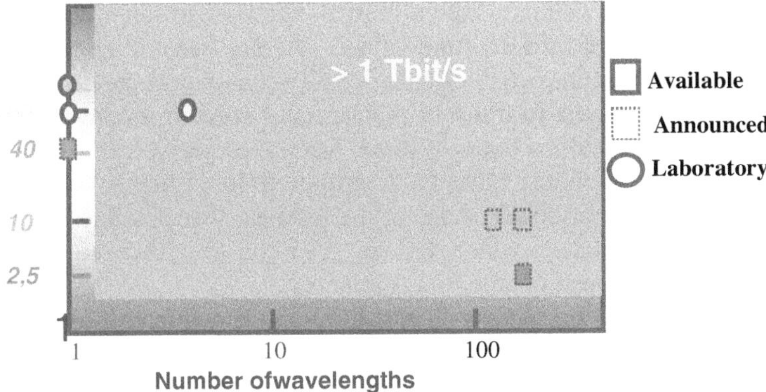

Figure 27 : State of the art in WDM technology.

3.3.8. What Future for Microwave in Telecommunications ?

First of all, electronic circuits associated with the aforementioned high capacity WDM optical systems are clearly microwave circuits.

Only transport networks have been discussed in this paper, and the future is clearly the all-optical high capacity backbone network, offering a capacity completely unattainable with radio systems. Let us just recall again that one satellite could provide a capacity of some tens of thousands of voice channels, while the most recent undersea lightwave systems will offer some millions of channels (2.5 Gbit/s corresponds roughly to 32000 voice channels).

In the access network, situation is much more contrasted. The bitrates to be transmitted are smaller and then compatible in a lot of cases with microwave technology, the copper infrastructure is existing and can be used (with ADSL, VDSL...techniques), and cost is a very important factor, because it cannot be shared among a so large number of users as in transport network, where traffic concentration is very high.

For these reasons, different technologies are present in the access networks, and microwaves is one of them. The radio loop can, for example, be preferred to a cable to connect a new subscriber if the cable has to be laid.

Radio is clearly the only solution each time mobility is required, and the importance of mobile services is well known, and can be observed everyday.

If satellites cannot compete with undersea optical systems and are no more actors for the major intercontinental permanent links, they remain necessary each time communication must be established rapidly and with points when no fixed infrastructure exists : satellite transmitters/receivers with very small antennas (for instance VSAT technology) are commonly used now by the media to transmit pictures and voice from any point of the world when something happens there. Several projects, with some tens of low orbiting satellites, have been proposed to insure a worldwide coverage for mobile users. It seems at the moment that possible business has been overevaluated, which led an operator to close the service, because of the too small number of customers.

Satellite can also be cheaper than cable for communications with low population density areas.

On the other hand, the importance of satellites in broadcasting is today very well known, and this is domain where they will keep a key role.

Finally, microwaves are also a candidate for indoor communications (communications with somebody moving within a building), and they compete in this domain with infrared devices.

4. WIRELESS SYTEMS

4.1. Wireless Systems Using Photonic Network Infrastructure

J.F. Cadiou, P. Jaffré, E. Pénard
France Telecom BD CNET/DTD/AEA
2, Avenue Pierre Marzin, 22307 Lannion Cedex - FRANCE
e-mail : jeanfrancois.cadiou@cnet.francetelecom.fr

Introduction

Photonic network infrastructure is an important topic for telecommunication operators because of its capacity to transmit very high bit rates. In the transmission network the use of such technologies is the only possibility to reach several Gbits/s over transatlantic distances for example. Although the subject of this paper deals with the access network, “the last mile deal”, where such technologies are still not completely deployed due to mainly two factors:

The price: Installing new optical infrastructures in the access could represent more than 60 % of the global access connecting prices and operators are trying to reuse the existing cable infra-structure to optimise their initial investments. That is the deal of the xDSL[2] technologies that are taking advantage of the enormous progress in signal processing and micro-electronics integration to overcome the problem of copper pairs band width by a powerful use of information coding.

The customers needs: For the business customers the needs in capacity is becoming more and more important for enterprise‘s sites interconnection (LAN interconnection), or connection to the supplier’s sites for example, bit rates are estimated today at a few 10 Mbits/s to a few Gbits/s tomorrow. In this case optical technologies are able to support those bit rates. For the residential customer the situation is less clear and for the moment services are essentially telephony, fast internet, poor quality images and broadcast services and ADSL techniques can easily give a first and fast answer with lower prices compared to FTTH techniques for example. Also the multi media era and the need for diversified service bundles are now becoming a reality and will boost the bit rate demand; optical technologies will then be the natural evolution of the access network for the delivery of service bundles unreachable for classical ADSL techniques.

In this paper we will first review the different access technologies to show their diversities and we will focus in the second part on fixed wireless access and show how it can be integrated over optical network trough the ACTS/FRANS[3] project example. The third part deals with the convergence between wireless cellular networks and optical networks.

4.1.1. Different Access Techniques

Basically three important public accesses network are present today, twisted pairs, coaxial and wireless (Mobile, satellite, fixed wireless). In the context of broad band network, the copper network has already given a first and powerful answer by the mean of xDSL techniques with down link bit rates in the case of ADSL up to 8 Mbits/s. Concerning fixed wireless LMDS[4] solutions are able to furnish up to 30 Mbits/s on the

---

[2] x Digital Subscriber Loop, A: Asymetric, V: Very high bit rate

[3] Advanced Communications technology & Services/Fibre Radio ATM Network & Services

[4] Local Multipoint Distribution System

down link and 2 Mbits/s on the up link per client. Satellite access is essentially a broadcast approach even if we talk actually of Internet access via satellite networks. The problem is to have a low cost return link and the intermediate solution is the use of the PSTN[5] via a classical 56 k V90 modem for this purpose. Coaxial cables networks are able to deliver to the customers broadcast video services and Fast Internet accesses. For example VoIP[6] services could be delivered via this media and are representing a serious concurrent to the classical PSTN.

If the demand evolve toward higher bit rates, FTTx[7] technologies could take an important place in the future access network. The layout of such an infrastructure is given below:

There are three different kinds of optical accesses depending on the end point of the fibre. The cab concept stops at the cabinet point and copper pairs are used for the drop line to bring analog services and new multimedia services using VDSL technology. In the FTTC and FTTB contexts the end point is located very close or at the building. The final step is the FTTH approach where ONU[8] is located at the customer premises and becomes the network termination (ONT[9]). Costs considerations are in this case a very hot topic for the equipment in the house.

If we have a look further, those architectures could evolve, by introducing the very well known WDM[10] technique in the optical access (Figure 29), or by coupling the optical network with wireless techniques that are actually in a tremendous growing phase. The last point is the HFR[11] concept shown figure 30.

[5] Public Switched Telphone Network

[6] Voice over Internet Protocol

[7] Fibre To The x : B : Building, H : Home, C : Curb, Cab : Cabinet

[8] Optical Network Unit

9 Optical Network Termination

[10] Wavelength Division Multiplexing

[11] hybrid Fibre Radio

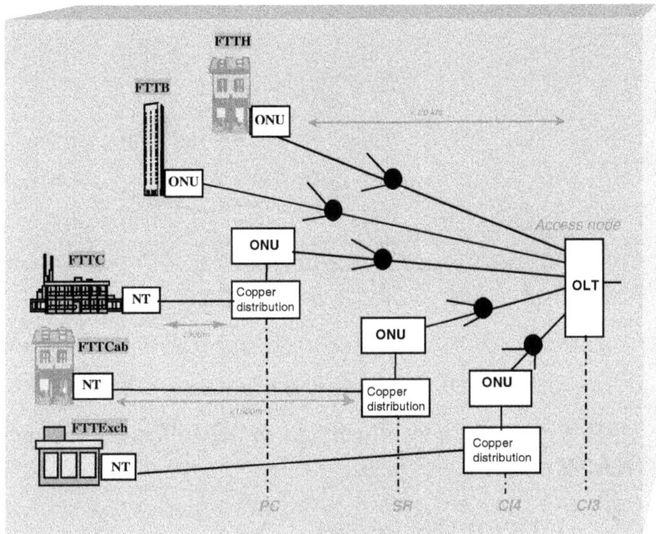

Figure 28 : FTTX approaches.

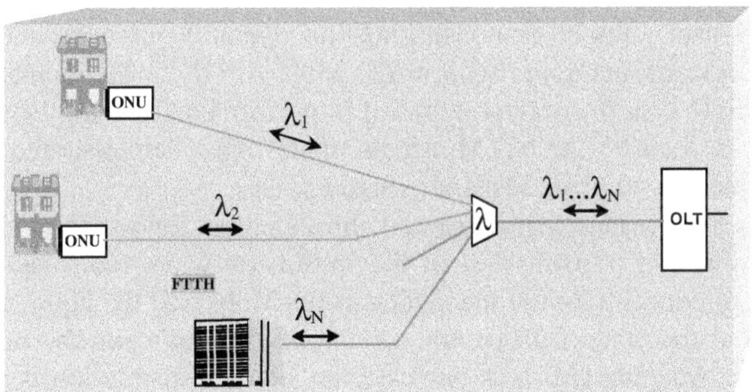

Figure 29 : WDM in the access.

WDM in the access has the following advantages

- allowing dynamic wavelength reallocation
- keeping the optical sharing
- enhancing the bit rate per client on demand

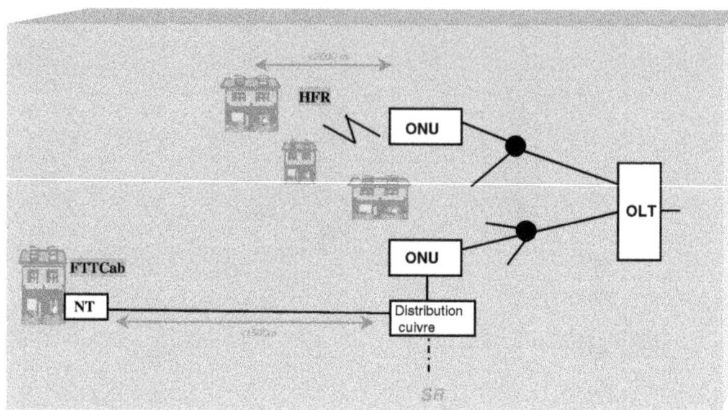

Figure 30 : HFR approach.

The HFR concept will be illustrated in the following paragraph by the ACTS/FRANS project.

4.1.2. The ACTS FRANS Project

4.1.2.1. Main Objectives of the Trial

This trial aims at demonstrating the operation of a broadband HFR access system integrated with a 622 Mb/s APON[12]. The management of the overall system (optical + radio) is performed from the central office where is located the ATM access node. The demonstrated upstream technique is a 40 Mb/s TDMA[13] shared access.

Another concept demonstrated through that field-trial is the optical generation and transmission of the millimetre-wave radio signal, which allows to concentrate the management functions and the equipment of the system at the central office and to simplify the optical/radio base station (Remote Antenna Unit). In our case the signal transmission is performed at 27.875 GHz on the fibre. Within that concept, the base station behaves like a transparent optical/radio interface.

A services demonstration is also one of the objectives of the trial : the system has been connected to a services platform for the delivery of VOD[14] and on-line services (high speed internet, visiophony, teleworking, teleteaching). The connection to PSTN has also been realised in order to provide telephony services.

---

[12] Asynchronous Tranfer Mode Passive Optical Network

[13] Time Division Multiple Access

[14] Video On Demand

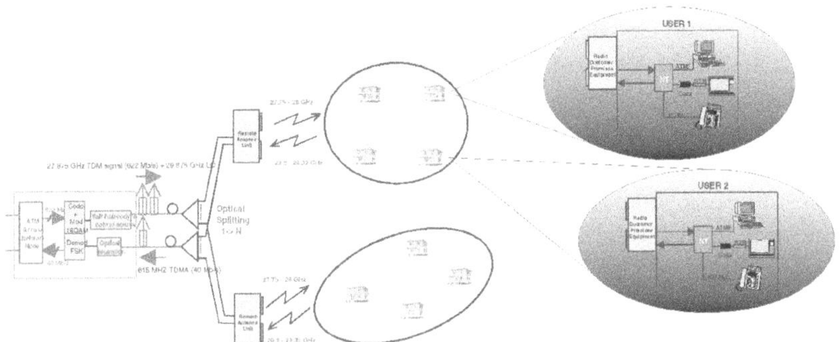

Figure 31 : Description of the field-trial.

The configuration of the field-trial is represented on figure 31. The figure represents the trial such as it was initially planned, with the following different subsystems :

- the 622 Mb/s ATM access system (Optical Line Termination), which performs the multiplexing of services in an ATM 622 Mb/s frame and the insertion of synchronous services (telephony) in the ATM frame. It uses 2 input interfaces : an optical STM1[15] interface for the VOD and on-line services, and a 2Mb/s G703 PCM[16] interface for the telephony services. The output interface is a parallel 8 x 78 Mb/s interface. The ATM access system is able to manage services towards up to 16 customers.
- the 622 Mb/s ATM customer's premises network termination, which provides the customer 3 kinds of services : VOD, on-line services and telephony. The network termination is connected to a PC by a twisted pair through an ATMF25.6 interface and to a set-top box by coaxial cable (proprietary solution).
- the Forward Error Correction Reed-Solomon encoder and decoder, which use a 8 x 78 Mb/s parallel processing : their function is to enhance the robustness of the system.
- a 16 QAM[17] 622 Mb/s modulator, which includes the main functions of mapping, channel filtering and modulation. The digital Nyquist filtering is performed in CMOS technology ICs allowing a processing speed of more than 300 MHz. The modulator can operate either in full-Nyquist or half-Nyquist mode.
- a 16 QAM 622 Mb/s demodulator, which performs the analog demodulation, the clock and carrier recovery, the analog-digital

[15] Synchronous Transfer Mode

[16] Pulse Code Modulation

[17] Quadrature Amplitude Modulation

conversion and the Nyquist filtering and equalisation. It is based on high speed signal processing (155 Mbauds but a useful eye width of 500 ps only in 16 QAM modulation). The demodulator includes complex looped functions and the balance of I and Q signals has revealed itself difficult to maintain in time. In front of the impossibility to interface that subsystem successfully with the 16 QAM modulator, it was decided to have recourse to a fall-back configuration represented on figure 35. A successfull achievement of the 622 Mb/s demodulator would require to reconsider the whole design of the subsystem, but the risk would remain. To have recourse to lower data rates would be a more reliable track.

• a self-heterodyne optical source, performing the photonic generation of the radio downlink signal at 27.875 GHz and of the local oscillator at 29.875 GHz. Both are transmitted on the optical fibre and photodetected inside the base station. The local oscillator is used inside the base station in order to downconvert the radio uplink signal from 29.26 GHz to the Intermediate Frequency 615 MHz . The optical transmission of the uplink signal is then performed at 615 MHz. The detail of the optical/radio transmission part of the trial is represented on figure 32.

•

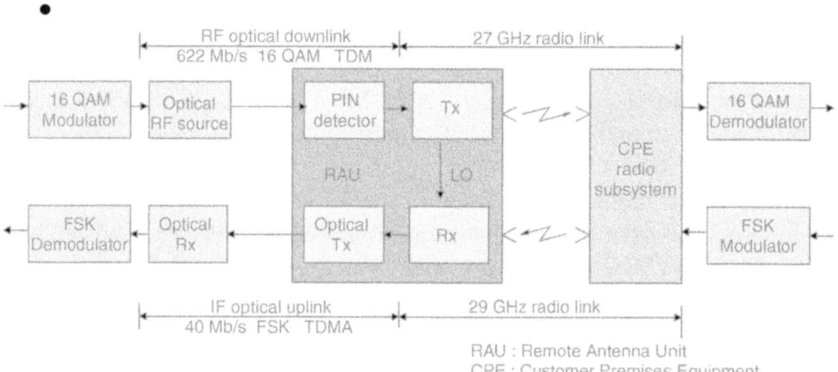

Figure 32 : Building blocks of the trial.

• a RAU[18] or base station, performing the photodetection and the amplification of the downlink signal. The local oscillator at 29.875 GHz is also separated from the signal. The antenna unit also performs the radio reception, the downconversion to 615 MHz and the optical transmission of the uplink signal. The antenna unit is based on a softboard substrate integration technology which allows the integration of the different functions of the antenna unit in a compact subsystem. The air interface is provided through 2 patch antennas with respective azimuth and elevation apertures of 45° and 5°.

---

[18] Remote Antenna Unit

• a radio CPE[19], performing the functions of reception and downconversion to 1.6 GHz of the radio downlink signal. On the uplink an FSK[20] modulation is applied. The CPE also includes an implementation of the TDMA function, which determines the duration during which the CPE is allowed to transmit. The air interface is provided through 2 planar directionnal cross-polarization antennas.

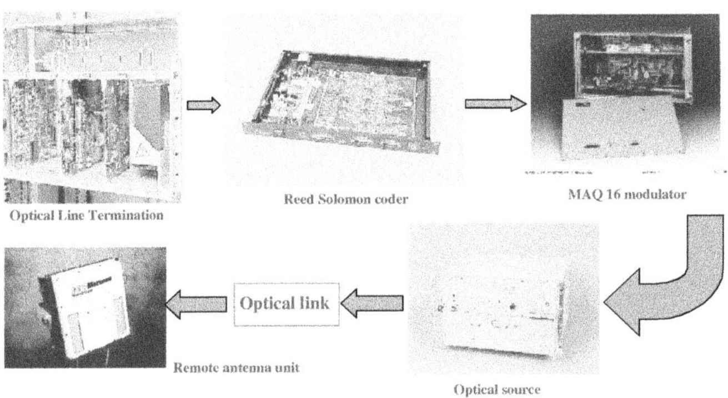

Figure 33: Central and ONU equipment.

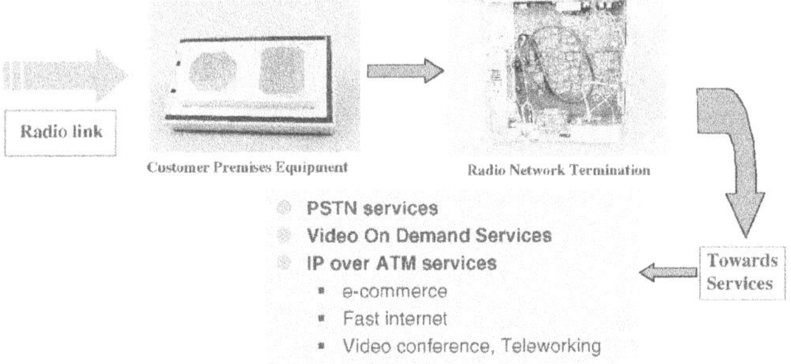

Figure 34 : CPE equipment.

4.1.2.2. Results of Integration and Operation

As mentioned above, the field-trial was implemented with the fall-back configuration represented on figure 35. In that configuration all the subsystems are integrated except the 16QAM modem and the RAU and

[19] Customer Premises equipment

[20] Frequency Shift Keying

CPE are common to the downlink and the uplink. On the downlink the Hybrid Fibre-Radio concept is demonstrated with a DVB-S[21] 40 Mb/s signal transmodulated into a 64 QAM modulated signal ; the downlink is independent from the ATM access system. The integration of the uplink with the ATM access system is demonstrated with the services which were initially planned ; the ATM access system is looped through the encoder and decoder.

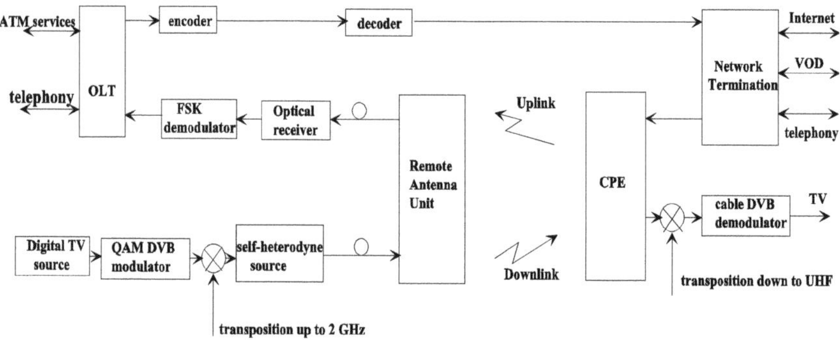

Figure 35 : Lannion modified test bed.

In that configuration, the system has been operating continuously in real trial conditions (RAU and CPE in line of sight in outdoor) since mid-April 1999 with a good stability of the quality of services. The distance between the radio base station and the radio customer's termination is 80 m whereas the system was specified for a maximum cell range of 500 m. Typical transmitted powers (EIRP[22]) from the radio base station and termination are +33 dBm and +26 dBm (with antenna gains of 18 dB and 24 dB respectively). The optical transmitted power from the self-heterodyne source is +11 dBm and the received power on the base station is 0 dBm (the optical budget including 15 km of optical fibre). The maximum drive level on the self-heterodyne source is +10 dBm without distorsion penalty.

The quality of the received 40 Mb/s 64 QAM signal could be assessed : a signal to noise level (S/N) of 32 dB was obtained after transmission through the optical/radio system. The corresponding constellation diagram is represented on figure 36. The extrapolation to a 622 Mb/s signal can be only theoretical.

---

[21] Digital Video Braodcast- Satellite

[22] Equivalent Isotrope Radiated Power

The performance of the uplink has also been evaluated in terms of BER[23] versus the signal to noise ratio ; the results are represented on figure 37 for the reference modulator on demodulator in PRBS[24] mode and for the overall uplink in ATM burst mode : a BER of 10^{-10} is achieved for a S/N of 13 dB.

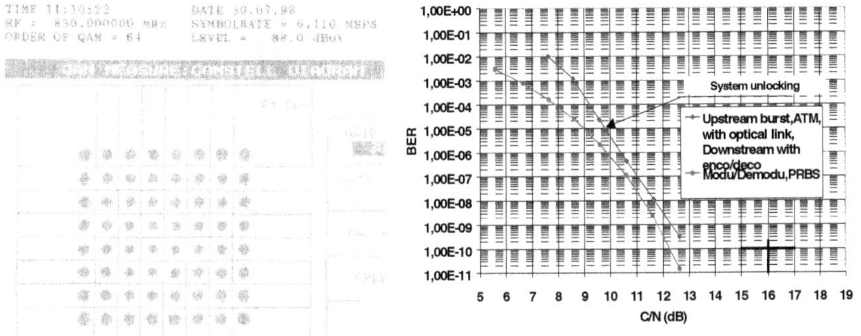

Figure 36 : 64 QAM constellation.

Figure 37 : Up link BER.

4.1.2.3. Conclusion

The main features of the field-trial, whether the matter is the transport on the optical fibre of a mm-wave signal (28 GHz) matched to the transmission in radio on the last drop of the network, the high data traffic (622 Mb/s downstream, 40 Mb/s upstream) processed by the system, the overall control of operation and of sharing of the resources centralised at the access node or the wide range of services (synchronous, asynchronous) connected to the demonstrator, all these features made of that field-trial a real challenge in the domain of application of the radio technology in the access network.

In practice the overall integration of the field-trial required to take into account the different constraints related to the real deployment of an ATM hybrid fibre/radio distribution system, from the connection to the services to the design of the customers premises equipment : in that way the trial induced some significant cumulated experience in the domain of the distribution of radio services on optical fibre for the access network.

---

[23] Bit Error Rate

[24] Pseudo Random Binary Sequence

4.1.3. The Example of DECT[25] Cellular System Integrated on an Optical Network

4.1.3.1. The DECT System : A Brief Overview

DECT system is a radio access technology, well adapted for low cost wireless applications. Its main characteristics are listed below.

- No cellular planning required
- High traffic capacity (small cells, powerful DCA[26])
- Unlicensed frequency
- Advanced services (voice, data, ISDN[27])
- Open interfaces to other networks

Its applications are basically for residential & SOHO[28] wireless systems, semi public and business WPABX[29] systems (coverage of campus, business area) and PCTM[30] systems (Coverage of entire town). The application aimed in this paper is the WLL[31] "avoiding cabling the last mile". The figure below is showing a DECT typical configuration and is fixing the terminology.

The physical layer is in the 1880-1900 MHz frequency band with up to 10 RF carriers (spacing: 1.728 MHz). The access technique is based on the MultiCarriers combined with TDMA (2*12 time slots). The duplexing technique is TDD[32] with DCA. Modulation format is the classical GMSK[33], with a sensivity of –83 dBm @ BER=10^{-3} and a nominal power of 24 dBm. The range for indoor and outdoor mobility applications is between 50 and 150 M, and for fixed access application between 350 and 5000 M depending on the antenna gains.

Supported services are typically voice telephony, voice band modem (4800 bits/s), wireless internet access, wireless LAN[34], wireless ISDN access, and WLL. As mentionned above we will mainly focus on the integration of DECT system over FTTx approaches and three different

---

[25] Digital European Cordless Telephone

[26] Dynamic Channel allocation)

[27] Integrated Services Digital Networks

[28] Small Office/Home Office

[29] Private Automatic Branch eXchange

[30] Public Cordless Terminal Mobility

[31] Wireless Local Loop

[32] Time Division Duplex

[33] Gaussian Minimum Shift Keying

[34] Local Area Network

networks configurations will be examined as a function of the DECT system building blocks positions in the network.

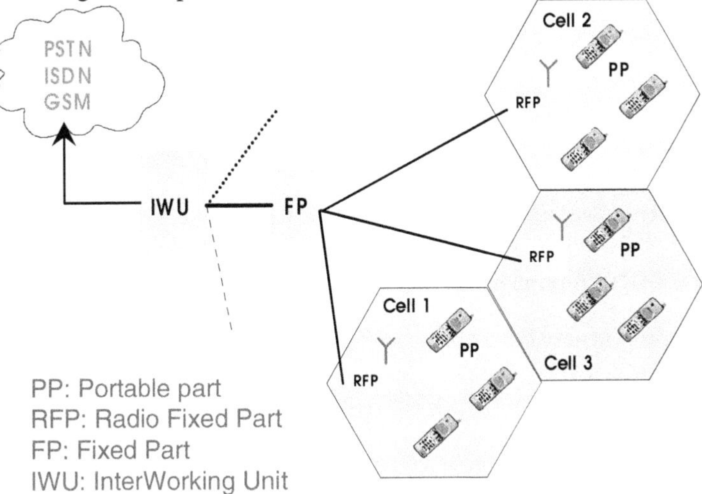

Figure 38 : DECT configuration.

4.1.3.2. FPs and RFPs are Localised at the ONU Level

This configuration is shown on the figure 39.

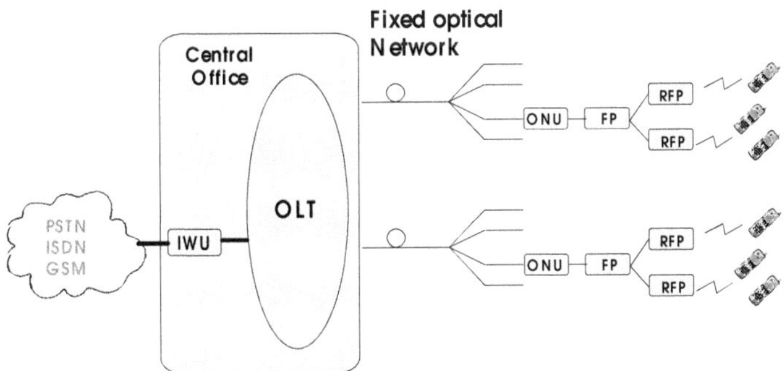

Figure 39 : FP and RFP are located at the ONU level.

This configuration is favourable to the synchronisation of RFPs to the same FP, although we have identified several limits.

- DECT system has a hand over management system and a mobile can change its RFP while it is moving. As it can be seen on the figure 39, RFP are not necessarily connected to the same ONU. The problem occurs because the data switching decision for the ONU is taken at the CO[35], while the channel reallocation occurs in the RFP.

- Moreover we need special functions in the CO to integrate in the frame information for PP authentication, location and channel allocation.

Finally the structure at the ONU is bulky and there is no simplification of the radio part by optical remote.

4.1.3.3. FPs Localised at CO whereas RFPs the ONU Level

This configuration is shown on the figure 40.

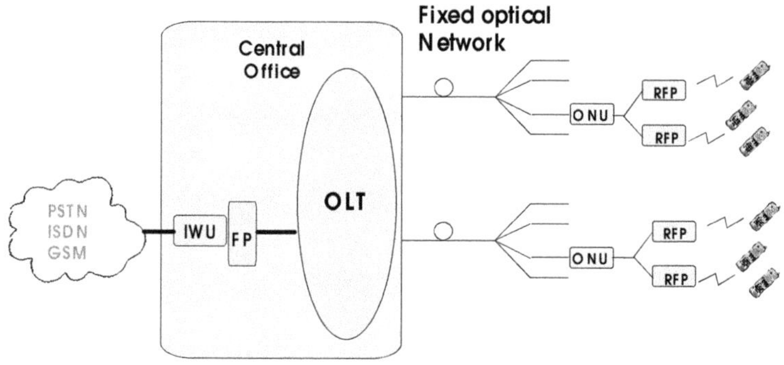

Figure 40 : FPs at the CO and RFs at the ONU.

This configuration is favourable for a global management of the mobility, but some problems of synchronisation may occur between FPs and RFPs due the dispersion of the optical branches and the resulting delay dispersion. A synchronisation signal may be carried on the fibre to the different RFPs connected to the same FP. Nevertheless the RFPs connected to the same FP are roughly in the same area of distribution and delay differences are attenuated.

4.1.3.4. FPs Localised at CO whereas RFPs the ONU Level

This configuration is shown on figure 41.

All layers of DECT system (MAC[36], DLC[37], network) are located at the CO and this configuration is completely favourable to the centralised

[35] Central Office

[36] Medium Access Control

[37] Digital Link Control

management of the system. The classical ONU is simply replaced by a DECT ONU, a RF/optical interface that restores the DECT channels.

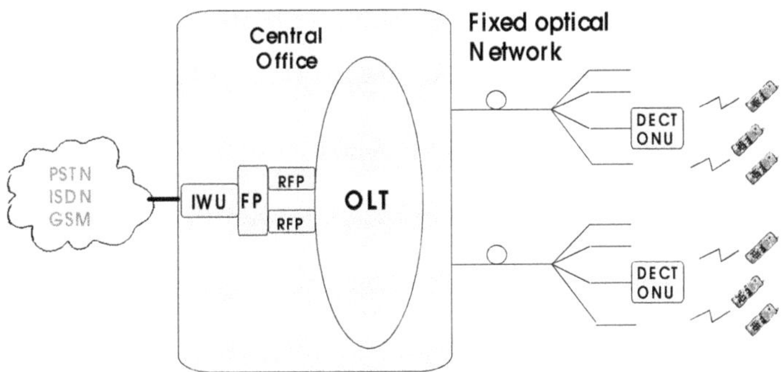

Figure 41 : FPs and RFPs are located at the CO.

This configuration allows the optimisation of the installed infrastructure with the possibility of RFP reallocation in case of peak traffic, and the simplification of installation, exploitation and maintenance. This solution overcomes the transmission of management signals between CO and RFP that are transmitted over expensive LL[38], resulting in a reduction of deployment costs.

4.1.3.5. Conclusion

The RF technology over fibre network allows a seamless management of the system and the centralised configuration simplifies installation, exploitation and maintenance operations. This study represents an example of a complete integrated optical-radio network (fixed-wireless convergence), and some results could be extrapolated to the next mobile generation: the UMTS system.

4.1.4. Conclusion

We have shown two examples of coupling between wireless and radio systems over fibre technologies. It shows the capacity of fibre to transmit RF signals from the low frequency band (< 2 GHz) up to the millimetre wave band. HFR type solutions are complementary to FTTcab approach by avoiding the cost of optical termination, by keeping the broadband aspect with a great facility of deployment. Moreover the centralised approach (RF over fibre solution) keep the transparency of the fibre to the

---

[38] Leased Lines

wireless system, simplifies the network termination and preserves the maximum of intelligence in the CO side.

4.2. Broadband Access Networks : The Opportunities of Wireless

G. Kalbe
European Commission DG-INFSO-F1, Future & Emerging Technologies, Rue de la Loi 200, B-1049 Brussels, BELGIUM
GUSTAV.KALBE@CEC.EU.INT

Introduction

The 1st January 1998 the telecommunications market in the European Union was completely liberalized. The provision of telecommunication services and the commercial exploitation of telecommunication infrastructure was no longer restricted to the national monopolies. Within a year the number of licensed public voice operators went up to more than 300 local and more than 500 national operators in the EU. This resulted in a fierce competition between the incumbent operators, i.e. the former monopolies, and the new entrants.

Competition meant for the incumbent a drastic change in their way of doing business. To survive, the prizes charged to the customers had to be cost oriented and could not be artificially high as in the past. Profit margins were coming down and the incumbent had to adopt business models as any other private sector company. Also, the new entrants have a competitive advantage as they are not burdened by legacy networks contrary to the incumbents. Both have to invest in the most modern available technology to either provide services for particular needs or to upgrade the existing infrastructure. With less cash to spend and reduced investment cycles, investments now have to be planned much more carefully, fulfil a market need and have to be financially justifiable.

The impact is particularly pronounced in the access network, the part of the network where the costs can not be shared between many customers. Although the equipments became affordable, the cost to deploy new access networks or to upgrade existing networks is still primarily determined by the huge installation costs. Therefor the migration of photonics into the access network is very slow, compared with the transport networks where DWDM is commonplace now, and there has been hardly any deployment of new access networks.

More than a decade ago the idea of providing a fibre to the home was a popular idea to provide customers with broadband applications. Soon however it became evident that neither the economics nor the available applications at that time would justify the investments. The high-bandwidth applications developed to justify the investments, like video on demand, were either not profitable or not needed by the customers. On the other hand the widespread use of the internet has since then pushed up slowly but steadily the customer demand for broadband services. Restricted by the existing infrastructure and the huge upgrade costs intermediate solutions have been developed, delivering broadband services while reusing to a maximum the existing infrastructure, like for example xDSL (Digital Subscriber Line). In addition alternative network infrastructures are exploited. Besides the copper twisted-pair network of the telephone companies, there are the networks of the cable companies, based on coax cables; the power lines from the electricity grid and the wireless solutions.

In the following we will give two examples of alternative access networks, extracted from ACTS[39]. Both are based on a combination of an optical feeder network with a wireless drop. Both examples implement an original approach to solve a particular problem typical to hybrid fibre radio (HFR) networks. The first example, the AC083 FRANS[40] project, demonstrates a clever way of integrating the optical and wireless network layers to reduce the network complexity and costs. The second example, the AC249 PRISMA[41] project, demonstrates how photonics can help to solve the hot-spot problem, typical to wireless networks. All the information extracted from the project documentation is reproduced with the kind permission of the project consortia.

---

[39] The 4th european framework for research, "Advanced Communication Technologies and Services", 1994 – 1998. For more information see http://www.cordis.lu/en/src/f_002_en.htm

[40] Fibre Radio ATM Network and Services. mmittric@rcs.sel.de

[41] Photonic Routing of Interactive Services for Mobile Applications. a.m.j.koonen@tue.nl

4.2.1. Direct RF-Wave Modulation of an Optical Carrier (FRANS[42])

The main objectives of the AC083 FRANS project were to demonstrate in a field trial the delivery of broadband services over a hybrid-fibre-radio access network. This included the development of new transmission concepts and the necessary components. We will not detail here the entire project, but concentrate on the different original approaches to directly transmit the RF-wave signals over an optical carrier, as demonstrated by the project.

In the classical approach the data is send from the switching centre to the antenna base station over a fibre by a STM-x link (e.g. STM-1 of SDH, Synchronous Digital Hierarchy, at 155 MBit/s). There, the optical signal is converted to an electric format, then the binary signal is modulated, added to a RF-wave carrier and finally sent over the air. This results in a high base station complexity, yielding a high number of different components and multiple conversions between signal formats. Thus the basic idea behind FRANS is to reduce the functionality of the optical-radio interface by shifting the radio-dependant equipment to the upper end of the optical feeder segment. Now, since the active equipment can be centralised at the head end, in general an accessible and spacious location, system upgrades may be achieved by changing head-end equipment without modification of the base station. A reduced base station complexity and bulk reduce furthermore the operation and maintenance costs, while preserving the low installation costs associated with a radio drop. Note, in combination with a passive optical network (PON) a dynamic service allocation using ATM (Asynchronous Transfer Mode) becomes possible.

Three solutions were studied in order to shift the radio-dependant equipment to the upper end of the optical feeder segment, transmitting the RF signal directly over the fibre:

- the RF (radio frequency) approach,
- the IF (intermediate frequency) approach,

---

[42] Project partners: Alcatel SEL (D), Centre Commun d'Etudes de Telediffusion et Télécommunication (F), Centre National d'Etudes des Télécommunications (F), Comatlas S A (F), Commissariat a l'Energie Atomique (F), CRITT Electronique (F), Dassault Automatisme et Telecom (F), Dassault Electronique (F), Fraunhofer Gesellschaft (D), GEC Marconi Ltd Research Centre (UK), GEC Marconi Materials Technology (UK), Ingenieurschule HTL Chur (CH), Institut d'Electronique et de Microelectronique du Nord (F), Institut fuer Kommunikationstechnik (CH), National Technical University of Athens (GR), Technical University Budapest (HU), Thomson CSF LCR (F), University College London (UK), University of Aveiro (P), University of York (UK)

- the optical PLL (phase lock loop) approach.

Figure 42 shows the principle of the RF optical feeder system. In the head-end, the downlink equipment pre-processes the digital baseband input signal in accordance to the characteristics of the radio segment including scrambling, interleaving, forward error correction, mapping, baseband shaping, modulation and up-conversion to an intermediate frequency. This signal is used to drive the data modulator of an optical millimetre wave source providing the final upconversion to RF and simultaneous electrical to optical conversion. The RF optical output signal, which can be a multi carrier, multi service signal, is transported via fibre to the remote antenna unit. The latter carries out the optical to electrical signal conversion, filtering, amplification and radiation.

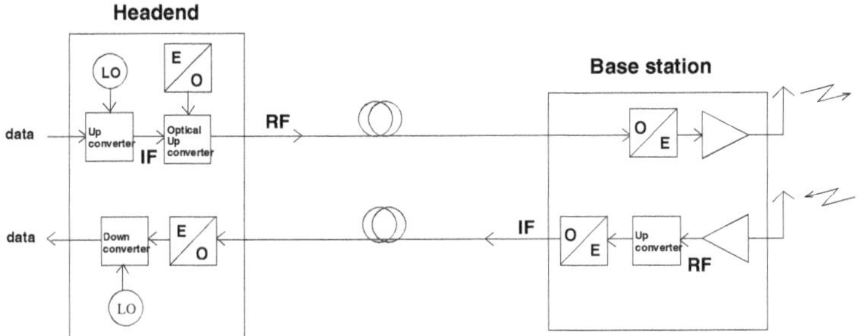

Figure 42 : Block diagram of a RF-based HFR system.

In principle the RF feeder technique is also applicable for the uplink. However, the boundary conditions in terms of temperature range, available power supply and volume, operation and maintenance effort as well as the cost target are different for head-end and remote antenna unit, and may require a different solution. In particular, from the system point of view it is not necessary to make an uplink RF signal available in the head-end. The signal has to be converted to digital baseband for switching since it is a signal composed of multiple contributions from multiple users, each contribution having another destination. Consequently it is a question of costs where the first down conversion from RF to IF is carried out, in the remote antenna unit or in the head-end.

The desired carrier frequency in the upper microwave/millimetre wave frequency range requires extremely high speed optical components in both the transmitter and the receiver. While laser modulation bandwidths of more than 30 GHz and external modulator bandwidths of 75 GHz have been demonstrated there are several detrimental effects which must be considered. Assuming a conventional amplitude modulation technique and the reuse of installed standard single mode fibre, the reach of an RF optical feeder is severely limited by chromatic dispersion of D=17ps/(nm·km) in the wavelength range of 1550 nm, and by attenuation at 1300 nm.

The alternative, most attractive solution is an RF feeder concept which is based on a self-coherent technique. Here, at the head end two optical carriers are generated transporting the information over the optical network to the base stations, where by coherent mixing a mm-wave signal is generated. This technique overcomes the impact of chromatic dispersion at 1550 nm and allows utilisation of erbium doped fibre amplifiers to compensate the insertion loss of the link and of the source components.

The RF approach reduces the technical function and complexity of the components at the cell site. This concept leads to the lowest volume, power consumption and maintenance effort in the outdoor-located base station, which mainly consists of a PIN photodiode followed by a stage providing power amplification. In contrast to the alternative IF transport solution, no oscillator for the up-conversion is needed. The base station offers broadband operation and transparency enabling a flexible upgrade and reconfiguration of services and capacity.

Within the RF transport scheme the mm-wave source located at the head end represents a high performance and very complex sub system where most of the electronic system equipment is concentrated. Due to the centralisation, several system parameters can be efficiently controlled at the head end, e.g. the operation mm-wave frequency can be change very flexible without the need of exchange of a LO (local oscillator) within the base station which would be necessary by use of IF transport schemes. In combination with broadband operating base stations the optical source allows the potential of system up-grade with regard to bit rate, capacity, mm-wave frequency at higher bands, number of frequencies transported/radiated at the antenna. Nevertheless, further mm-wave signals can be optically provided which can be used as reference signal at the base station/subscriber.

In summary, the main advantage of the RF source configuration is given by the centralisation of equipment and the available control mechanism. Both will lead to reduction of the operation and maintenance costs of such systems.

In contrast, the generation of optical carriers bearing the data signals and an efficient delivery of high quality mm-wave signals requires a considerable technical and economical effort. Presently, appropriate optical RF sources have been realised using several optical and electro-optical components. The volume of these configurations, as well as their high environment sensitivity (e.g. temperature effects), do not allow operation under outdoor conditions. The effort spent per source might be tolerable in the head end in a pure distribution system using passive optical splitting and radiating identical signal in each radio cell. Here, the costs for the source are shared by several base stations and even more customers. New mm-wave generation schemes and improvements of the degree of integration of the optical RF source and receiver are necessary to realise an economic point to point link which is considered to be the optimum long term solution.

In the IF approach (Figure 43), the downlink equipment in the head-end pre-processes the digital baseband input signal in accordance to the characteristics of the radio segment including scrambling, interleaving, forward error correction, mapping, baseband shaping, modulation, up-conversion to an appropriate intermediate frequency and electrical to optical signal conversion. This optical IF signal which can be a multi carrier, multi service signal is transported via fibre to the remote antenna unit. The latter carries out the conversion from optical to electrical signal, final up-conversion to the RF level, filtering, amplification and radiation.

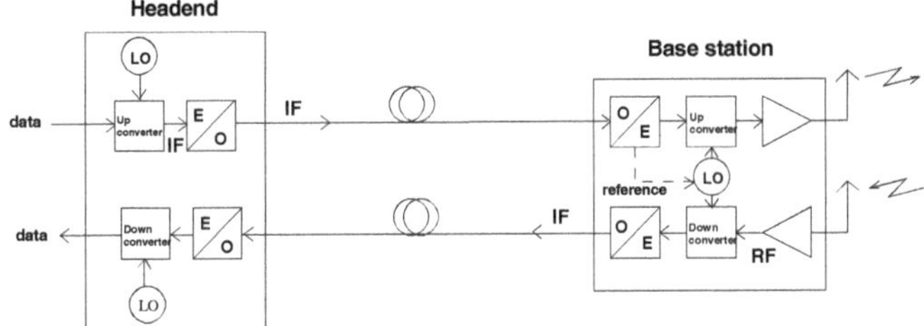

Figure 43: Block diagram of an IF based HFR system.

In uplink direction the signal is also transported optically on IF level, the remote antenna unit and the head-end carries out the inverse functions of the downlink direction. The local oscillator can be shared by both up- and downlink. As in the case of the RF approach, from the system point of view it is not necessary to make an uplink RF signal available in the head-end, as long as a signal has to be switched in the digital baseband. An RF optical uplink feeder would only make sense if it is cheaper than an IF solution.

The moderate upper band limit of the IF signal enables direct laser modulation, avoiding the need for an external modulator with it's corresponding non linear behaviour and high drive level. Together with a proper choice of the IF band lower limit the intermodulation limitations are significantly relaxed compared to the RF option. At the detector side, medium bandwidth photo diodes can be applied. The components required for the IF solution have a relatively high maturity and are significantly cheaper than the mm-wave optical components required for the RF approach.

Since the optical output signal of an IF source does not contain spectral components of the desired millimetre wave carrier frequency, it is only moderately subject to dispersion, since the effects are proportional to the square of the carrier frequency. This applies for signal deterioration due to chromatic dispersion, for noise power induced by phase de-correlation of the optical carriers increasingly with fibre length and for the impact of polarisation mode dispersion.

The main disadvantage of the IF approach is the local oscillator and mixer function, which has to be moved from the head-end to the remote antenna unit, increasing the complexity and the operation and maintenance effort. Here the up/down-converter has to be operated under outdoor conditions which increases the requirements in terms of frequency accuracy and phase noise. In particular for low bandwidth channels the provision of a frequency reference is mandatory to maintain the frequency error at an acceptable level. Nevertheless the IF approach offers the same transparency as the RF option in terms of channel bitrate, channel load, modulation format and frequency allocation within the overall bandwidth.

The optically-supported phase-locked loop (OPLL) is an extension of the IF approach. The OPLL approach addresses the problem of frequency stability and phase noise of the local oscillator used for remote up/down conversion by providing an optically-transmitted reference for a remote PLL in the base station. The mm-wave PLL effects a multiplication of the reference frequency while preserving frequency and phase stability. With a suitable multiplication factor, the reference signal can be transported at an IF frequency within the direct modulation bandwidth of a semiconductor laser. Reference and data signals may be multiplexed and demultiplexed at the base station, and the data upconverted by mixing, using the PLL output as LO. A block diagram of the OPLL approach is shown on figure 44.

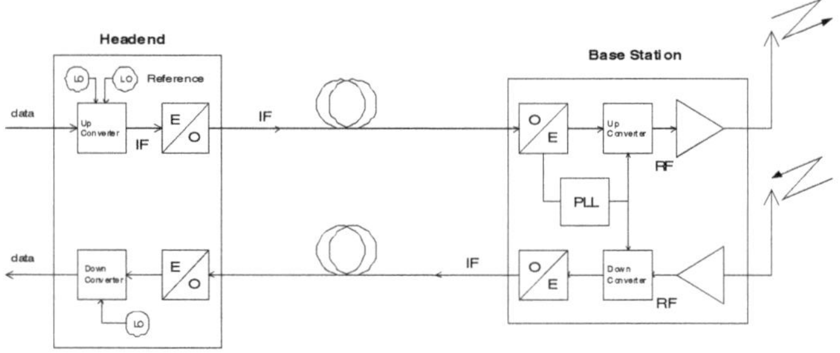

Figure 44 : Block diagram of the OPLL system.

The head-end is identical to the IF head-end with the addition of the reference signal multiplexed with the IF data signal. The base station provides the same functionality as for the IF base station, with a PLL providing the LO signal for upconversion. The PLL and mixer may be implemented as MMIC (mm Integrated Circuit) devices together with the subsequent amplifiers and filters, providing potential for reduced cost, reduced size and increased reliability.

In principle the OPLL approach provides similar system flexibility to the IF approach. Multiple carriers can be supported, with transparency to modulation format. Differences in the systems arise from the requirement of the OPLL that an optical reference signal is transported together with the data signal. Further, the OPLL approach will allow MMIC integration to simplify the base station, which is the key aim of generic radio-over-fibre techniques. Below we focus on the significance of these features for system operation and performance.

The OPLL approach allows implementation of a potentially simple optical link between head-end and base station. Since IF frequencies are utilised, direct laser modulation can be employed without significant dispersion penalty at 1550 nm. For fibre spans of several tens of kilometres and modest splitting losses, optical amplification is not required. If commercial off-the-shelf DFB (Distributed Feedback) lasers and photoreceivers can be employed, the link can be expected to be low cost and highly reliable.

The OPLL approach achieves considerable simplification of the optical link compared to the RF system, but shares with the IF approach the disadvantage of increased base station complexity. A PLL must be integrated in the base station, together with demultiplexer to separate IF data and reference signals, a mixer to perform upconversion, and further filtering to remove unwanted mixing terms. To provide significant advantage compared to the IF approach it seems clear that the PLL should be implemented using MMIC technology

All three techniques outlined above provide a good deal of flexibility for system deployment. Each can support multiple carriers and are transparent to data format. Employing the self-heterodyne technique for the RF approach overcomes the effect of fibre chromatic dispersion, allowing all three techniques to utilise installed standard singlemode fibre.

The RF link is clearly technologically more challenging. The most serious effect results from the interaction between laser phase noise and chromatic dispersion, producing an increase in carrier noise. The main complexity associated with the RF approach is centred on the mm-wave source. The self-heterodyne source designed for the field trial provides good performance, but is a complex subsystem requiring a large assembly effort. The high-speed photodiode appears to present much less risk than the source sub-system, the main issue is that future demand will be sufficiently high to push down component prices.

The IF approach provides significant simplification of the optical link, but shifts complexity to the base station. In this case component and assembly costs for the base station are expected to dominate. The OPLL approach provides an avenue to provide significantly reduced costs by allowing monolithic integration of the local oscillator and upconverter in the base station. This approach achieves a good compromise between base station complexity and optical link requirements. However, the technology required is not sufficiently mature to accurately predict techno-economic performance. Preliminary measurements of the OPLL illustrate the difficulty of achieving good phase noise performance. The VCO (Voltage Controlled Oscillator) phase noise will be critical in determining system capacity.

4.2.2. Dynamic Capacity Reallocation in a HFR Network (PRISMA[43])

Every network has to satisfy a certain capacity demand, which has to be determined during the planning phase, before the network is actually built. The installation costs of a network are proportional to the dimensioning of the network. Overdimensioning it, resp. underdimensioning it, i.e. providing more, resp. less, capacity than is actually required, leads to a capital overspending, resp. to a high upgrade cost and to revenue loss. Notice, once the hardware is installed following the planned design, the network is fixed and a redesign comes at a high cost. The implemented solution is then always a tradeoff, i.e. a network providing enough capacity in most circumstances with a determined call blocking probability.

In the case of a wireless network the dimensioning issue is relaxed compared to a fixed network, where the bulk of the installation costs are determined by the installation of the cables. Furthermore, a capacity upgrade in the wireless network is simpler, as long as the increased number of customers or their relocation stays within the radio coverage.

In an optical network an upgrade can be done in several ways, depending on the causes. When the fibres get exhausted new fibres or cables can be added on the same route, provided there is the physical space, or a new fibre route can be installed. If there is no space and if the

---

[43] Project partners: Lucent Netherlands (NL), Corning SA (F), Intracom (GR), KPN Research (NL), Norcontel Ireland (IRL), University of Limerick (IRL), CTIT – University of Twente (NL), IMEC (B)

amount of additional capacity justifies it, the network can be upgraded to WDM (Wavelength Division Multiplexing).

In a wireless network, once the capacity gets exhausted in a cell, the size of the cell has to be reduced and new cells have to be added. This requires a careful reengineering of the spectrum allocation to avoid interferences. Once the new base stations are installed the network configuration is fixed again. The figure 45 shows the cell structure of a wireless network. Although the cell is in reality a circle it can be represented to a good approximation as a hexagon. At the centre of each cell is the base station (BTS) radiating at a particular frequency. To avoid interferences, neighbouring cells have to use different frequencies. All the base stations are linked by cables to the switching centre (BSC). Cells not necessarily have the same size, depending on the required geographic coverage and capacity demands.

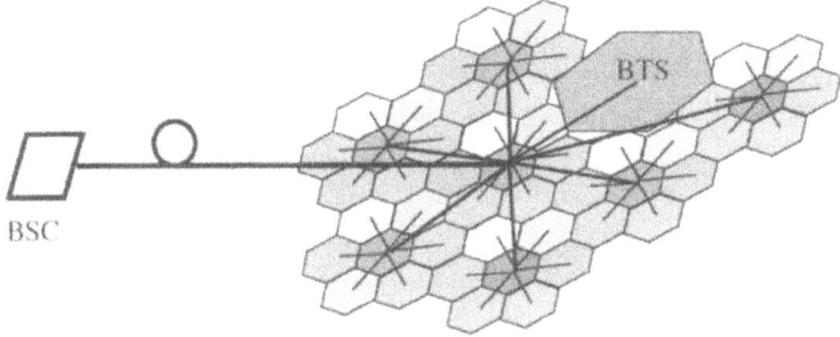

Figure 45 : Cell structure of a wireless network.

The great advantage of the wireless networks, the user mobility, is also their weakness. Contrary to fixed-line users, the wireless users are free to move around and the occupation ratio of the radio cells shifts accordingly over daytime. For example, during office hours the traffic would be concentrated in the business district, whereas outside these hours it would shift towards the residential areas. Inevitably hot-spots occur, i.e the traffic concentration in a cell saturates the allowable capacity, since the network can not dynamically reallocate the capacity where needed and when needed. This situation even worsens with the introduction of broadband mobile applications, like for example UMTS. The PRISMA project solves the hot-spot problem, allowing a dynamic reallocation of capacity for broadband mobile applications, utilizing WDM.

In every PON (Passive Optical Network) there are several splitting points in the optical network. The splitting can be a power splitting, a spatial splitting, a TDM splitting (Time Division Multiplex), a wavelength splitting (WDM based), or a mix of those. Each approach has its distinct advantages but the WDM approach, adopted by PRISMA, allows a dynamic reconfiguration of the network. Flexible wavelength routers located at the splitting centres allow to assign a radio cell or a group of radio cells with a distinct optical wavelength or a set of wavelengths, as illustrated by the example of the figure 46. It now becomes possible to shift the capacity around the network to provide it where it is most needed. When in a network cell there is a sudden high demand for capacity, an extra radio carrier is added in this cell and extra feeder capacity is offered by rearranging the wavelength channel distribution among the base stations. The cell extension can also be modified in this way. Thus, the operator can reconfigure the network at will, from the BSC, without the need to intervene on site at the base stations. Of course, this adds some complexity to the base station equipment.

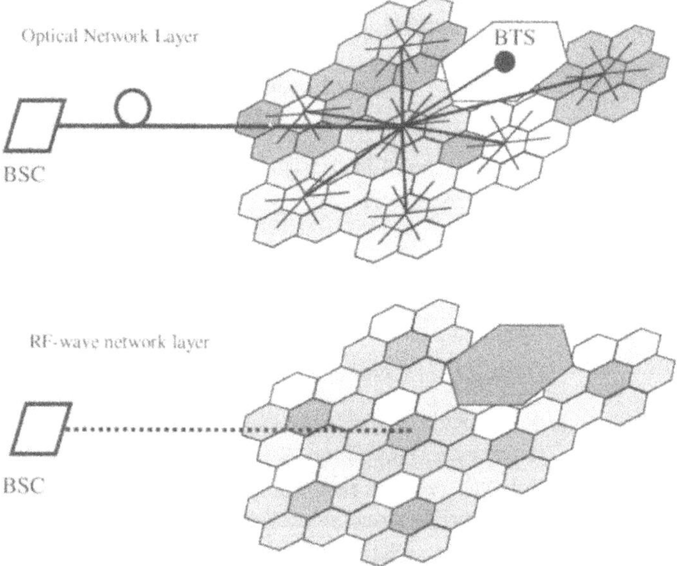

Figure 46 : Matching optical wavelengths to groups of RF cells.

In the classical configuration, all base stations share the same optical carrier wavelength. The total number of cells or the total radio capacity of the network is ultimately limited by the available optical capacity that has to be shared between base stations. An overloading of demand in a particular cell limits the usage of the other cells. This situation is difficult to solve as long as the connections are not terminated and capacity released to be reused in another cell. Currently the GSM operators are trying to deal with this situation employing sophisticated routing algorithms but the dynamics of the system do not allow to react on time.

With the introduction of WDM it becomes possible to solve this problem as it arises. The capacity exhaust of the optical wavelength is solved by allocating the capacity requesting base station to a different optical carrier. This is shown in the example of the figure 47. This approach becomes particularly interesting with the introduction of broadband services.

Another advantage of this WDM HFR approach is the allocation of different wavelengths to different operators or services. In deed, a network operator could lease its network to competing mobile service providers, assigning each one its own optical carrier. Alternatively different wavelengths could be used to carry in parallel different services, like for example GSM and UMTS.

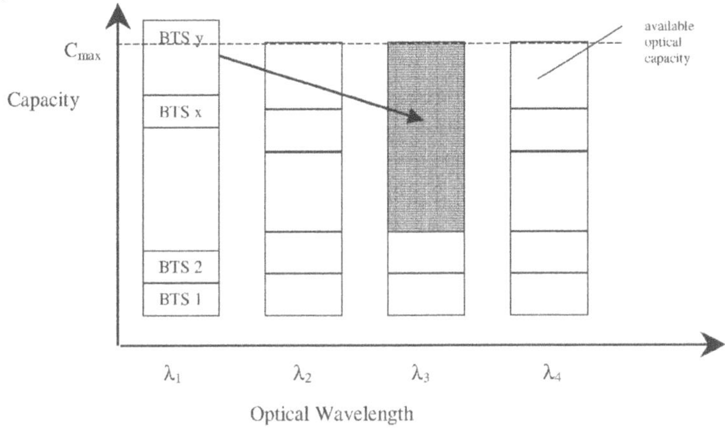

Figure 47 : Reallocating RF bandwidth to a an empty optical carrier.

A cost comparison between the PRISMA, the space division multiplexing, the power splitting, and fixed WDM approaches, revealed the attractivity of the chosen approach, providing more network flexibility at only a minor additional investment. The economics in the PRISMA case are highly dependent on the number of wavelength splitting points in the network. This is a design parameter that has to be carefully selected to balance the network granularity and costs. Replacing the power splitters closer to the OLT (Optical Line Termination) requires less WDM components and results in a larger granular size. However, a large granular size decreases the flexibility to assign a large capacity to a small group of cells, as the capacity is shared with other cells. The location of the wavelength splitters is then a trade-off between the required granularity and the economics. In a real situation the wavelength splitters will most probably be located close to the OLT and migrate closer to the base stations as the customer base grows, and the number of cells multiplies.

5. ANTENNA – BEAM FORMING

5.1. Planar Antenna Technology for Microwave-Optical Interactions

Y. Qian, W. R. Deal, T. Itoh
Electrical Engineering Department, University of California, Los Angeles
405 Hilgard Avenue, Los Angeles, CA 90095, USA
e-mail: yqian@ucla.edu

Abstract

Antennas are electromagnetic transducers between different forms of electromagnetic signals and energy. Since many advanced optical systems carry microwave signals, effective interaction of the microwave signals with free space is important. Although there are a number of antenna structures available, some of them are more suitable for RF photonics environment. Since the advanced form of RF photonics will benefit from planar technology, the antennas are desired to be planar as well. This lecture critically reviews a number of available and proposed planar (and quasi-planar) antenna structures. The evaluation of these antennas is carried out from fundamental and practical aspects as well as compatibility with photonic systems. An antenna selection guide will be provided.

5.1.1. Introduction

The field of RF photonics has expanded enormously in recent years, leading to a number of important applications including antenna remoting for cellular and micro-cellular radio using analog fiber links, cable-TV signal distribution, and optically assisted phased-array antennas [55, 56]. A typical analog fiber link, by definition, includes a pair of E/O (modulator) and O/E (photodetector) devices at the two ends of the low loss, low dispersion optical fiber, as well as any passive impedance-matching circuits to match the modulation device and photodetector impedance to the RF signal. The antenna, which is also an indispensable part of the complete link, is usually treated as a separate component with 50 Ω interfacing impedance in a conventional design approach

As wireless applications move towards higher microwave and millimeter wave frequencies, photonic device designers are faced with the great challenge of achieving the maximum bandwidth and saturation power product within each of the popularly adopted design topologies [57]. Traveling-wave photodetectors with bandwidth of 190 GHz and efficiency of up to 0.45 A/W has been reported [58]. At UCLA, a velocity-matched distributed photodetector (VMDP) consisting of an array of small MSM-diodes connected to each other via velocity-matched optical and electrical waveguides has been developed [59]. Detector currents of 56 mA have been experimentally demonstrated at 850 nm in a 49 GHz VMDP with three active photodiodes. In comparison, the typical maximum photocurrent of a commercial 50-GHz photodetector is about 1-2 mA [56].

In addition to developing innovative photonic devices which satisfy the ever demanding requirements for higher power and higher frequency for current and future applications, we believe that the system performance of an RF photonics-based link can be greatly enhanced by appropriate integration of the photonic devices and antennas. The authors' group has put a lot of efforts in developing novel active integrated antennas, which have shown great promise in designing modern microwave and millimeter-wave architectures with desirable features such as compactness, light weight, low cost, low profile, minimum power consumption, and multiple functionality [60, 61].

As a proof-of-principle study, we recently demonstrated the integration of a tapered slot antenna (TSA) with our velocity-matched distributed photodetector (VMDP). The concept was confirmed by successful photomixing and antenna measurement with an X-band prototype [62]. It was also realized, however, that the TSA is not the ideal antenna structure to be used in a VMDP-based antenna array, both because of its large electrical size and because of the relatively high cross polarization radiation when high dielectric-constant substrate (GaAs, InP, etc.) is to be used eventually for MMIC implementation. Therefore, we need to find a better antenna candidate to optimize our design, or to develop a totally new antenna structure if such a candidate does not exit.

This paper intends to give an overview of planar antenna structures which we believe may find important applications in modern RF photonics systems. Most of the antennas we describe here are based on microstrip, CPW and CPS designs, thus compatible with modern planar fabrication technology. The evaluation of these antennas is carried out from fundamental and practical aspects as well as compatibility with photonic systems. An antenna selection guide will be provided at the end of the paper.

5.1.2. Design Considerations for Planar Antennas

Antennas, which serve as the important interface between guided waves in RF circuits and free space radiation, can be generally classified as 1D (wire), 2D (planar) and 3D structures. Some of the most popularly used wire antennas include dipole, monopole, loop and helical antennas. Planar antennas, on the other hand, are best represented by patch, slot, ring and tapered slot (Vivaldi) antennas. 3D antennas, such as horn, lens and paraboloid, are usually realized by more complicated machining processes. It should also be noticed that both 1D and 3D antennas can be modified into planar versions so that they can be more easily realized with cost-effective printed circuit technology. For example, printed dipole and spiral antennas have been developed and used in a number of medium- to large-scale antenna array applications. The bow-tie antenna, on the other hand, is another example of reducing the order of a 3D antenna (biconical) to 2D for easier fabrication.

One driving factor that makes planar integrated antennas so desirable is the ease of integration of these antennas with microwave or millimeter-wave circuit components. For this reason, planar integrated antennas must be compatible with these technologies, which, at microwave and millimeter-wave frequencies, are typically microstrip or Coplanar Waveguide (CPW) based. These transmission lines have several advantages, including ability to integrate 3-terminal devices, mechanical and heat-sinking capabilities due to metallic ground planes, as well as simplified packaging issues. Therefore, it is essential that these types of transmission lines can directly or indirectly feed the planar integrated antennas. Examples of direct feeding include the patch antenna and slot antenna, which are easily integrated with microstrip or CPW, respectively. Indirect feeding can include transitions or various forms of EM coupling. The method of feeding is critical and can affect antenna cross-polarization, patterns, bandwidth as well as possible array architectures.

However, the dielectric substrates on which microstrip and CPW compatible antennas are fabricated will support surface waves, and can propagate energy away from the antenna, thereby lowering its efficiency. While the losses are small at lower frequencies, this can be a major problem at microwave and millimeter-wave frequency where many new applications are targeting planar antennas. The thickness of the substrate, permittivity and frequency of operation determine the amount of surface wave losses. Several methods have been developed to reduce this, as will be briefly discussed later.

Different classes of planar antennas are capable of a broad variety of radiation characteristics. The most common classes, patch and resonant slot antennas, demonstrate broad, low gain patterns making them excellent for use in multi-element beam-forming arrays. Additionally, some of these antennas can be easily modified for dual-linear or circular polarization. More sophisticated classes demonstrate higher gains and some are capable of frequency scanning. Another important parameter is the frequency bandwidth of the antenna, which usually poses a severe design trade-off among size, dielectric constant, number of substrate layers, as well as possibly degradation in other parameters such as cross polarization and front-to-back ratio.

5.1.3. Microstrip Patch Antennas

The microstrip patch antenna has a broadside radiation pattern that allows it to be integrated into two-dimensional arrays, with desirable features including low profile, low-cost, conformability and ease of manufacture. Additionally, various feeding schemes can be used to achieve linear or circular polarization. Feeding is extremely important with the patch antenna. Patches with direct microstrip feed (either inset or offset feeding) have very narrow bandwidths, almost invariably less than 5%. Other feed mechanisms have been used to increase bandwidth, including proximity coupling and aperture coupling, both of which require multi-layer fabrication. A review of this technology is discussed in [63].

An alternative approach to increase the bandwidth of patch antennas is to use electrically thicker substrate, thus effectively lowering the Q factor of the antenna cavity. However, high levels of TM_0 surface waves can result and therefore reduce the radiation efficiency as well as degrade the radiation pattern. The problem of electrically thick substrate is also a common one for high frequency antennas on high permittivity substrates such as Si, GaAs or InP. A lot of research efforts have been devoted recently to solve this radiation inefficiency problem, including the use of the latest micro-machining technique [64].

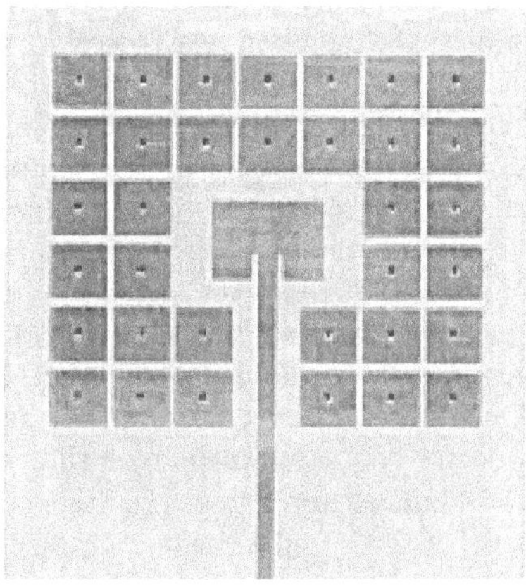

Figure 48 : A Ku-band microstrip inset-fed patch antenna integrated with PBG for surface wave suppression.

More recently, the photonic band-gap (PBG) concept has also been used for this purpose. In this case, a periodic array of perturbations is used to suppress the surface wave mode, such as the structure shown in figure 48. The PBG lattice consists of capacitive pads on the top plane connected to the ground by inductive shorting pins. Full-wave analysis demonstrates that this periodic structure is indeed effective at eliminating the TM_0 surface wave. When integrated with a patch antenna on RT/Duroid substrate, it was found that the gain of the antenna could be increased by 1.6 dB when compared with a reference patch without PBG [65].

5.1.4. Resonant Slot Antennas

The slot antenna, consisting of a narrow slit in a ground plane, is a very versatile antenna. With modification, it is amenable to waveguide, coplanar waveguide (CPW), coaxial, slotline or microstrip feeding schemes and has found application in all aspects of wireless and radar applications. Planar microstrip-fed slot antennas have been reported in the early 1970s [66].

The resonant half-wavelength slot antenna is a desirable choice in many cases because of its compact size, but has large input impedance, typically larger than 300 Ω, which makes it unattractive to match to. This can be circumvented by using an offset microstrip feed or the folded-slot antenna, which stems directly from the folded dipole by Booker's relation. In this case, the slot is folded in upon itself. The overall length of the antenna remains approximately a half-wavelength, but increasing the number of folds reduces the radiation resistance.

The CPW version of the folded-slot has been investigated extensively [67, 68]. This antenna requires no input matching which makes it an inexpensive and compact candidate for direct integration with microwave circuits. The broad radiation pattern also makes this antenna an excellent candidate for wireless communications systems, which are currently pushing into the microwave regime.

The folded slot antenna can also be easily fed by using microstrip lines, as shown in figure 49 [69]. The folded-slot is etched in the ground plane of the substrate. One of the inner metalizations of the slot is connected to the microstrip conductor on the top plane by a shorting pin. A 50 Ω input impedance is easily obtained for a two-fold slot on a RT/Duroid with a relative permittivity of 2.33 and substrate thickness of 31 mils. This antenna has been found to have a very broad frequency bandwidth in spite of a simple, single layer design. The bandwidth for the prototype shown in figure 49 is measured to be from approximately 1.5 to 2.9 GHz (BW=61% for S11<-10 dB). One complication with this sort of antenna, however, is its bi-directional radiation pattern, which may not be suitable for some applications.

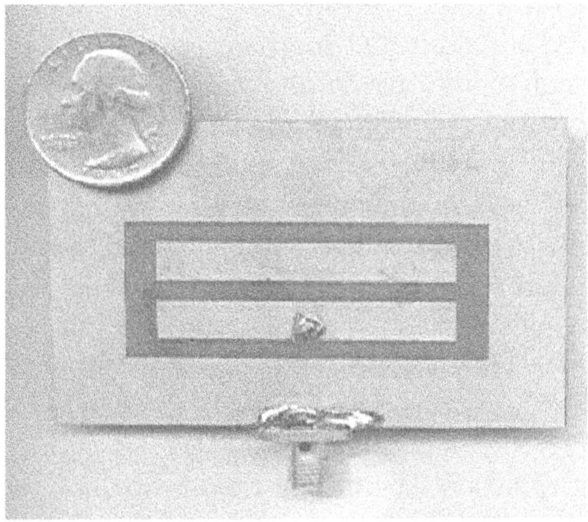

Figure 49 : A microstrip-fed folded slot antenna with broadband characteristics.

5.1.5. Tapered Slot Antennas

Both the patch and slot antennas described in the previous sections are resonant-type (or standing wave) structures. These types of antennas typically have modest gain radiation patterns, and narrow bandwidth unless special broadband techniques are employed. Another class of antennas, travelling wave or non-resonant antennas, is comprised of structures that radiate as the wave propagates. These structures are electrically large when compared to resonant type structures and typically demonstrate higher gain and may demonstrate broad-bandwidth performance.

One important category of the traveling wave antennas is the tapered slot antenna, or TSA. Extensive reviews can be found in [70, 71]. These antennas are completely planar, can be easily printed on dielectric substrates, have endfire radiation patterns, and are capable of obtaining high directivity and/or bandwidth. Proposed applications include millimeter-wave imaging, power combining and use as an active integrated antenna element. The TSA is usually etched into the metalization on one side of a dielectric substrate, and takes one of the three popular configurations, i. e., the Vivaldi (exponential taper), the linear taper (LTSA) and the constant width slot antenna (CWSA). Radiation of a particular frequency will occur where the slot is a certain diameter. Slot width should reach at least one half wavelength for efficient radiation to occur. Therefore, maximum and minimum widths roughly determine the bandwidth of the structure.

There are several important design concerns with this type of antenna. TSA are usually built on thin, low permittivity substrates if they are to achieve good radiation patterns and maintain good radiation efficiency which will be reduced by TM_0 surface wave losses. As with the patch antenna, micro-machining has been used to reduce the electrical thickness of millimeter-wave TSA antennas by reducing the effective permittivity. This also allows a physically thicker substrate, which is essential for mechanical support for this kind of structure at millimeter wavelength.

Additionally, when used as an integrated antenna, a transition from the transmission line of choice (microstrip or CPW typically) must be used. The bandwidth of the transition may limit the bandwidth of the structure. A CPW-fed TSA slot with exponential taper is shown in figure 50. The antenna is fabricated on 25-mil substrate with 10.2 relative permittivity. The broadband CPW-slotline transition allows broadband response; measured bandwidth is greater than 70% centered at 13 GHz [72].

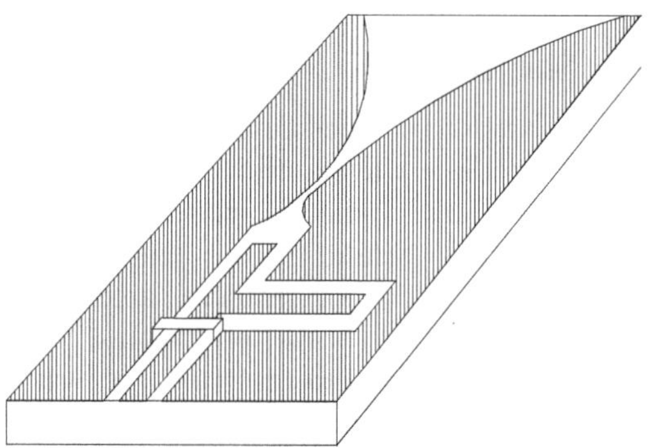

Figure 50 : Schematic of a uniplanar CPW-fed Vivaldi antenna.

5.1.6. Uniplanar Quasi-Yagi Antenna

An ideal planar antenna, in the authors' opinion, should have the following desirable features: (1) It has high-quality radiation characteristics including well-defined pattern, low cross-polarization and good front-to-back ratio; (2) It has a broad frequency bandwidth, especially in the present case where the ultrawide instantaneous bandwidth of photonic devices is to be exploited; (3) It has to be as compact as possible for easy integration with active circuits and array implementation; (4) It has to be simple and low-cost in fabrication. Unfortunately, most existing planar antenna structures developed so far fail to meet at leat part of these requirements.

Our intensive searching for such an "ultimate" planar antenna has led to the recent invention of a novel uniplanar quasi-Yagi antenna, as shown in figure 51. The antenna uses the truncated microstrip ground plane to replace the reflector dipole in a traditional Yagi-Uda configuration (thus the name "quasi-Yagi"), and uses a unique microstrip-CPS transition as balanced feed to the driving dipole [73]. After optimization, the antenna has achieved a record frequency bandwidth of 50 % (VSWR<2), using only one single layer of high dielectric substrate (Duroid, ε_r=10.2) [74]. The quasi-Yagi antenna radiates an end-fire beam, with a front-to-back ratio typically greater than 15 dB, and cross polarization level below –12 dB across the entire frequency band. A very low mutual coupling level of below –22 dB has been measured for a two-element array with $\lambda_0/2$ separation.

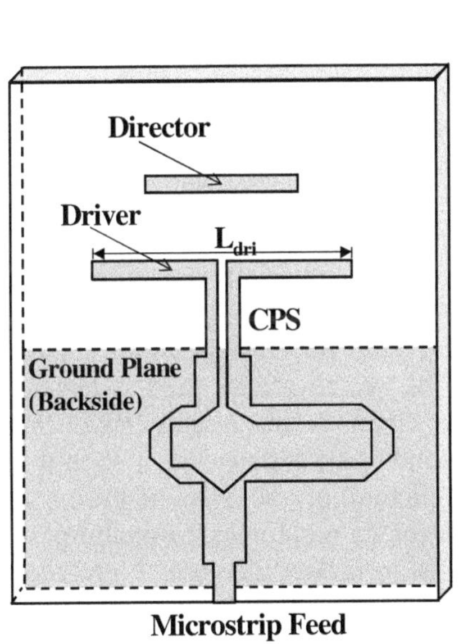

Figure 51 : Schematic of the quasi-Yagi antenna.

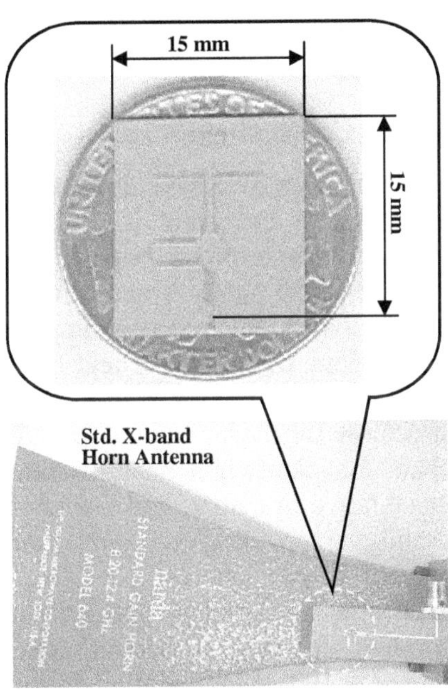

Figure 52 : X-band prototype of the quasi-Yagi. The simple, compact antenna demonstrated a record 50 % bandwidth ever achieved with a printed antenna on a single piece of high dielectric constant substrate.

As can be seen in figure 52, the quasi-Yagi antenna is at least two orders smaller in volume than a standard horn antenna for the same frequency coverage. The antenna gain is also found be relatively flat within the frequency band. Although a single element quasi-Yagi has a moderate gain of 5~6 dB, they can be easily implemented into "card-type" arrays which, subsequently, can be stacked to realize a two-dimensional array. This new structure offers several distinctive advantages over the venerable tapered slot antenna array configuration, including compactness, reduced cross polarization and mutual coupling, as well as much easier integration with active circuits since the quasi-Yagi can be readily interfaced with a 50 Ω microstrip line.

We have also found that the quasi-Yagi antenna can be scaled linearly to any frequency band of interest while retaining the wideband characteristics. In fact, a C-band prototype we have simulated and fabricated on 1.27 mm thick Duroid (ε_r=10.2) has measured a similar 50 % frequency bandwidth (4.17 to 6.94 GHz) [75]. FDTD simulation for a millimeter-wave version indicates that a single quasi-Yagi antenna works from 41.6 to 70.1 GHz (51 % bandwidth), which covers part of Q- and most of the V-band.

5.1.7. Concluding Remarks

In this paper we have discussed a variety of classes of planar antennas which might be useful in RF photonic system applications. Although serious research efforts in planar antennas dated back to the early 1970s, there is still great room for improving both their fundamental characteristics and practical implementations, especially for commercial applications where simple and low-cost mass production is almost a pre-requisite. To facilitate readers in identifying the best possible antenna structures for their particular applications, we have appended a selection chart for planar antennas. While the chart is more qualitative than quantitative in nature, and does not intend to be exhaustive, it should be useful in providing a rough, global picture of the state-of-the-art of modern planar antennas. As the technology continues to mature, we expect the functionality of these antennas to further increase. With recent wireless applications moving to higher frequencies, planar antennas should attract more attention from RF designers in order to realize high performance systems which can remain a strong competitive edge in a rapidly evolving market.

5.1.8. Acknowledgment

This work was supported by ONR MURI N00014-97-1-0508.

Appendix: Planar Antenna Selection Chart

		Pattern	*Directivity*	*Polarization*	*Bandwidth*	*Comments*
Patch		Broadside	Medium	Linear/Circular	Narrow	Easiest design
Slot	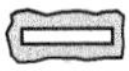	Broadside	Low/Medium	Linear	Medium	Bi-directional
Ring		Broadside	Medium	Linear/Circular	Narrow	Feeding complicated
Spiral		Broadside	Medium	Linear/Circular	Wide	Balun & Absorber
Bow-Tie		Broadside	Medium	Linear	Wide	Same as Spiral
TSA(Vivaldi)		Endfire	Medium/High	Linear	Wide	Feed transition
LPDA	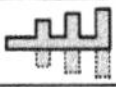	Endfire	Medium	Linear	Wide	Balun, Two Layer
Leaky-Wave		Scannable	High	Linear	Medium	Beam-steering, Beam-tilting
Quasi Yagi	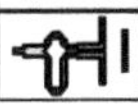	Endfire	Medium/High	Linear	Wide	Uniplanar, Compact

5.2. Antenna Applications of RF Photonics

J.J. Lee
Raytheon Systems Co.
R2-V541, PO Box 902, EL Segundo CA 90245-0902, U.S.A.
e-mail : jjlee8@west.raytheon.com

This part reviews system applications of RF photonics. Two examples will be given to discuss the requirements, benefits, and design approaches of photonics for phased array antennas. The first example illustrates the optical control of phased arrays by fiber-optic links for RF and data remoting and a photonic time-shift network for wide instantaneous bandwidth. The second example illustrates how photonics can be used to form a wide band beamforming network for multibeam arrays.

5.2.1. Optical Control of Phased Array

5.2.1.1. Background

To begin with, the design and the wide band performance of an L-band 96-element array controlled by photonics are discussed. In recent years there has been a growing interest in applying the photonic technology to phased arrays [76-81]. Significant progress has been made in the reduction of RF to optics conversion loss. The unique features of an optically fed array are wide instantaneous bandwidth, low transmission loss for data remoting, and reduction in size and weight as a long term goal. The 2-D array developed for technology demonstration[44] consists of 96 wideband elements, grouped into 24 columns, with each column steered by an 11-bit time shifter. The L-band array is capable of transmitting and receiving over ± 60° scan in the azimuth plane. It is controlled by RF and digital fiber optic links from a remote site. The design parameters of the photonic array are:

Aperture size	~ 1 x 2.7 m, conformal, 3 m radius
Frequency	L band, 850 - 1400 MHz
Bandwidth	50% at 1125 MHz center frequency
Radiation element	printed "bunny ear" elements
No. of elements	4 x 24 (96)
Element spacing	10.7 cm AZ, 21.3 cm EL
Directivity	~ 25 dBi (midband)
Beamwidth	~ 5° AZ, 15° EL (midband)
Scan limit	±60° AZ, no scan in EL
No. of T/R modules	24
Time shifters	5 bits photonic, plus 6 bits electronic
Radiated power	~ 30 W
Peak sidelobe	Transmit -13 dB AZ & EL
Receive	- 25 dB AZ, -13 dB EL

5.2.1.2. Photonic Beamformer

The system concept for a wideband array is quite simple and form the beam by group delays instead of phase combining. Each column was supported by a T/R module with a 6-bit electronic microstrip delay line. Every three columns were combined to form one subarray, which was controlled by one 5-bit photonic time shifter. The photonic time shifter provides the long delays for the whole group of subarray, while the electronic time shifter refines the delays within one nanosecond for each column.

The system block diagram shows how the phased array is controlled by RF and digital fiber optic links from a remote site. On transmit a laser light is modulated by the RF source and transmitted through the fiber to the antenna site. The RF signal is photo-detected and amplified before it is distributed to eight subarrays. After going through the programmable 5-bit time shifter, the signal is further divided into three ways, one for each column of four elements.

The building block of the wideband beamforming network is the photonic time shift module. The programmable time shifters provide the

---

[44] Program funded by DARPA/Rome Lab under Contract No. F30602-91-C-006

coarse delay steps ranging from 0.25 ns to 7.75 ns for the subarrays, while the electronic delay lines in the T/R module provide fine differential delays ranging from 0.01 ns to 0.5 ns. The physical dimension of the 5-bit photonic time shift module is 10.5 x 12 x 6 cm. Key components inside the time shifter are four semiconductor pigtailed lasers, one 4x8 fiber coupler, and two 1x4 detector arrays with FET bias switches.

During transmission, the microwave signal goes though a 1:4 RF switch and modulates one of the four lasers. The laser converts the microwave signal into light which is coupled into the 4x8 fiber coupler. After splitting by the coupler, the light is incident on all the detectors in the array. By switching on one of the 8 detectors with the bias switch, the modulated light is routed through one of the 32 preset delays before recovering the RF signal.

The RF signal is then post amplified and divided into three ways with each feeding another 6-bit time shifter in the T/R module. In the receive mode, the signal path is reversed except that the signal must be routed through two transfer switches in the photonic module so that the signal can go through the non-reciprocal 5-bit time shifter in the same direction.

In the 5-bit time shift module most of the insertion loss is incurred in the 4x8 optical coupler between the lasers and the photodetector array. The internal fan-out loss is 18 dB plus 2 dB excess loss. Further, the input impedance of the laser is only a few ohms, while the output impedance of the detector is very high, on the order of several kilo-ohms. These mismatches contribute to additional losses. To overcome these losses, matching circuits have been developed and a preamplifier and a post amplifier are usually included in the circuit to make the link appear to be transparent. Significant progress has been made by many researchers, and it is expected that this conversion loss will be further reduced in the future.

5.2.1.3. Noise Figure and Dynamic Range

Several key issues were examined in the system design and tradeoff study. The most obvious question is the impact of the high conversion loss of the time shift element on the overall noise figure of the receive system. As in a single-channel case, the overall noise figure of an array with differential weightings is primarily set by the noise figure of the LNA in each channel. Also, minimizing the front-end loss is most important because it directly affects the noise figure. This is accomplished by placing a high gain LNA right behind the radiating element. If the gain is not sufficiently large, the overall noise figure will be affected to some extent by the losses after the LNA. This is especially true for a photonic array, where the downstream loss is significant and the effect of the beamforming network can not be overlooked.

The conversion loss of a photonic 5-bit time shifter is on the order of 40 dB without wide band input and output impedance matching. To overcome this loss, the LNA gain must be at least 45 dB or higher to reduce the overall noise figure to a reasonable level. For example, with a nominal front-end loss of 1.5 dB and a noise figure of 2 dB for the LNA, the overall system noise figure can be maintained at the level of 2.5 dB if the LNA gain is 40 dB.

The use of high gain LNA is not without limitations in terms of feedback and leakage. A lesson learned from this development is that the transfer switches in the T/R module for the transmit and receive operation must be specially designed with very high isolation, 60 dB or more each. This results from the fact that the output of the time shifter is on the order of 0 dBm, and the power amplifier must provide 35 dB gain to boost the radiated power to two watts level specified for this application. Thus, with a 45 dB LNA, the loop gain in the T/R module is close to 80 dB, which tends to cause oscillations if the transmit and receive paths are not sufficiently isolated from each other.

There are different definitions of the dynamic range in the calculation of the radar performance. In this case, the spur free dynamic range, defined as the third order intermodulations not to exceed the noise floor, is used. Based on the beamforming network discussed, a signal to noise ratio analysis using spread-sheet program was carried out to estimate the dynamic range of the receive path.

When two amplifiers are cascaded in series, the 3rd order intercept point is somewhat degraded. To maximize the dynamic range in a cascaded system, the overall gain should be distributed properly at different stages. Lumping all the gain at the front-end is not optimal. This is especially true in the photonic array where three stages of amplification were required to overcome the loss in the receive path. In this system, an LNA in the T/R module was used to support each column; the combined output of the subarray was pre-amplified in front of the photonic time shifter, which is followed by a post-amplifier to offset the insertion loss.

Using actual device parameters in the analysis, we can optimize the dynamic range of the system to exceed 95 dB by properly distributing the gains of the amplifiers at different stages along the signal path. A high gain LNA at the front-end tends to improve the overall noise figure but reduce the dynamic range. On the other hand, a lower LNA gain will boost the dynamic range, but degrade the noise figure somewhat. Thus a tradeoff is needed to optimize the performance so that a balance on the noise figure and the dynamic range can be achieved.

Note that if the noise bandwidth increases by 3 dB the dynamic range is reduced by 2 dB, which is the two-thirds rule intrinsic to the spur free dynamic range definition. On the other hand, if the number of subarrays goes up by 15 dB, the dynamic range increases by 10 dB because the signal to noise ratio is enhanced by as much due to coherent signal combining.

Compared to an active aperture array using conventional phase shifters in the T/R modules, the optically fed array suffers about 13 dB degradation in the dynamic range due to the additional loss in the time shift module supporting each subarray. However, in spite of the high conversion loss of the photonic time shifter, many array systems can benefit from the insertion of this emerging technology to achieve wide instantaneous bandwidth, and reasonably high dynamic range at the expense of a slightly degraded system noise figure.

Now consider the effects of the conversion loss on the transmit path. The power limitations of photonic devices preclude their substitution for conventional cables or waveguides for high power distribution in the phased arrays. The concern here is how much gain is required to boost the power up to the radiation level needed for typical array applications. Note that the maximum input power for the laser is about 10 dBm, so the input to the remote link and the time shifter is limited to this level. In the demo system, three stages of amplification were required. A post amplifier of 35 dB gain was used at the end of the remote link, followed by another post amplifier of 37 dB at the output of the time shifter. In addition, a power amplifier of 38 dB gain was used to produce one Watt radiated power for each column.

For other applications where a single channel photonic link is used with no RF fan-out loss, the transmit path will require a 10 dBm input power for the fiber optic link, followed by 30 dB post amplifier, and then a 35 dB power amplifier to produce one Watt power level at the aperture.

5.2.1.4. Array Performance

Antenna patterns of a nine-column test array over the specified frequency range with the beam scanned to broadside, ±30°, and ±60° showed that the beam did not squint over the bandwidth by using a true-time-delay beamforming network. A conventional array with phase shifters could not have achieved this performance.

The bandwidth of the array was studied by a new technique performed in the time domain. The basic concept is to inject a 2 ns pulse into each column of the array through a series feed and wideband couplers, and then record the waveform after the pulse propagates through all the components in the transmit or receive path. By examining the pulse shape, magnitude, and the relative time delay, we can determine the insertion loss, time delay setting, and the status of the components in each channel. This time domain reflectometry type measurement can not be used for conventional band limited arrays, but it is most suitable for a photonic time shift system.

The impulse response of a 3-column subarray consisting of 3 T/R modules and a 5-bit photonic time shifter verified that the system has a 550 MHz bandwidth, which corresponds to a range resolution of 30 cm. The pulse propagated through all the RF and optical components in the receive path, so the pulse shape revealed the true frequency characteristics of the antenna system.

The antenna patterns of the 24-column photonic array at 850, 1000, 1250, and 1450 MHz will be shown. The average peak sidelobe on transmit is -15 dB, and the level on receive is -20 dB. A 10 dB edge taper was imposed on the 8:1 power combiner for the receive patterns, which produced lower sidelobes than the transmit case.

5.2.2. Multi-Beam Photonic Array Feed

5.2.2.1. Background

Photonics can also be used to support a multibeam wide band feed for array antennas. The main advantage is to reduce the complexity of the array front end. This is accomplished by replacing multiple sets of discrete phase shifters at the array element level with a simplified fiber optic Rotman lens supplemented with a RF heterodyne technique for fine scan. The feed "engine" can be used for both transmit and receive operations. On receive, the signal across the aperture is conjugatedly matched at the front end by the phase gradient produced by the transmit network.

This development was motivated by the need to reduce the number of antennas on many airborne and shipboard platforms. Conventional techniques to achieve multiband and multibeam capabilities are impractical because of the size, weight, packaging density, and high cost of the beamforming networks. Packaging is difficult because of the small element spacing required for a typical 3:1 bandwidth array. It is a major challenge to package multiple sets of phase shifters, drivers, and control lines in the space available behind each element. Also, phase shifters are usually lossy, complex, and expensive to fabricate. In addition, heat dissipation imposes a heavy burden on the mechanical and thermal designs needed to achieve dense module packaging. Thus, innovative multibeam feed and independent beam scan concepts are needed.

5.2.2.2. Multi-Beam Wideband Beamformer

The new beamforming system uses a simplified Rotman lens configuration supplemented with a RF heterodyne system to provide continuous scan. The configuration consists of a few feed ports to point the beams in the general directions over a ±60° range, and the heterodyne system scans the beam over a small region around the discrete offset angles. This phase-locked RF mixing feed combines the signal distribution and beam scan unit into one beamformer for both transmit and receive operations, thereby replacing the multitude of phase shifters, drivers, and beam control circuitry conventionally used. Heterodyne approaches had been studied before, but none has focused on the aspect of wide band and multibeam applications [82-91].

The basic architecture of the multifunctional, wide band beamformer will be discussed, using a 16-element array with five feed ports as an example. Each port covers a 30-degree sector over the ±60° scan range in the azimuth plane. Within any sector, each beam is steered by a heterodyne phase-locked loop, which constrains two frequencies ω_1 and ω_2 to produce a constant beat frequency, ω_0, radiated by the linear array. Frequency ω_1 is fed into the constrained lens as a reference signal. To a first order approximation, this signal provides a uniform amplitude and certain phase distribution across 16 elements along the pick-up side of the Rotman lens. The second frequency, ω_2, from an offset feed port supplies the desired phase gradient along the same 16 elements to steer the beam in the desired direction when the phase front is transferred to the array aperture. These two frequencies will mix to produce the constant ω_0. However, the spatial phase gradient is not affected by the heterodyne process. By varying ω_2 and ω_1 the phase gradient along the aperture and, hence, the beam direction, can be changed. By exciting other feed ports, one can use the same heterodyne process to generate multiple beams with different pairs of RF frequencies.

5.2.2.3. Bandwidth and Beam Broadening

It can be shown that, to a first order approximation, the amount of beam squint normalized to its local beamwidth for a given bandwidth Δf_0 is given by

$$\frac{\Delta\theta_o}{BW} = \frac{N}{2}\frac{\Delta f_o}{f_{MAX}}\frac{d_R}{d_o} Sin\theta_R\left(1-\frac{f_2}{f_o}\right) \tag{16}$$

where N is the number of elements, f_{MAX} is the highest operating frequency of the antenna, and f_o is the current operating frequency. The element spacing is assumed to be $d_o = \lambda_{MIN}/2$ where λ_{MIN} is the wavelength corresponding to f_{MAX}. A criterion to define the bandwidth is to restrict the squint (absolute value) to be less than one quarter or one half of the beamwidth. So one can set

$$\frac{\Delta\theta_o}{BW} = \frac{N}{2}\frac{\Delta f_o}{f_{MAX}}\frac{d_R}{d_o} Sin\theta_R \left(1 - \frac{f_2}{f_o}\right) < 0.5 \tag{17}$$

to calculate the maximum size of the array (N) for a given bandwidth ($\Delta f_o/f_{max}$) in terms of feed angle θ_P, relative element spacings, and vice versa. Note that when f_2 is equal to f_0 without heterodyne, the system degenerates into a conventional Rotman lens with infinite bandwidth, consistent with the definition of a true time delay beamformer. Also, when f_2 is varied to go above or below f_o the beam will deviate from the normal setting θ_o, scanning to the right or left depending on the frequency variation. This is the basic principle of the heterodyne beam scan system.

5.2.2.4. Fiber Optic Implementation

The space feed can be replaced by bundles of fibers precisely cut to produce perfect wave fronts for the directions associated with the feed ports. The fiber version of the feed makes the system compact and foldable. Using light sources of different colors will provide high isolation between independent beams. A special case of the system was described in [91], in which one set of equal-line-length fibers represents the central reference port and another set of unequal-line-length fibers of incremental length ΔL generates the phase gradient required to scan the beam by frequency control through a phase-locked loop. Multibeam operation is achieved by using laser light of different colors to carry control signals for each beam while sharing the same fiber feed system. This sharing is made possible by the use of optical wavelength division multiplexers (WDMs), which allow light signals of different wavelengths to be combined, passed through the common feed system, then separated at the output to generate independent, noninteracting beams.

5.2.2.5. Receive Operation

The transmit (TX) manifold can be used for receive by producing a conjugate phase front to mix with the incoming wave. The TX "engine" produces an outgoing wave with a slightly offset ω_0 to heterodyne with the receive signal by another set of mixers. The IF outputs at the elements can then be added in phase with a summing network and sent to the remote site by digital photonics for further filtering and processing. Again, multiple beams can share the same beamforming manifold to reduce cost and complexity. The receive signal does not go through the entire beamforming manifold in the reverse direction. Hence, the overall noise figure is not degraded by the total loss of the beamformer in the transmit path. This is especially significant when the photonic conversion loss is still high. With the new design, the receive path by passes most of the transmit components so the noise figure is limited only by the front end loss and the noise figure of the LNA. This eliminates the most severe drawback encountered in other competing designs where a conventional photonic beamformer is used.

In summary, the wide band beamformer is a low-loss, compact system for simultaneous multibeam, multiband, and wide scan operation. Multiple beams can share the same optical feed manifold without duplicating the complex network of phase shifters, drivers, and beam-control data lines of a conventional feed system. Continuous beam scan by the heterodyne process eliminates the problem of gain ripple (crossover between beams) encountered in a conventional Rotman lens. Phase shifters are replaced by Wide band mixers at lower cost and less system complexity.

5.3. Microwave/Photonic Feed Networks for Phased Array Antenna Systems

R. A. Sparks
ANRO Engineering, Inc., 63 Great Road, Maynard, MA 01754 USA
e-mail : r.sparks@ieee.org

Abstract

Linear and planar arrays may consist of tens, hundreds or thousands of antenna radiating elements, each of which must be interconnected to the system transmitter and receiver by means of a feed circuit, beamformer or power distribution network. A review of the requirements and design considerations for several microwave and photonic circuits that perform this function are presented.

Several examples of constrained and unconstrained microwave feed networks for phased arrays are described, including a brief description of the component building blocks and their circuit properties. Equivalent photonic beamformer implementations in fiber or integrated optical waveguide that have been demonstrated or proposed for radar, electronic warfare and communication system applications are discussed. The potential insertion of photonic beamformers in operational systems is briefly examined.

5.3.1. Introduction

The development of array antenna systems dates back to the earliest days of radio and wireless transmission experiments near the beginning of the 20th century. Hansen [92] briefly discusses the background for the modern era of phased arrays, and includes an annotated description of several currently available references. The body of knowledge pertaining to microwave feed systems and power distribution networks for contemporary antenna arrays is covered from various perspectives that have been published in a number of books and periodicals. The approach that has been adapted for this presentation follows closely the format outlined by Patton [93]. For the purposes of this discussion, "a feed network is that part of an antenna that distributes power from the transmitter to the array elements in a prescribed manner, and collects the power captured by the array elements with some desired weighting for transfer to the receiver" [94]. Antenna arrays may be classified according to their principal features as noted in Table 1.

Linear
Planar
Mechanically Scanned
Electronically Scanned
Passive
Active
Photonically Controlled

Table 1 : Classifications of antenna arrays.

Some photographs of antenna array types are illustrated in figure 53. These are fairly representative of operational systems in current use. In the paragraphs that follow a description of some of the key components used in feed network and beamformer construction is presented. The microwave implementations will be introduced first, and where appropriate, equivalent photonic embodiments will be described. The organization of array feed types is described and examples of several types are presented.

a) Electronically Scannable Linear Array.

b) Mechanically Scanned Array.

(c) Planar Passive Array.

(d) Planar Active Array.

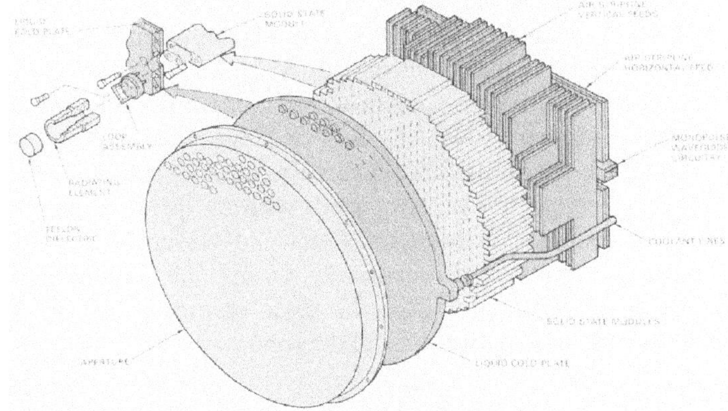

(e) Exploded View, Solid State Active Array.

Figure 53 : Antenna Arrays.

5.3.2. Antenna Feed Selection Criteria

It should be apparent from the variety of antenna array types illustrated in figure 53 that the associated microwave feed and power distribution networks vary significantly in design with respect to each other. The radar, electronic warfare or communication system of which the array is a part largely dictates the kind and complexity of the feed network that becomes an integral member of the aperture design. The exploded view of the solid state active array in the lower photo shows explicitly the feed networks, but the others are integral within the structures. The feed configuration is dependent on the functions the antenna system is designed to perform, which in turn is related to the overall system application and its requirements. Mechanical considerations of antenna size and weight including provisions for mounting will often constrain the choice of transmission line media and feed network materials. The system bandwidth, frequency of operation and peak and average power to be transmitted are major design parameters that enter early in the trades-off that must be performed. Many of the critical electrical, mechanical and environmental performance requirements [93] that influence the choice of techniques to be employed in the phased array antenna feed selection process are identified

5.3.3. Microwave and Optical Feed Components

In this section a brief departure from the main theme is provided to consider some of the component building blocks that are employed in the design of antenna array feeds. A few of the key microwave circuit elements [95] are described first, and where appropriate, equivalent optical counterparts are illustrated.

The distribution and collection of power in an antenna feed network is accomplished by interconnecting in-series and/or in-parallel, combinations of passive transmission line components that are described as directional couplers, power dividers or waveguide junctions. A partial listing of some types to be described are noted in Table 2. There may be three, four or more ports associated with a given component. The properties of couplers, divider/combiners and waveguide junctions have been studied and analyzed extensively during the past 50 years. New design variations with tailored performance parameters are invented and fabricated for special applications all the time.

Directional Couplers
Power Dividers/Combiners
Hybrid Couplers
Tees
Magic Tees
Rat Races
Short-Slot Sidewall/Topwall

Table 2 : Microwave/optical feed components

A directional coupler, figure 54, is a four port device that taps off some fraction of the power flowing through a primary waveguide into a secondary waveguide. The ratio of the power coupled to the auxiliary waveguide to the power input to the main guide is defined as the coupling coefficient, and can vary from 3 dB to 60 dB. A power divider is a specific type of coupler in which the primary waveguide may be split into N equal or unequal parts. 3 dB couplers, or hybrid junctions, are four port devices with two inputs and two outputs, each output voltage being a different linear combination of the two input voltages. There are several types of hybrid couplers having unique performance properties that have found application in microwave feed networks.

One of the best known examples of a hybrid coupler is the symmetric waveguide 'magic tee" shown in figure 55. Under perfectly matched conditions, power applied to port 1 is split equally and in phase to ports 3 and 4; port 2 is completely isolated. If power is applied to port 2 it splits equally between ports 3 and 4, but the output phase differs by 180 degrees, and port 1 is isolated. These interesting junction properties have been exploited in a number of subsystem applications, such as microwave mixers, modulators, high power ferrite circulators, feed networks, etc.

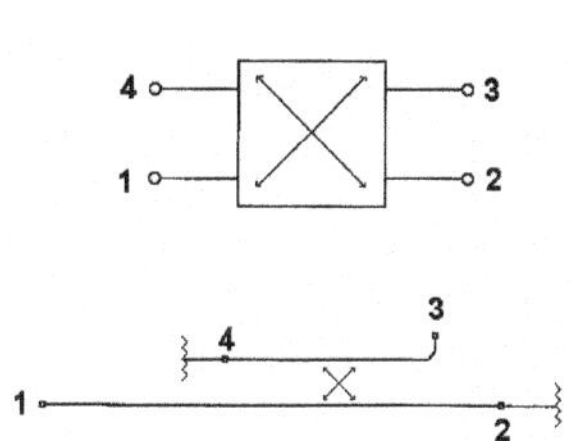

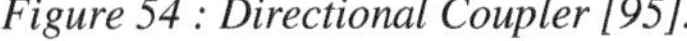

Figure 54 : Directional Coupler [95].

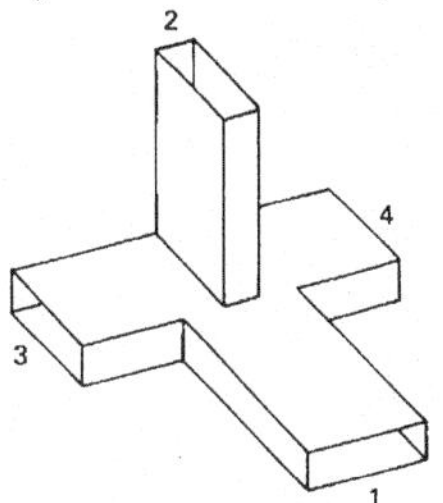

Figure 55 : Magic Tee [95].

A related planar junction, the hybrid ring or 'rat-race', often implemented in a microstrip or strip transmission line medium, is illustrated in figure 56. Power applied to port 2 will be split equally and in phase between ports 1 and 3, with port 4 isolated. Power applied to port 1 will be split equally, but 180 degrees out of phase, between ports 2 and 4, and port 3 will be completely isolated.

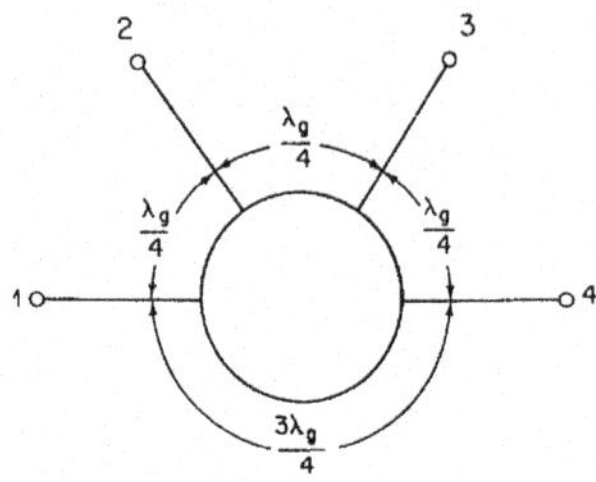

Figure 56 : Rat Race [95].

The power dividing properties of certain optical components bear a very close similarity to these microwave devices, some fabricated in single mode optical fiber and others processed in optical waveguide. Figure 57 illustrates a very common fused biconical taper fiber optic coupler, and figure 58 shows a typical optical power divider fabricated in silica glass waveguide. Both types have been used in optical beamformers that are described later.

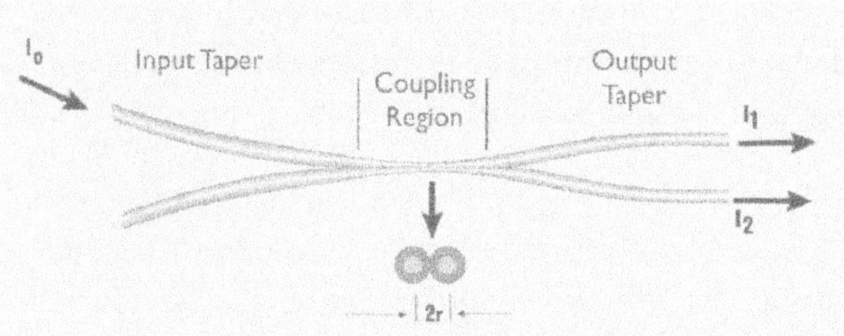

Figure 57 : Fused Biconical Taper Fiber Optic Coupler [96].

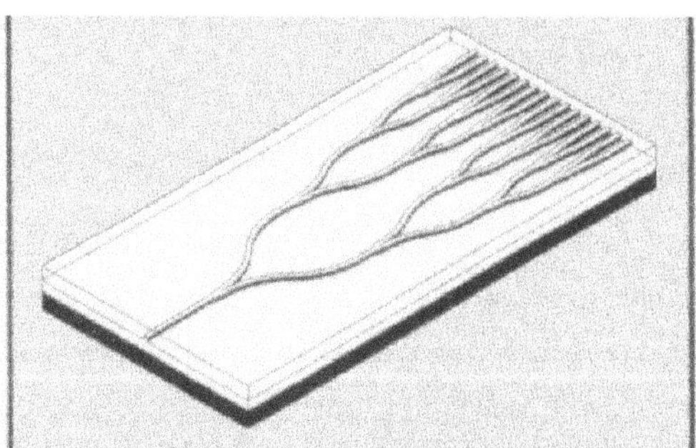

Figure 58 : Silica Glass Waveguide Power Divider [97].

5.3.4. Array Feed Techniques

Examination of figure 53 provides a small sample of antenna array configurations that have been developed, each with its own particular feed implementation. One organization [93] of microwave array feeds into three categories is listed in Table 3 . Constrained feeds are defined as those that limit the electromagnetic energy to travel along surfaces defined by the conductors of the transmission medium. An unconstrained feed refers to those techniques where the energy is launched from an open horn or radiator into free space, without conducting boundaries, to a collection of receiving elements, each of which becomes an output port for the feed. The semi-constrained feed category is less well defined, and may include some aspects of the other two in one or more dimensions. Only the first two categories, each of which is subdivided further, are addressed in the following discussion.

Figure 59 illustrates several parallel feed techniques, and figure 60 shows a number of series feed networks that have been used in antenna array power distribution manifolds. The two principle types of unconstrained feed examples, the transmission lens and reflect array technique, are depicted in figures 61 and 62.

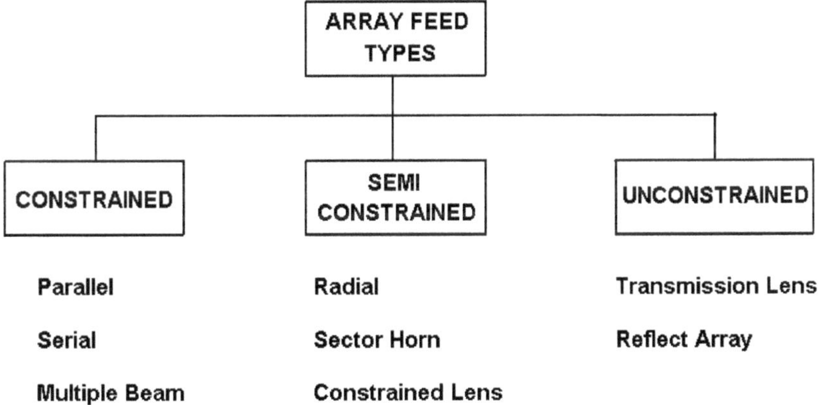

Table 3 : Categories of array feeds

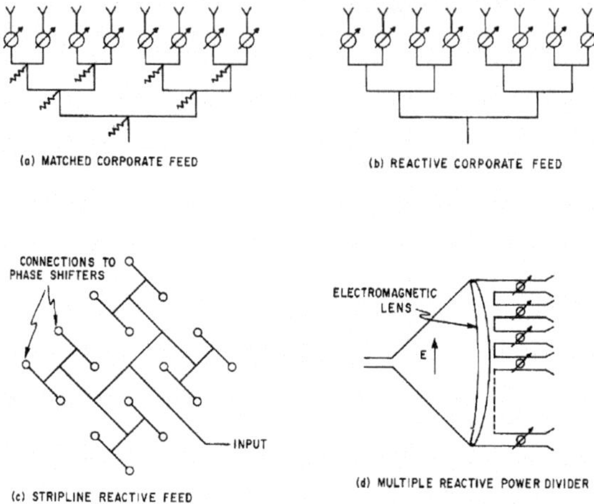

Figure 59 : Parallel feed networks [102].

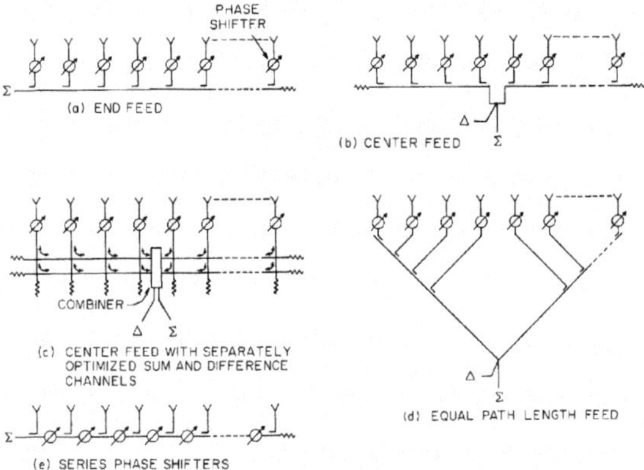

Figure 60 : Series feed networks [102].

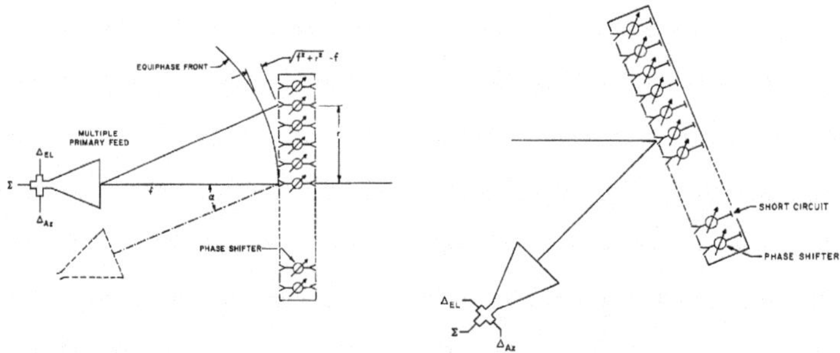

Figure 61 : Transmission Lens Feed [102]. *Figure 62 : Reflect Array Feed [102].*

5.3.5. Multiple Beam Feed Techniques

The Monopulse Comparator Network, (MCN), is probably the most widely used multi-beam feed technique that is incorporated in almost every modern tracking radar system. An MCN circuit schematic is shown in figure 63 connected to an antenna aperture that has been divided into four similar quadrants. The transmitted power is radiated from all four sub-apertures corresponding to an input to the sum channel port, and on receive the energy reflected from the target is processed by the network of magic tees to produce a sum signal, an elevation error difference signal and a traverse or azimuth error difference signal. The two error signals are fed back through tracking loops to point the peak of the antenna beam in the direction of the target. A novel fiber optic multi-mode feed for monopulse laser radar antenna applications was reported [98] several years ago.

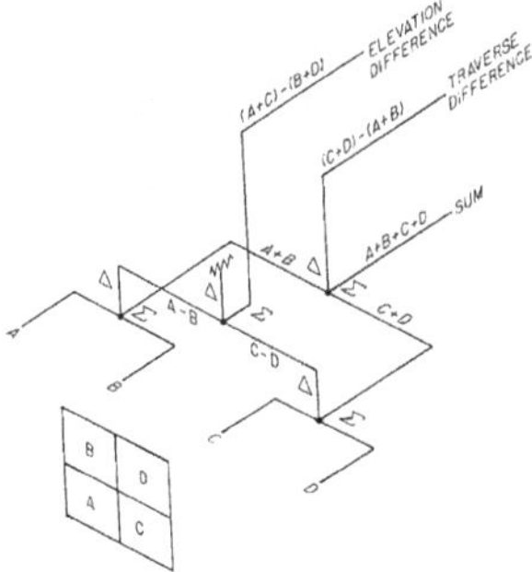

Figure 63 : Monopulse Comparator Network [94]

The one-dimensional Butler matrix [99] illustrated in figure 64 was introduced almost 40 years ago to allow multiple, simultaneous, independent, overlapping beams that correspond to the number of elements in the array. It uses 2^{n-1} hybrid couplers at each of its n levels, or $n(2^{n-1})$ couplers in total. Fixed phase shifters are inserted in some of the internal coupler outputs to provide the proper phase relationships at the beamformer array ports. Two-dimensional beam steering can be realized by stacking several matrices vertically, and interconnecting them to a similar stack of horizontal matrices. An integrated optic Butler matrix [100] has been investigated for angle-of-arrival applications with a circular array.

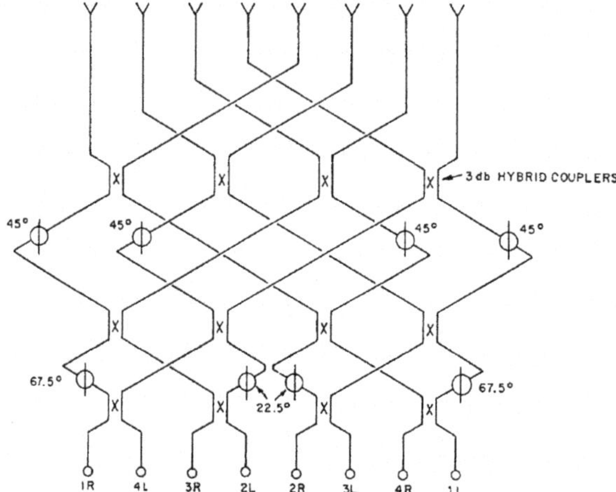

Figure 64 : Butler Matrix [93], [99].

The Rotman-Turner Lens [101] shown in figure 65 consists of a planar waveguide structure bounded by two nearly circular contours. There are three perfect foci on the right hand beam-side contour, one on-axis at port 4, and two symmetrically located at ports 4 and 7. Intermediate points are nearly focussed, and can be improved with contour corrections. On transmit, feed ports along the beam-side surface launch microwave energy into the planar medium where it propagates with a true time delay and is absorbed by the array-side probes. Specified lengths of transmission line interconnect the probes to each radiator forming a wavefront that travels in a direction corresponding to the beam port excited. The R-T lens is capable of very broadband, multioctave frequency operation, generally limited only by the radiating elements.

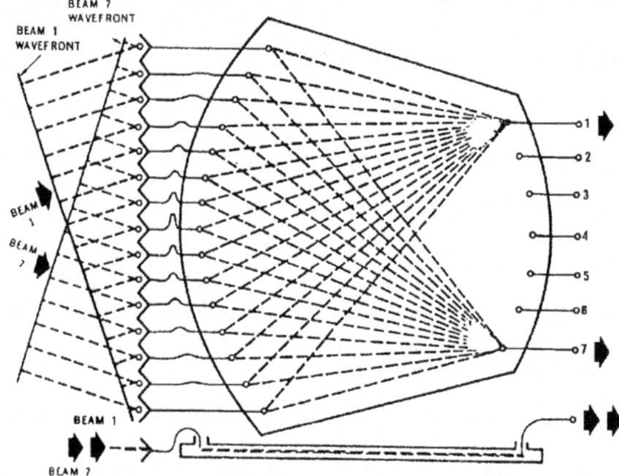

Figure 65 : Rotman-Turner Lens.

An equivalent 8-beam fiber optic Rotman lens, figure 66, has been demonstrated [103] with a 16 element 2-4 GHz linear array antenna. The 128 fixed time delays between each beam port and the corresponding array ports are fabricated from precisely cut lengths of single mode fiber optic cable.

Figure 66 : Fiber Optic Rotman Lens [103].

5.3.6. Conclusions

This brief review of antenna feed and beamforming networks for phased arrays has addressed most of the commonly implemented microwave techniques. Several circuit components that comprise the building blocks for many of the feeds have been described, together with equivalent optical counterparts. The literature of this decade is replete with photonic beamformers that have demonstrated squint-free, array beam steering under laboratory controlled conditions. However, their insertion into practical system applications has been lagging, due in large part, to their inherent complexity and high component costs. Any replacement or upgrade with photonic hardware into an existing system must satisfy rigorous criteria of form, fit and function, and insure a seamless exchange between all electrical interconnections and mechanical interfaces. Furthermore, it must demonstrate clear improvements in performance or cost over the incumbent microwave hardware implementation. The passive optical Butler matrix or fiber optic Rotman lens would seem to offer the greatest near term promise in this regard.

5.4. Photonics and Phased Array Antennas

J. Chazelas, D. Dolfi*
Microwave Photonics Department, Thomson-CSF Detexis**,
1 Bld Jean Moulin, Elancourt, France,
e-mail : jean.chazelas@fr.thalesgroup.com
Central Research Laboratory***, Domaine de Corbeville,
91404 Orsay Cedex FRANCE,e-mail : daniel.dolfi@thalesgroup.com
**NOW:THALES AIRBORNE SYTEMS ***NOW: THALES RESEARCH &1 TECHNOLOGY

Introduction

Phased array antennas offers many advantages over conventional antennas especially for steering and beam pointing accuracy, low sidelobes according to the phase and amplitude control of each array element. In future generation phased array radars, signal distributions will have to fulfill strict performance criteria. These include high isolation from both electromagnetic interference and crosstalk between module or subarray feeds; analog frequencies of operation into the millimeter-wave range with bandwidths approaching one octave; dramatic reduction in size and weight regarding present fielded radars; and performance compatible with growing requirements.

New avenues are opened for controlling many thousand array elements together with handling the wide bandwidth of shared aperture antennas through the marriage of photonics and microwave technologies. Photonics technologies will provide an interconnect solution for future airborne phased array radar antennas, which have conformality, bandwidth, EMI immunity, size,and weight requirements increasingly difficult, if not impossible, to meet using conventionnal electrical interconnect methods. The simultaneous requirements of wide bandwidth and large scanning angle emphasize the need for True Time Delay steering techniques and optical distribution of microwave signals. Future system requirements are reported together with a review of demonstrated approaches for True Time Delay and photonic switching architectures. An optical architecture for processing of the radar receive mode is also proposed and discussed.

5.4.1. Phased Array System Requirements

Airborne radars, for fighter aircrafts, are presently all equipped with antennas, operating in X band, which are planar devices fitted in the nose cone of the aircraft. This situation cannot change as long as the high performance level required from the radar system implies either an optical reflector antenna or a planar slot array, as with present technology. New phased array technologies, passive or active, associated with advanced signal processing, will lead to antenna systems with more flexible shapes which can possibly follow, more or less closely, the shape of the aircraft body.

It is understood in the following that only **active arrays** can be implemented conformally on modern aircraft : constraints due to RF power distribution at high level on transmit and RF collection at very low level on receive make implementation of passive (phase shifter) conformal arrays very difficult, especially when they are distributed across the aircraft body .

Three main applications have been identified for conformal arrays in aircraft radars :

- **Side-looking arrays :**

These arrays are disposed on the aircraft body can be desirable for backwards and / or up-downwards visibility.

It appears that the curvature radiuses at these points of the body are quite large (around half a meter) in the body section and can be assimilated to straight lines along the aircraft axis. Then degenerated conformal arrays are to be implemented, which can be made by stacking linear rows of T/R Modules.

- **Wing-edge array**

These arrays are disposed onto the leading edge of the aircraft wings or vertical fin can be desirable to increase the transverse or vertical dimension of the front array, giving access to higher separation capability between targets.

In the following example, two arrays are fitted (one on each wing), one the leading wing edges (mobile flaps) at a distance TBD (say, 1 meter) from the aircraft body. They are supposed hereafter to be operated in the same frequency band (X-Band) as the main array. Rx only is implemented, with only one channel, as array part is to improve front antenna resolution on reception of pulses transmitted by the front array. Antenna width is comparable to front antenna diameter or smaller (around 40 cm), height is smaller (10 cm or so), depth of radiating face is around 15 cm (figure 67).

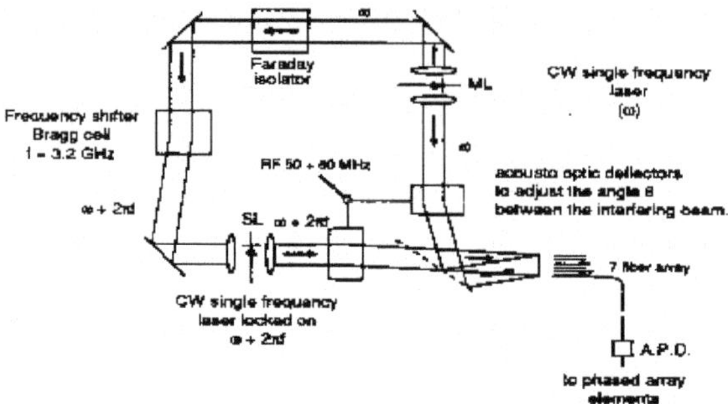

Figure 67 : NRL Goldberg.

Technology is nearly accessible, as curvature radius along the wing length is quite large (unwrappable antenna).

- **Nose-cone antenna**

This concept is more advanced than the previous one, it requires a 3D active array with full polarisation control on Tx and Rx (except in circular polarisation).

5.4.1.1. Benefits Expected from Optical RF Distributions

In this context, the main benefits of optical distributions are :

- a high layout flexibility, with easy 3-D, little interference with mechanical structure, small required real estate, ability to separate the location of the splitter from the array back-face;
- high performance, independent from mechanical shape complexity, with losses only at interfaces, EMC / EMS immunity, quality of RF interfaces (VSWR etc...) insensitive to mechanical layout.
- shape flexibility
- electromagnetic compatibility / susceptibility

They can be made better than corporate feeds for higher complexity or more complex shapes, performance may be lower for very simple distributions.

- possibility to easily mix RF and lower frequencies or digital signals on the same fibers
- easy remoting of signal splitting
- compatible with true time delays, ie very wide bandwidths.

The techniques of true time delay shifting which allow into the beamforming system to increase intrinsic array bandwith.

Optical based radar beam control promises should allow significant improvement and enhancement capability in terms of lage, agile bandwith, increased instantaneous bandwith, simultaneous multiple beams, multiple independant beams on independant frequencies, during either transmit or receive operation.

5.4.1.2. Constraints and Drawbacks On / From Optical Distributions

This chapter deals with constraints on / from Optical Distributions for conformal arrays and possible drawbacks, compared with corporate RF feeds for planar arrays as a reference.

Optical distributions have drawbacks :

- they are not bi-directional (reciprocal); optical fiber can be used in both ways, but separation at ends is mandatory with some complexity involved;
- very low power levels are available at interfaces, requiring distributed gains. Then T/R modules with higher gains (Tx and Rx) are required, with related higher DC power consumptions and thermal dissipations as consequences (plus possible energy feeding / extraction support).
- components are required at both ends for light modulation / demodulation, with resulting power consumption / dissipation
- flexible, cheap optical interfaces for several hundreds of T/R modules must be developed according to antenna layout and constraints : environment, maintenability - dismountability, reliability, loss dispersion control including optical fiber connectors

5.4.2. True Time Delay Optical Beamforming Networks

As pointed out True Time Delay (TTD) beamforming is required when wide band operation is combined with significant beam steering offset. In this case there is a need of low loss transmission links allowing the remote control of the antennas and the distribution of large bandwidth microwave signals. This need is fulfilled today by microwave optical links, owing to an increase in the modulation bandwidth and the dynamic range of optical emitters and detectors. Furthermore, optoelectronic architectures, because of their inherent parallel processing capabilities, bring attractive perspectives for radar signal control and processing.

According to these considerations, a large number of Optical Beamforming Networks (OBFN) have been proposed during the last decade. One can classify these architectures according to five generic approaches :

- switched delay lines
- laser/photodiode switching
- dispersive delays/Bragg grating delays
- 2D optical delay lines.

- coherent OBFN

For each approach we will detail in the following a typical demonstration that already includes a built array. This overview is completed with related published references.

5.4.2.1. Coherent Beamforming Network

This approach is based on the generation of the phase delays to be distributed onto the antenna, through the use of dual frequency optical carrier of the microwave signal.

This concept can be illustrated by the experiment performed by Tamburrini & al (cf figure 67). The operating principle is shown in the following figure.

Two mutually coherent, frequency offset optical beams are obtained by injection locking a slave laser (SL) with the emission of a master laser (ML), which was frequency shifted by a Bragg cell operating at frequency f = 3.2 GHz. These thus beams interfere and give rise to a moving interference pattern. A regularly spaced array of multimode fibers is used to spatially sample the moving pattern and to transmit the optical signals to the antenna plane. The light intensity coupled into each fiber varies at the beam frequency f with a microwave phase depending linearly of the fiber position and of the angle between the interfering beams. Phased array beam steering is achieved by changing the angle between the beams, thereby changing the spatial period of the interference pattern. The far field pattern of a 7 element linear antenna was characterised.

This concept was revisited (see references) but all these different approaches are based on the optical control (integrated optics, free space,...) of the microwave signal by changing the relative phase of the optical components of a dual frequency beam. It provides generally simple structures but does not permit a large frequency bandwidth operation of the antenna since these architectures only perform phase scanning.

5.4.2.2. Switched Delay Lines

This approach is based on the optoelectronic switching of fiber delay lines. This switching provides a digital control on the path lengths experienced by an optical carrier microwave signal and thus permits a true time delay control of a phased array antenna. This concept can be illustrated by the experiment performed by Goutzoulis et al (cf figure 68). The operating principle is shown in this figure.

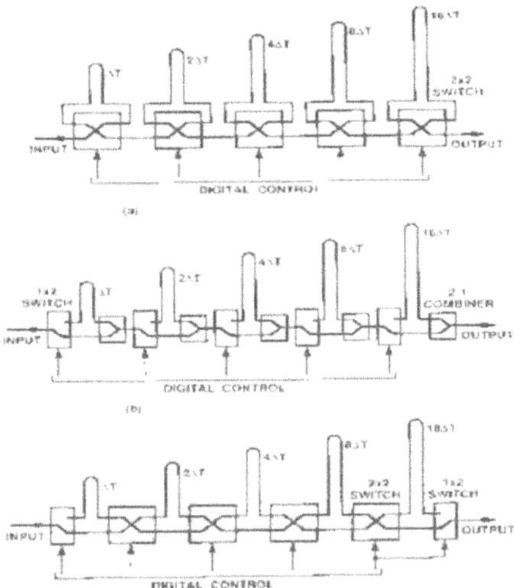

Figure 68 : Switched fiber delay lines.

In the binary fiber optic delay line (BIFODEL) architecture, the optical carrier of the microwave signal is optically routed through N fiber segments whose lengths increase successively by a power of 2D. The required fiber segments are addressed using a set of N 2x2 optical switches. Since each switch allows the signal to either connect or bypass a fiber segment; a delay T may be inserted which can take any value, in increments of ΔT, up to a maximum value T_{max} given by :

$$T_{\max} = (2^N - 1)T$$

For each reading element or subarray of a phased array antenna it is necessary to implement such a BIFODEL. It yields that this technique is very well adapted to a TTD control of a subarrayed antenna.

The performances of this concept can be extended, mainly for an antenna divided in subarrays, according to the use of optical wavelength multiplexing. This approach is the one proposed by Westinghouse in its proof of concept demonstration (Goutzoulis et al, cf figure 69).

The partioned phased array concept can be implemented using optical WDM in conjunction with all optical programmable delay lines. Furthermore it is reversible since the hardware can be used for both transmit and receive modes.

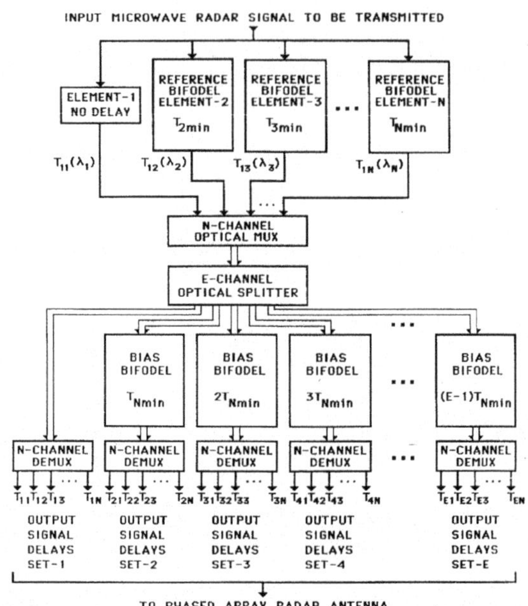

Figure 69 : BIFODEL architecture.

In the transmit mode M-1 BIFODELs with outputs at wavelengths λ_2, ..., λ_M are driven in parallel radar signal. The M-1 BIFODEL outputs, along with an undelayed output at wavelength, λ_1, are multiplexed via an M-channel multiplexer (MUX), the output of which is divided into E channels via a 1:E- channel optical divider.

All but one of the divider outputs independently drive a bias BIFODEL, each of which is followed by an optical M-channel demultiplexer (DEMUX) output will also contain M wavelengths, λ_1, λ_2, ..., λ_M. The outputs of the nonbiased DEMUX contain the M progressively delayed signals required for the set 1.The outputs of each the remaining DEMUXs contain a similar set of signals but they are further delayed via the bias BIFODELs.

Similar wavelength outputs drive similar location elements in each set. All BIFODELs must have N = $\log_2 R$ cascaded segments and different time resolution T_{imin}. The latter is determined by the location of the specific element, the antenna geometry, the radar characteristics, and so on. Similar comments apply to the bias BIFODELs, which have time resolutions (j-1)T_{Mmin}. In the receive mode, the same architecture is used, but in reverse.

Here, the output of each element of the phased array drives an LD of a different wavelength. Elements with similar locations in different sets drive LDs of the same wavelength.

The experimental demonstration of this concept was performed at Westinghouse for a 16 element linear antenna (16 elements for the transmit mode, 8 elements for the received mode). The far field pattern was characterised for both modes over the frequency range 600-1500 MHz for the distribution of the microwave signals. The antenna is divided in 4 subarrays. The microwave signal is first divided in parts, 3 of them can be electrically delayed (from 8 ps to 1500 ps with a 1.5 ps accuracy). The output of the non delayed line end of the 3 delayed lines are used to feed 4 directly modulated semiconductor lasers at different wavelengths.

5.4.2.3. Laser / Photodiode Switching

In this approach (originally proposed and demonstrated by Hughes Aircraft), the delay path of the optical carrier of the microwave signal is defined, for each radiating element or subarray, by selectively turning on a laser and detector located respectively at the beginning and end of an analog optical link (see figure 70).

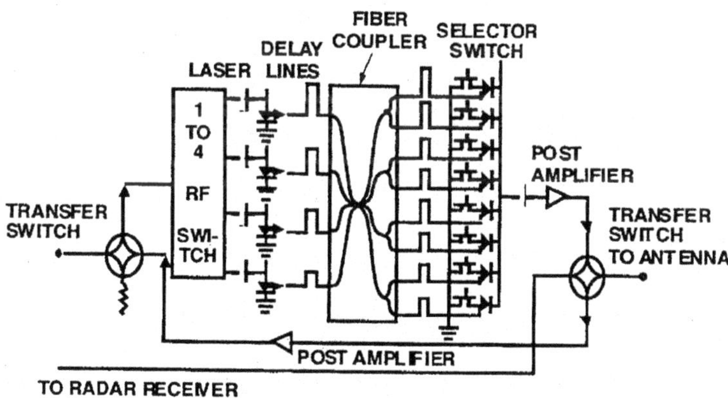

Figure 70 : 5-bit time shifter.

Combining laser and detector switching the network of the above figure provide 32 delay options (i.e 5 bits of resolution). This 5 bit time shift module is the building block of a wide band feed network. The programmable time shifters (8) provide the coarse delay steps ranging from 0.25 ns to 8 ns (5bits) for the 8x3 subarrays of a 96 element antenna. Electronic delay lines in the T/R modules provide fine differential delays ranging from 0.01ns to 0.5 ns (6 bit precision).

According to this concept both transmit and receive modes of a 2D conformal antenna were characterized over the frequency bandwidth 850 – 1400 MHz. This proof of concept is, at the moment the most achieved demonstration of optical remote control and beamforming of a large bandwidth antenna.

5.4.2.4. Dispersive Delays / Bragg Grating Delays

5.4.2.4.1. Dispersive Delays

In this approach, the time delays experienced by optically carried microwave signals are provided by the use of one or several tunable wavelength lasers in conjunction with a wavelength selective material. This material is either an optical fiber including permanent Bragg gratings (Lembo et al. from TRW, Smith et al. From GEC-Marconi) or an optical fiber used in its dispersive region (Frankel et al. from the Naval Research Lab.).

In the following we will detail the NRL approach, since it is already demonstrated with a radiating antenna.

The microwave signal driving the antenna elements is transmitted on a single wavelength-tunable optical carrier via a bank of dispersive fiber optic links. The TTD function is realized by tuning the carrier wavelength to vary the group velocity of the propagating signal. Each fiber link feeding an array element incorporates an overall amount of dispersion proportional to the element position. A set change in the carrier wavelength provides the necessary proportional time delay for all array elements with a single wavelength control input.

This approach seems to be very well adapted to linear antenna, with a number of elements in the range 10-100. It was experimentally demonstrated, for the transmit mode, with a very large bandwidth antenna (2 – 18 GHz) of 8 radiating elements. Receive mode operation is also possible with this concept when the optical beamformer is used to generate a properly phased local oscillator. In this case, a tunable laser is used in the module.

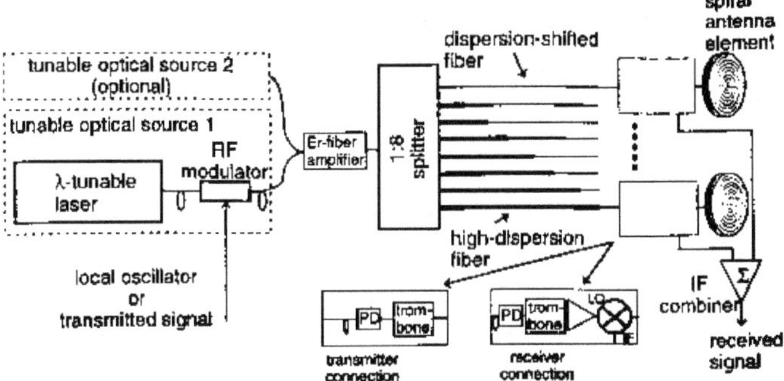

Figure 71 : dispersive delays.

5.4.2.4.2. Bragg Grating Delays

Several laboratories have investigated the use of Bragg fibre gratings to provide true time delay beam steering in optically controlled phased array antennas. These studies have considered the performance of single channel discrete multi grating arrays (C.Edge & I Bennion) as shown in figure 72, chirped grating beamformers (see reference (a) below) and full transmit /receive antenna systems (see ref (b) below).

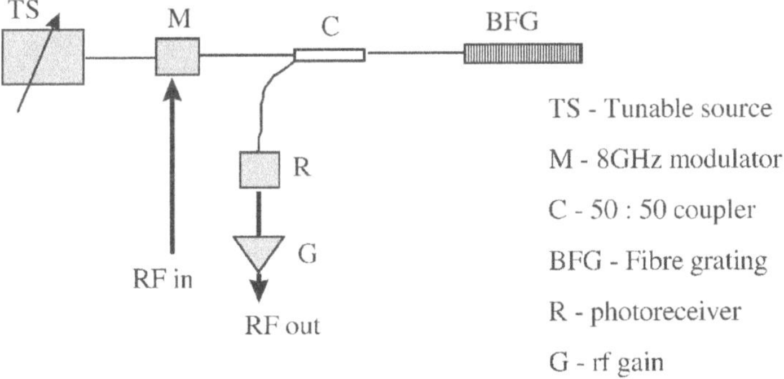

Figure 72 : 8 element (3-bit) Bragg fibre grating test configuration.

Generation of TTD using multi element or chirped gratings is advantageous since all of the required delays for a single antenna element can be provided on one fibre rather than the more complex switched time delay modules described in the previous section. There are significant disadvantages however including the manufacturing reproducibility of fibre gratings, the requirement for highly wavelength stable tunable laser sources (only currently available in bench top form and with slow tuning speeds) and the ability to achieve suitable close-to-carrier phase noise performance within a system (B.Smith and M.Nawaz).

5.4.2.4.3. 2D Optical Delay Lines

A 2D optical delay lines has been implemented and demonstrated (cf Dolfi & Riza).

In this approach the time delays are provided by free space delay lines, switched using 2D spatial light modulators (SLM), cf figure 73.

A dual frequency laser beam is the optical carrier of the microwave signal. This beam is expanded and travels through a set of SLMs whose number of pixels (pxp) is the number of radiating elements of the antenna. M_O is a parallely aligned nematic liquid crystal (LC) SLM. It controls the phase of the microwave signal by changing the relative optical phase of the cross polarized components of the dual frequency beam.

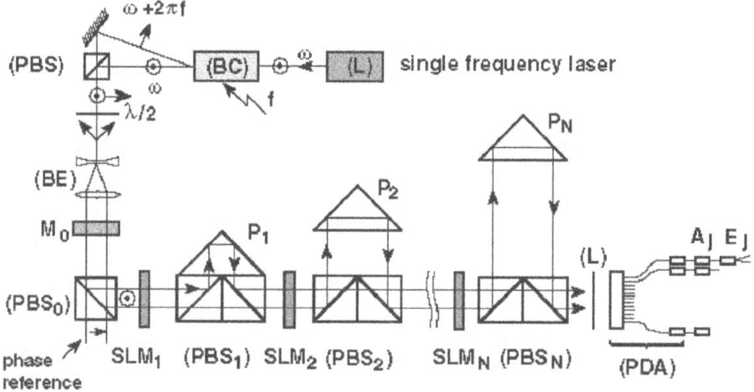

Figure 73 : 2D optical delay lines.

At the output of M_O, the linearly polarized dual frequency beam intercepts a set of spatial light modulators SLM_i, polarizing beam splitters PBS_i and prisms P_i. They provide the parallel control of the time delays assigned to the antenna. The beam polarization can be rotated by 0° or 90° on each pixel. According to the polarization, PBS_i is transparent (and the light beam intercepts the next SLM_{i+1}) or reflective (and the microwave signal is delayed). The collimated beam travels through all the (PBS_i) and is focused by an array of microlenses (L) onto an array of pxp fiber pigtailed photodiodes (PDA).

For a given photodiode, the phase of the microwave beating signal is determined by the applied voltage on the corresponding pixel of M_O and by the choice of the PBS_i on which the reflections occur. Since the positions of prisms P_i provide delay values according to a geometric progression (t, 2t, 4t..), the beating signal can be delayed from 0 to $(2^N-1)t$ with step t.

Experimental demonstration of an optically controlled phased array antenna, operating between 2.5 and 3.5 GHz. The 2D architecture is implemented with 6 SLMs of 4 x 4 pixels. It provides 32 delay values (5 bits), an analog control of the phase [0,2p] and permits the control of a 16 element phased array antenna. Figure 69 displays an example of the microwave beam deflection using time delays. Furthermore, when far field patterns at different frequencies are superposed for a given scan direction, one can notice the absence of any beam squint.

5.4.2.4.4. Thomson-CSF Receive Mode Approach

Future active radars, based on solid state transmit/receive modules, will provide new capabilities, in terms of reliability and jamming robustness. The far field pattern of such phased array systems is controlled by the relative phase of the microwave signals emitted by the modules. Instantaneous frequency bandwidth (up to 30 %) of these multifunction systems will require the implementation of time delay beamforming networks (TDBFN).

This paragraph reviews recent work on TDBFN, performed at Thomson-CSF/LCR, in collaboration with the radar divisions.

The operating principle of our 2D optical architecture is detailed in figure 74. A dual frequency expanded laser beam is the optical carrier of the microwave signal. It travels through a set of SLMs (spatial light modulators) whose number of pixels (pxp) is the number of elements of the antenna. M_O is a parallely aligned liquid crystal SLM. It controls the phase of the microwave signals by changing the relative optical phase of the cross polarized components of the dual frequency beam. At the output of M_O, the now linearly polarized optical carrier intercepts N spatial light modulators SLM_i, polarizing beam splitters PBS_i and prisms P_i. They provide the parallel control of the time delays assigned to the antenna. On each pixel the beam polarization is rotated by 0° or 90°. That is, PBS_i is transparent (and the light beam intercepts the next SLM_{i+1}) or reflective (and the microwave signal is delayed). The channelized beam is then detected by an array of pxp fiber pigtailed photodiodes (PDA). The phase of the microwave signal delivered by each photodiode is determined by the applied voltage on the corresponding pixel of M_O and by the choice of the PBS_i on which the reflections occur. Since time delay values are set according to a geometric progression (τ, 2τ, 4τ..), the beating signal can be delayed from 0 to $(2^N-1)\tau$ with step τ.

We have completed the experimental demonstration of an optically controlled phased array antenna, operating around 3 GHz. The architecture, implemented with 6 SLMs of 4 x 4 pixels, feeds a 16 element antenna with 32 delay values and an analog control of the phase. The far field pattern of this antenna was characterized, using optical phase and time delay switching, without any beam squint over the 2.5-3.5 GHz range.

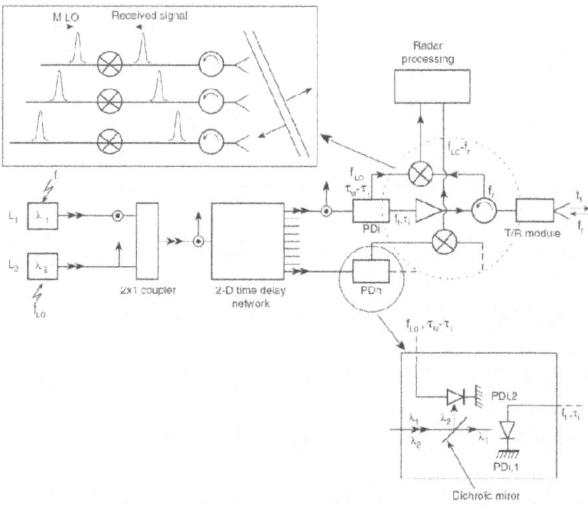

Figure 74 : Optical architecture for receive mode operating principle.

Analog optical processing of the receive mode is possible, reversing the previous architecture. For a radar detection in the same direction as for emission, the optically carried transmitted and received signals, have to travel through the same TDBFN. It permits in-phase addition, over a large frequency bandwidth, of the microwave signals received by the antenna. In this case, since these signals are spread over a dynamic range as large as 120 dB, the implementation of highly linear optoelectronic links is necessary (spurious free dynamic range $\approx$ 80 dB/MHz$^{2/3}$). It is still difficult to obtain, over large bandwidth, with currently available optoelectronic components.

In order to overcome this limitation, we proposed an original architecture in which a channelized microwave local oscillator (LO), optically carried, is used for mixing with the received microwave signals. There is no more microwave-to-optic conversion of the received signals. The transmit/receive architecture which operates in a way similar to optical phase conjugation is shown. In this architecture, two cross-polarized optical beams at wavelengths λ_1 and λ_2 are modulated by microwave signals at frequencies f_{tr} (signal to be transmitted) and f_{lo} (local oscillator). They travel through a 2D switching network similar to the one of figure 73. The delays experienced by signals at f_{tr} permit the control of the emitted far field pattern. The delays experienced by Los at f_{lo} are chosen to be complementary to the one experienced by f_{tr}, using a remarkable property of an optical architecture based on polarization switching. When two cross-polarized beams travel along the same channel, their polarizations remain orthogonal and they experience complementary paths. In this case, when the carrier of the frequency f_{tr} is delayed by τ_k along channel k, the cross-polarized carrier of frequency f_{lo} is delayed by τ_{max}-τ_k, where τ_{max} is the maximum available time-delay. On each channel, a dichroïc mirror switches the carriers on two different photodiodes which provides the signal to be emitted and a perfectly matched microwave LO, respectively.

Because of jamming, frequency f_{lo} must stand out of the radar bandwidth Δf. The ideal homodyne processing must be replaced by an heterodyne detection, where $f_{lo} \neq f_r$ (f_r is the frequency of the received signals). It can results, for large Δf, in prohibitive phase errors. In order to minimize these errors, the LO is generated in two successive steps. The first step provides a channelized LO with $f_{lo}=f_{tr}$. Then, those signals are mixed with in-phase microwave signals at an intermediate frequency $f_i > \Delta f$. It provides, on each channel k, a LO with phase

$$\varphi_k^{LO}(t)=2\pi(f_{tr}-f_i)t-2\pi f_{tr}(\tau_{max}-\tau_k),$$

which is mixed with the corresponding received signal.

A proof of concept was recently completed, which consists of two transmit/receive modules and two delay blocks. On each channel, a laser beam is modulated at frequency f_r or f_{lo}. Phase differences equivalent to 10 ps delays at f_i = 700 MHz were measured that permits in-phase addition of the received signals (received signals were generated using time delays that simulate reflection from the target).

In addition, one can notice that these approaches could greatly benefit from a holographic backplane scheme, in order to solve the problems of compactness, reliability and scaling up to 10^3 transmit/receive modules as for large array systems.

5.4.3. Conclusion

According to the requirements for increasing the instantaneous bandwidth of radar systems, numerous groups are involved in the research of optical implementation of true time delay beamforming architectures.

In addition, developments of future shared systems will require the optical control and switching of wideband conformal array. In order to overcome the present limitation in the dynamic range of available optoelectronic components, an original optical architecture for the analog processing of the receive mode has been proposed based on the mixing of a matched local oscillator and the microwave signals.

5.4.4. Acknowledgments

We thank DGA /STTC for their partial support. We acknowledge JP Huignard for its contribution to this review and S.Formont, G. Granger, T. Merlet and O. Maas (Thomson-CSF/Airsys) are acknowledged for their contributions to this work.

6. PHASE NOISE DEGRADATION IN NONLINEAR FIBER OPTIC LINKS DISTRIBUTION NETWORKS FOR COMMUNICATION SATELLITES

A.S. Daryoush
Dept. of ECE,
Drexel University, Philadelphia, PA, 19104 USA
e-mail : Daryoush@ece.drexel.edu

Abstract

Large phased array antennas play a significant role in many wireless applications such as communication satellites. Optical beamforming networks is attractive because weight, size, and volume reduction over electrical beamforming networks. The challenge of realizing fiber optic network that is low cost while satisfying high spurious free dynamic range in the harsh space environment is met using directly modulated fiberoptic links in the T/R level data mixing. Phase noise degradation of the frequency reference distributed using fiber optic links is analyzed and comparison with electrical distribution are made. Dynamic response of a Fabry-Perot laser diode is altered by adding an external feedback, resulting in a resonance peak. Electrical injection locking of this resonance frequency results in stable oscillation frequency. Using a monolithically integrated electro-absorption modulator with a long F-P laser diode both injection locking and mode-locking are demonstrated resulting in over $\approx$90dB.Hz$^{2/3}$ spurious free dynamic range without a significant degradation of close-in to carrier phase noise.

6.1. Introduction

Future communication satellites are designed to operate at frequencies of Ka-band using active phased array antennas to simultaneously generate as many as 100 radiated beams. To reduce size and weight optical beamforming networks are proposed as a viable solution [128]. Many beamforming networks are reported, however any viable beamforming should meet low cost, superior performance over electrical distribution under harsh space environment. The large temperature variation (-150 to +150 C) and ionized radiation (electrons and protons) of space influence operation of both passive and active optical components. Radiation hardened optical fiber are required to minimize fiber darkening for extended radiation exposure. Moreover, high temperature cycling results in micro-bending loss in optical fibers, if one does not select proper fiber jacket and coating. Optical modulators based on insulator integrated optic waveguides suffer from pyro-electric and photorefractive effects. The impact is significant in LiNbO3 Mach-Zehnder modulators and $\Delta\beta$ couplers. On the other hand, semiconductor laser diodes suffer from increased threshold current under radiation which could be compensated for when its output light power is monitored using a monitor photodiode. The impact of radiation on photodiodes is increase in dark current.

From cost performance view point, externally modulated FO links are more expansive than directly modulated links. Therefore, it appears that from both cost and its hardness to harsh space environment, directly modulated fiber optic links are the most appropriate candidate. Even though the reported spurious free dynamic range (SFDR) performance of the externally modulated fiberoptic links are higher than the directly modulated one [129], however, this higher SFDR performance is only attained in the case of a high power Nd:YAG laser as the optical source. On the other hand, when a semiconductor laser diode is used as a source for the externally modulated FO links, the SFDR performance superiority of the externally modulated FO links over the directly modulated ones disappears.

Unfortunately, the physical limitations of present laser structures have restricted practical system applications of the directly modulated fiber optic links to the frequencies of a few gigahertz. To overcome this limitation, the carrier and data signals are proposed to be separately distributed using T/R level data mixing where a higher SFDR is achieved [130]. However, questions of frequency coherency and the amount phase noise degradation of the frequency reference requires to be addressed. Since subcarrier modulated fiber optic (FO) links require only a relatively narrow bandwidth about a high microwave carrier, Lau [131] has proposed a number of laser structure modifications to apply directly modulated FO links up to millimeter wave frequencies.

This paper reviews the sources of phase noise degradation in directly modulated FO links and compare overall performance of fiber optic distribution to the electrical ones in terms of dynamic range, phase noise degradation, phase fluctuations, size, and weight. Next a novel F-P laser structure is reviewed where efficient transmission of data and carrier signals with high SFDR and low phase noise degradation are achieved.

6.2. Carrier Signal Generation

Fiber optic (FO) links are employed for distribution of frequency reference in distributed systems where phase and frequency coherency of the individual receivers are important for coherent integration of signals [131]. The FO distribution link contributes residual phase noise to the reference signal, which is a function of operation frequency. The phase noise of the reference signal could be degraded if residual phase noise is too close to the signal noise floor level. Therefore, an appropriate selection of reference frequency is necessary to avoid being degraded significantly after passing through the FO link. For example as shown in figure 75, to generate a 12 GHz local oscillator (LO) at front-end, a reference signal at frequency of 100 MHz (UHF), 4 GHz (C-band), and 12 GHz (X-band) can be sent through FO link. Since the phase noise contributions for FO links are different at these frequencies, an optimum frequency for reference signal can be found to have the least phase noise degradation due to FO link.

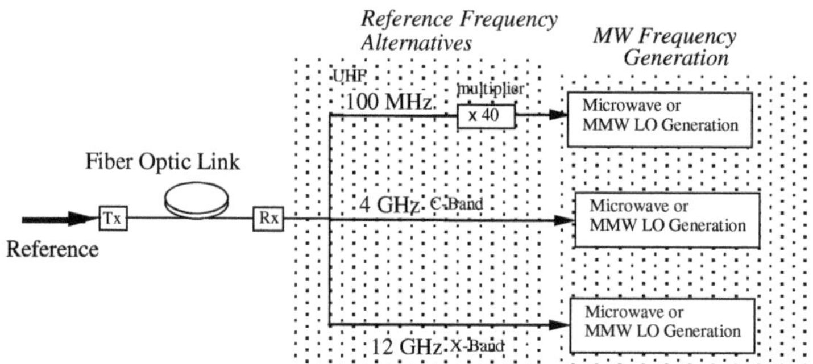

Figure 75 : Various regime of operation for the frequency reference used in a FO synchronization network.

The optical spectra of a modulated optical signal is expressed as :

$$P_{opt}\{1 + m\cos(\omega_m t + \delta\phi_m(t)) + n_{RIN}(t)\}\cos(\omega_{opt} t + \phi_{opt}(t) + \delta\phi_{opt}(t)),$$

where P_{opt} is the averaged optical power, **m** is the optical modulation index at modulating microwave carrier, ω_m. The focus of present work is $\delta\phi_m$, the residual phase noise added to the microwave carrier from the laser diode noise source. n_{RIN} is the relative intensity noise; ω_{opt} is the optical frequency; ϕ_{opt} is the optical phase signal due to side modes and modulation, and $\delta\phi_{opt}$ is the phase noise of the optical signal. Since most fiber-optic links for antenna remoting applications use intensity detection, only the noise signals in optical intensity affect the microwave carrier signal, namely, $\delta\phi_m$ and n_{RIN}. The n_{RIN} could contribute to the FM noise of the reference signal through nonlinear AM/PM conversion [132].

The laser diode SSB phase noise of the nth harmonic of the modulating signal, $£_{out}$, has contributions from three noise terms: i) the input reference signal phase noise, $£_{in}$; ii) the low frequency noise of laser diode up-converted to the carrier frequency , $£_{up}$; and iii) the RIN noise at the offset microwave carrier, $£_{RIN}$. This behavior is quite analogous to microwave systems [133]. Therefore at angular offset carrier frequency of Ω, $£_{out}$ can be approximately expressed as [134] :

$$£_{\mathrm{out,n}\omega}(\Omega) = n^2 £_{\mathrm{in},\omega}(\Omega) + n^2 £_{\mathrm{up},\omega}(\Omega) + £_{\mathrm{RIN,n}\omega}(\Omega) \tag{18}$$

The factor of n is the harmonic order of the modulation signal, if any nonlinearity of laser diode is exploited to generate the nth harmonic. The subscript ω indicates the modulation frequency. The up-conversion factor of LF RIN to phase noise is $C_{up,\omega} = 1/2(\partial\theta(\omega)/\partial P_0)^2 P_0^2$, which θ is phase of optical signal at the modulating frequency ω. The dependence of phase on the optical output power is through "a" parameter [135], and is a function of modulation frequency and averaged optical power.

Since in semiconductor laser diodes RIN noise are strong up to 100 MHz because of mode partition noise, it is predicted that the spectral purity of the UHF reference signal is greatly degraded, resulting in a higher FM noise. Moreover, the X-band modulating signal is close to the relaxation oscillation frequency where RIN is peaked. The best frequency for reference signal distribution through DMFO link is the C-Band signal as depicted in figure 76, where the phase noise of the 12 GHz LO signal is generated from the reference signal through the above mentioned DMFO link. Clearly, the signal generated from a C-band signal has the best phase noise performance. The signal from UHF reference degrades greatly because the residual phase noise of the FO link is higher than the reference phase noise at offset frequency higher than 100 Hz.

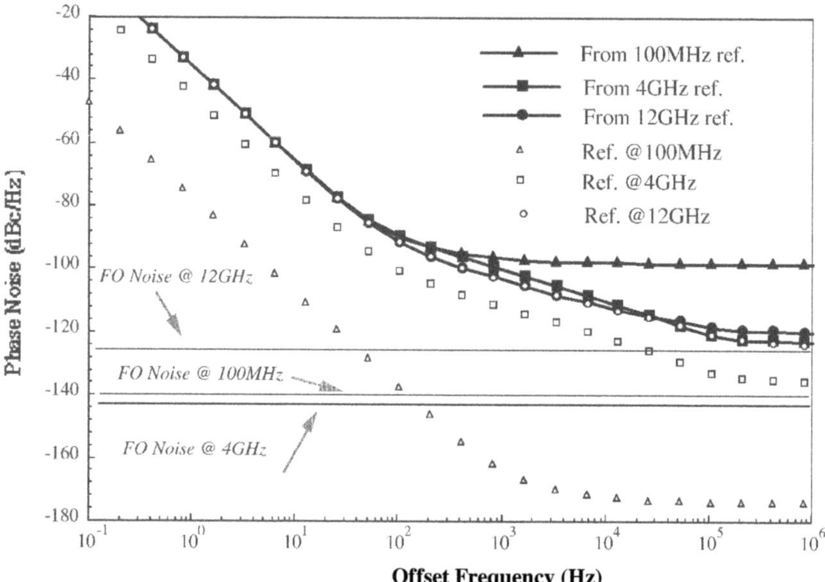

Figure 76 : The simulated phase noise of a LO signal at 12 GHz, which is generated from different reference signals through a directly modulated FO link.

6.3. Comparison of Optical and Electrical Distribution

Distribution networks are needed to link the CPU to the remote front-ends to be able to control, monitor, and exchange information. A comparison between fiber optic and electrical distribution in all aspects is very necessary to study how advantageous fiber optical link is over microwave cable in the system design. This section compares the two distribution approach in terms of dynamic range, phase noise performance, size, and weight as a function of distance-frequency product.

6.3.1. Dynamic Range Comparison

The dynamic performance of electrical cable and fiber optic link is simulated and shown in figure 77 as a function of operation frequency and length. The coaxial cable from Gore Incorporation is used [136], and different state-of-the-art fiber-optic links are also presented [137, 138, 139]. A 10 dBm signal within 20 MHz band width is used to examine the signal to noise ratio. The required S/N is around 85 dB.

The signal to noise ratio of coaxial cable is determined by the cable loss, which is primarily dominated by the loss caused at central and outer conductor. Therefore, at high frequency, the skin depth of the conductor is smaller and loss is higher, which can clearly be seen in figure 77. Since the electrical loss is proportional to the cable length, signal to noise ratio decreases greatly for long length cable. On the other hand, fiber optic link loss is not length dependent for a short length. The noise in laser and optical detector determines the noise floor of the link. In addition, fiber optic link has upper limit for input power because of the nonlinearity of laser diode and external modulator, but usually the 1 dB gain compression point is larger than 10 dBm at input. Therefore, for the FO links used in our application, noise floor is the dominant factor to the signal to noise ratio.

From figure 77, the FO link using external modulator provides the best performance and meets the system requirement. The reason is that, in the case of using external modulator, high power solid-state laser is used, which has very low RIN at microwave and MMW band. However, since high speed semiconductor laser used in directly modulated links has relatively higher RIN at microwave frequencies and contributes to the noise floor of the whole link. As shown in figure 77, as laser band width increases, the RIN at low frequency end will decrease, henceforth, the signal to noise ratio can be improved [140].

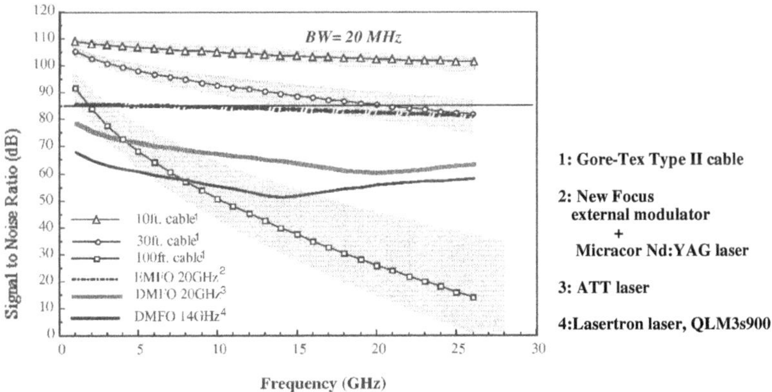

Figure 77 : Dynamic performance comparison as functions of frequency and length. Signal to noise ratio as a function of operating frequency. EMFO 20 GHz represents an externally modulated fiber optic link with bandwidth at 20 GHz, and DMFO 20 GHz and DMFO 14 GHz stands for the directly modulated links with laser band width at 20 and 14 GHz. The shadow regions presents the variation range as function of cable quality.

The directly modulated optical link has the advantage over externally modulated optical link in terms of simple system and low cost. It is better to use directly modulated FO (DMFO) instead of externally modulated FO (EMFO) link, if possible. To meet the signal to noise ratio requirement, a novel technique can be used to improve laser's RIN up to 20 dB[141] by simply adding an external optical feedback. Simulation of this system at

18 GHz as a function of distance reveals that coaxial cable has better performance only at low frequency and short distance. For a length at 100 feet, the S/N of the EMFO and DMFO links are taking over coaxial cable at frequencies of 2 GHz and 5 GHz, respectively. The cross over point of length between the optical and electrical link in terms of S/N at 18 GHz. For EMFO and DMFO link, the cross over point is around 30 feet and 50 feet, respectively.

6.3.2. Comparison of Phase Noise Degradation

For a reference signal distribution, the most important thing is to maintain the quality of the signal through either electric cable or fiber optic link. In the coaxial cable link, the influence on the phase noise performance of reference signal is from the cable loss. However, in the DMFO link, laser RIN is high and the RIN could be converted into phase noise of the reference signal under large signal modulation, since laser diode is a highly nonlinear device. To simulate the phase noise contribution from laser diode in DMFO link, we have developed a large-signal phase noise model for laser diode based on the internal noise force and nonlinear characteristic within semiconductor laser cavity [132]. Figure 78 shows the residual phase noise performance to a reference signal sent through different links.

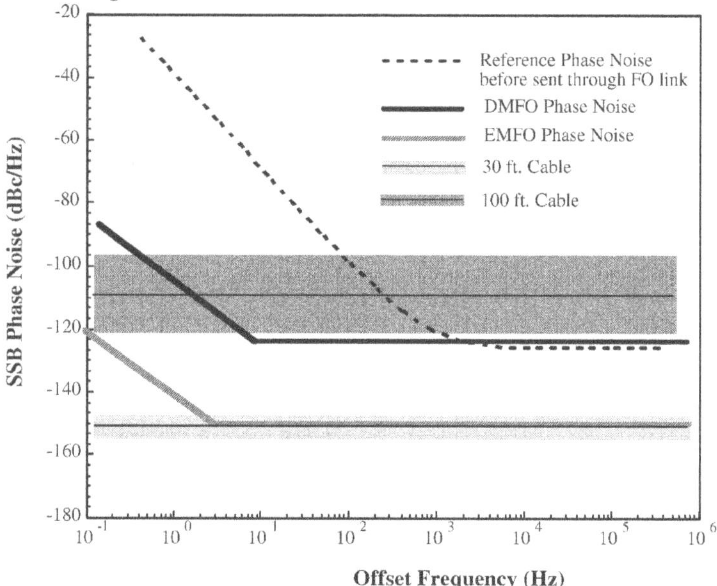

Figure 78 : Phase noise degradation to a 12 GHz reference signal sent through different electrical and optical links. The dash presents the phase noise of a reference signal before it is sent to the link.

The coaxial cable's contribution to the phase noise increases as the cable length increases. However, the residual phase noise of DMFO and EMFO are not length dependent. EMFO link has a very small residual phase noise and is much smaller than the original phase noise in the reference. DMFO link has high phase noise contribution because of the RIN noise and the AM/PM noise conversion under large signal condition. At offset frequency higher than 1 kHz, and residual phase noise is higher than reference phase noise and then will degrade the phase noise. The slope change of residual phase noise as a function of offset frequency is caused by the intrinsic low frequency fluctuation of property of the laser diode and the Mach-Zehnder modulator.

6.3.3. Comparison of Phase Fluctuation

The phase and amplitude balance between front-ends is important in the direction finding system. However, due to the temperature variation, the phase of information signal and reference will be changed. In the electrical path, the phase variation is primarily caused by the thermal expansion of the conductors [136]. On the other hand, the fiber has much smaller thermal expansion factor than metal. Therefore, FO link has a much lower phase variation rate than coaxial cable, as shown in figure 79. The phase variation due to the thermal expansion is proportional to the operating frequency. For example, for a signal at 450 MHz, the phase variation in both optic and electric links will be 40 times smaller than that shown in figure 79.

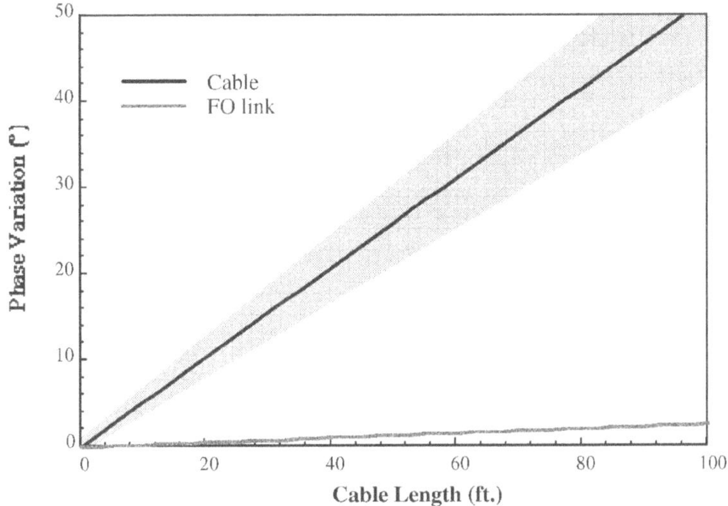

Figure 79 : Comparison of maximum phase variation due to temperature variation for an 18 GHz signal as a function of length for the electrical and optical link.

6.3.4. Comparison in Terms of Weight and Size

Even though it is clear that FO link is generally much better than the coaxial cable in terms of weight, size and EMI, it is still important to know quantitatively how much and under what condition the FO link is better than electrical cable in our application. Figure 80 shows the size comparison between FO links and coaxial cable. Clearly, FO links have very small size, and directly modulated link is the smallest in size. The EMFO is relatively bulky because the big solid-state laser source and mach-zender modulator. It can be predicted that EMFO and DMFO is smaller than the coaxial cable at length longer than 12 and 5 feet, respectively. Figure 81 compares the weight of cable and FO links [136-138]. The shadow region for cable presents the variation from different quality cables. For a high quality cable, it usually has large size of conductors which can make the link very heavy. EMFO and DMFO is lighter than the coaxial cable at length longer than 14 and 4 feet, respectively.

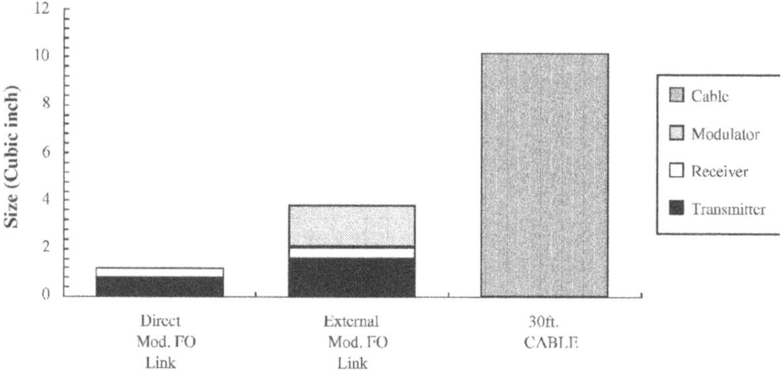

Figure 80 : Size comparison. The cable is Gore-Tex type II cable.

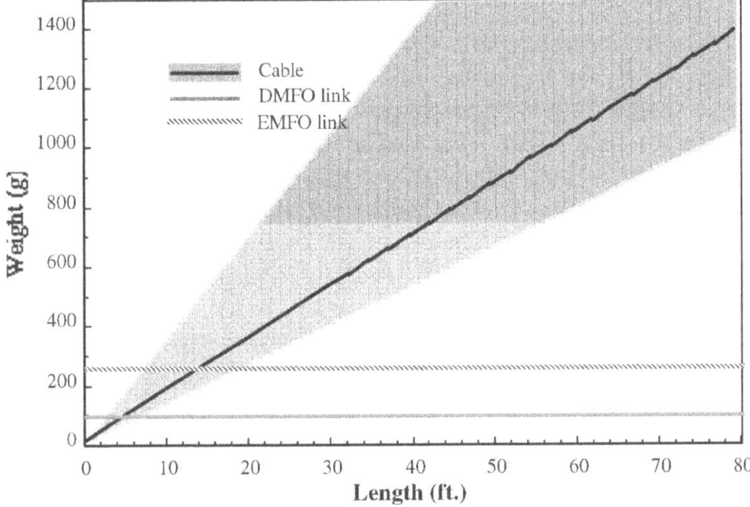

Figure 81 : Comparison of weight for electrical and optical links.

6.4. Novel Semiconductor Laser Structure

It is attractive to combine coherent distribution of the reference and sub-carrier data signals using a single optical transmitter (E/O), as shown in figure 82. This performance is cost saving and provides the highest SFDR and power consumption efficiency. First the performance of F-P laser with optical feedback using external cavity is reviewed in section 6.4.1. Our approach is to use a monolithically integrated electro-absorption (EA) modulator with a long F-P laser. This device performance is discussed in 6.4.2.

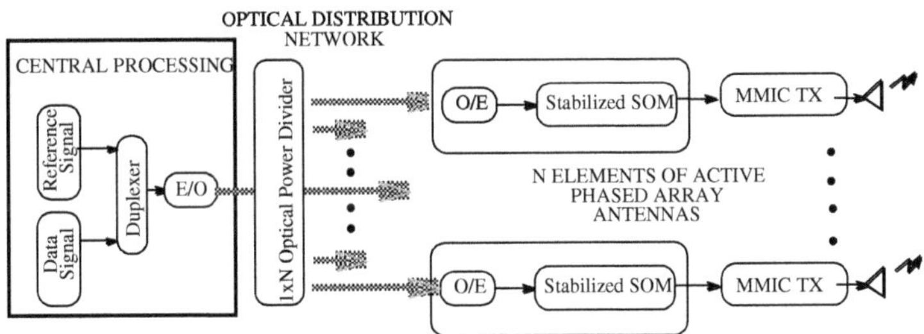

Figure 82 : Conceptual representation of optical distribution of a frequency reference and data signals to N elements of a Ka-band phased array antennas. The reference signal is used to synchronize the local oscillators in the SOM. Data signal is mixed with the stabilized SOM to generate an RF signal. (E/O and O/E are optical transmitter and receiver modules. MMIC TX represents a MMIC based transmitter module).

6.4.1. Microwave Carrier Phase Noise Measurements

Numerous measurements were conducted to assess the FM noise degradation of the frequency reference in the laser with and without external cavity. These measurements were conducted at the carrier frequency of 5.08 GHz. The measured degradation was the standard multiplication by four effect (20 Log(4) = 12 dB) indicating that the laser's RIN converted to the microwave FM noise is lower than the synthesizer noise spectra. Therefore, we have proceeded to measure the residual FM noise level to identify the laser diode's contribution to the FM noise floor of the microwave carrier.

The single side-band (SSB) residual phase noise of the microwave carrier signal is measured using the set-up shown in figure 83. The measurement system phase noise level (i.e., without the laser and detector) is depicted as the baseline in figure 84. This measured SSB system noise level is lower than the noise floor of signal analyzer. The phase noise measurement repeatability of the whole system is about ±3dB.

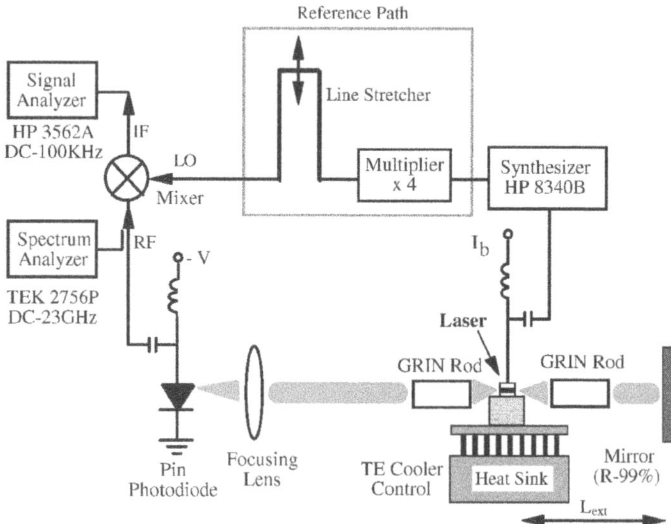

Figure 83: Experimental setup for measuring the close-in to carrier phase noise of laser diode with the external feedback.

The laser is modulated at 1.27GHz by an input power level of 0 dBm (corresponding to a current modulation index of m=0.9). The measured residual phase noise of the fourth harmonic signal at 5.08GHz for the laser with and without external optical feedback is shown in figure 84. The offset carrier frequency is from 1Hz to 100kHz. As seen in this figure, the phase noise are very close for both with and without feedback cases at offset carrier frequencies less than 1kHz, which is primarily dominated by the 60Hz sub- and super- harmonics of the ac power-line. A crossover of residual phase noise at about few hertz is observed for the laser with and without feedback. It is speculated that this increase level of the residual phase noise is due to the light scintillation in the external optical cavity and is avoided in the monolithically integrated version. At offset carrier frequency larger than 1kHz, the difference in the phase noise levels are distinct for these two cases. In fact, the phase noise level of the feedback case is reduced by 20dB from -115 to a level of -135dBc/Hz at far-away offset carrier frequency.

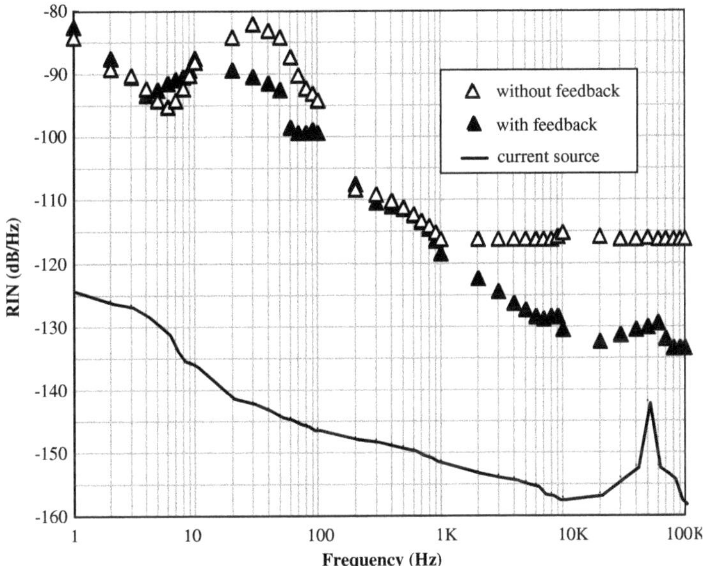

Figure 84 : The measured residual phase noise at the fourth harmonic (5.08 GHz) for the laser with and without optical feedback. The modulation frequency is 1.27 GHz, and the input power level is 0 dBm.

The basis for this reduction in phase noise can be explained using forced oscillation process. Considering that the optical feedback creates a low Q oscillation at the round trip frequency, one can consider that the oscillation signal in the optical cavity is injection locked by the harmonic of the modulation signal. This process is similar to subharmonic injection locking of microwave oscillators. Using the subharmonic injection locking theory in regular microwave oscillator, the phase noise of the locked oscillation signal in light intensity at n_ω can be expressed as [142] :

$$\pounds_{\text{out},n\omega}(\Omega) = \frac{\left[n^2 \pounds_{\text{in},\omega}(\Omega) + n^2 \pounds_{\text{up},\omega}(\Omega)\right]\Delta\omega_{Lock}^2 + \Omega^2 \pounds_{\text{osc},n\omega}(\Omega)}{\Omega^2 + \Delta\omega_{Lock}^2} \tag{19}$$

The first term in the numerator represents the contribution from the modulation signal, which is degraded by up-converted laser LF noise $£_{up,\omega}$. The locking range is a function of the oscillating optical power density, P_{osc}, and the harmonic signal in photon density, $P_0 I_n(a)/I_0(a)$, which is generated by laser's nonlinearity :

$$\Delta\omega_{Lock} = \frac{n\omega}{2Q}\frac{P_0 I_n(a)}{I_0(a)P_{osc}} \tag{20}$$

On the other hand, adopting the phase noise expression of a microwave oscillator, the phase noise of the free-running laser with coherent feedback (an oscillation without any external force) is written as [141, 142] :

$$£_{\mathrm{osc},\,n\omega}(\Omega) = \left\{\left(\frac{n\omega}{2Q}\right)^2 \frac{1}{\Omega^2} + 1\right\}\left\{£_{\mathrm{up},\,n\omega}(\Omega) + £_{\mathrm{RIN},\,n\omega}(\Omega)\right\} \tag{21}$$

In Eq. (21), the RIN contribution can be neglected for close-in offset frequency since it is much lower than the up-conversion noise because of high oscillation power. Substituting Eqs. (21) and (20) into Eq. (19) and recognizing $\Omega << \Delta\omega_{Lock}$, one obtains an approximation for the residual phase noise in the enhanced harmonic signal, $£_{r,n\omega}$:

$$£_{\mathrm{r},\,n\omega}(\Omega) = £_{\mathrm{out},\,n\omega}(\Omega) - n^2 £_{\mathrm{in},\,\omega}(\Omega) - n^2 C_{up,\omega} RIN(\Omega) + C_{up,n\omega}\left(\frac{I_0 P_{osc}}{P_0 I_n}\right)^2 RIN(\Omega) \tag{22}$$

The first term of the residual phase noise in Eq. (22) is as result of the up-converted RIN, whereas the second term is controlled by the injection power level and is dominated by the ratio of $I_0(a)/I_n(a)$, where a is related to the modulation index of laser. The predicted results matches well with the measured results [135].

6.4.2. Monolithically Integrated Laser Diode

A monolithic version of laser with external cavity is realizable using semiconductor fabrication process. Figure 85a shows a schematic drawing of the monolithic laser with an integrated EA modulator. Stacked structure consisting of two MQW layers, a MQW for laser diode (MQW-LD) and a MQW for EA modulator (MQW-MD) are employed. The details of this structure and the fabrication process are described in [143]. The F-P cavity length for our experiment is cleaved approximately for a length of 2170μm. This total length is composed of 1970 μm long gain section, 150 μm long modulator, and a 50 μm long separation region. The facet of the modulator section is coated with high reflective film (R=85%). The facet of the gain section is as cleaved. The laser is mounted in a high-frequency package. The schematic diagram of the long FP laser with integrated EA modulator is shown in figure 85b. RF inputs to the gain and EA modulator sections are through K connectors' transition to microstrip lines. The 50Ω microstrip lines are realized on alumina substrates. The gain section is resistively matched or reactively matched to 50Ω. The EA modulator is left unmatched.

The gain section of the laser diode is forward biased at different bias currents and the EA section is reverse biased by different voltage levels. The natural frequency response of the laser is measured at various laser bias currents, ranging from I_b=80mA up to I_b=150mA for bias voltage of V_m=0V. A resonance peak is observed that is associated with the longitudinal mode separation in the long FP laser. The longitudinal mode separation is calculated as $\Delta f=c/2nL$=19.3GHz, where c=300mm.GHz is speed of light in free space, n=3.5 is index of refraction of waveguide, and L=2.17mm is the F-P cavity length. This resonant frequency has a frequency tuning sensitivity of 1MHz/mA.

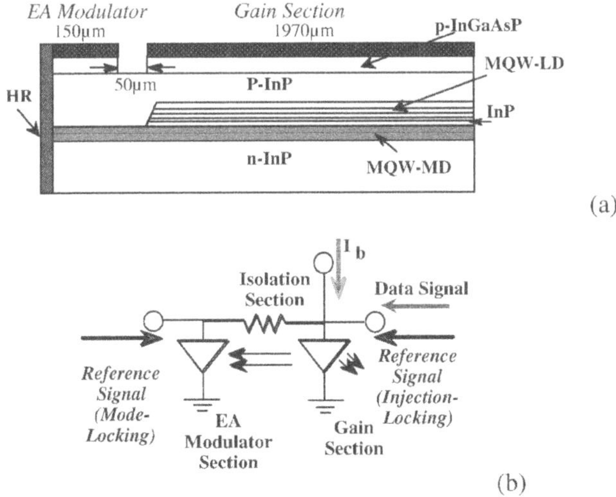

Figure 85 : a) Conceptual representation of the long F-P laser integrated with an electro-absorption modulator b) mechanical fixture used for mounting of the laser diode for injection locking, mode locking and opto-electronic conversion evaluations.

One could stabilize the optical oscillations using injection-locking. As the gain section is modulated by a synthesized frequency reference (HP83640A) of $P_m \geq$-1dBm at f_m=19.258GHz, a single oscillation peak appears. The familiar one sided injection locking spectra is observed outside the injection locking range and the close-in to carrier phase noise is significantly reduced within the locking range. The measured close in to carrier phase noise degradation at 100Hz offset carrier is depicted in figure 86, where 31dB and 6dB degradation are measured for the injected power of P_m=+0.5dBm in the resistively- and reactively-matched modules respectively. However for injected power level of +4.5dBm, a close-in to carrier phase noise identical to the reference source is measured for the reactively-matched case.

On the other hand one could use EA modulator and by pumping as high as 15 dBm rf power at 19.3 GHz, mode-lock the optical side modes to one another. The close-in to carrier phase noise can be compared to the injection locked case using gain modulation. In comparison, at 100Hz offset carrier for the P_m=+15dBm and V_m=-4.5V, a phase noise of -73dBc/Hz and -70dBc/Hz is measured at I_b=140mA and I_b=100mA respectively using an HP83640A as the reference source. Note also that the minimum electrical power required to achieve mode-locked pulses of 6mW or 12mW using EA modulation is 15dBm at V_m=-2.5V. On the other hand, estimated electrical power of P_m =12dBm and +20dBm is required to generate mode-locked pulses of 6mW and 12mW using gain modulation for V_m=0V. Therefore, electrical injection locking provides a low power consuming alternative to the active mode-locking to stabilize the intermodal oscillations.

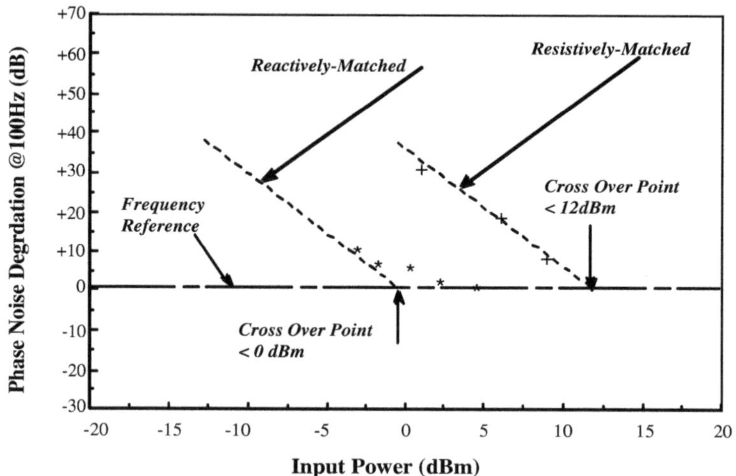

Figure 86. Measured FM noise degradation at 100Hz offset carrier for the injection locked intermodal oscillation. The measured injection locking power of the long F-P laser diode are also depicted at the cross over points for the resistively- (+) and reactively-matched (*) lasers.

Since this stabilized signal has much cleaner close-in to carrier phase noise than the free-running oscillation, it could be employed as the LO signal. Next the gain section of this laser is modulated by S-band signals (2.2 GHz ± 50 MHz). Strong nonlinearity of the mode-locked laser at the

LO signal of 19.3 GHz up-converts the S-band signals to 17.1 GHz and 22.5 GHz. The data modulation power level is changed over a wide range. An optical conversion loss is defined as the ratio of the generated mixed RF signal (19.3 ± 2.2 GHz) to the IF signal (2.2 GHz). The optical conversion loss is as low as 1.4 dB resulting in the electrical conversion loss of 2.8 dB. The opto-electronic conversion loss for the lower side-band (LSB) at 17.1 GHz is higher than the upper side-band (USB) of 22.5 GHz by 1.3 dB (i.e., 2.6 dB electrical) [144].

On the other hand, a modulation loss greater than 51 dB is measured when the the gain section is directly modulated by the RF signal at 17.1 GHz. The spurious-free dynamic range (SFDR) of this opto-electronic mixer is also evaluated. The intermodulation distortion (IMD) measurements are conducted for two modulating tones which are 5 MHz apart (e.g., f_1 = 2.200 GHz and f_2 = 2.205 GHz). Both tones are up-converted by stable LO signal of 19.360 GHz and IMD of the up-converted RF signals are measured at LSB and USB frequencies. Based on the mode-locked laser IMD and RIN noise measurement results for the up-converted RF tones, SFDR LSB and USB RF signals are 88 $dB.Hz^{2/3}$ and 89 $dB.Hz^{2/3}$ respectively [144].

6.5. Conclusions

Future generation communication satellites are envisioned to have as many as 100 beams covering various part of globe. Optical beamforming networks are considered as a viable solution, however cost and reliability constraints restrict the distribution to directly modulated FO links. Both carrier and data signals are to be distributed, whereas the trade off in terms of phase noise degradation, dynamic range, weight and size is presented for both electrical and optical distribution networks. A new device structure has shown also investigated which is based on the concept of a long F-P laser that is monolithically integrated with an EA modulator. The analytical models indicate that stabilized LO signal at 19.3 GHz can be attained using injection locking and mode-locking. The achieved close-in to carrier phase noise of the stabilized LO signal is lower in the case of injection locking than mode-locking for the same modulating power level. The opto-electronic mixing of 19.2 GHz LO and S-band data is also feasible using this device with SFDR as high as 90 $dB.\ Hz^{2/3}$.

6.6. Acknowledgment

The author wishes to acknowledge the contribution of many of my students, particularly Dr. Tsang Der Ni, Dr. Xiangdong Zhang, and Dr. Joong Hee Lee. The experimental results on the monolithically integrated

EA modulator with long F-P laser diode were conducted as part of my sabbatical leave in NTT Wireless Communication laboratories and I would like to acknowledge of assistance of many colleagues such as Dr. Kenji. Sato and Dr. Hiroyo Ogawa.

7. REFERENCES

[1] E. I. Ackerman
personal communication
1997

[2] G. E. Betts
Linearized Modulator for Suboctave-bandpass optical analog links
IEEE Trans. on Microwave Theory and Techniques, 1994, vol.42, no. 12, pp 2642-2649.

[3] G. E Betts
personal communication
1996

[4] W. B. Bridges and J. H. Schaffner
Distortion in Linearized Electrooptic Modulators
IEEE Trans. on Microwave Theory and Techniques, 1995, vol.43, no. 9, pp. 2184-2197

[5] J. L. Prince
personal communication
1998

[6] D.D. Marcenac, D. Nesset, A.E. Kelly, M. Brierley, A.D. Ellis, D.G. Moodie and C.W. Ford
40 Gbit/s transmission over 406 km of NDSF using mid-span spectral inversion by four-wave-mixing in a 2 mm long semiconductor optical amplifier
Electronics Letters, vol.33, pp.879-880, 1997

[7] A.H. Gnauck, A.R. Chraplyvy, R.W. Tkach, J.L. Zyskind, J.W. Sulhoff, A.J. Lucero, Y. Sun, R.M. Jopson, F. Forghieri, R.M. Derosier, C. Wolf and A.R. McCormick
One terabit/s transmission experiment
OFC '96, vol.2, pp. 407-410, 1996

[8] J.V. Collins, I.F. Lealman, P.J. Fiddyment, C.A. Jones, R.G. Waller, L.J. Rivers, K. Cooper, S.D. Perrin M.W. Nield and M.J. Harlow
Passive alignment of a tapered laser with more than 50% coupling efficiency
Electronics Letters, vol.31, pp. 730-731, 1995

[9] A.E. Kelly, I.F. Lealman, L.J. Rivers, and S.D. Perrin
Low noise figure (7.2 dB) and high gain (29 dB) semiconductor optical amplifier with a single layer AR coating
Electronics Letters, vol.33, pp. 536-538, 1997

[10] T.R. Chen, P.C. Chen, J. Ungar, J. Paslaski, S. Oh, H. Luong and N. Bar-Chaim
Wide temperature range linear DFB lasers with very low threshold current
Electronics Letters, vol. 33, pp. 963-965, 1997

[11] O. Kjebon, R. Schatz, S. Lourdudoss, S. Nilsson, B. Stalnacke and L. Backbom
30GHz direct modulation bandwidth in detuned InGaAsP DBR lasers at 1.55μm wavelength
Electronics Letters, vol.33, pp. 488-489, 1997

[12] K. Kato
Long-wavelength photodetectors for ultrawide-band systems
IEICE Trans. on Electronics, vol.E79-C, pp. 14-20, 1996

[13] J.S. Paslaski, P.C. Chen, J.S. Chen, C.M. Gee and N. Bar-Chaim
High power microwave photodiode for improving performance of rf fibre optic links
Proc. SPIE, vol.2844, pp. 110-119, 1996

[14] R.G. Walker
GaAs/AlGaAs travelling-wave modulators for mm-wave frequencies
IEE Colloquium 'Towards Terabit Transmission' (Digest No.1995/110), 1995

[15] F. Devaux, P. Bordes, A. Ougazzaden, M. Carré and F. Huet
Experimental optimisation of MQW electroabsorption modulators with up to 40 GHz bandwidths
Electronics Letters, vol.30, pp. 1347-1348, 1994

[16] www.ortel.com

[17] Anacom Systems Corp., 100 Jersey Ave., New Brunswick, New Jersey 08901

[18] www.mikom.com

[19] www.intec.rug.ac.be/research/projects/horizon/projects/frans/frans.html

[20] J. J. O'Reilly, P. M. Lane, R. Heidemann, and R. Hofstetter
Optical generation of very narrow linewidth millimetre wave signals
Electronics Letters, vol. 28, pp. 2309- 2310, 1992

[21] L. D. Westbrook, and D. G. Moodie
Simultaneous bi-directional analogue fibre-optic transmission using an electroabsorption modulator
Electronics Letters, vol. 32, pp. 1806-1807, 1996

[22] L.D. Westbrook, L. Noël and D.G. Moodie
Full duplex, 25km analogue fibre transmission at 120Mbit/s with simultaneous modulation and detection in an electroabsorption modulator
Electronics Letters, vol. 33, pp. 694-695, 1997

[23] D. Wake, D. Johansson and D.G. Moodie
Passive picocell - a new concept in wireless network infrastructure
Electronics Letters, vol. 33, pp.404-406, 1997

[24] G.H. Smith, D. Novak and Z. Ahmed
Technique for optical SSB generation to overcome dispersion penalties in fibre-radio systems
Electronics Letters, vol. 33, pp.74-75, 1997

[25] D. Wake, C. R. Lima and P. A. Davies
Optical generation of mm-wave signals for fibre-radio systems using a dual-mode DFB semiconductor laser
IEEE Trans. on Microwave Theory and Techniques, vol. 43, pp.2270-2276, 1995

[26] F.N. Timofeev, S. Bennett, R. Griffin, P. Bayvel, A.J. Seeds, R. Wyatt, R. Kashyap and M. Robertson
High spectral purity millimetre-wave modulated optical signal generation using fibre grating lasers
Electronics Letters, vol. 34, pp. 668-669, 1998

[27] L. Noël, D. D. Marcenac, and D. Wake
Optical millimetre-wave generation technique with high efficiency, purity and stability
Electronics Letters, vol. 32, pp. 1997-1998, 1996

[28] L. Noël, D. Wake, D.G. Moodie, D.D. Marcenac, L.D. Westbrook, and D. Nesset,
Novel techniques for high capacity 60 GHz fibre-radio transmission systems
IEEE Trans. MTT, vol.45, pp.1416-1423, 1997

[29] J. Aprille
Filtering and Equalisation for Digital Transmission
IEEE Comm. Magazine, March 1983, vol 21 n°2

[30] K. Bhargava
Forward Error Correction Schemes for Digital Communications
IEEE Comm. Magazine, January 1983, vol 21 n°1

[31] B. Sidar
A structured overview of digital communications. A tutorial review
IEEE Comm. Magazine, August, 1983 vol 21 n°5 and october 1983 vol 21 n°7

[32] A. Glavieux, M. Joindot
Communications Numériques, Collection Pédagogique de Télécommunication
Masson, 1996

[33] A. Siller
Multipath Propagation
IEEE Comm. Magazine, February 1984, vol 22 n°2

[34] D. P. Taylor P. R. Hartmann
Telecommunications by Digital Radio
IEEE Comm. Magazine, August 1986, vol 24 n°8

[35] J. K. Chamberlain, F. M. Clayton, H. sari, P. Vandamme
Receiver Techniques for Microwave Digital Radio
IEEE Comm. Magazine November 1986 vol 24 n°11

[36] Special Issue on Mobile Communications
IEEE Comm. Magazine, June 1987, vol 25 n°8

[37] PCS : the second generation
IEEE Comm. Magazine, December 1992, vol.30, n°12

[38] Optical Wireless : the story so far
IEEE Comm. Magazine, December 1998, vol.36 n°12

[39] M. Zeng, A. Annamalai, V. K. Bhargava
Recent Advances in Cellular Wireless Communications
IEEE Comm. Magazine September 1999 vol.37 n°9

[40] Le guide d'ondes circulaires.
Numéro Spécial des Annales des Télécommunications, Tome 29, n°9-10 septembre-octobre 1974

[41] R. D . Ehrbar
Undersea cables for telephony
IEEE Comm. Magazine, August 1983, vol 21 n°5

[42] P. R. Trischitta, W. C. Maria
Applying WDM technology to undersea cable systems
IEEE Comm. Magazine, February 1998 vol.36 n°2

[43] Special Issue on Satellite Communications
IEEE Comm. Magazine, March 1984, vol 22 n°3

[44] D. Chakrabarty
VSAT Communication Systems : an Overview
IEEE Comm. Magazine, May 1988, Vol. 26 n°3

[45] G. P. Agrawal
Nonlinear fiber optics
Academic Press New York, 1989

[46] I and M. Joindot and 12 coauthors
Les Télécommunications par fibre optique
CTST, Dunod, Paris, 1996

[47] E . Desurvire
Erbium Doped Fiber Amplifiers. Principles and Applications

Wiley, New York, 1994
[48] I. W. Stanley
A tutorial review of Techniques for Coherent Optical Fiber Transmission Systems
IEEE Comm. Magazine, August 1985 n°8
[49] A. Linke
Optical Heterodyne Communication Systems
IEEE Comm. Magazine, October 1989, vol.27 n°10
[50] M. Potenza
Optical Fibre Amplifiers in Telecommunication Systems
IEEE Comm. Magazine, August 1996, vol 36 n°8
[51] Undersea Communication Networks
IEEE Comm. Magazine, February 1996, vol 34 n°2
[52] P. Kaiser, J. Midwinter, S. Shimada
Status and Future Trends in Terrestrial Optical Fiber Systems in North America Europe and Japan
IEEE Comm. Magazine, October 1987, vol.25 n°10
[53] Deployment of WDM Fiber Based Optical Networks
IEEE Comm. Magazine, February 1998, vol.36 n°2
[54] Multiwavelength Fiber optic Communication
IEEE Comm. Magazine, December 1998, vol.36 n°12
[55] C. Cox III, E. Ackerman, R. Helkey and G. E. Betts
Techniques and performance of intensity-modulation direct-detection analog optical links
IEEE Trans. Microwave Theory Tech., vol. 45, pp. 1375-1383, Aug. 1997
[56] M. Y. Frankel, P. J. Matthews and R. D. Esman
Fiber-optic true time steering of an ultrawide-band receive array
IEEE Trans. Microwave Theory Tech., vol. 45, pp. 1522-1530, Aug. 1997
[57] M. Alles, U. Auer, F. J. Tegude and D. Jager
Distributed velocity-matched 1.55 um InP travelling-wave photodetector for generation of high millimeterwave signal power
IEEE MTT IMS Dig., pp. 1233-1236, 1998
[58] K. S. Giboney, M. J. W. Rodwell and J. E. Bowers
Traveling-wave photodetector theory
IEEE Trans. Microwave Theory Tech., vol. 45, pp. 1310-1319, Aug. 1997
[59] L. Y. Lin, M. C. Wu, T. Itoh, T. A. Vang, R. E. Muller, D. L. Sivco and A. Y. Cho,
High-power high-speed photodetectors – design, analysis and experimental demonstration
IEEE Trans. Microwave Theory Tech., vol. 45, pp. 1320-1331, Aug. 1997
[60] J. Lin and T. Itoh
Active integrated antennas
IEEE Trans. Microwave Theory Tech., vol. 42, pp. 2186-2194, Dec. 1994
[61] Y. Qian and T. Itoh
Progress in active integrated antennas and their applications
IEEE Trans. Microwave Theory Tech., vol. 46, pp. 1891-1900, Nov. 1998
[62] B. S. Ke, T. Chau, Y. Qian, M. Wu and T. Itoh
Tapered slot antenna integrated with velocity-matched distributed photodetector
1998 IEEE MTT IMS Dig., pp. 1241-1244, June 1998
[63] D. M. Pozar
Microstrip antennas
Proc. IEEE, vol. 80, no. 1, Jan. 1992
[64] G. P. Gauthier, A. Courtay and G. M. Rebeiz
Microstrip antennas on synthesized low dielectric-constant substrates

IEEE Trans. Antennas Propagat., vol. 45, pp. 1310-1314, Aug. 1997

[65] Y. Qian, R. Coccioli, D. Sievenpiper, V. Radisic, E. Yablonovitch and T. Itoh
Microstrip Patch Antenna Using Novel Photonic Band-Gap Structures
Microwave Journal, pp. 66-76, January 1999

[66] Y. Yoshimura
A microstrip slot antenna
IEEE Trans. Microwave Theory Tech., pp. 760-763, Nov. 1972

[67] T. M. Welller, L. P.B. Katehi and G. M. Rebeiz
Single and double folded-slot antennas on semi-infinite substrates
IEEE Trans. Antennas and Propag., vol. 53, no. 12, Dec. 1995 pp. 1423-1428

[68] H. S. Tsai, M. J. W. Rodwell and R. A. York
Planar amplifier array with improved bandwidth using folded-slots
IEEE Microwave and Guided Wave Lett., vol. 4, no. 4, April 1994

[69] W. R. Deal, V. Radisic, Y. Qian and T. Itoh
A broadband microstrip-fed slot antenna
1999 IEEE MTT-S International Topical Symposium on Technologies for Wireless Applications Digest, pp. 209-212, Feb. 1999

[70] K. S. Yngvesson, D.H. Schaubert, T. L. Korzeniowski, E.L. Kollberg, T. Thungren and J.F. Johansson
Endfire tapered slot antennas on dielectric substrates
IEEE Trans. on Antennas and Propag., vol. 33, no. 12, Dec. 1985

[71] K. S. Yngvesson, T.L. Korzeniowski, Y.K. Kim, E.L. Kollberg and J.F. Johansson
The tapered slot antenna- A new integrated element for millimeter-wave applications
IEEE Trans. on Microwave Theory and Tech., vol. 37, no. 2, Feb. 1989

[72] K.P. Ma, Y. Qian and T. Itoh
Analysis and application of a new CPW-slotline transition
IEEE Trans. on Microwave Theory and Tech., vol. 47, no. 4, April 1999

[73] Y. Qian, W.R. Deal, N. Kaneda and T. Itoh
Microstrip-fed quasi-yagi antenna with broadband characteristics
Electronic Letters, vol. 34, no. 23, Nov. 1998

[74] Y. Qian, W. R. Deal, N. Kaneda and T. Itoh
A uniplanar quasi-Yagi antenna with wide bandwidth and low mutual coupling characteristics
1999 IEEE AP-S International Symposium, Orlando, FL, July 1999

[75] Y. Qian, W. R. Deal, J. Sor and T. Itoh
A novel printed antenna with broadband circular polarization
1999 Asia-Pacific Microwave Conference, Singapore, Dec. 1999

[76] W. Ng, A. Walston, G. Tangonan, J. J. Lee and I. Newberg
Optical steering of dual band microwave phased array antenna using semiconductor laser switching
Electronics Letters, Vol. 26, Page 791, 1990

[77] R. Tang, A. Popa, J. J. Lee
Applications of photonic technology to phased array antennas
Proc. IEEE AP-S Symp., Vol. II, Page 758, 1990

[78] W. Ng, A. Walston, G. Tangonan, J. J. Lee, I. Newberg
The first demonstration of an optically steered microwave phased array antenna using true-time-delay
IEEE Journal of Lightwave Technology, Vol. 9, p. 1124, 1991

[79] A. P. Goutzoulis, D. K. Davies, J. M. Zomp
Hybrid electronic fiber optic wavelength-multiplexed system for true time-delay steering of phased array antennas

Optical Engineering, Vol. 31, p. 2312, 1992
[80] H. Zmuda, E. N. Toughlian
Photonic Aspects of Modern Radar
Artech House, 1994
[81] J. J. Lee, R. Y. Loo, S. Livingston, et al
Photonic Wideband Array Antennas
IEEE Trans. Antennas and Propagation, Vol. 43, pp. 966-982, Sept., 1995
[82] W. R. Welty, US Patent No. 3,090,928, 1963
[83] H. Shnitkin
Electronically scanned antennas
Microwave Journal, p. 57, January, 1961
[84] E. Pels, W. Liang
A method of array steering by means of phase control through heterodyning
IRE Trans. Antennas & Propagation, p. 100, January, 1962
[85] J. L. Butler
Variable I.F. scanning
Microwave Scanning Antennas, Vol. III, Ed. R. C. Hansen, Academic Press, 1966
[86] K. Aamo
Frequency controlled antenna beam steering
IEEE MTT-S Digest, p. 1549, 1994
[87] R. Benjamin, C. Zaglanikis and A. J. Seeds
Optical Beamformer For Phased Arrays with Independent Control of Radiated Frequency and Phase
Electronics Letters, 1990, Vol 26, No. 22, p. 1853
[88] A. J. Seeds
Optical Beamforming Techniques for Phased-Array Antennas
Microwave Journal, Vol. 35, No. 7, July 1992, pp 74
[89] D.K. Paul
Optical Beamforming and Steering Architecture for SATCOM Phased Array
IEEE Antennas and Propag. Symp., 1996, p. 1508
[90] A. S. Daryoush and M. Ghanevati
True Time Delay Challenge in Optically Controlled Phased Array Antennas
IEEE Antennas and Propag. Symp., 1997, p. 732
[91] J. J. Lee, et al
Multibeam Arrays Using RF Mixing Feeds
IEEE Antennas and Propag. Symp., 1997, Vol. 2, p. 706
[92] R. C. Hansen
Phased Array Antennas
John Wiley & Sons, Inc., 1998
[93] W. T. Patton
Array Feeds
Lecture 6, Practical Phased Array Antenna Systems, E. Brookner, Editor, Artech House, Inc., 1991
[94] S. M. Sherman
Monopulse Principles and Techniques
Artech House, Inc., 1984
[95] G. P. Kefalas & J. C Wiltse
Transmission Lines, Components and Devices
Chap. 8, Radar Handbook, M. I. Skolnik, Editor, McGraw-Hill Book Co., 1970
[96] 1999 Gould Fiber Optic Component Catalog
[97] 1997 Photonic Integration Research, Inc. Silica Optical Waveguide Devices Data Sheet

[98] Novel Fiber Optic Multimode Feed for Monopulse Lidar Antenna Application, Final Report
Contract No. DASG60-89-C-0067, U.S. Army Strategic Defense Command, October 1991

[99] J. L. Butler and R. Lowe
Beam Forming Matrix Simplifies Design of Electronically Scanned Antennas
Electronic Design, pp.170-173, Vol. 9, 12 April 1961

[100] J. T. Gallo and R. DeSalvo
Experimental Demonstration of Optical Guided-Wave Butler Matrices
pp. 1501-1506, Trans. IEEE MTT-S., Vol. 45, No. 8, Part II; August 1997

[101] W. Rotman and R. F. Turner
Wide-Angle Microwave Lens for Line Source Applications
pp. 623-632, Trans. IEEE AP-S, Vol. 11; November 1963

[102] T. C. Cheston & J. Frank
Phased Array Radar Antennas
Chap. 7, Radar Handbook, M. I. Skolnik Editor, 2nd Edition, McGraw-Hill Pub. Co., 1990

[103] R. A. Sparks, et al
Eight Beam Prototype Fiber Optic Rotman Lens,
1999 IEEE International Topical Meeting on Microwave Photonics, Melbourne, Australia, Nov. 1999

[104] R. J. Mailloux
Phased Array Antenna Handbook
Artech House, Inc., 1994

[105] 1996 IEEE International Symposium on Phased Array Systems and Technology Digest, Boston, MA, 15-18 October 1996

[106] C. Pell
Phased Array Radars
Editor, MEP Ltd., 1988

[107] G. W. Stimson
Introduction to Airborne Radar
2nd Edition, IEEE Press, 1983

[108] M. I. Skolnik
Introduction to Radar Systems
McGraw-Hill Book Co., Inc., 1962

[109] M. Tamburrini, M. Parent, L. Goldberg, D. Stillwell
Optical feed for a phased array microwave antenna
Electron. Lett., 23, 680, 1987

[110] J.L. Guggenmos, R. L. Johnson, B. D. Seery
Optical distribution manifolds for phased array antennas
Proc. SPIE, vol . 886, 214, 1988

[111] E. N. Toughlian, H. Zmuda, Ph. Kornreich
Deformable mirror based optical beam focusing system for phased array antenna
IEEE Photon. Tech. Lett., 2, 444, 1990

[112] M. J. Wale, C. Edge, M. G. Holliday, R. G. Walker, N. S. Birkmayer, B. Hösselbarth, T. Jakob
Optoelectronic beamforming for a space borne phased array antenna
Proc. SPIE, vol. 1703, 368, 1992

[113] M. F. Lewis
Coherent optical RF beamforming
Proc. IEEEE Microwave Photonics'97, 23, 1997

[114] Goutzoulis, K. Davies, J. Zomp, P. Hrycak, A. Johnson

Development and field demonstration of a harware compressive fiber optics true time delay steering system for phase array antennas
Appl. Opt., 33, 8173, 1994

[115] K. Horikawa, I. Ogawa, T. Kitoh
Silica based integrated planar lightwave true time delay network for microwave antenna applications
OFC'96 Technical Digest, 100, 1996

[116] T. Sullivan, S. D. Mukherjee, M. K. Hibbs-Brenner, A. Gopinath, E. Kalweit, T. Marta, W. Goldberg, and R. Walterson
Switched time delay elements based on AlGaAs/GaAs optical waveguide technology at 1.32 μm for optically controlled phased array antennas
SPIE, vol. 1703, 264, 1992

[117] G. A. Magel and J. L. Leonard
Phosphosilicate glass waveguides for phased-array radar time delay
SPIE, vol. 1703, 373, 1992

[118] E. Ackerman, S. Wanuga, D. Kasemset, W. Minford, N. Thorsten, and J. Watson
Integrated 6-bit photonic true-time delay unit for light weight 3-6 Ghz radar beamformer
MTT-Symposium Digest (Institute of Electrical and Electronic Engineers, New-York, 1992, pp. 681-684

[119] W. Ng, A. Walston, G. Tangonan, J.J. Lee, I. Newberg, N. Bernstein
The first demonstration of an optically steered microwave phased array antenna using true time delay
IEEE J. Lightwave Technol., LT9, 1124, 1991

[120] J.J. Lee, R.Y. Loo, S. Livingston, V.I. Jones, J.B. Lewis, H.W. Yen, G. Tangonan, M. Wechsberg
Photonic wideband array antennas
IEEE Trans. Antenna Propag., 43, 966, 1995

[121] Frankel, R.D. Esman
True time delay fiber optic control of an ultrawideband array transmitter/receiver with multibeam capability
IEEE Trans. Microwave Theory and Tech., 43, 2387, 1995

[122] M. Wickham, L. Lembo, L. Dozal, J. Brock,
Fiber optic grating true time delay generator for broad band RF applications
Proc. SPIE, vol. 2844, 175, 1996

[123] Smith, M. Nawaz
Evaluation of Bragg fibre grating as TTD elements in optical phased array systems
Proc. IEEEE Microwave Photonics'97, 205, 1997

[124] J.E.Roman et al
Time Steered Array with a Chirped Grating Beamformer
Electron. Lett., Vol. 33, No.8, pp. 652-653, April 1997

[125] H. Zmuda et al
Photonic Bramformer for Phased Array Antennas using a Fiber Grating Prism
IEEE Photonic Technology Letters, Vol. 9, No. 2, pp. 242-243, Feb 1997

[126] D. Dolfi et al
Optically controlled true time delays for phased array antenna
Proc. SPIE1102,152 (1989)

[127] N.A.Riza
Transmit/Receive time delay beam forming optical architecture for phased array antenna
Appli.Opt. 30, 4594 (1991)

[128] D.K. Paul

Optical Technology in Communication Satellites
in The Froehlich/Kent Encyclopedia of Telecommunications, Volume 14, pp. 71-110, Marcel Dekker, Inc., 1997

[129] A. S. Daryoush, E. Ackerman, N. Samant, S. Wanuga, and D. Kasemset
Interfaces for High-Speed Fiberoptic Links: Analysis and Experiment
IEEE Trans. on Microwave Theory and Techniques, vol. 39. no. 12, pp. 2031-44, December 1991

[130] E. Ackerman and A. S. Daryoush
Antenna remoting in Comunication Satellites
IEEE Trans. Microwave Theory and Technique, Vol. 45, no. 8, pp. 1436-1442, August 1997

[131] K. Lau
Short-Pulse and High-frequency Signal Generation in Semiconductor Lasers
IEEE Journal of Lightwave Technology, Vol. 7, No. 2, pp. 400-419, Feb. 1989

[132] X. Zhang, T.D. Ni, and A.S. Daryoush
Laser Induced Phase Noise in Optically Injection Locked Oscillator
Digest of IEEE 1992 MTT Symposium, pp.765-768, 1992

[133] X. Zhang and A.S. Daryoush
Bias Dependent Low Frequency Noise Up-Conversion in HBT Oscillators
IEEE Microwave and Guided wave Letters, vol. 4, no. 12, pp. 423-425, Dec. 1994

[134] E. Bava, G.P. Bava, A. Godone. G. Rietto
Transfer function of amplitude and phase fluctuations and additive noise in varactor doublers
IEEE Trans. Microwave Theory and Technique, Vol. 27, no. 8, pp. 753-757, August 1979

[135] T.D. Ni, X. Zhang, and A.S. Daryoush
Experimental Study on Close-in carrier Phase Noise of Laser Diode with Coherent Feedback
IEEE Trans. Microwave Theory and Techniques, vol. 43, no. 9, pp. 2277-2283, Sept. 1995

[136] Product Catalog, "Gore Microwave Assembles", 1990

[137] Product Specification of QLM3S900 microwave laser, Lasertron, 1993

[138] P. Morton, T. Tanbun-EK, R. Logan
Ultra-high speed long wavelength MQW lasers
1993 Conference on Laser and Electro-Optics, p 55, May, 1993

[139] Product Specification of MicraChip Laser for Electro-Optic Systems and Instrumentation, Micracor, 1993

[140] C. Harder, J. Katz, S. Margalit, J. Shacham, and A. Yariv
Noise Equivalent Circuit of a Semiconductor Laser Diode
IEEE Journal of Quantum Electronics, vol. QE-18, no. 3 pp. 333-337, 1982

[141] Ni, Zhang, S. Daryoush
Harmonic enhancement of modulated semiconductor laser with optical feedback
IEEE Microwave and Guided wave Letters, vol. 3, no. 5, pp.139-141, 1993

[142] D. J. Sturzebecher, X. Zhou, X. Zhang, and A. S. Daryoush
Optically Controlled Oscillators for Millimeter-wave Phased Array Antennas
IEEE Trans. Microwave Theory and Techniques, vol. 41, no.6/7, pp.998-1004, 1993

[143] K. Sato, K. Wakita, I. Kotaka, Y. Kondo, and M. Yamamoto
Monolithic Strained-InGaAsP Multiple-Quantum-Well Lasers with Integrated Electroabsorption Modulators for Active Mode locking
Appl. Phys. Lett., Vol. 65, no. 1, pp. 1-4, July 1994

[144] A.S. Daryoush, K. Sato, K. Horikawa, and H. Ogawa

Efficient Opto-electronic Mixing at Ka-Band using a Mode-locked Laser IEEE Microwave and Guided wave Letters.

CHAPTER 5 : ALL OPTICAL PROCESSING OF MICROWAVE FUNCTIONS

1. INTRODUCTION

In this chapter we discuss of novel and innovative photonic techniques for optical signal generation and processing that are required in ultra-wide band systems.

The all-optical processing of microwave signals has been the subject of much research in the past decade and synthesis of such structures is now possible. We have tried to give in this chapter a glance on recent results of this field.

We have divided the chapter into four groups of general subjects.

The first group presents a general overview on microwave functions enable by photonics and industrial requirements as well. Original methods for frequency conversion and tunable optical filters are also presented.

The three last group concern semiconductor devices for optical processing, analog and digital convertors, and the very promising field of terahertz optoelectronics.

2. PHOTONIC BASE MICROWAVE FUNCTIONS

2.1. Microwave Functions Enabled by Photonics

R.D. Esman

Naval Research Laboratory, Photonics Technology Branch, Code 5650, Washington, DC 20375, USA

Esman@nrl.navy.mil

Abstract

Numerous microwave functions have been demonstrated utilizing photonics. This paper highlights the functions that are difficult or impossible without utilizing photonics. The true power of photonics for a particular application is not limited to one benefit, but typically includes many of both the well known advantages: size, weight, immunity to EMI, low differential loss, and large time-bandwidth product; and the not-so-well-known advantages: flexible harness, non-intrusive, non-conducting, submarine and subterranean, and lightning safe.

2.1.1. Introduction

In the past few years there has been significant progress in fiber optic components and technology applicable to microwave (analog) systems. Notable advances include high-power lasers and amplifiers, reduced-$V\pi$ modulators, and higher-power photodetectors, all well suited for >16 GHz operation. The corresponding order-of-magnitude increase in capability has strengthened interest in fiber optics as a viable alternative to conventional electrical techniques (e.g., coaxial cable). In the following we present some of the current capabilities and limitations of fiber optics (primarily for wideband applications) for microwave signal processing (including beamforming, filtering, switching, variable delay, and microwave circuit control).

In particular, there is some concentration on a primary application for photonics--that of time-delay beamforming and signal distribution for wideband arrays. Wideband arrays and photonics appear as a natural match for numerous reasons. First, there does not appear to be any microwave-based alternative for generating controllable time delay of numerous RF signals. Second, size and weight comparisons give photonics a significant advantage over the conventional cable alternative. Third, photonics offers multiplexing of RF signals. Fourth, active array T/R modules are amenable to low input levels. Fifth, fiber optics offers substantial immunity to electromagnetic interference. Additionally, fiber optics offers flexible harnessing, is non-intrusive, and reduces communication to the array backplane. But one of the most compelling reasons for using photonics is the low incremental signal loss in fiber optics. Most advantages of fiber optics are summarized in the table below—the obvious size and weight advantages are often *the* driving motivation for using fiber optics.

Feature	**Fiber Optics**	**Coax**
Size	250 μm diameter	2 ~ 6 mm diameter
Weight	0.073 g/m	130 ~ 300 g/m
Isolation	>80 dB	freq. dependent
Differential Loss (dB/km)	0.4 ~ 0.8	1000 ~ 1500
Differential Loss (dB/μs)	0.08 ~ 0.16	200 ~ 300
Time Bandwidth Product	1,000,000	100

2.1.2. Photonic-Microwave Links

The simplest function of photonics is that of a microwave link-delivering a microwave signal either from point A to point B or from time 1 to time 2 (i.e., introducing time delay or serving as memory). In the microwave domain, this may be accomplished using a cable or radio link. Surface acoustic wave devices have also been use for narrowband, small time delays. For photonics, we are simply converting the RF signal to an optical signal (by impressing the signal on an optical carrier) and then converting the optical signal back to RF. Many of the techniques of RF carrier modulation apply to optical carrier modulation. That is, an optical carrier may be amplitude (optical intensity) modulated, phase modulated, or frequency modulated—of course, each technique would require a different optical-to-RF conversion process. Concepts such as optical mixing (heterodyne) and single side band modulation have also been transplanted to optical systems. To consider the optical-microwave link, we examine each of the component parts: Electrical to Optical Converter (EOC), the transmission and amplification of the optical carrier, and then the Optical to Electrical Converter (OEC). Due to the availability of components (as related to the telecommunications industry) the optical wavelength of is typically 1550 nm where there is a minimum in optical attenuation in fiber or near 1300 nm where the standard single mode optical fiber exhibits zero dispersion.

2.1.2.1. Electrical to Optical Converter (EOC)

Generating microwave signals on an optical carrier can be accomplished by direct or external modulation or by optical heterodyne. While direct modulation offers small size, simplicity, and respectable modulation bandwidths (~18 GHz available and 37 GHz demonstrated), the (often acceptable) drawbacks include higher noise, nonlinearities, and limited optical power. The most advanced DFB devices have demonstrated operation >17 GHz and offer lower chirp, lower noise, and new flexibility for wavelength multiplexing. In general, external modulation offers better linearity, higher modulation frequencies (typically <20 GHz but 100 GHz has been demonstrated), higher output power, narrow spectral linewidth, and lower noise; but suffer from polarization dependence, high drive power requirements, and higher cost. Modulator linearization, electroabsorption modulators, and efficiency improvement are topics of recent attention. Lastly, optical heterodyne provides large dynamic range, no harmonic content, full modulation depth, high optical power, and high difference frequencies; however, system implementation is complex and is typically narrow band.

2.1.2.1.1. Direct Modulation

Direct modulation refers to directly modulating the optical source as opposed to external modulation, where the optical carrier is fed to a modulator external to the optical source (Figure 1). The only viable optical source for direct modulation is the semiconductor laser diode (SLD). There are numerous laboratories that have fabricated and packaged SLD with flat frequency response to beyond 20 GHz. Currently, devices with 3-dB bandwidths of 12 GHz are available with certain limitations on delivery time and emission wavelength. Hero devices have exhibited bandwidths up to 37 GHz. These devices are small (< 1 cc. packaged) and are relatively simple to operate—the output optical power is proportional to the drive RF current.

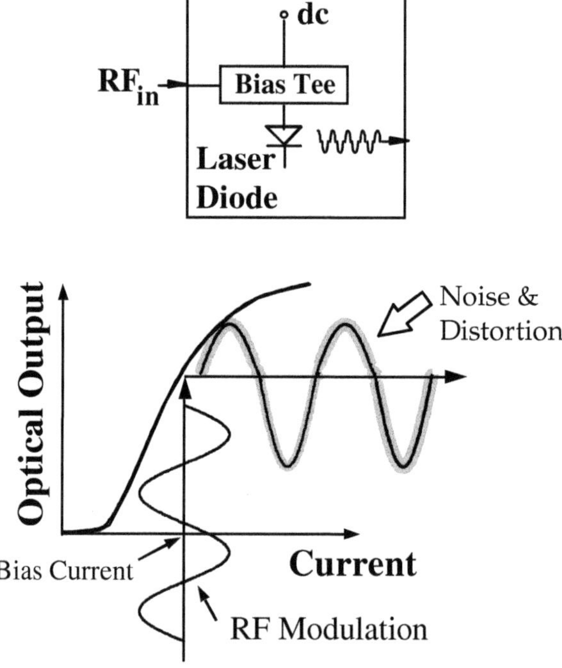

Figure 1 : Direct modulation.

The drawbacks associated with direct modulation include higher RIN, nonlinearities, limited wavelength selection, wider optical linewidth, optical wavelength shift (chirp) with modulation, and limited optical power. Directly modulated lasers can be either intensity or frequency modulated, however, due to the added complexity of the receiver and due to the large frequency noise of the laser, intensity modulation is preferred and most widely used. The relative intensity noise, then, is a measure of the intensity noise performance of the laser and is given by the ratio of the RMS fluctuation to the average intensity. Typical RIN performance of an SLD is from -130 to -160 dB/Hz, which (for a typical ~0.6 mA average photocurrent or -20 dBm electrical) corresponds to an equivalent input noise of about -110 to -140 dBm/Hz. Hence, with a typical input RF power for 1-dB compression of +15 dBm, there is a compression dynamic range of about 125 to 165 dB·Hz (from taking +15 dBm +140 dBm/Hz). Unfortunately, the spur-free dynamic range is much less due to the SLD nonlinearities and is approximately 110 dB·$Hz^{2/3}$.

2.1.2.1.2. External Modulation

For the microwave frequency range, modulators rely on the electro-optic effect that exists in several crystals. Similar to direct modulation, an intensity modulation scheme is preferred due to the complexity of the receive system. One of the most common and most attractive means to intensity modulate an optical carrier is using a Mach-Zehnder Modulator (MZM). In the MZM the input optical signal is split into two paths, one or both paths are (optical) *phase* modulated while it travels through an electro-optic material where an electric field (RF signal) is applied. Then, the two separate optical signals are recombined (typically in the same crystal). Since the two signals are optically coherent they either constructively or destructive interfere depending on their relative optical phase difference between the two paths—variation in the relative phase depends on the applied voltage. An important performance parameter with MZM's is the voltage required to switch from completely constructive (maximum optical output) to completely destructive (minimum optical output) interference. This switching voltage is referred to as $V\pi$. Other external schemes include electroabsorption (electric field dependent absorption) and directional couplers (electric field dependent cross coupling). In general, as compared to direct modulation, external modulation offers better linearity, higher modulation frequencies (typically <20 GHz but 100 GHz has been demonstrated), higher output power, narrow spectral linewidth, and lower noise. These advantages mainly arise because the modulator and laser can be optimized separately: the laser for high power and low noise and the modulator for high-speed and efficiency. On the other hand, since the electro-optic effect usually depends on the polarization of the optical wave, the MZM (and other external modulators) suffer from polarization dependence. Also, despite great improvements (100-fold) in recent years in the switching voltage (down to ~4 V for 20 GHz), the sensitivity of MZM still remains poor in relation to available signal levels. Other detractions of external modulation are the increase part count, increase complexity, and nonlinearities. Since the output is that of two interfering optical beams, the intensity is sinusoidally related to the input voltage (Figure 2). A topic of recent and increased interest is that of modulator linearization via predistortion, feedforward, or multiple modulator schemes. Performance of the externally modulated link will be addressed in greater detail below after the receiver (photodetector) has been discussed.

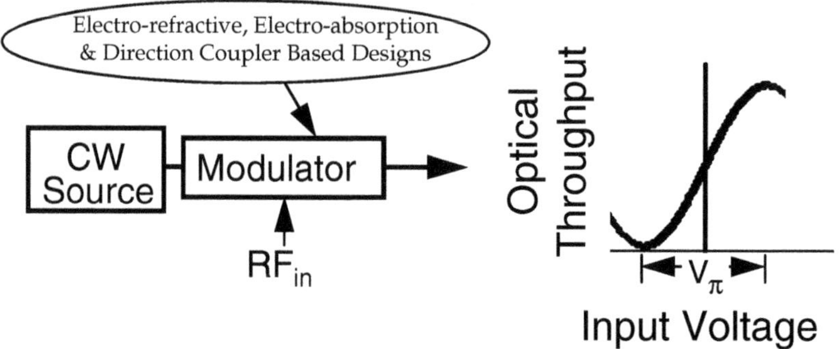

Figure 2 : External modulation.

2.1.2.3. Optical Heterodyne

A third approach to generating a microwave-modulated optical carrier is optical heterodyning (Figure 3). Optical heterodyne is similar to electrical (RF) mixing--two optical signals (optical fields) are combined and mixed in a photodetector (PD). If the two signals are separated in optical frequency by ω, then the optical intensity (the square of the optical fields) will vary sinusoidally at the frequency ω. Then since the PD generates a current proportional to the optical intensity, the PD will generate an electrical signal at frequency ω. One can view this as one optical signal and one optical local oscillator and the RF signal produced by the PD as the difference or intermediate frequency (IF).

Lasers used for optical heterodyning can be very spectrally pure (narrow linewidth, low phase noise), individually tuned, and exhibit low intensity noise. As a result optical heterodyne provides large dynamic range, high optical power, full modulation depth, high difference frequencies, and *inherently* no harmonic content. The large dynamic range is related to the freedom to use low-noise lasers (that cannot be modulated at microwave frequencies). As for the RF frequency capability, the frequency separation of any two lasers is enormous indeed. But practically speaking, since optical heterodyning produces an RF signal linewidth with the geometric mean of the two input lasers, typically narrow linewidth lasers (e.g., Nd:YAG) lasers are used. In addition, if the linewidth is narrow enough (as with the Nd:YAG lasers), then a phase lock loop can be formed to lock the optical phase of one laser to the other. Fortunately these lasers also exhibit plenty of in-fiber power levels—exceeding 300 mW, more than an external modulator or photodetector can handle. Another related advantage to optical heterodyning is the fact that the RF phase is related to the phase of both the input optical signals and a change is optical phase corresponds directly to a change in the RF phase. This proves to be useful in generating phase-controlled RF signals from optical signals. These numerous advantages are only shadowed by increased part count, increased system complexity, and a limitation in RF frequency band coverage, i.e., tuning the RF difference frequency is limited by the tuning speed of the lasers.

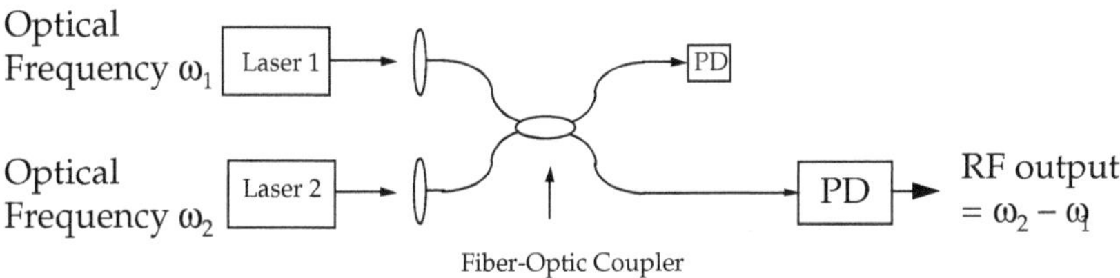

Figure 3 : Optical heterodyne.

2.1.2.2. Transmission

Having addressed generation of the modulated optical carrier, a brief examination of the medium to deliver the optical signal is in order, namely, the optical fiber itself and optical amplifiers.

2.1.2.2.1. Optical Fiber

The optical fiber transmission media has many attractive features. First, the fiber itself only introduces about 0.25 dB/km optical attenuation (~0.5 dB/km electrical). Second, the small size and low weight compares favorably to metallic alternatives. Third, optical fiber is flexible for ease of routing through the platform structure. Fourth, greater phase stability over coaxial cable feeds. For the short distances of interest here (< 300 m), other more exotic deleterious effects including chromatic and polarization dispersion, stimulated Brillouin scattering, Raman scattering, and self-phase modulation are not evident.

2.1.2.2.2. Optical Amplifiers

Another advantage to optical fiber systems is the availability of optical amplifiers. Optical amplifiers fall into two categories: doped-fiber based and semiconductor based. In either case, the amplifier is essential for overcoming losses associated with signal distribution and insertion losses of other optical components. Both EDFA and SOA amplifiers exhibit low distortion when the RF modulation frequency is greater than about 1 GHz and 1 kHz, respectively.

a. Fiber amplifiers

Fiber amplifiers can be made to operate at both the 1300 and 1550 nm wavelengths, with the 1550 nm band (1525 - 1600nm and based on the Erbium dopant) being technologically much easier to fabricate and exhibiting much better performance. Erbium-doped fiber amplifiers (EDFA's) are readily available and have demonstrated gains of up to 50 dB—corresponding to 100 dB electrical gain—a truly phenomenal amount of gain for a microwave engineer. The EDFA is polarization independent and can be built to saturate at high optical output powers (~5 W). The EDFA is also fairly low noise with several demonstrations of noise figures near the 3 dB quantum limit.

b. Semiconductor optical amplifiers

Semiconductor optical amplifiers (SOA's) are simply semiconductor lasers without mirrors. They suffer from high fiber-to-chip losses, less net gain, polarization dependence, and higher noise figures (>5 dB worse than EDFA's). One advantage, however, of the SOA is the possibility of 1 GHz signal modulation (whereas EDFA's can only be modulated at kHz frequencies). Hence, SOA's can be used to route signals, as will be discussed in relation to controlled time delay below.

2.1.2.3. Optical to Electrical Converter (OEC)

For microwave signal frequencies, depletion layer photodetectors (PD's) are the preferred optical to electrical converter. For operation well within the 3 dB bandwidth a simple current generator model for the photodiode can be used where the output current is proportional to the detected optical power.

2.1.2.3.1. The Pin Photodetector

PD's based on a simple *p-i-n* structure and exhibiting 3-dB bandwidths in excess of 20 GHz were first reported c. 1987 (J. Schlafer, *et al.*, Appl. Phys. Letts., **50**, 1260, 1987). Since then, reports of PD bandwidths in excess of 100 GHz are fairly routine. Hero PD's have extended to frequencies >500 GHz, with monolithic balanced receivers demonstrated to 12 GHz, and RF-amplified receivers to 50 GHz. The basic PD, however, is commercially available to 60 GHz, with 20 GHz devices available in packages amenable to RF system integration. As a photonic link component, the PD is the most developed and best understood. The efficiency of most PD's is near 80% of the quantum limit. Since the average photocurrent sets both the photonic link insertion loss and output RF power, the only real concern that exists with PD's is the high-power handling capability.

2.1.2.3.2. PD Limitations

A current topic with PD's is, therefore, performance with large incident optical power. The physical mechanisms, modeling, symptoms, and remedies of PD nonlinearities are beyond the scope of this text. To extend the power handling capability, several groups have proposed and demonstrated waveguide PDs. Others have improved upon conventional pin PDs with operation to 15 mA for 17-GHz devices and >140 mA for 200-MHz devices. (For more information, see IEEE Trans. On Microw. Theory and Techn., MTT-45, No. 8, Part II, Special Issue on Microwave and millimeter-Wave Photonics, August, 1997.)

2.1.3. Typical Link Performance

As microwave photonic links enter serious consideration for analog interconnects in state-of-the-art wideband microwave functions such as multifunction phased array beamforming, it becomes crucial to develop an understanding of the baseline capabilities of such links in terms of gain, sensitivity, distortion, and dynamic range in order to determine whether or not they are suitable for particular applications. Existing assessments of link performance do not sufficiently address the effect of broader bandwidths, higher photocurrents, lower RIN, modulator half-wave voltage, and external preamplifiers on the total performance of a photonic link (PL). Here, the basic performance model presented heretofore is expanded to account for multi-octave and balanced detection PLs.

In order to establish a baseline for evaluating microwave photonic links, we begin by focusing on the performance of unconditioned PLs as opposed to links involving linearization schemes such that distortion products can actually add out of phase. Since state-of-the-art dynamic range values are currently achieved using externally-modulated links, we will consider these rather than laser diode direct modulation techniques. There are a variety of ways in which optical signals can be heterodyned in order to recover the baseband RF signal; some involve indirect detection through a squaring rectifier or a frequency discriminator. To simplify further, we look only at intensity-modulated direct-detection (IMDD) links where the output photocurrent is the baseband signal.

2.1.3.1. Link Noise

Three different effects typically dominate noise power in IMDD links: thermal noise, shot noise, and intensity noise. The thermal noise power received is a combination of that created at the output of the link and that created at the input to the active portion of the link amplified by the gain of that active portion. The relations describing shot and intensity noise have a characteristic dependence on the detected optical power, the detection scheme used, and photodetector (PD) impedance matching. Mach-Zehnder modulators (MZMs) used for intensity modulation may be single-output (Y-coupled) or dual-output (X-coupled). Both outputs of the X-coupled MZM carry the same IM signal but are 180° out of (RF) phase thus allowing dual-channel balanced detection. Because intensity noise variations in each arm of the balanced PL are correlated, the intensity noise power from each PD of the balanced detector coherently subtracts, assuming the path lengths from modulator to detector and the amplitude responses of the detectors are well matched.

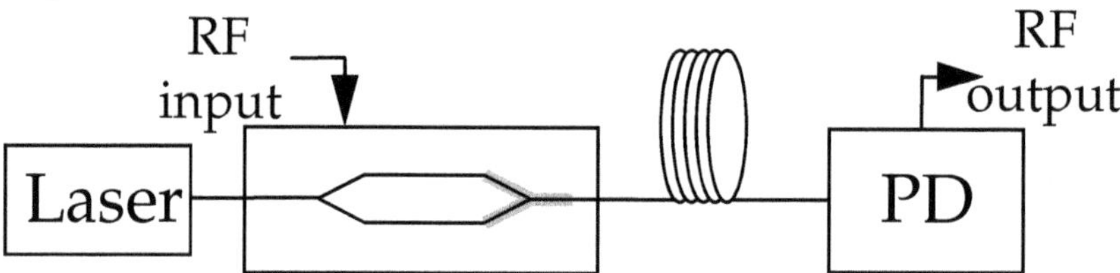

Figure 4a : Y-Coupled modulator.

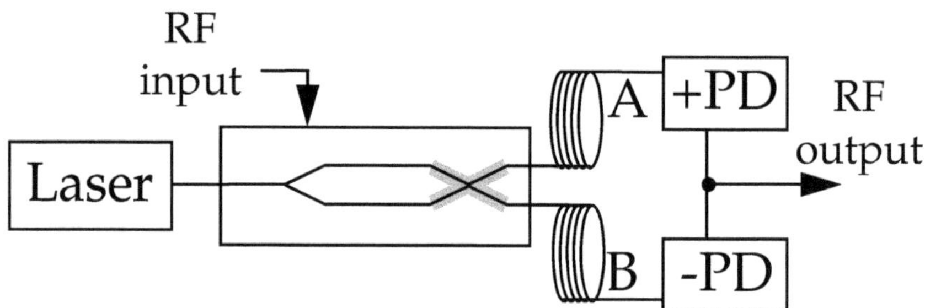

Figure 4b : X-Coupled modulator with balanced detection.

2.1.3.2. Linearity

To quantify the linearity of the link, we first assume an equal-power two-tone input $\phi_m(t) = \phi(\cos\omega_1 t + \cos\omega_2 t)$

Where : $\phi = \pi\sqrt{2k_{in}R_l P}/v_\pi$, P is the input RF power to the MZM, and v_π is the voltage swing required to shift the relative phase by 180°. Then, with the help of Kolner and Dolfi (Appl. Opt., 26, pp. 3676-3680, 1987), we can derive numerous relationships relating to the nonlinear distortions if limited to the sinusoidal nonlinearity of the modulator (Figure 5).

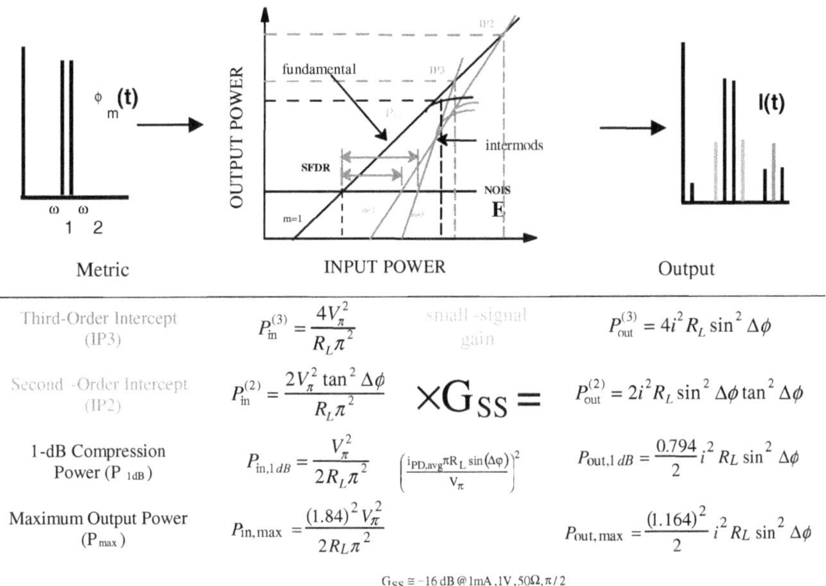

Figure 5 : Intermodulation definitions and formulas.

2.1.3.3. Link Noise Figure

The importance of RIN noise in these considerations is shown in Figure 6 where Y-coupled links are compared to a balanced detection link at quadrature bias. Clearly RIN limits the PL sensitivity that can be obtained by increasing photocurrent into the tens or hundreds of mA's. Since the current values indicated in this figure and all that follow correspond to i, the total average photocurrent at quadrature, the performance of the X-coupled PL at a particular value should always be compared to the Y-coupled PL performance at half the current (the same optical source power); visually, this means shifting the Y-related curves toward higher photocurrents by a factor of 2. For instance, the noise figure of a quadrature-biased Y-coupled PL with a -170 dB/Hz RIN source and 10 mA photocurrent can actually be improved by 4 dB using balanced detection. It should also be noted that the X-coupled link data assumes ideal RIN cancellation.

Practically, balanced detection offers at least 20 dB of RIN suppression so that, at higher photocurrents, the performance indicated by a particular Y-configuration curve in figure 6 can be achieved by an X-coupled PL using an optical source with roughly 20 dB greater RIN. Alternatively, at higher modulation frequencies where PDs are limited to lower photocurrents, balanced detection is not as essential.

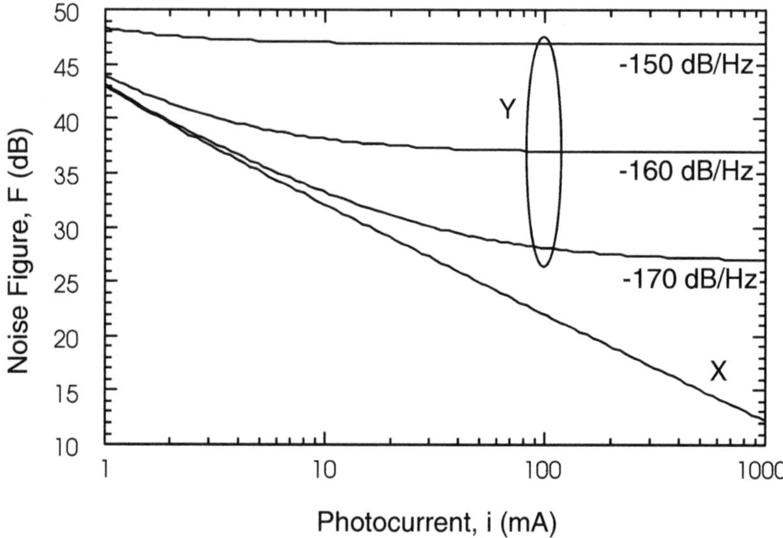

Figure 6 : Noise figure comparison of quadrature-biased X- and Y-coupled IMDD PLs for optical sources with varying RIN. Assumptions: input and output matched, $R_l=50\ \Omega$, and $V_\pi=10$ V.

2.1.3.4. Example High Performance Link

A state of the art link was demonstrated by Williams et al. (Electron. Lett., 33, pp. 1327-1328, 1997), which utilized balanced detection and high power photodetection. The experimental arrangement, link gain, noise figure and distortion measurements are shown in figure 7.

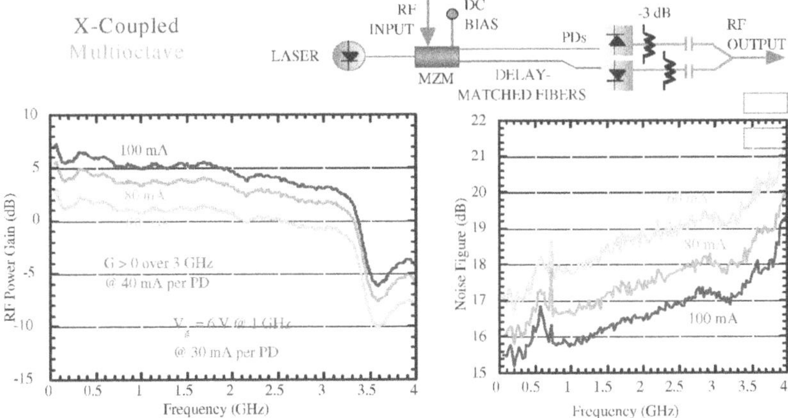

Figure 7a : High performance 3 GHz link.

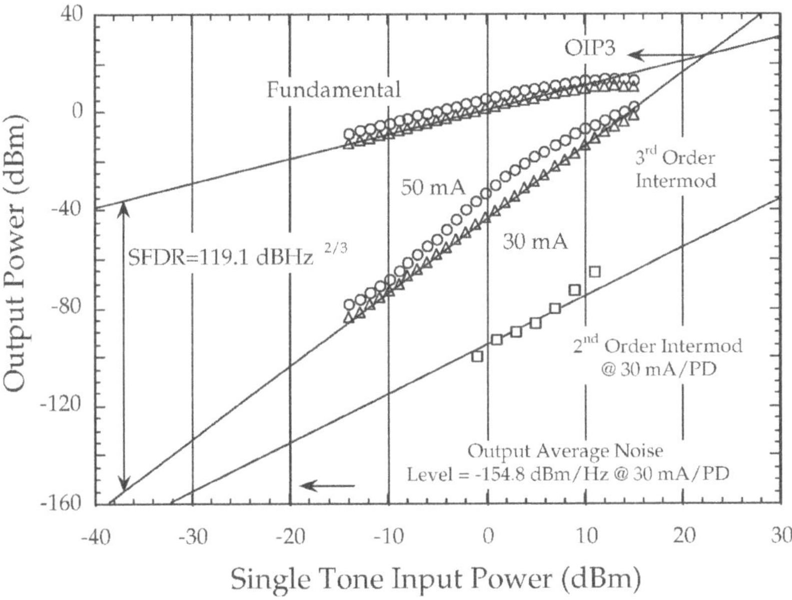

Figure 7b : Distortion measurements at 1 GHz fundamentals.

2.1.3.5. Photonic Link Conclusions

The newest photonic links have performance levels comparable to microwave components and subsystems. Certainly for use in long delay lines or transmission links, the photonic link will outperform any microwave system. Recent results presented here also show that photonic links and exhibit a spur free dynamic range greater than wideband microwave amplifiers; so long distances or delays are no longer prerequisites for utilizing photonic links.

2.1.4. Microwave Functions in Photonics

Several microwave functions are preferred in or are strictly limited to the fiber optic (photonic) domain. Most notable is time-delay beamforming for wideband, wide-aperture arrays. Using the fiber-optic dispersive prism approach, squint-free, multiple-beam transmission has been demonstrated over three octaves (2-18 GHz) and is reviewed in the next section. Downconverting of a microwave signals can be performed by a series cascade two of Mach-Zehnder modulators and has recently been demonstrated with 11 dB loss and 34 dB noise figure. Continued interest in fiber-optic transversal filters has resulted in new capabilities, including tunable and high-pass filters.

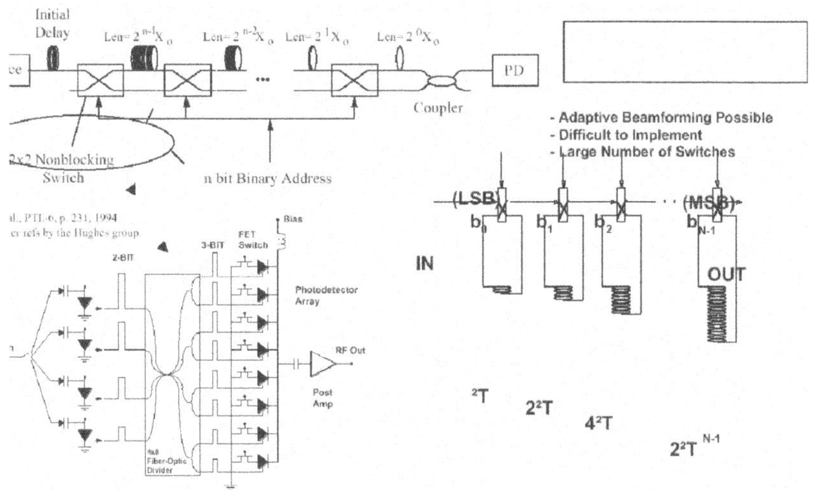

Figure 8

2.1.4.1. Controlled Time Delay

A controlled time delay can be introduced to an RF signal on an optical carrier by using one of many possible switched architectures. In the parallel delay line scheme the optical signal is split or switched between varying lengths of optical fiber. The signals are then combined or switched to a common output port. Two such architectures are shown in figure 8. The figure (lower left) shows a structure that varies the delay of the signal by sending the RF signal to all of the semiconductor lasers but only turning on the laser that corresponds to the delay needed. Likewise, only the photodetector that corresponds to the delay needed is activated. In this way a 5-bit variable delay line is formed. In the upper left, each bit of delay through the system is controlled swapping of extra path lengths (lengths of fiber); the difference in length for each bit will be twice the previous bit. Routing the light could be by using nonblocking switches (as shown) or by splitting the light to both channels (fibers) and then using an optical blocking switch to allow either one or the other channel to pass (utilizing an optical amplifier, for example). Drawbacks include the discrete nature of the incremental delay, complexity, size, and speed.

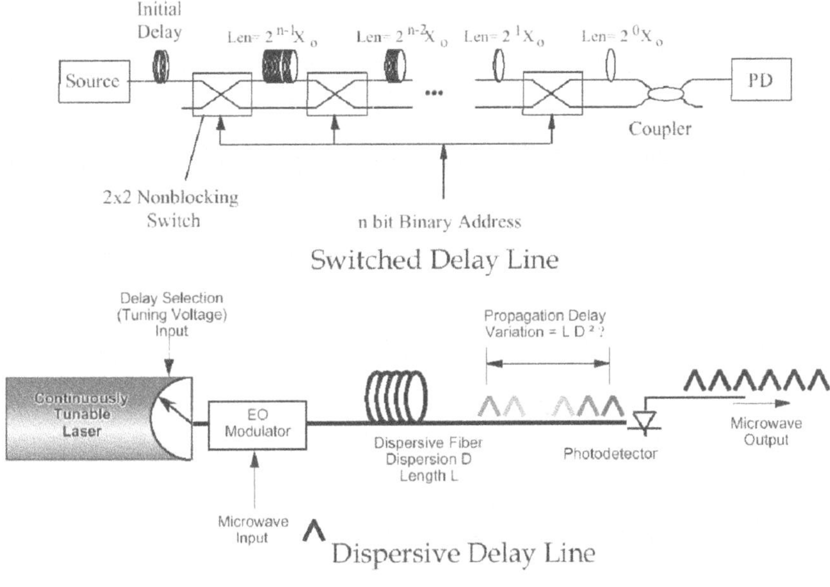

Figure 9

Another technique to implement variable delay (see Fig. 9) relies on optical dispersion. The technique provides a continuously variable time delay without significant additions to a fiber-optic system and is controlled by low-voltage signals. This TTDM technique utilizes a tunable narrow-band optical source that can be modulated to microwave frequencies and an optical fiber with chromatic dispersion. Simply put, modulation of the source wavelength modulates the velocity and, hence, the delay of the optical signal.

This dispersion-based technique modifies a standard externally modulated link in two ways: dispersive optical fiber and a widely tunable laser are used. The signal (RF, microwave, millimeter wave) to be variably delayed intensity modulates the continuous wave optical signal from the tunable laser. The modulated laser signal propagates along the fiber at a velocity determined by the wavelength of the laser. The photodetector converts the optical intensity back into an electrical signal.

2.1.4.2. Frequency Translation

The ability to remote microwave antenna systems with optical fiber offers attractive possibilities for new semi-nonintrusive RF receiving systems. These systems must however compete with electrically mixed RF-to-IF systems, which can exhibit noise figures below 10 dB in the multi-gigahertz regime. A simple fiber-optic downconverting technique has achieved low conversion loss (< 20 dB). However, the configuration suffers from high noise figures (~50 dB). A modified configuration based on the same dual in-line modulator approach includes optical amplification for reduced loss, lower noise, and distribution capability and includes a balanced receiver for canceling laser intensity noise and added EDFA noise. The system exhibits a >16 dB improvement in noise figure—with further improvements possible.

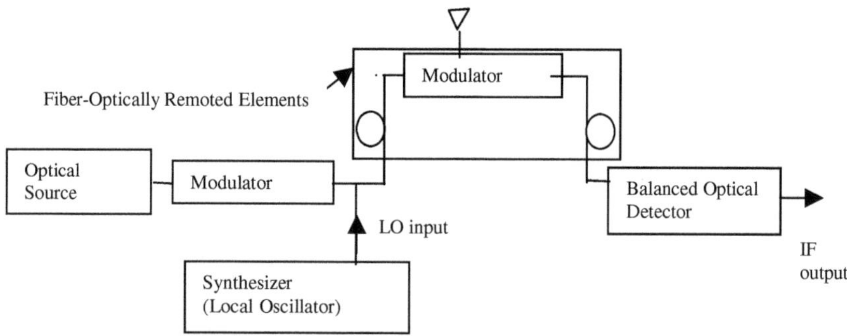

Figure 10

The optimized configuration (see Fig.10) comprises a 50 mW single-frequency diode-pumped 1550-nm Erbium-doped glass laser coupled to a Mach-Zehnder modulator (MZM) with local oscillator (LO) feed, the output of which is optically amplified with a polarization maintaining Erbium doped fiber amplifier (EDFA). The optical signal can then be distributed remotely to several receive antenna elements. A received RF signal is fed directly to a dual-complementary-output MZM with both outputs connected to a balanced photoreceiver. The dual in-line optical modulation effectively results in the translation of the RF input by the LO signal applied to the first MZM and results in an IF=|RF-LO| signal out of each of the photodetectors (PDs) of the balanced photoreceiver. The complementary MZM outputs provide IF (160 MHz) signals 180° out of phase, whereas the EDFA added noise are common (in phase) to both outputs and so cancel electrically in the balanced receiver. Additional noise sources include the laser relative intensity noise (RIN), both at the IF frequency and low-frequency RIN (<1 MHz) which is up- and down-converted. However, laser RIN contributions to IF noise are negligible since 1) solid-state lasers inherently exhibit low-RIN at the IF (for our laser the RIN is < -165 dB/Hz for freq > 10 MHz), 2) IF RIN is canceled in the balanced receiver, and 3) potentially significant RIN levels at low frequencies (<1 MHz) are easily eliminated with simple laser feedback circuits. Thus, with sufficient common-mode noise rejection (amplitude and phase balance), the principal noise contributor—the EDFA added noise—is rejected by the photoreceiver, leaving only shot noise and thermal noise at the IF output.

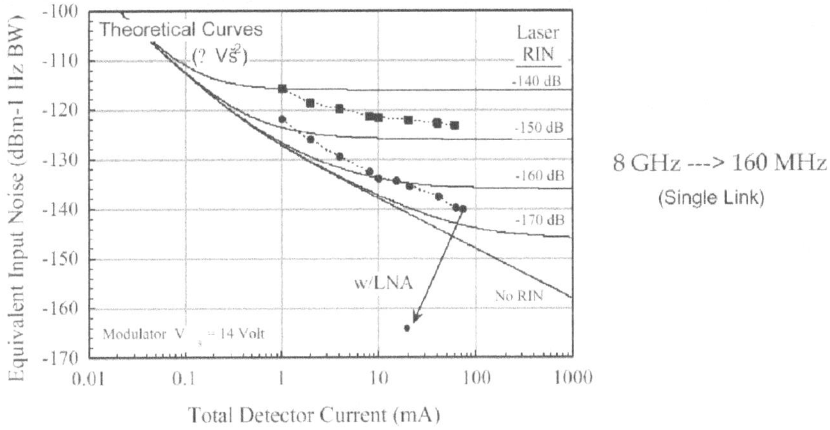

Figure 11

Figure 11 shows the resulting equivalent input noise (EIN) obtained from the link configurations (balanced and unbalanced) as measured with both LO and RF turned on. Also plotted is the non-downconverting EIN that is expected for a 12-Volt modulator $V\pi$ and various laser RIN levels. The EIN of the balanced configuration displays a general decrease with increasing current (minus a 7 dB downconverting loss) with a slope equal to the No-RIN curve, which displays shot-noise-limited performance. Shot-noise-limited results are obtained with the balanced receiver configuration because laser RIN and EDFA intensity noise are effectively rejected. This results in a link EIN of -140 dBm/Hz (34 dB noise figure). By contrast, the EIN for unbalanced configuration is 6 to 16 dB higher due to the uncorrelated EDFA noise which cannot be balanced out.

The insertion loss (RF to IF) of the balanced downconverting link configuration at 72 mA and 8 GHz was 11.0 dB. Although this is only

9 dB better than the conversion loss previously reported using single-output modulators ($V\pi = 9$ V), the EIN has been reduced 16 dB from -124 dBm/Hz to -140 dBm/Hz. With present-day technology, a 4-Volt $V\pi$ dual-output 18 GHz MZM would result in an EIN of –150.5 dBm/Hz (~24.5 dB noise figure). Further noise figure improvements are possible with improvements to MZM maximum output optical power, since additional or improved PDs could be used to increase the maximum detectable current.

2.1.4.3. Filtering / Signal Processing

The have been several demonstrations of the use of photonics for signal processing and, in particular, for signal filtering. Figure 12 shows the use of fiber gratings to effectively implement a finite impulse response filter.

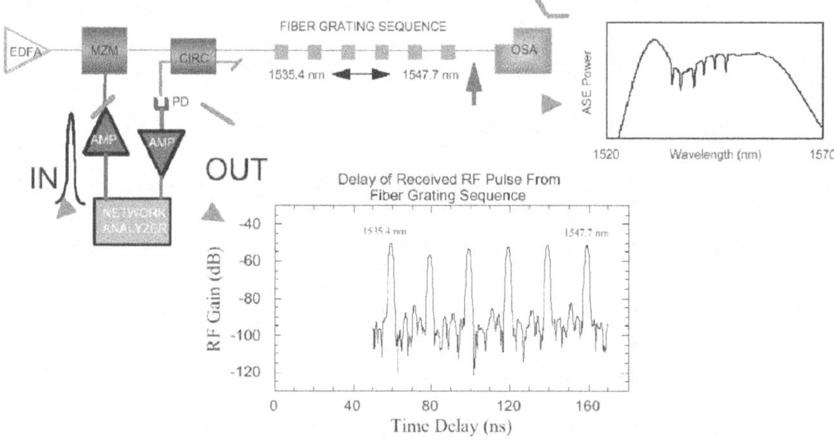

Figure 12

The electro-optic modulator modulates a broadband optical source and multiple replicas of the RF input are generated at the photodetector corresponding to multiple reflections from different path lengths. The measured response of a similar 29-tap filter is shown in figure 13 and shows a center frequency of 2 GHz and a Q of about 10.

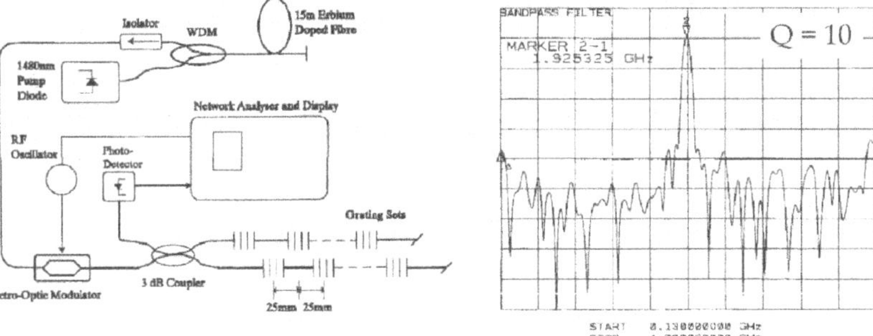

Experimental Setup

Measured Frequency Response of 29-tap Transversal Filter with Tapered Weighting

Figure 13

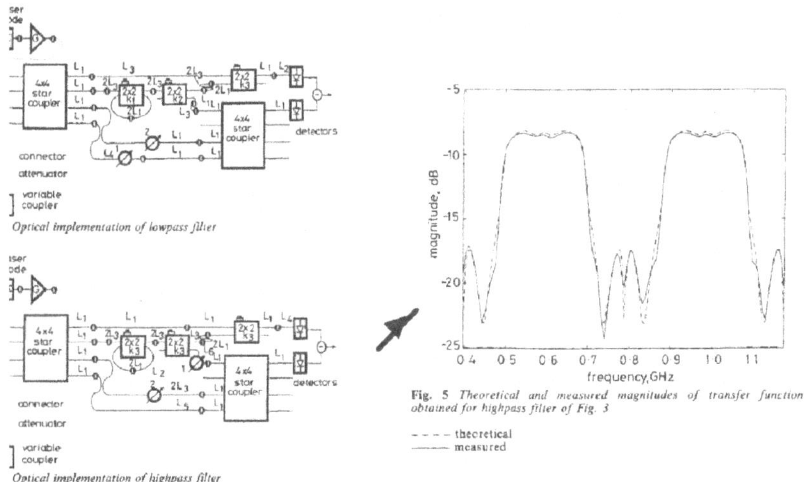

Figure 14

Sales, *et al.*, [S. Sales, *et al.*, Electronics Letters, **31**, 1095, 1995.] have also developed tapped delay line filters but appear to be the first to implement negatively biased photodetectors (see Fig. 14) to generate inverted RF signals. This additional degree of freedom allows additional processing architectures including a high-pass filter. At NRL we have utilized the dispersive fiber concept to implement a tunable filter (OMW 19 & 20) based on a tunable laser. With a single input control, the filter is able to tune from 8.9 to 18.2 GHz. It is interesting to note that the Q (of 30) is constant as the filter is tuned, which is characteristic of a tine-delay tune filter.

2.1.4.4. Control of Microwave Devices

An alternative to distributing controlled RF signals to various points in a system is to use optical control of a microwave device. Optical control of oscillators using IMPATTs, bipolar transistors, MESFETs and HEMTs has been demonstrated. However, the control range achieved has been small due to the low optical responsivities of standard microwave devices. For example, the modulated output light from a laser diode was used to injection lock a microwave oscillator. This dramatically improves the phase noise of the oscillator. Further, to demonstrate phase modulation, the laser diode average output power is varied. An example of phase deviation is shown in figure 15 where the phase at 7 Ghz is seen to vary by 180°. Phase modulation by optical injection has been demonstrated to 1 MHz and should be possible to beyond 10 MHz.

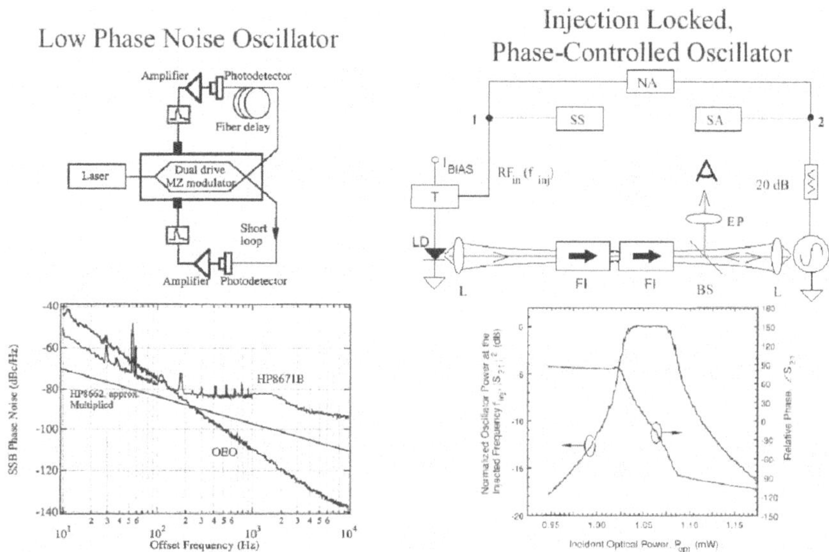

Figure 15

2.1.5. Beamforming

Photonic control and beamforming of phased-array antennas has seen active recent development. The development efforts are mostly driven by the expected benefits over conventional all-electronic phased-array antenna control methods in the areas of size and weight reduction, interference immunity, remoting capability, etc. In addition, photonics has held out a promise of being an enabling technology for true time-delay beam steering, which permits wide instantaneous bandwidths and squint free operation. A variety of photonic techniques have been proposed for obtaining true time-delay capability. Among others, these include electronically-selected optical delay lines, switched optical delay lines, optimized schemes combining both optical and electronic time-delay switching, schemes based on optically-coherent control of arrays, acousto-optically based delay lines, fiber Bragg grating-based delay lines, and schemes based on fiber-optic dispersive delay lines. However, most of these techniques have not progressed beyond conceptual laboratory demonstrations, as they are hampered by the demands for precisely matched optical elements, excessive power losses, instability, or specialized component development. The exceptions have been the switched delay line techniques developed by and for MHz through 3 GHz frequency ranges, and a dispersive delay line technique developed by NRL for under 2 to over 18 GHz frequency ranges. Recently, a true time-delay fiber-optic control of an eight-element one-dimensional transmitter array was demonstrated, which exhibited squint-free ±50° azimuth steering over a 2 to 18 GHz frequency range. Some details (below) of the beamformer are in order and are representative of some of the issues facing Rf beamformers and their photonic control.

In brief, the microwave signal driving the antenna elements is transmitted on a single wavelength-tunable optical carrier via a bank of dispersive fiber-optic links. The TTD function is realized by tuning the carrier wavelength to vary the group velocity of the propagating signal. Each element requires a time-delay proportional to its relative position within the array. Thus, each fiber-optic link feeding an individual array element incorporates an overall amount of dispersion that is proportional to the element position. A set change in the carrier wavelength provides the necessary proportional time-delay for all array elements with a single wavelength-control input. This system is illustrated in figure 16.

Our bank of fiber-optic links was based on a combination of high-dispersion (HD) fiber (D_{hd}~-70 ps/nm·km) and dispersion-shifted (DS) fiber (D_{ds}~0 ps/nm·km). Then, for a signal of wavelength $\lambda=\lambda_o-\Delta\lambda$ we can compute the time-delay *change* relative to the delay at a reference wavelength λ_o. Assuming a link containing *n* units of HD fiber of length l_{hd} and $l_{link}-nl_{hd}$ meters of DS fiber, the time-delay *change* is given by

$$\Delta\tau_{link} = \ell_{link}\left(D_{ds}\Delta\lambda + D'_{ds}(\Delta\lambda)^2\right) + n\ell_{hd}\Delta\lambda\left((D_{hd} - D_{ds}) + \Delta\lambda(D'_{hd} - D'_{ds})\right)$$

where the dispersion may vary linearly with wavelength as $D_{hd}+D'_{hd}(\lambda-\lambda_o)$ and $D_{ds}+D'_{ds}(\lambda-\lambda_o)$ for the HD and the DS fiber, respectively. The link $\Delta\tau_{link}$ is given by a sum of one term which is common to all links (and may be ignored) and another term, which is proportional to the link number *n*. For this system, $l_{hd} \approx 46$ m. The consecutive links in this system contained nominal sections of 0, 138, 276, 414, 598, 782, 1104, 1334 m of HD fiber and were length-equalized to 1350 m with DS fiber.

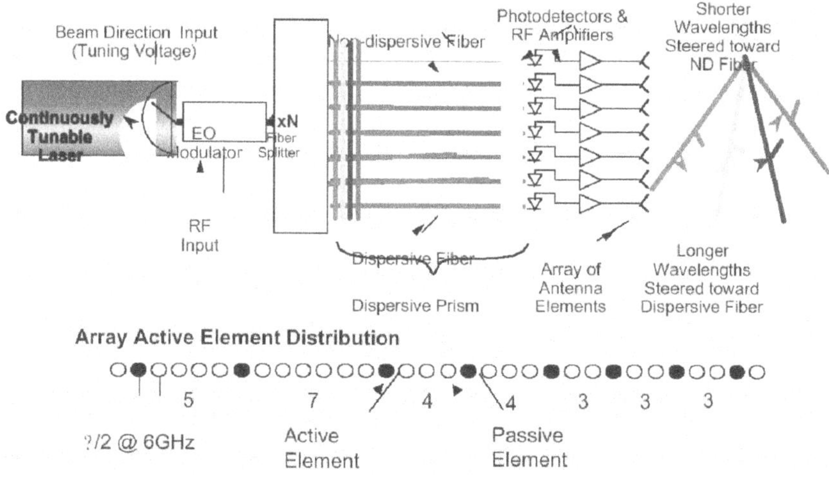

Figure 16

The HD fiber length was measured with a mechanical counter to within ±0.25 m. This error results in under a ±1 ps *relative* delay error among the links. The overall delay was equalized by trimming the DS fiber to within ±0.5 cm of the required length at a center wavelength of $\lambda_0 = 1558$ nm. The final delay equalization to within ±5 ps was accomplished via microwave trombone delay lines.

The laser driving the beamformer was a fiber-optic tunable laser. The laser was tunable from under 1530 nm to over 1580 nm with a 0.06-nm linewidth (implying more than 800 resolvable wavelength settings) and single-polarization output. The laser wavelength tuning speed was limited to the millisecond range by the Fabry-Perot etalon filter, but microsecond tuning speeds have been demonstrated with lasers including an acousto-optic filter. The laser output was amplitude-modulated by a Mach-Zehnder (M-Z) modulator, amplified in an Er-fiber amplifier, and split into eight fiber-optic dispersive links. The outputs of the links were fed to individual photodetectors (PDs) followed by microwave trombones (not shown in figure 16) for time-delay calibration. A second (optional) tunable optical source can be added to implement multiple simultaneous beams.

The amplitudes of the RF signals from each link were matched by introducing controlled bend loss into the fiber to provide frequency-independent attenuation.The RF spectral amplitude uniformity among all the links was acceptable over the complete 2 to 18 GHz frequency range. In addition to the customary small variability in performance (~±0.2 dB) due to microwave component differences, there were two additional deleterious effects present. The first is an abnormally large ~2-3 dB gain and ~15° phase ripple across the whole band due to the mismatch between the high output impedance of the PDs and the 50Ω impedance of the following microwave elements. This ripple could be eliminated in an improved system by either properly terminating the PDs or by inserting broadband isolators after the PDs. The second error is intrinsic to dispersive fiber-optic links and originates from microwave sideband phase walk-off at high modulation frequencies. This leads to a microwave loss on the order of ~3 dB at 18 GHz for the most dispersive link in our system, but is much smaller at lower frequencies and in less dispersive links. Therefore, the array pattern distortion is believed to be dominated by the array element coupling and by the PD mismatch ripple.

The array antenna patterns were measured in an anechoic compact radar range. A network analyzer under computer control was used to drive the M-Z modulator and to measure the received signal power and phase as a function of the antenna mechanical azimuth.

Figure 17 (top) shows a comparison between ideal calculated (dashed) and measured (solid) patterns at 6 Ghz, with the beam steered to ~-24° by simply tuning the laser by -10 nm from its nominal broadside wavelength of 1558 nm. This comparison emphasizes that the interelement amplitude and phase mismatch errors do not significantly affect the array pattern for this system. The main lobe steer angle is –24° off broadside and is independent of frequency in the 2 to 18 GHz range (Fig. 17 middle), as expected for a time-steered beamformer. In general, we observe a narrowing main lobe, and sidelobes converging towards the main lobe with increasing frequency, as expected. The side lobe level is over 10 dB below the main lobe in its vicinity and rises up to ~5 dB below far off the main lobe. The measured nulls are as deep as 40 dB, indicating a reasonable amplitude and phase uniformity across the array. The grating lobes are absent in the data up to 12 GHz but appear in the higher frequency patterns. The sidelobe levels, beamwidth, and grating lobes are all in agreement with the sparsely-populated unequal-spacing array design. For comparison, a phase-steered PAA (Fig. 17 bottom) yields main lobe squint from ~8° to over 35° as the frequency is scanned from 18 GHz to under 4 GHz.

This fiber prism technique has since been shown to transmit multiple simultaneous beams and has been extended to a two-dimensional arrangement that operates over the full 6-18 GHz band.

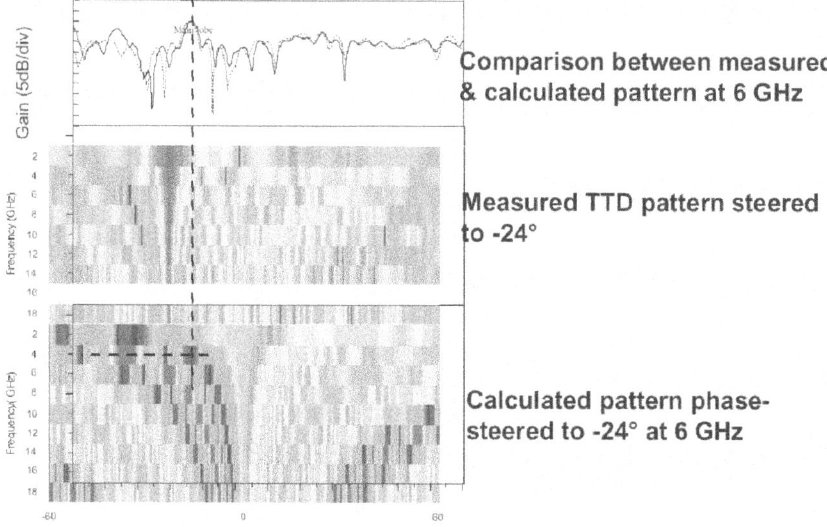

Figure 17

2.1.6. Summary and Issues

Photonics *now* plays a key role in many microwave (analog) applications. Numerous advantages include size, weight, link loss, bandwidth, EMI immunity, and new functions. Certainly the ability to multiplex several wide bandwidth, high-frequency signals on a single strand of optical fiber is an attractive alternative for multifunction systems.

The main challenges of the photonic link are the output RF power, linearity, noise and cost. The issue with output RF power is that of a high-power high-speed photodetector, as mentioned in section 2.1.2.3 ; several new ideas have appeared in the literature for improving the power handling capability of these photodetectors. The issue of linearity and noise was introduced in section 2.1.4 above. The issue of cost is becoming less important as the telecommunications and cable TV industries continue to advance the component and manufacturing technologies. In addition, the added utility and the multiplexing of signals leads to reasonable amortizing of photonic costs.

Since about 1995, most components have become commercially available for photonic links and networks suitable for RF systems and, in particular, for phased arrays at frequencies throughout the microwave region. The use of optical amplifiers is attractive in distribution networks for large arrays and has been utilized in the fiber-optic beamformer described in section 2.1.5. above. Dynamic range and sensitivity requirements for the return of received signals are much harder to meet and it appears that further advancement will be required on component sensitivity and linearization to make optical links attractive for this function

Also, it is interesting to note that, as demonstrated by the dispersive fiber prism beamformer, many systems now are limited by the associated *microwave* components.

Acknowledgment: This work was supported by the Office of Naval Research.

2.2. Industrial Requirements to Photonic Generation of Microwave Signals

S. Gevorgian[1,2], L.R Pendrill[3], A. Alping[2]
[1]Chalmers University of Technology, 412 96 Gothenburg, SWEDEN
[2]Ericsson Microwave Systems, 431 84 Moelndal, SWEDEN
[3]Swedish National Testing & Research Institute, Box 857, S-501 15 Borås, SWEDEN
spartak@ep.chalmers.se

Abstract

Generation of microwave/millimeterwave signals by optical heterodyning of laser light has reached a state of maturity, where the development of microwave oscillators may be considered from the industrial point of view. The cost, especially for commercial applications, is one of the main driving forces. It is shown that short wavelength (λ<0.9 µm) Vertical Cavity Surface Emitting Lasers and standard low frequency (f< 10-20 GHz) Silicon or Silicon/Germanium Monolithic Microwave Integrated Circuits are a promising combination for the development of commercial photonic microwave oscillators.

2.2.1. Introduction

Wireless Communication Systems move towards higher-speed/higher microwave frequencies to handle audio, video, and high bit rate data transmission. It seems that the optical technology may substantially improve both cost and performance of these systems. The battle for the market in commercial wireless networks makes the infrastructure cost one of the most critical issues. Currently new cost effective microwave fibre optical links and other photonic microwave devices are being developed. The main components of these systems are electrooptic and optoelectronic transducers. For many systems photonic microwave sources should be based on long wavelength laser diodes (1.3 and 1.55 μm) to match standard optical fibres. On the other hand, in many microwave systems, commercial (communication, automotive, sensor etc,) and special (military, radio-astronomy etc.) there is a need in microwave (fixed and frequency tuneable) low cost oscillators. Attempts are being made to fabricate such oscillators in standard silicon technology for frequencies below 10 GHz. Today only GaAs or InP based oscillators are capable to operate at higher frequencies. Development of oscillators operating at shorter millimetre and submillimeterwaves is still a challenging research problem. Photonic microwave generators seem to be one of the most promising approaches for such applications. They may have both optical (modulated at microwaves) and electrical (microwave) outputs. In the latter case they may have a wide application in any microwave system, while the former may also be used in microwave photonic systems, including fibre optical.

Several methods of generation of continuos wave (CW) microwave/millimeterwave signals by heterodyne mixing of CW optical illuminations with different wavelengths are proposed in the past. A brief review covering the topic is given in [1]. To reduce the phase noise of the output microwave signals injection locking of two slave lasers by a master laser is proposed [2]. Recently Optical Phase Locked Loops (OPLL, see for example [3]), or Optical Injection Locking Phase Locked Loop (OPILL, [4]) are proposed to reduce the phase noise of the output microwave signals. These methods are based on expensive edge emitting laser diodes and complex phase looked loops. They are not cost effective for commercial applications.

Vertical Cavity Surface Emitting Lasers (VCSEL) based oscillators seems to be favorable in terms of cost. Long wavelength VCSELs are still under development, and photonic microwave generators based on long wavelength VCSELs will not be addressed here, while short wavelength VCSEL (λ<0.9 μm) are commercially available but need substantial improvement in terms of linewidth and noise. However, short wavelength VCSELs already are quite promising and attractive in many microwave photonic applications. Particularly, recent experiments show that VCSELs may be useful in photonic microwave oscillators, provided that special measures are undertaken to reduce the noises in optical and/or microwave domains [5], [6], [7], [8].

The objective of this work is to review the possibilities of the development of cost effective architectures of frequency tunable photonic microwave oscillators based on short wavelength VCSELs. The emphasize is made on electrical rather than optical mode locking methods aimed on the reduction of the phase noise of generated frequency tunable microwave signals. The main aim is to propose an architecture, which provides high integration levels based on standard silicon IC fabrication technology.

2.2.2. The Optical Heterodyne Principle

We will focus only on one particular microwave photonic device-photonic microwave oscillator. Generation of micro/millimeter wave signals is a well-known problem, and application of optical technology for this purpose seems to be quite promising. Mixing of two optical signals with different frequencies, figure 18, generates Microwave/millimeterwave signals with a frequency :

$$f_m = \nu_{o1} - \nu_{o2} = \frac{c_o(\lambda_{o2} - \lambda_{o1})}{\lambda_{o2}\lambda_{o1}} \quad (1)$$

where f_m is the frequency of generated microwave signal ν_{o1} (λ_{o1}) and ν_{o1} (λ_{o1}) are frequencies (wavelengths) of two optical signals, and c_o is the velocity of the light in free space.

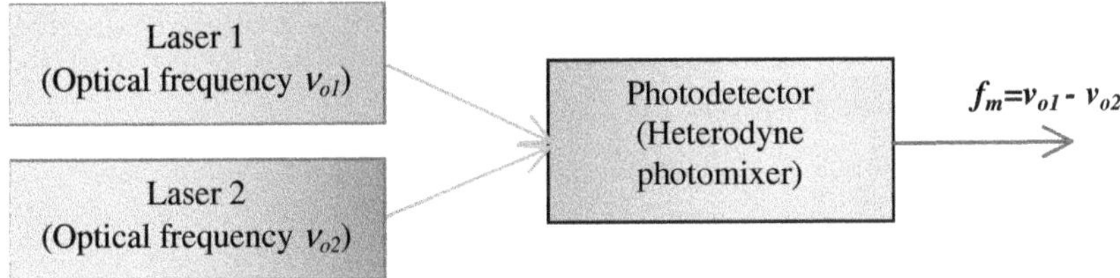

Figure 18 : Generation of micro/millimeter wave signals by optical heterodyning.

One has to remember that two optical signals do not interact in a free space, i.e. mixing two lightwaves in free space will not result in micro/millimeter waves propagating in the same free space. Generation (detection) of the difference frequency takes place only in a medium, where the lightwaves transform the dielectric or electric properties of the medium in a non-linear way, in this case a semiconductor photodetector, figure 18.

The main requirement in this case is that the generation and recombination of free carriers should be fast enough. A simplified theory of generation of microwave signals by heterodyne mixing of two optical waves may be given as follows. The electric fields of two optical signals are represented as:

$$E_{o1}(\nu) = E_{o1} \exp j(2\pi\nu_{o1}t + \varphi_{o1}) \quad (2)$$

$$E_{o2}(v) = E_{o2} \exp j(2\pi v_{o1} t + \varphi_{o2}) \tag{3}$$

The total electric field of the optical signals at the input of a photodetector:

$$E(v) = E_{o1}(v_{o1}) + E_{o2}(v_{o2}) = E_{o1} \exp j(2\pi v_{o1} t + \varphi_{o1}) + E_{o2} \exp j(2\pi v_{o2} t + \varphi_2)$$

The output current from the photodetector (the medium where the optical signals interact) is proportional to the square of the total optical electrical field absorbed by the photodetector :

$$\begin{aligned} I_{ph} &\propto \left|E\ (v\ )\right|^2 = \left|E_{o1}(v) + E_{o2}(v)\right|^2 = \\ &\left|E_{o1} \exp j(2\pi v_{o1} t + \varphi_{o1}) + E_{o2} \exp j(2\pi v_{o2} t + \varphi_{o2})\right|^2 \end{aligned} \tag{4}$$

$$= E_{o1}^{\ 2} + E_{o2}^{\ 2} + 2E_{o1}E_{o2} \cos[2\pi(v_{o1} - v_{o2})t + (\varphi_{o1} + \varphi_{o2})]$$

+ sum-frequency terms

The optical power collected by the detector is the product of the intensity and the detector area :

$$P = P_{o1} + P_{o2} + 2\sqrt{P_{o1}P_{o2}} \cos[2\pi f_m t + (\varphi_{o1} - \varphi_{o2}] \tag{5}$$

In reality the laser light consist of a spectrum (not a single frequency, as discussed above). Illustrative spectrums of the two lasers and the heterodyned microwave signal are shown in figure 19.

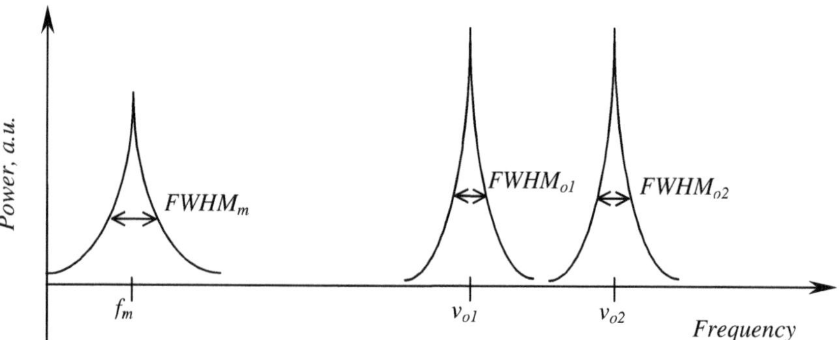

Figure 19 : Spectrums of laser (optical) and heterodyne (microwave) signals.

If the Full Width Half Maximum (FWHM) linewidths of the two laser signals are known, the FWHM linewidth of the microwave signal can be calculated by :

$$\Delta f_{0,FWHM} = \sqrt{(\Delta f_{1,FWHM})^2 + (\Delta f_{2,FWHM})^2} \tag{6}$$

where $FWHM_m$ is the linewidth of the microwave signal, $FWHM_{o1}$ linewidth of the first laser, and $FWHM_{o2}$ linewidth of the second laser. Thus, only lasers with small linewidth can be used as optical sources for generation of low phase noise (linewidth) microwave signals.

2.2.3. Components - State of the Art

The main components used in photonic microwave generators are lasers and photodiodes. In industrial/commercial applications the basic driving force is the cost of these components and more importantly, the cost of the complete generator in the form of a single chip or multichip module. The package, if required, may be the main part of the cost. In what follows we will briefly discuss laser and photo diodes on the cost/performance background.

2.2.3.1. Laser Diodes

It seems that among different laser diodes VCSEL is the best low cost high quality laser [9], [10], [11], [12]. VCSELs also offer injection modulation at frequencies up to 25 GHz and easy on chip and Multichip Module (MCM) integration possibilities. The main limiting factor regarding application in photonic microwave oscillators is the large linewidth, usually more than 100 MHz. Linewidth and noise reduction of the individual lasers is known to be achieved by optical or electrical injection locking.

Optical injection locking – is based on the injection of optical signals inside the main cavity of the laser. The simplest optical injection locking is achieved by using external mirrors to reflect a part of the laser beam back into the internal cavity, as it is shown in figure 10a. By proper choice of the length L of the external cavity and the amount of the reflected back power one can achieve a substantial narrowing of the laser linewidth. To achieve a better locking the external mirrors may be replaced by Fabry-Perot etalons [13]. Optical injection locking is also achieved in Master-Slave configuration, where the wavelength of one laser (slave) is synchronized to the wavelength of the other (master), figure 20b. The Optical phase lock loops [14], may are further improvements of the optical injection locking technique to achieve better performance.

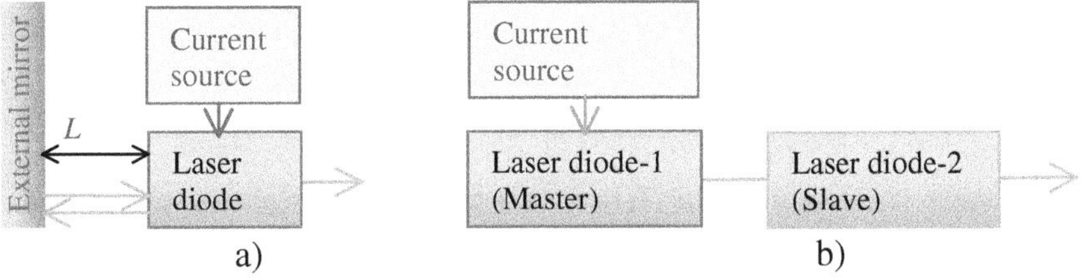

Figure 20 : Optical injection locking using feedback (a) and Master-slave arrangement.

<u>Electrical injection locking</u> – In this case the electric currents injected into the diode, in contrast with the optical injection. In case of feedback type arrangement, figure 21, a wavelength sensitive device, such as a Fabry-Perot cavity [15] or a narrow band Bragg filter (e.g. fiber grating [16]) is used to detect optical frequency fluctuations and inject the detected electrical current fluctuations back in the laser diode. The wavelength fluctuations are compensated if the feedback (current injection) is negative. Electrical stabilization (locking) of the wavelength is also possible by external RF signals applied to lasers, figure 21b, where the frequency of the RF signal coincides with a sub-harmonic of the round trip frequency [17]. In a similar experiment with a dual mode laser the frequency of the RF signal was chosen to be equal to a subharmonic of the free-running mode beat frequency of two modes of the laser [18]. Last two experiments have been used to generate low phase noise millimeterwave signals.

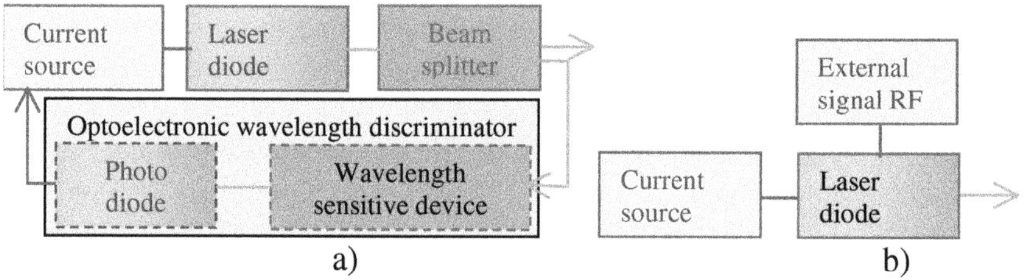

Figure 21 : Electrical injection locking. Feedback (a), and external (b) injections.

2.2.3.2. Photodetectors

Most of the high speed/microwave photodetectors are based on GaAs or InP [19], useful for long wavelength (1.3 and 1.55 μm) analogue (frequency>10-20 GHz) and high speed (broad band) digital fiber optical communication systems. On the other hand silicon or Silicon /Germanium based photodetectors are more desirable in microwave oscillators based on short wavelength (λ< 0.9 μm) VCSELs. For Silicon the maximum quantum efficiency is at 0.8 μm, [20]). This wavelength compatibility makes it possible to consider integration of Silicon photodetectors with advanced Silicon MMICs. Today the cut-off frequency of Si based bipolar transistors is more than 50 GHz in the mass production lines. For Si/Ge transistors it is even higher [21]. Moreover, 1-2 GHz Silicon MMICs are already available, and X-band MMICs based on these transistors (technologies) are about to appear in the market. The basic question is whether or not one can use high performance microwave Si and Si/Ge transistors made in standard technology as high speed/microwave phototransistors? First of all heterojunction bipolar transistors in high speed/microwave photodetectors have been successfully demonstrated [19], [22], [23], [24]. It has been shown that HBTs have better performances in terms of signal/noise ratio in comparison with FET, HEMT and JFET devices. At the same time it is known that Si/Ge HBTs have even better noise performances in comparison with the A_3B_5 (i.e. GaAs) based HBTs [21]. The brief discussions above show that standard Si or Si/Ge technology based bipolar transistor or varactors are quite promising for applications in photonic microwave oscillators. They may also be useful for short wavelength VCSEL based optoelectronic microwave frequency converters [25] wide band microwave amplifiers (such as [26]), and other photonic microwave devices, where the optical wavelength is not limited by standard optical fibers (1.3 and 1.55 μm).

Bipolar standard Si or Si/Ge technology based transistors may be arranged (as in [27]) to achieve high linearity (dynamic range) and to handle high optical saturation powers. The latter is important where high microwave powers are required at the output of photonic microwave oscillators. Chip level integration of bipolar photodetectors with other components in an SiMMIC is beneficial from the matching (microwave) point of view and offers additional advantages of reduction of circuit parasitics where the generated microwave signals are used for feed-backing (negative [15] or positive [23]) or additional reduction of noise using balanced photodetector architecture of [27].

2.2.4. VCSEL Based Photonic Microwave Oscillators. Industrial Perspectives

The optical and electrical injection locking techniques discussed above may be combined to have an additional flexibility in the development of photonic microwave oscillators. Moreover, it might be possible to use the generated microwave signals in feedback architectures to improve the performance of the lasers and microwave oscillator itself. Particularly for industrial applications, it may help to win in the performance vs. cost battle where both the performance and the cost are critical issues. In this regard, we will focus on generators based on VCSELs only, since they are the low-costest high performance lasers available commercially today. In contrast to GaAs the standard silicon technology offers higher degree of integration. Extensive use of electrical rather than optical locking seems to be beneficial, since higher integration densities may be achieved in comparison with the optical integration.

No monolithically integrated microwave generators incorporating both VCSELs and photodetectors are reported so far. Integration of VCSELs and photodetectors (with a VCSEL type structure) on the same chip demonstrated in recent publications [28] may be regarded as prototypes of fully integrated generators. Such a generator, if integrated with a low noise microwave amplifier, may be useful at high speeds/frequencies (f>20 GHz).

Below we will discuss optical microwave generators where VCSEL and photodetectors may be integrated in a single chip or multichip module to produce microwave signals (electrical domain). Such a generator may also have an optical output for use in short distance (free space or optical fiber) optoelectronic microwave interconnects.

Experimentally linewidth reduction in VCSELs using external cavities has been demonstrated in [29]. The first experiment where VCSELs are used for optical heterodyne generation of microwave signals date back to 1991 [5]. In [6] two cost effective VCSELs are used in combination with an external cavity based on a fibre-loop mirror, to make the linewiths of the lasers narrower and hence to reduce the phase noise of microwave signals. The current tuning rate was measured to be approximately 54 GHz/mA. The single mode operating regime of the lasers used in this experiment allows the generation of beat frequencies up to 150 GHz by tuning the current. The beat-note microwave signal has a minimum free running (without optical feedback) linewidth of about 120 MHz. By implementing optical feedback, beat-note linewidth of less than 200 kHz have been observed up to frequencies limited by the photodetector (25 GHz). An improvement in linewidth of about 600 times! The minimum linewidth has been found to be dependent on the laser bias current. In this case the optimum bias current was about 12 mA (26 % above the threshold current). The side-mode suppression was typically greater than 15 dB. The optical feedback power level for stable operation was between –37 and –34 dB. The phase noise was –75 dBc/Hz at 1 MHz offset, limited by measurement system noise. Application of the fiber-loop mirror makes the system both expensive and not useful for integration of all components in a single, small size commercial unit. Additionally, the phase noise of the generated microwave signal still has to be improved.

A new architecture of VCSEL based microwave generator for commercial application is proposed in [7], [30]. The center of gravity in this frequency tunable generator is moved toward the maximum use of low cost electronic rather than optical components to achieve low noise performance. A simplified equivalent circuit of the proposal is shown in figure 22.

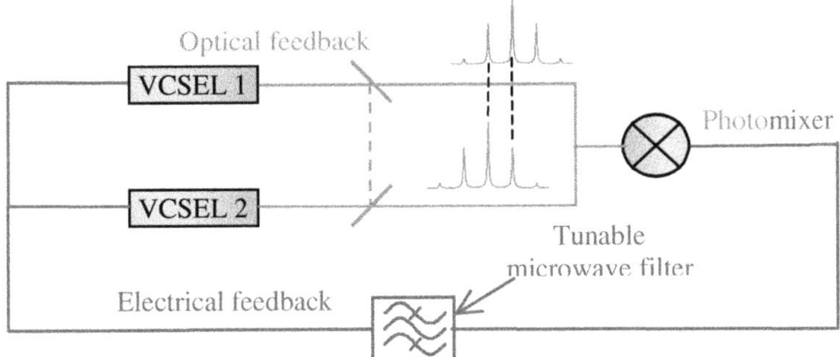

Figure 22 : Simplified equivalent circuit of a frequency tunable photonic microwave generator [30].

In this architecture both optical and electrical feedback are combined and integration possibilities are envisaged. The main features of the proposed architecture are as follows. Two VCSELs are equivalent, preferably made on a single semiconductor wafer to ensure similar temperature conditions for both of them [7]. First laser acts as a mirror (external cavity) for the second one, and, similarly, the second laser acts as external mirror for the first one. Two optical injection locking options may be used. In the first case the wavelengths of two lasers are outside the locking ranges the VCSELs act as passive mirrors for each other. Each of the lasers may be regarded as being external cavity optically injection locked. This case is used where the frequency difference between the laser diodes is in submillimeter and THz range. At near-millimeter-wave and microwave frequencies, where the wavelengths of two lasers are within locking ranges of each other, additional locking is achieved by external (second VCSEL as an extrinsic source) optical injection. In this case both laser act as masters and slaves at the same time. A portion of the illumination from the one laser entering in the cavity of the second one optically injection locks this laser. Since the mirrors of the external cavities are Distributed Bragg Reflectors (DBR) of the internal cavities of the VCSELs the phase of the reflected signal is not sharply defined and hence the optical length of the external cavity is not critical to the geometrical length. This makes the mechanical tolerances less critical and acceptable for mass production. No additional mirrors are required to form external cavities for the lasers. The optical length of the external cavity for achieving phase control is adjusted by changing the refractive index of the beam splitter(s) or any other low optical loss material (phase controller) inserted in the external cavity. The change may be achieved by heating or any other way (e.g. by applying electric field to an electro-optic material). In both optical injection cases mentioned a stabilization of the polarization of both lasers may be achieved by using a polarizer in the external cavity, if the lasers do not have stable polarization. Beam splitters and an optical attenuator inserted in the external optical cavity set the level of the optical injection. A spatial filter inserted in the external cavity may be used to select one mode in case where multimode VCSELs are used.

A part of the generated microwave power with a frequency given by $f= |\nu_{01}-\nu_{02}|$, taken from the output of the photodetecting mixer (or low noise amplifier) is used as a feedback signal, figure 22. Three different electric feedback schemes are possible. In the first case, which is the basic requirement for the proposed architecture, figure 22 to figure 25, it is used to make amplitude modulation of VCSELs and produce optical sidebands in the laser spectrums given by $\nu_{n1}=\nu_{01} \pm n \bullet f$, $\nu_{n2}=\nu_{02} \pm n \bullet f$, $n=1,2,\ldots$. One of the first order sidebands, n=1, of one of the lasers exactly coincides in frequency with the main lines of the second laser. Modulation spectrums with main and sideband lines for two lasers are shown in figure 22. The frequency of such a microwave generator may be stabilized if a transistor with internal gain is used as a photodetector. To stabilize the output microwave frequency a part of the microwave signal is applied back to phototransistor, as it is shown in figure 21. In the simplest case the tunable narrow band filter may be a varactor controlled LC lumped element resonator [21]. Such a positive feedback makes the photodetector highly frequency selective. In this way both lasers are electrically injection locked, which reduces further the phase noise of the microwave signals at the output of the system. In contrast to purely electronic transistor oscillators the positive feedback is not critical for getting microwave oscillations. Furthermore, in a self oscillation region of the transistor it will be inherently optically injection locked.

For long term temperature stabilization a part of microwave signal may be detected at the output of the band pass filter and superimposed to the bias current VCSELs, figure 24. Note that the photodetector with microwave filter and rectifier form an electronic wavelength discriminator, which may be compared with the optical analogue shown in figure 21a. In contrast to figure 21a no optical wavelength sensitive component is used here. All electronic components used may be realized in the same chip as the photodetector, leading to lower sizes and cost.

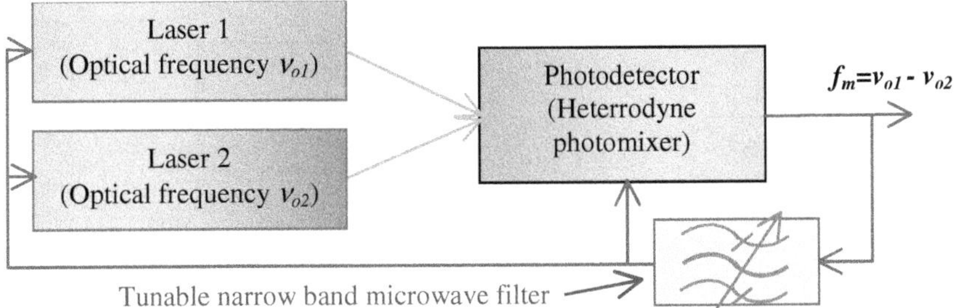

Figure 23 : Microwave feedback with microwave stabilization of frequency.

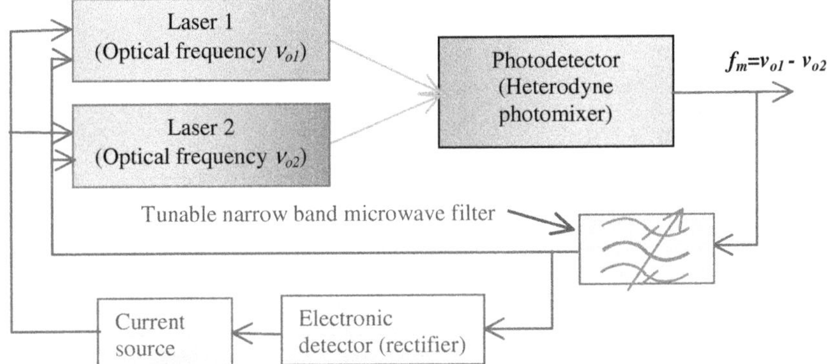

Figure 24 : Low frequency stabilization of the bias current by electrical feedback.

The photonic microwave generator may be used as frequency synthesizer, as it is shown in figure 25, where the output microwave frequency is compared (possibly after downconversion/division) with an external signal (e.g. low frequency quartz oscillator), and the differential signal is superimposed on the bias currents of the lasers. In figure 22 to figure 25 microwave matching networks will be required to provide impedance matching between output of the filter and input impedance of the lasers and also to de-couple DC power supply and low frequency (figure 22 and figure 23) feedback networks. Optionally an optical power divider before the photomixer to facilitate the system with an additional optical output, which could be coupled to an optical fiber. Additionally, VCSEL/photodetector integration in a multichip model is envisaged as possible way to bring the cost down. A simple integration example is shown in figure 26, where the chip with VCSELs is flipped on a standard SiMMIC substrate.

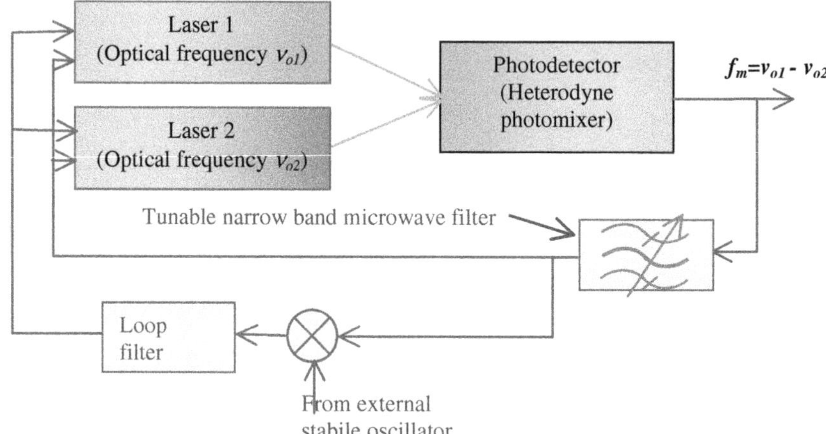

Figure 25 : Photonic microwave frequency synthesizer.

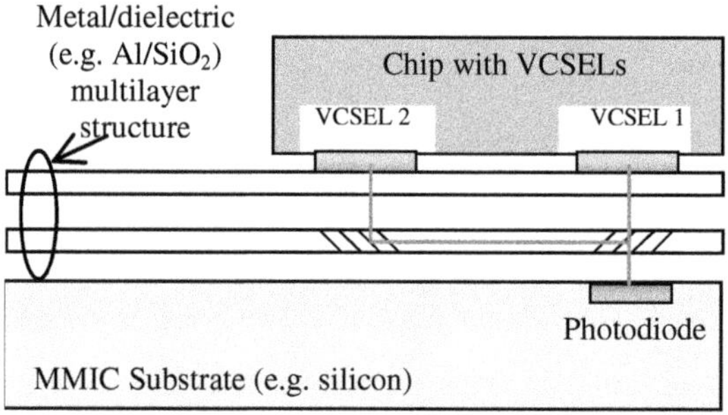

Figure 26 : Possible integration example with optical feedback as discussed above.

A set of preliminary experiments have been carried out to check microwave and optical performances of the VCSELs planned for application in a frequency tunable photonic microwave generator [7]. Similar experiments are available in the literature [31]. Figure 27 shows the current tuning rate of the VCSELs used in the experiments and the spectrum of the generated microwave signal. The problem which arises in VCSEL based frequency tunable photonic microwave oscillator is high current tuning rate of VCSELs, Figure 27a. This implies rather strict requirements to the tuning electronics. For similar applications new VCSEL with reduced Current Tuning Rate (i.e. dependence of the optical frequency on laser current, CTR) should be developed.

Figure 28 shows time dependencies of microwave frequency and microwave $FWHM_m$ for [7].

The data from preliminary experiments (the level of the fed back optical power, effects of polarization instability, parasitic reflections from the other optical components into the VCSEL's) without temperature stabilization and feedback (optical, electrical) is required where the feedback circuits and/or a phase locked loop in a photonic microwave oscillator are synthesized. Different combinations of the feedback arrangements may be used to achieve the desired phase noise specification.

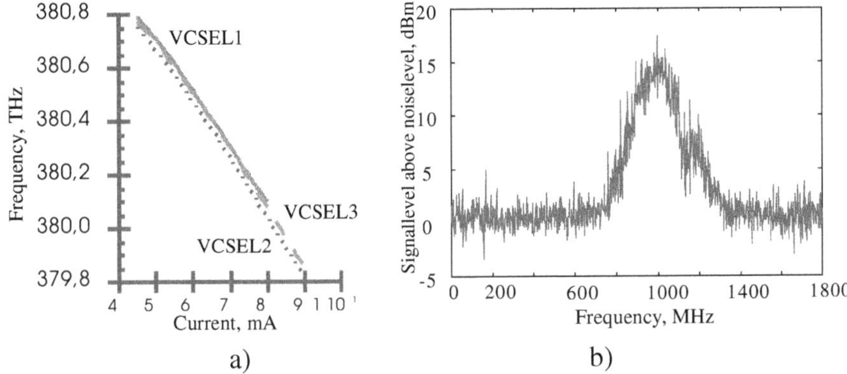

Figure 27 : Current tuning rate (a) and spectrum of microwave signal (b). No feedback is used.

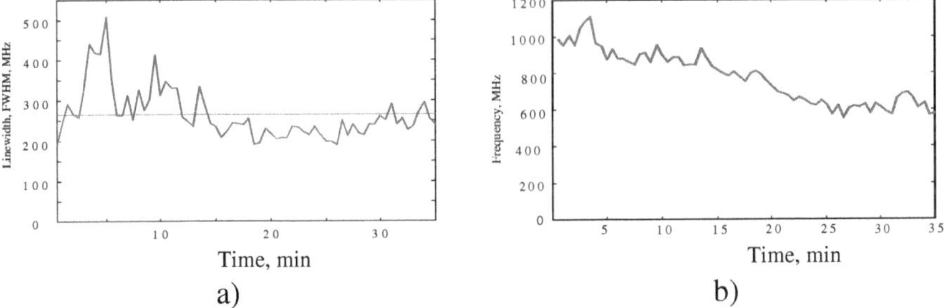

Figure 28 : Time dependence of frequency (a) and FWHM$_m$ (b). No feedback is used.

2.2.5. Conclusions

The discussion above show that low phase noise microwave generation is possible by using cost effective VCSELs. No temperature stabilization (e.g. Peltier cooler) or complex optical phase lock loops are required. The other advantages of the proposed generator [7], [30] is that the linewidth reduction of VCSEL's by optical feedback is done without use of any optical phase locking loop. Most of the components used for the phase noise improvement are compatible with the standard semiconductor fabrication technology, more specifically with a silicon technology. Thus a further improvement of the system is possible by full integration, on a single semiconductor substrate, both laser/photo diodes and microwave (passive and active), and in some cases digital circuits. To a large extent the hybrid integration of short wavelength VCSELs with SiMMICs may be practical up to 40 GHz and above. Photonic microwave oscillators may have a large range of applications in advanced microwave systems. Particularly, a stepwise (balanced) change in bias currents of one or both VCSELs will result in frequency hopping of the output microwave signals with a dynamic electrical/optical injection locking of the system to the new output frequency. In this case modulating baseband (useful) signal may be applied at the microwave output of the system. In an alternative system modulating baseband signal may be applied as a difference between bias currents of lasers, which will result in a frequency/phase modulation of the output microwave signal.

2.3. OPTICAL GENERATION OF MICROWAVE FUNCTIONS

B. Cabon, V. Girod, G. Maury
LEMO /ENSERG, UMR 5530 INPG-UJF-CNRS, B.P.257 38016 Grenoble Cedex 1, FRANCE
cabon@enserg.fr

Abstract

The progress achieved in performing optoelectronic components makes feasible the generation of microwave functions using all-optical devices. The application concerns signal processing at very high frequencies, which is usually difficult to perform in the microwave range.

The principle of using optical devices for microwave processing is described for both optical coherent and non-coherent regimes. Optical components are addressed in terms of microwave-optical S parameters.

Filtering microwave signals with optical delay lines is addressed and the experimental realization using fibers or integrated optics is explained. Generation of microwave mixing is explained as well. Then the generalization of these techniques to WDM and digital systems is presented. Applications to cellular radio on fiber systems and radar systems are discussed.

2.3.1. Introduction

Processing microwave signals on the optical link, directly in the optical domain, avoids intermediate conversions from optics to electronics (O/E) and then back again from electronics to optics (E/O). New functions are generated at microwave frequencies by processing the RF modulating signal of the optical carrier. This is not achieved before emission or after detection in the electrical domain, but after optical modulation, by optical components inserted in the link.

The bandwidth for processing is limited. In the microwave domain, the limitation comes from the bandwidth of components that achieve processing, and by other propagation and electromagnetic effects : attenuation and dispersion of planar integrated waveguides and cables and radiation of components and waveguides. The limitation is on the order of 10 GHz in hybrid microwave integrated circuits. In the optical domain, if we neglect dispersion, the limitation comes neither from the processing itself, nor from propagation since the bandwidth of fibers and optical guides is extremely wide (on the order of THz) and attenuation is negligible. The frequency bandwidth limitation comes here again from the components, but only those that are placed at each end of the optical link : the modulator and the photodetector. For commercially available components, direct modulation is limited to about 20 GHz, external modulation typically to 40 GHz, while photodetectors are limited typically to 60 GHz.

By using ordinary passive optical components like interferometers, single mode fibers as optical delay lines and couplers, some interesting microwave functions can be generated. The interference of the microwave envelope and/or the optical carrier generates filtering and mixing of the microwave subcarriers. These functions are not frequency limited. They are periodic, and valid to infinite frequency. While these passive (filtering) and active (mixing) functions might be achieved in the microwave range by power consuming and frequency limited components, they can be achieved in the optical domain with few power consumption and with common and low-loss components. Consequently, it is interesting to process optically those functions at higher frequency, where they are not easily realized in the microwave domain.

2.3.1.1. All Optical Processing of Microwave Functions by Insertion of an Optical Passive Component

In figure 29 is shown the point to point optical link with direct modulation (figure 29a), or external modulation (for example, electro-optical modulation using a Mach Zehnder interferometer, figure 29b). Optical passive devices can be inserted between planes P_1 and P_2 for achieving optical processing of the microwave subcarrier. They can be delay lines, fibers, or a passive interferometer made either of fibers and couplers or integrated optical waveguides as shown later on.

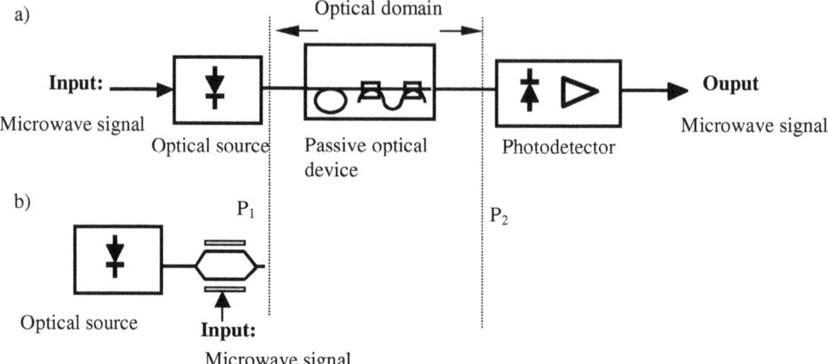

Figure 29 : Insertion of optical components in the microwave link for all-optical signal processing.

2.3.1.2. Electrical Power Gain of the Link

The link transducer power gain G at RF frequency can be expressed as the product of three separately determinable parts:

$$G(RF) = \frac{P_{out}}{P_{RF,a}} = \frac{P_{mod}^{2}}{P_{RF,a}}.t_o^{2}.\frac{P_{out}}{P_{od}^{2}} = F_{E/O}.t_o^{2}.F_{O/E} \tag{7}$$

where P_{out} is the RF power delivered to the load at the link output, $P_{RF,a}$ is the available power from the RF source at the link input, P_{mod} is the fiber coupled optical power from the modulating device, t_o is the link optical transfer efficiency, $t_o = P_{od}/P_{mod}$, and includes all factors affecting the transmission of light in fiber from the modulating device to the photodetector, P_{od} is the modulated optical power received by the photodetector.

The first term of the product, $F_{E/O}$, concerns E/O conversion, the second , $F_{O/E}$, concerns O/E conversion.

The optical powers are squared because for the optoelectronic devices, the optical power varies linearly with current and the RF power delivered to the load of the photodetector is proportional to the square of the detected current. Considering also a laser diode for direct modulation, P_{mod} is related to the driving current I_L by the slope efficiency : $\eta_L = \frac{dP_{mod}}{dI_L}$. For a photodetector, the same relationship exists $\eta_D = \frac{dI_D}{dP_{od}}$. This shows that when optoelectronic devices operate under a linear regime, G is proportional to the product $\eta_L^{2}.\eta_D^{2}$.

The relation giving G explains why 10 dB of optical loss produces 20 dB of electrical loss at the detected end.

2.3.1.3. External and Direct Modulation

2.3.1.3.1. The External Modulation

The external modulation (e.g. electro-optic modulation) is a phase modulation $\Delta\Phi(t)$ operated by the RF electrical field, which is converted by the interferometer into an intensity (power) modulation IM, at the output of the interferometer.

The expression of the modulated optical intensity, when the interferometer is biased in the linear regime, is:

$$P_{mod}(t) = \frac{P_0}{2}\left[1+\cos(\Delta\Phi(t))\right] \approx \frac{P_0}{2}\left[1+m\cos(\omega_m t)\right], \qquad (8)$$

m is the modulation index, m<<1 (small RF signal), P_0 is the optical power delivered from the optical source without modulation, and ω_m is the microwave (RF) modulation angular frequency.

The expression for the optical modulated field at the output of the external modulator is, in push-pull configuration :

$$E(t) = \frac{E_0}{\sqrt{2}}\left[1+\frac{m}{2}\cos(\omega_m t)\right].\cos(\Omega_0 t)\,, \qquad (9)$$

where Ω_0 is the optical angular frequency.

It is a pure amplitude modulation (AM), without frequency modulation of the optical carrier. The optical spectrum of E(t) is composed of the central carrier at frequency $\nu_0 = \Omega_0/2\pi$, and two lateral peaks at a distance of $\pm f_m = \pm\omega_m/2\pi$, of the carrier, so the spectral width is $Lu=2f_m$. The two peaks apart from the carrier are visible on an optical spectrum analyzer and one peak at the f_m offset frequency can be measured on a RF spectrum analyzer. Then, the value of m can be determined.

2.3.1.3.2. The Direct Modulation

The direct modulation of the driving current of the LD produces AM modulation plus FM modulation of the optical field. Due to the chirp effect, the optical frequency is modulated. This leads to broadening of the optical spectrum of the impulse emitted by the laser diode, and has the same consequence as chromatic dispersion of the fibers when the impulse propagates along the fiber. Normally, this effect is undesirable in telecommunications where single mode pulse and the narrowest optical spectrum are desired. But in some cases like in this presentation, for purposes of signal processing, advantages can be taken from this FM modulation. This will be discussed in section 2.3.3.

Without frequency modulation, the directly modulated laser diode (modulation index m, IM) would emit the modulated power

$P_{mod}(t) = P_0[1 + m\cos(\omega_m t)]$ under a linear operation and the electrical field would be $E(t) = E_0\sqrt{1 + m\cos(\omega_m t)}.\cos(\Omega_o t)$.

Introducing FM modulation (chirp), the optical wavelength becomes :

$$\lambda\ (t) = \lambda_0 + \Delta\lambda\cos(\omega_m t) \quad (10)$$

where $\Delta\lambda$ is the amplitude of wavelength variation, and is proportional to the linewidth enhancement factor α (or Henry Factor), proper to the LD.

The instantaneous optical frequency is:

$$\nu(t) = \nu_0 - \Delta\nu\cos(\omega_m t) \quad (11)$$

where $\Delta\nu$ is of the order of 100-200 MHz per mA with common LD in the bandwidth of the LD, $\nu_0=\Omega_O/2\pi$. Taking into account this chirp effect, the optical field is :

$$E(t) = E_0\sqrt{1 + m\cos(\omega_m t)}.\cos(\Omega_o t - \beta\sin\omega_m t)\ . \quad (12)$$

β is the FM modulation index and $\beta = \dfrac{\Delta\nu}{f_m}$. (13)

The ratio β/m is correlated to α by : $\left|\dfrac{\beta}{m}\right| = \dfrac{\alpha}{2}\sqrt{1 + (\dfrac{\omega_c}{\omega_m})^2}$ (14)

where ω_c is a cutoff angular frequency depending on the laser propertis and on its bias current.

For frequencies lower than the resonant frequency of the laser diode and beyond the cutoff frequency, $\left|\dfrac{\beta}{m}\right| \approx \dfrac{\alpha}{2}$. (15)

Measured with a high resolution optical spectrum analyzer, the combined AM (amplitude modulation) and FM (frequency modulation) effects give peaks at distances of ±fm, ±2fm,... from the optical carrier ν_0. The optical spectral width of E(t) is broadened and is related to β by :

$$Lu = 2f_m(1 + \beta)\ . \quad (16)$$

Due to combined AM+FM, the amplitude of the peaks are not symmetrical, regarding the central frequency ν_0, and are in direct relation with β and m. The detected power of the photocurrent can be measured on a spectrum analyzer, and m can be derived.

The optical modulated field (effects of β and m) can be measured on an optical analyzer, and by inverse modelling, β is derived.

2.3.1.4. Measurement of the Optoelectronic Scattering Parameters

We consider now external modulation, the optical intensity is modulated at a microwave frequency (modulation index m).

It is convenient to define the optoelectronic S parameters as in the microwave range. The optical scattering waves (a_i,b_i) at each port, i , of the optical device, related to modulated optical intensity (envelope of the carrier) are :

$a_i = I_{ai} m \cos(\omega_m t + \varphi_{ai})$ and $b_j = I_{bj} m \cos(\omega_m t + \varphi_{bj})$, where φ_{ai} is the phase of the microwave envelope at port i, related to the incident wave a, while φ_{bj} relates to the emergent wave b. I_{ai} is the incident optical intensity at port i, I_{bi} the emergent intensity at this port.

The optoelectronic S parameters of the optical device are then :

$$S_{ij,\,opt} = \frac{b_i}{a_j} = \frac{I_{bi}}{I_{aj}} \exp j(\varphi_{bi} - \varphi_{aj}) . \tag{17}$$

These S parameters defined with the optical intensity can be measured with a microwave Vector Network Analyzer and its lightwave extension (Figure 30).

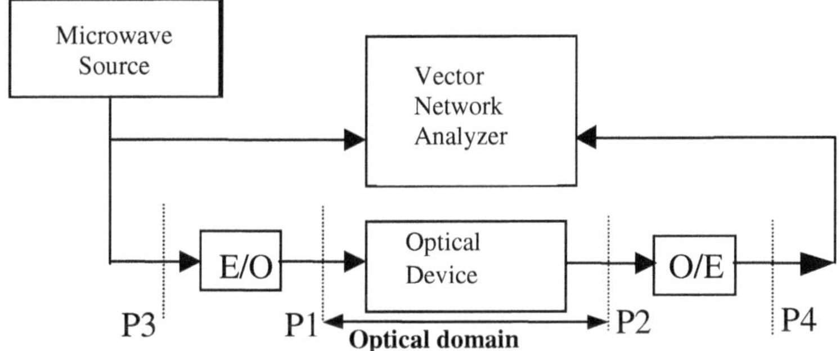

Figure 30 : Measurement of optoelectronic S parameters.

The intensity of the light emitted by a DFB laser source (1300 nm) is modulated at frequency f_m up to 20 GHz by a MZ external modulator and a microwave source (E/O conversion). At the output of a rapid photodetector (O/E conversion), a synchronous detection is operated by the Vector Network Analyzer. The photocurrent detected by the photodetector is compared to a reference signal. The optoelectronic S parameters are then obtained. The microwave frequency response, the insertion loss, the group delay, etc. are finally derived.

Assuming a linear operation of E/O and O/E converters, then the optoelectronic transfer function $S_{21,\,opt}$ (f_m) of the optical device in the planes P1-P2 can be obtained by measurements :

i) of the global transfer function T_{global} in the planes P3-P4 and

ii) of the E/O and O/E responses by a preliminary calibration.

This transfer function becomes :

$$S_{21,opt}(f_m) = \frac{T_{global}(f_m)}{T_{E/O}(f_m).T_{O/E}(f_m)}. \tag{18}$$

2.3.2. Generation of Microwave Filtering

The solution presented is based on interference (coherent regime and incoherent regime) using a directly modulated DFB (Distributed feedback) LD and a passive unbalanced Mach-Zehnder (UMZ) interferometer [33]. Other solution has been presented that uses cascaded passive Mach-Zehnder (UMZ) interferometers [34].

The microwave filter consists of a single optical unbalanced Mach-Zehnder interferometer, which is composed of two optical directional 3dB-couplers separated by two unequal optical paths, when fibers are used (Figure 31a). It can also be realized in integrated optics on glass, with two Y junctions separated by two optical integrated waveguides of different lengths : one straight, the other curved (Figure 31b).

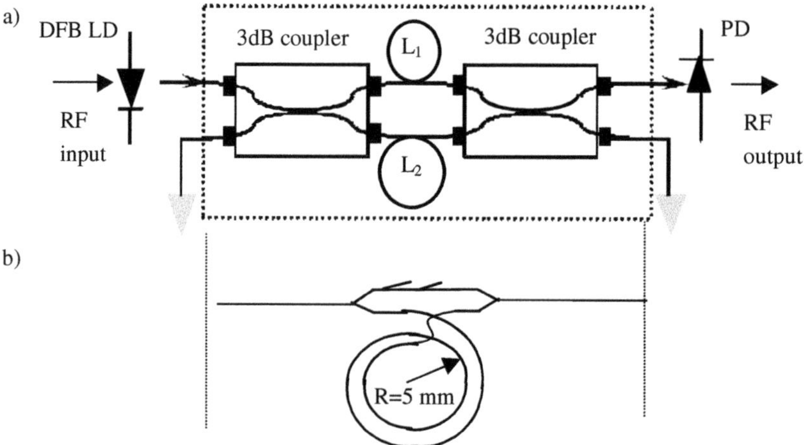

Figure 31 : UMZ: Unbalanced Mach-Zehnder used as a microwave filter a) fibered b) integrated optical waveguides on glass.

Because of the path difference $\Delta L = L_2 - L_1$ between the arms of the UMZ interferometer, two interference figures can occur at the output of the interferometer :

- microwave interference on the envelope : when two non-coherent optical pulses arrive at the same time at the output of the UMZ. This is the case when the coherent length Lc of the source is shorter than ΔL. The intensity at the output is the sum of optical intensity $I_1 + I_2$ on each arm.

- microwave interference on the envelope plus optical interference on the optical carrier, when the laser coherence length is greater than the path difference, Lc>ΔL.

In each case, the RF frequency period of the transfer function $S_{21}(f_m)$ equals the FSR (Free Spectral Range) of the interferometer, which is defined as :

$$\frac{n_{eff}}{c}(L_2 - L_1) = \frac{1}{FSR} = \tau. \quad (19)$$

The optical intensity at the output of the UMZ interferometer is

$$I(t) = I_1(t) + I_2(t-\tau) + 2\sqrt{I_1(t).I_2(t-\tau)\cos(\Omega_0\tau)}.V(\tau)\,. \quad (20)$$

Loss of coherence in the two waves propagated on the two arms of the interferometer is illustrated by $V(\tau)$, which equals 0 in incoherent regime and is approximately 1 in the coherent regime.

2.3.2.1. Non-Coherent Regime

The interference of the modulated intensity waves in the two arms produces a periodic transfer function $S_{21}(f_m)$ with minima and maxima. The microwave frequency of the minima is an odd integer multiple of FSR/2, that of maxima an integer multiple of FSR.

The UMZ acts as a frequency rejection filter over a large frequency range. Filtering is periodic, period equals FSR that can be set by adjusting ΔL. A rejection ratio (maximum divided by minimum of $|S_{21}|$) greater than 25 dB optical or 50 dB electrical can be obtained.

It is worth noting that such a similar value for the rejection ratio could not be obtained in the microwave range, and moreover periodically, up to infinite frequencies if there were no limitation in the frequency response of the optoelectronic components at emission and detection sides.

2.3.2.2. Coherent Regime

This regime exists when the coherence length of the source is high enough to be greater than ΔL. The coherent optical interference suffers from some additional effects :

- influence of the optical phase of each arm and influence of the refraction index. The temperature of the component must be controlled accurately
- influence of the linewidth of the laser, and of the wavelength emitted (by accurate control of the DC bias and temperature of the LD)
- influence of polarization of the light, when the optical waveguides used in the UMZ are birefringent.

2.3.2.3. Experimental Results with Passive UMZ Interferometer Integrated on Glass

The example of a passive optical UMZ interferometer integrated on glass substrate by Tl+/Na+ ion exchange is presented here. The design allows a FSR of 3 GHz (ΔL=6.3 cm). The layout of the UMZ is shown in figure 32a, and is similar to the schematic of figure 31b. When excited with a coherent DFB laser diode source emitting at 1300 nm (coherence length ≈ 5m) and modulated by an external modulator (20 GHz of bandwidth), the optoelectronic transfer function $S_{21}(f_m)$ can be obtained, as shown in figure 32b.

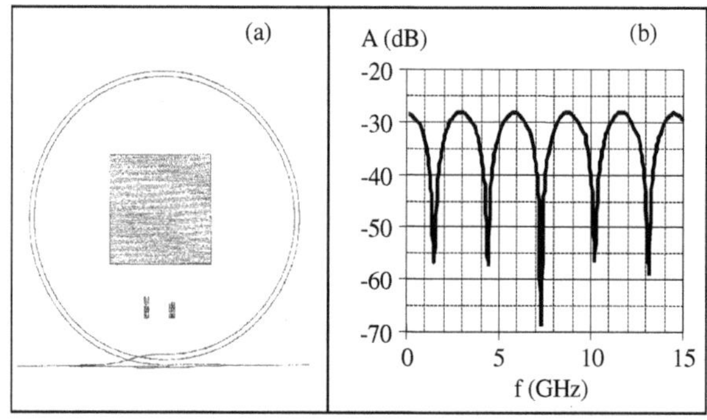

Figure 32 : layout of the integrated UMZ (a) and the corresponding measured frequency response (b).

2.3.3. Generation of Microwave Mixing

2.3.3.1. Definition of Microwave Mixing and Processing by Photonics

Mixing is a common function used in electronics for converting the frequency of a microwave signal. A pumped non-linear device generates a spectrum of frequencies (intermediate frequency IF) based upon the sum and difference of the harmonics of the signal (RF) and local oscillator (LO) frequencies. The LO acts as a pump of non-linearity. The mixer conversion factor (conversion gain or loss) is defined as :

$$\text{Conv} = 10\log\left[\frac{P_{out}(\text{IF})}{P_{RF,a}}\right]. \qquad (21)$$

In addition to the primary frequencies, $f_{LO}+f_{RF}$, $|f_{LO}-f_{RF}|$, one can find mainly the fundamentals f_{RF}, f_{LO}, plus harmonics of f_{LO} at the output of the mixer. But, for better conversion, the fundamentals and their harmonics must be rejected as much as possible in the output spectrum.

In microwaves, active mixers are necessary to obtain a conversion gain and not a conversion loss. But this requires a complex matching circuitry with DC bias for the active elements, which is power consuming.

Processing mixing in the optical domain can be achieved with active devices as well as with passive devices. Non-linear figures are required, and processed either by

- using a non-linear transfer function (that of an LD, or of an external modulator biased at Vπ) [35,36]
- or by multiplication of two linear transfer functions (that of an a external modulator biased at Vπ/2) [37]
- or by using non-linearity of the interference figure. In the latter case, a passive component (UMZ) can be used [38].

Photonics mixing can be generated either at the emission side or at the reception side of the optical link. At emission, mixing is generated by LD, modulators or interferometers as explained above. At the reception side, it is generated by photodetectors.

2.3.3.2. Example of Mixing Generation at Emission Side : Utilization of a Passive UMZ Interferometer [38]

A non-linear function is here produced by coherent interference. It is the non-linearity of the optical power detected as a function of the optical frequency ν_o. This is obtained at the output of a passive optical UMZ interferometer as :

$$P_{out}=1/2.P_o(t).[1+\cos(2\pi\nu_o\tau) \quad (22)$$

]where P_o is the optical intensity at the input of the UMZ, τ=1/FSR is the optical delay between the two arms of the UMZ.

When this property is used for generation of mixing, the non-linearity required for the mixer is neither due to the DFB LD (operated in the linear region $I_L(P_0)$ nor to the non-linear $I(P_o)$ detection. The field at the input of the UMZ modulated by two harmonic signals at frequency f_{m1} and f_{m2} is similar to expression (12) and is :

$$E_{in}(t)=E_0\sqrt{1+\sum_{i=1,2}m_i\cos(\omega_{mi}t)}.\cos\left[\Omega_o t+\phi(t)-\sum_{i=1,2}\beta_i\sin(\omega_{mi}t)\right] \quad (23)$$

and can be written as $E_{in}(t)=E_0m(t).\cos\left[\Omega_o t+\phi(t)-\sum_{i=1,2}\beta_i\sin(\omega_{mi}t)\right]$

Eq. (23) shows that the optical frequency ν_o is modulated (FM modulation) by the fundamentals $f_{m1}=f_{RF}$ and $f_{m2}=f_{OL}$. The optical power varies non-linearly with the modulated optical frequency as previously mentioned above in Eq. (22). Maximum non-linearity is obtained when the

UMZ operates at maximum or minimum of transmission. Consequently, intermodulation products are present in the optical modulated power detected by the PD.

The modulated field, at the output of the UMZ (before detection by the PD) results from coherent interference and is :

$$E_{out}(t)=1/2[E_{in}(t)+E_{in}(t-\tau)]. \quad (24)$$

The optical intensity I(t) is proportional to $E_{out}(t)^2$ and is similar to Eq. (20). The power detected by the rapid photodetector is the average of I(t) on several optical periods.

$$P_{od}(t)=\langle I(t)\rangle= \frac{E_0^{\,2}}{4}\left[2+\sum_{i=1,2} m_i\left\{\cos[\omega_{mi}t]+\cos[\omega_{mi}(t-\tau)]\right\}+2V(\tau)A(t)B(t)\right] \quad (25)$$

with $A(t)=m(t).m(t-\tau)$,

and $$B(t)=\cos\left[\Omega_o t-2\sum_{i=1,2}\beta_i\sin(\omega_{mi}\tau)\cos[\omega_{mi}(t-\tau/2)]\right] \quad (26)$$

This shows an intensity modulation (IM). The first term exists in the incoherent regime and results from AM. The second term results from AM+FM.

$$V(\tau)=\langle \exp j[\phi(t)-\phi(t-\tau)]\rangle \quad (27)$$

Thus, coherent interference converts FM modulation present in $E_{in}(t)$ into IM modulation at the photodetector side.

The interference of the envelope (AM) provides filtering of undesired fundamentals present in the output detected spectrum (cf. section 2.3.2), while coherent interference (AM +FM) produces the desired mixing products.

Figure 33 shows mixing products obtained experimentally and the effect of modulation of the optical frequency (via laser chirp present in the directly modulated LD) on the optical power.

The curve "P_{od}" shows the interference regime, since this optical power at the output of the UMZ interferometer is measured by a power-meter. It is presented as a function of the optical frequency which is controlled by the bias current of the LD. This is the experimental verification of Eq. (22).

The two other curves (quasi superimposed) show the mixing products via P_{mix} (P_{out} in Eq. 7) measured by a rapid PD, as a function of the optical frequency. The frequency 3 GHz corresponds to the difference f_{OL}-f_{RF} and 6 GHz to the sum f_{OL}+f_{RF} since f_{LO}=4.5 GHz and f_{RF}=1.5 GHz. FSR=3 GHz. Figure 33 shows that the maximum or minimum of transmission corresponds to the maximum power of mixing products. This is the first condition for best mixing.

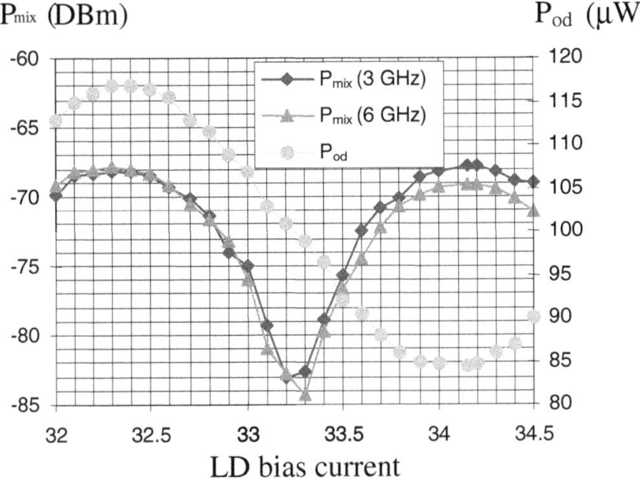

Figure 33 : Measured powers and effect of the interference regime on the mixing response of the passive UMZ.

The second condition for optimized mixing, with respect to the FSR of the UMZ is:

$$f_{OL} = (2k+1)\frac{FSR}{2},\ k \in N \text{ and } f_{RF} = (2k'+1)\frac{FSR}{2},\ k' \in N \qquad (28)$$

This condition is fulfilled f_{LO}=4.5 GHz and f_{RF} by f_{LO} and f_{RF}.

This shows that the fundamentals must be chosen according to a given FSR and vice versa. The fundamentals must be chosen periodically on the frequency range: this is well demonstrated by the simulations of figure 34. The power of the different mixing products is shown for α=5, P_{RF}= 15 dBm, P_{mod}=1mW, I_L-I_{th} =25 mA, η_D=0,5 A/W, t_o=1, FSR=1 GHz, and P_{LO}=3 dBm. The frequency f_{LO} is fixed and equals 8.5 GHz, which respects Eq. (28), while f_{RF} varies. The 3 dB RF bandwidth equals FSR/2. Rejection of the fundamentals and maximum power for the mixing products are obtained periodically and simultaneously when conditions in Eq. (28) are fulfilled, (f_{RF}=0.5, 1.5, 2.5,.. GHz) and at maximum of transmission of the interference. The difference in the two side-band products, $f_{LO} + f_{RF}$ and $f_{LO} - f_{RF}$ is due to combined effects of AM and FM.

The experimental mixing spectrum is shown in figure 35. P_{out} is measured by a rapid photodetector connected to a spectrum analyzer, where FSR=1 GHz, f_{LO}= 8.5 GHz and f_{RF} is swept from 2.7 to 4.2 GHz. P_{LO}=3 dBm, P_{RF}=-15 dBm. FSR=1 GHz. The choice for frequencies f_{LO} and f_{RF} is imposed by the frequency bandwidth of the DFB laser diode and by the condition of Eq. (28). Both the lower sideband f_{LO} - f_{RF} of the mixing product and the upper sideband $f_{LO}+f_{RF}$ are significantly increased by inserting the interferometer. The rejection of fundamentals is higher than 15 dB and 10 dB for the two sidebands. After preliminary measurements of the AM index m and the FM index at RF frequency, α is then deduced and is approximated at 6.8 . The calculation of α is necessary for estimating the mixing conversion gain. The higher α is, the higher the conversion gain is.

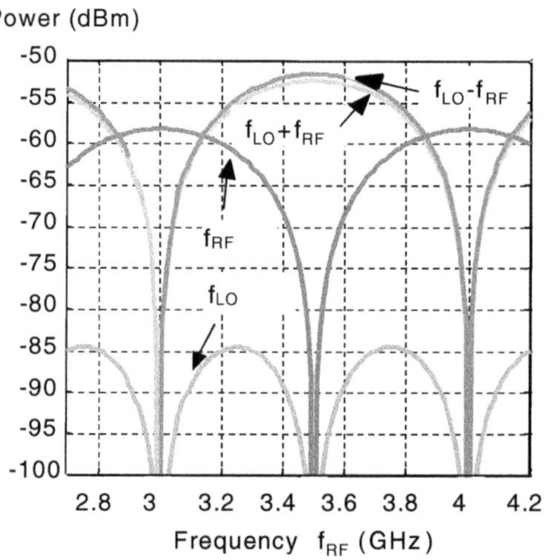

Figure 34 : Variation of f_{RF} with fixed f_{LO}=8.5 GHz, FSR=1 GHz,and effect on the mixing response of the passive UMZ.

In fact, the definition of the optimal optical conversion gain comes from Eq. (21) where $P_{RF,a}$ is, in the present case of an optical link, the optical power modulated at RF frequency and detected by the PD. Using Eq. (1), the normalized expression is derived where t_o no longer interacts :

$$\text{Conv, optical} = \frac{G(IF)}{G(RF)} . \tag{29}$$

Calculated for the present configuration of photonics mixing, at maximum of interference and when Eq. (28) is satisfied, the conversion gain Conv,optical is correlated to the linewidth enhancement factor α by :

$$\text{Conv, optical} \approx \frac{\alpha^2 0.58^2}{4} \tag{30}$$

and equals +6 dB for the experimental measured value α= 6.8.

In figure 35, a ratio G(IF)/G(RF) of -22 dB is obtained for f_{RF} equal to 1.7 GHz, i.e. closer to 1.5 GHz (FSR/2). But here, only G(IF) takes into account the coupling between fibers and integrated device.

This induces an optical loss of - 14 dB corresponding to -28 dB of electrical loss. So, +6 dB of conversion gain is obtained experimentally for this solution of mixing, which is in agreement with the theoretical value of this gain.

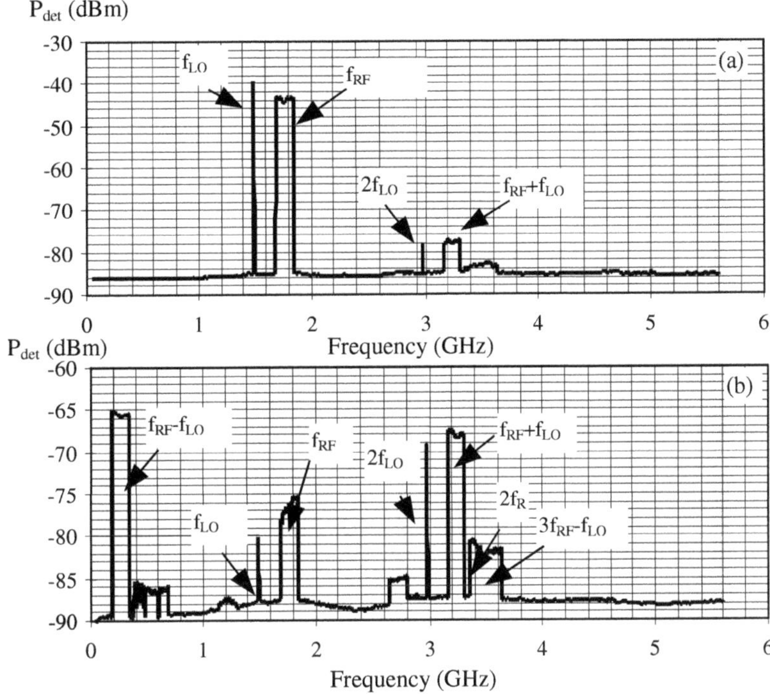

Figure 35 : Measured power spectra of detected signals at the input of the UMZ (a) and at its output (b).

2.3.3.3. *Mixing at Detection Side : Utilization of a Passive P-I-N Photodetector for Generating Mixing*

Mixing is realized at the reception side by a PD operated in the non-linear region of the curve I(V).

It can be, for example, a P-I-N photodetector, or a HPT heterojunction phototransistor. In the latter case, the performances of InP/InGaAs have been demonstrated [39]. The HPT is one of the most promising optical-electrical transducers for high-speed hybrid fibre radio (HFR) distribution networks. This device processes not only detection of the modulated optical signal, but also amplification of the microwave output signal. Optoelectronic mixing can be obtained also. Results are presented here for the P-I-N photodetector [40]. The optoelectronic system for optical-microwave mixing is shown in figure 36. Simultaneous injection of a LO microwave signal at the electronic port of a PD, via a circulator, and a RF signal at the optical port of the PD by IM external modulation, results in the mixing of the two signals.

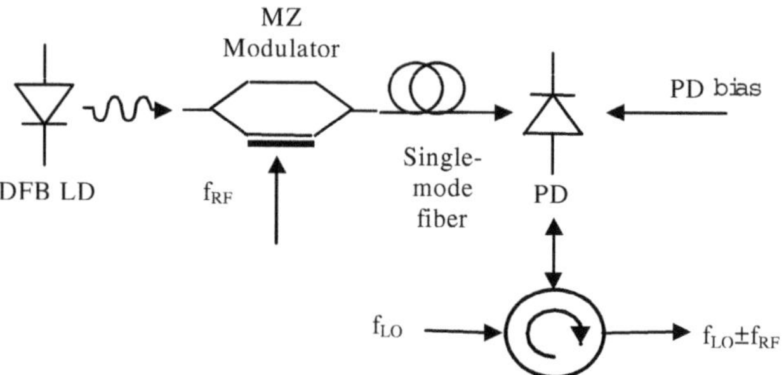

Figure 36 : System used for optoelectronic mixing using a P-I-N-photodetector.

The PD is biased close to 0V to obtain the most efficient mixing. LO frequency has to be within the frequency band of the circulator. Double sideband conversion is presented in figure 37, with f_{LO} = 8 GHz. The modulation frequency f_{RF} of the light is swept from 45 MHz to 850 MHz. The response is quite flat up to a 1.5 GHz modulation frequency. The influence of P_{LO} is demonstrated, and two optical modulation depths (OMD) of the light have been examined, OMD =25% (Min) and OMD =75%(Max). At low levels, the power of the mixing products is proportional to the OMD and to the LO power (P_{LO} < -5 dBm). At higher levels a saturation effect is observed ($P_{LO}\cong 0$ dBm). However, the frequency behavior in figure 37 is dependent on varying neither the OMD nor the LO power.

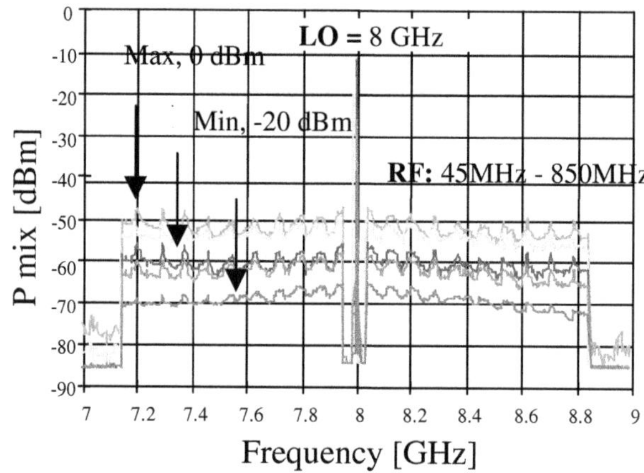

Figure 37 : PD mixing. P_{LO}= 0, -20 dBm, modulation depth of 25% (Min) and 75% (Max).

2.3.3.4. Comparison of the Two Methods

COMPARISON OF THE TWO MIXING METHODS	**UMZ / PD MIXING**	**PD MIXING**
Principle of operation	Passive optical device (UMZ)	Non-linearity of the PD
Frequency dependence	Periodic (period is a function of ΔL)	Flat response
Operation frequency	Limited by LD and PD responses	Limited by modulator, PD and circulator responses
Bandwidth	FSR/2	Limited by modulator, PD, circulator bandwidths
Conversion gain / loss	Gain is possible	
Conversion loss vs. P_{LO}	Curve with optimum	Curve with saturation
Rejection of the fundamentals	Complete rejection if mixing is optimal	Only by external filtering

Table 1 : Comparison of two techniques of optoelectronic mixing.

The principle of frequency conversion is different in the last two presented methods.

In the first case, the f_{LO} and f_{RF} signals operate FM+AM of the light, are mixed by the UMZ and transmitted. Applications are at emission side of a fiber-optic link. In the second case, mixing is realized at the reception of the optical link : only the frequency f_{RF} is transmitted by optical IM. A summary of the differences of the two methods is presented in table 1.

2.3.4. System Applications

2.3.4.1. Cellular Radio Systems

In cellular radio on fiber systems (Fig. 38), up-conversion is requested for transmission at millimeterwave frequency. For this purpose, up-converted subcarrier multiplexed signals could be up-converted and transmitted by the system LD-UMZ interferometer-PD that performs mixing. Conversion could be operated by a unique MZ interferometer at emission side. This avoids multiple conversions at the detection side, before transmission to the antenna connected to the base station.

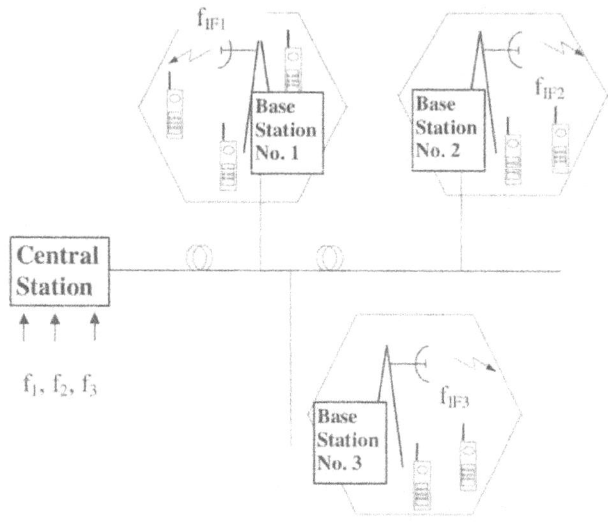

Figure 38 : Transmission of SCM signals with Radio on Fibre Systems.

Digital transmission

Digital high bit rate data may be transmitted on SCM up-converted signals, by the LD-UMZ-PD technique of mixing, as illustrated in figure 39. A sine wave (f_{RF}=4.45 GHz) is PSK modulated by a digital sequence (Fig. 39a).

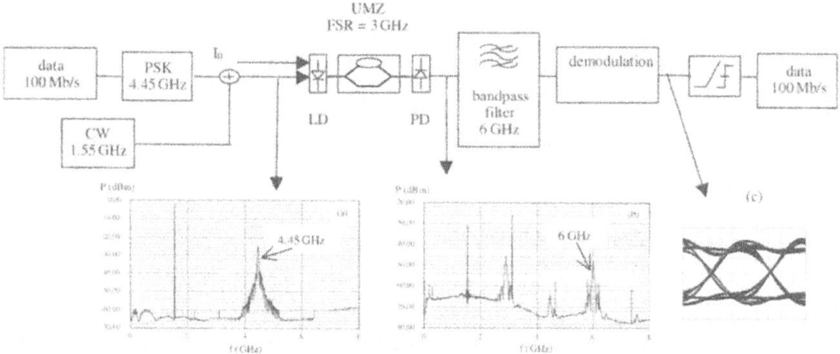

Figure 39 : Transmission of digital data with up-conversion of the sub-carrier microwave signal.

This microwave subcarrier at 4.45 GHz and a LO signal at 1.55 GHz modulate both the driving current of a DFB LD. Up-conversion is then achieved at the output of the UMZ. The mixing product at the frequency $f_{RF}+f_{OL}$=6 GHz (Fig. 39b) is filtered, finally demodulated. The digital sequence is well recovered at the system output, as shown with the opened eye-diagram of figure 39c.

WDM combined with photonics mixing

The demand for increase optical communications capacity has fueled great interest in wavelength-division multiplexing (WDM). Mixing microwave signals on WDM optical carriers is easy with the system LD-UMZ-PD [41]. Since several optical carriers and several RF ports are needed, other solutions for mixing, e.g. 2-port external modulators, cannot be utilized. Mixing microwave subcarriers multiplexed (SCM) signals with the system LD-UMZ-PD has the advantage over other solutions, like cascaded active external modulators, to use a unique passive optical component, the UMZ, to perform mixing on each channel.

2.3.4.2. Radar Systems

Photonics-microwave mixers presented previously can be utilized for down-conversion in radar systems.

One optical channel, with conversion of SCM signals

The main domain of application of optical-microwave mixing is in radar ground-based systems as well as airborne and future spatial radar systems. In optical architectures needed for beamforming, the down-conversion of microwave signals on one optical carrier allow the realization of the phase and delay synthesis on microwave IF (intermediate frequency), before coding and radar processing.

Several optical WDM channels, with conversion of SCM signals

The function of WDM does not exist in microwaves. Optics brings a unique advantage of mutliplexing channels, without interference, each channel bearing microwave SCM signals. Thus the capacity of the system is greatly increased.

Consider now the system LD-UMZ-PD with WDM optical channels. The main application of this photonics-microwave mixer on multi wavelengths could concern ultra wide band systems which are now needed in the electronic warfare. The N different beams present in a phased array antenna could be converted simultaneously with a unique photonics mixer when received.

2.3.5. Conclusions

Microwave functions can be generated with all optical components. Photonics-microwave rejection filters with high extinction ratio, and photonics-microwave mixers not limited in frequency range, have been presented. They both use a passive unbalanced Mach-Zehnder interferometer. Mixing with integrated amplification can be achieved with heterojunction phototransistors. Microwave SCM signals can be frequency-converted on different WDM optical channels, and digital signals can be transmitted with the SCM converted channels. This opens new possibilities to all- optical processing of microwave signals.

2.4. Optical Filtering for RF Signal Processing

J. Capmany, D. Pastor, B. Ortega, S. Sales
Optical Communications Group, IM002 Research Institute, Universidad Politécnica de Valencia - Camino de Vera s/n, 46022 Valencia, SPAIN
jcapmany@dcom.upv.es

2.4.1. Introduction

The progress of radiofrequency, microwave and millimetre wave technologies for telecommunications applications requires a coordinated effort in the development of signal processing techniques suitable for them. This is especially important as novel applications demand the use of increasingly higher frequency carriers and broadband signals. The traditional approach towards RF signal processing is illustrated in the upper part of figure 40. Here a RF signal originated at a RF source or coming from an antenna is fed to a RF circuit that performs the signal processing tasks either at the RF signal or at an intermediate frequency band after a downconversion operation. In any case, the RF circuit is capable of performing the signal processing tasks for which it has been designed only within a specified (often reduced) spectral band. This approach results in a poor flexibility since changing the band of the signals to be processed requires the design of a novel RF circuit and possibly the use of a different hardware technology. Furthermore, even if the RF carrier is not changed, the nature of the modulated signal might be, requiring from the processor more bandwith or sampling speed. This is especially true in the case where discrete time signal processing has to be carried over the Rf signal. These set of drawbacks are often termed in the optical communications technology literature as the *electronic bottleneck* . Being important it is by no means the only source of degradation, since electromagnetic interference (EMI) and frequency dependent losses can also be sources of important impairments.

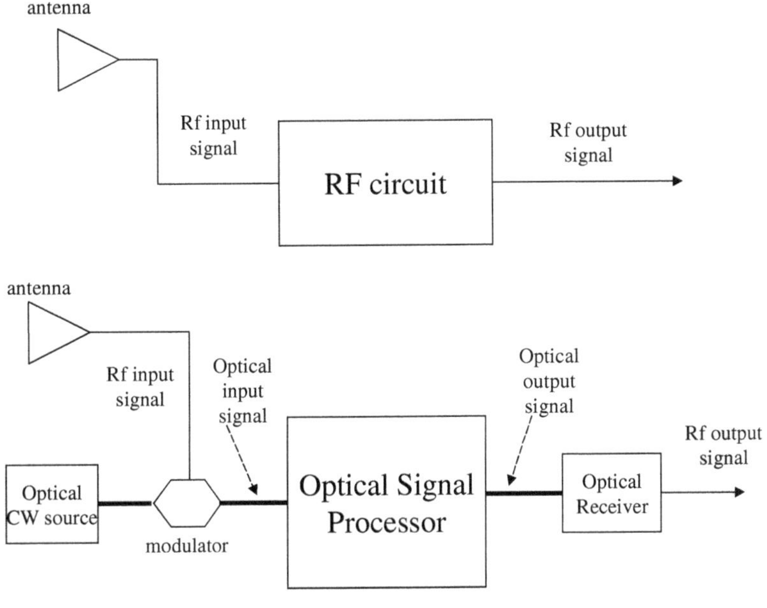

Figure 40

An interesting approach to overcome the above limitations involves the use of optical technology and especially fiber and integrated optics circuits to perform signal processing of RF signals conveyed by an optical carrier directly in the optical domain. We will refer to this as OPRFS[1]. This approach is shown in the lower part of figure 38. The Rf to optical conversion is achieved by direct (or externally) modulating a laser. The Rf signal is the conveyed by an optical carrier and the composite signal is fed to photonic circuit using optical delay lines for signal processing. At the output/s the resulting signal/s are optical to RF converted by means of an/various optical receiver/s.OPRFS has several advantages: optical delay lines have very low loss (independent of the RF signal frequency), provide very high time bandwidth products, are inmune to EMI, lightweight, can provide very short delays which result in very high speed samplig frequencies (over 100 GHz in comparison with a few GHz with the available electronic technology) and finally but not less important optics provides the possibility of spatial and wavelength parallelism using WDM techniques. .

A fundamental distinction must be made on the OPRFS operation regime in terms of the relationship between the coherence time τ_c of th optical transmitter and the basic delay T (time bewteen adjacent temporal

[1] OPRFS: Optical Processing of Radio Frequency Signals

samples provided by the structure). If $\tau_c >> T$ then the processor is said to work under *coherent* regime and its transfer function is linear in terms of the electric field thus depending on the optical phase shifts experienced by the carrier that conveys the RF signals. These are highly dependent on environmental parameters (i.e temperature, ..) and polarization, making their implementation quite difficult under realistic conditions. On the contrary, if $\tau_c << T$ the signal processor works under *incoherent* regime and the overall structure transfer function is linear in terms of the optical intensity (i.e power) and the effect of optical phase shifts can be discarded. Although work has been reported on both operation regimes, the majority of contributions focus on incoherent operation, since it is more prone to practical implementation.

This section aims to review the fundamental concepts, limitations, technologies and major milestones in RF filtering applications of OPRFS. Research contributions within this area extend over the last 25 years starting with the seminal paper of Wilner and Van de Heuvel [42] who noted that the low loss and high modulation bandwidth op optical fibers made then suitable for broadband signal processing Several contributions during the 70s addressed experimental work on OPRFS using multimode fibers [43,44]. An intensive theoretical and experimental research work on incoherent OPRFS using singlemode fiber delay lines was carried by researchers at the University of Stanford during the period between 1980 and 1990. Multiple congigurations, applications and potential limitations of these structures were considered and the main results of it can be found summarised in [45-47]. The technology status regarding optical fiber and integrated components was at the time at its infancy and therefore the OPRFS demonstrated had serious limitations arising from losses and lack of reconfiguration.

The advent of the optical amplifier at the end of the 80s and the development of optical components (variable couplers, modulators, electrooptic switches) and specific purpose instrumentation fueled the activity towards more flexible structures employing these components [48-75]. Most of these contributions present filters that still rely on the implementation of time delays by means of fiber strands. Yet, the availability of a novel component, the fiber Bragg grating has openned a new perspective towards the implementation of OPRFSs using this component which can lead to fully reconfigurable and tunable filters (see [35]-[62]).

2.4.2. Fundamental Concepts and Limitations

Any filter implemented using OPRFS tries to provide a system function for the RF signal given by:

$$H(z^{-1}) = \frac{\sum_{r=0}^{N} a_r z^{-r}}{1 - \sum_{k=1}^{M} b_k z^{-k}} \tag{31}$$

where z^{-1} represents the basic delay between samples. The numerator represents the finite impulse part (i.e non recursive or FIR) of the system function, whereas the denominator accounts for the infinite impulse part (i.e. recursive or FIR) of the system function. N and M stand for the order of the FIR and IIR parts respectively. If $b_k=0$ for all k, the the filter is non-recursive and is also known as transversal filter. Otherwise the filter is recursive and it is common to use the term recirculating delay line.
Figure 41 illustrates how (31) is implemented for the specific case of an N-order transversal[2] incoherent filter using a single optical source. Note that the impulse response corresponding to this situation can be directly derived from (31) yielding:

$$h(t) = \sum_{r=0}^{N} a_r \delta(t - rT) \tag{32}$$

Which convolved with the input Rf signal $s_i(t)$ yields the following output signal $s_o(t)$:

$$s_o(t) = \sum_{r=0}^{N} a_r s_i(t - rT) \tag{33}$$

The implementation of the OPRFS requires specific optical components to provide: a) Signal tapping, b) Optical delay lines, c) Optical weights and d) Optical signal combination. 2x2 and 1xN, Nx1 star couplers have been proposed for the implemetation of a) and b), variable 2x2 couplers, optical amplifiers (both EDFAs and SOAs), electrooptic and electroabsorption modulators can be used to implement c) Standard, high dispersion singlemode fiber coils and fiber Bragg gratings have been proposed for the implementation of b).

[2] The implementation of a recursive filter is similar but is not considered here due to space restrictions.

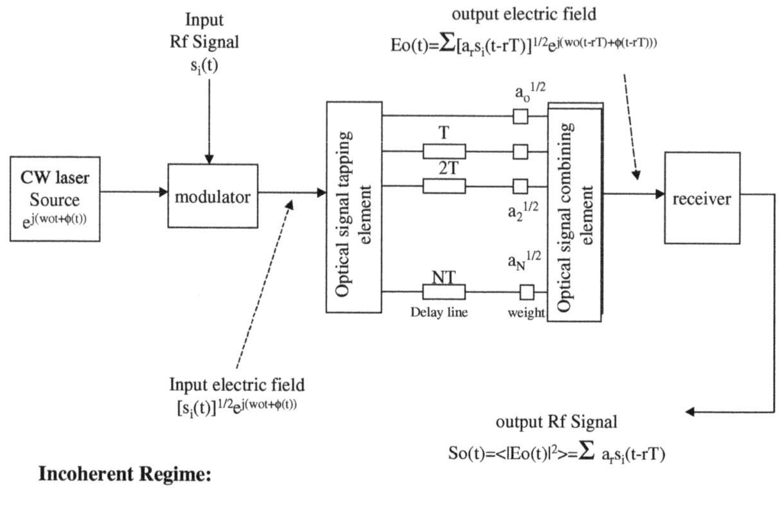

Figure 41

OPRFS must overcome a series of potential limitations prior to their practical realization as pointed out by various researchers. The main limitations arise from :

a) *Source coherence:* The source spectral characteristics must be carefully chosen attending to the desired working regime. While coherent operation provides the possibility of implementing any king of desired transfer function, these structures are very sensitive to environmental conditions [45]. Thus in the majority of cases, incoherent operation is employed since the filters are very compact and robust. Undesirable coherent effects may be overcome by the use of birrefringent fiber delay lines [104]
b) *Polarization:* Polarisation effects are only important under coherent operation [45]. However, it has been outlined and experimentally demonstrated that even under incoherent operation the filter can be sensitive to signal polarization [68], [105]. The main cause for this apparent contradiction is that some signal samples experience exactly the same delay within the filter leading to coherent interference between them even if a broadband source is employed [68].
c) *Positive coefficients*: Filters working under incoherent regime are linear in optical intensity, thus the coefficients of their impulse responses are always positive. This has two important implications as derived from the theory of positive systems [46]. The first one and more important is that the range of transfer functions that can be

implemented is quite limited. The second one is that regardless of its spectral period, the transfer function has always a resonance place at baseband. This is not a serious limitation since a DC blocking filter can be inserted at the optical receiver output. Nevertheless, incoherent filters with negative coefficients can be implemented by means of differential detection [46], [64] and cross gain modulation in a SOA [106].

d) *Limited Spectral period or FSR (Free Spectral Range):* OPRFSs are periodic in spectrum since they sample the input signal at a time rate given by *T*. Thus the spectral period or FSR is given by *1/T*. If the OPRFS is fed by only one optical source then the source coherence time (which is inversely related to the source linewidth) limits the maximum (minimum) value of the attainable FSR under incoherent (coherent) operation. This is depicted in figure 42. To overcome this limitation it has been proposed to feed the OPRFSs with source arrays [87].

e) *Noise*: As far as the optical source is concerned, passive OPRFSs behave as frequency discriminators and thus convert the optical source phase noise into intensity noise which materialises into RF baseband noise at the filter output [107]-[111]. This conversion is dependent on the operation regime. For incoherent operation the noise is periodic in spectrum showing notches at zero frequency and multiples of the filter FSR [107]-[110]. Under active operation (i.e. when incorporating optical amplifiers) new RF noise sources appear as a direct consequence of the beating between the signal and the spontaneous emission [55], [111]. It has been proved however that the converted phase noise is still the dominant noise source. The use of source arrays to feed the OPRFS is an attractive solution to overcome noise limitations [87]. This is due to the fact that signals recombining at the photodetector at different wavelengths will generate the intensity noise centered at the frequency resulting from the beats of the optical carriers. Since these have very high values they will be filtered out by the receiver.

f) Reconfigurability: This property refers to the possibility to dynamically change the values of a_r and b_k in (31). Passive structures are incapable of this functions. Several solutions have been proposed to overcome this limitation including the use of optical amplifiers [48]-[50], modulators [51], [75], fiber gratings and laser arrays.

g) Tunability: This property refers to the possibility to dynamically change the position of filter resonances or notches. To provide tunability it is necessary to alter the value of the sampling period T. Solutions that include the use of switched fiber delay lines [55], high dispersion fibers [71] and fiber Bragg [76] gratings have been proposed. In the last two options a tunable source is required.

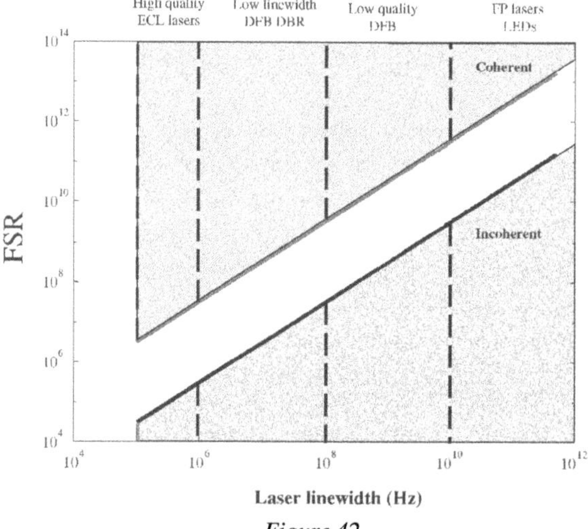

Figure 42

To date none of the two main approaches that have been followed by most of the research groups throughout the world has been able to address successfully all the above limitations. These main approaches are:

a) Implementation of OPRFS using fiber coils as delay lines, single source illumination and signal tapping combination and weighting by means of discrete fiber or integrated optics components. We will refer to these as FDLFs (Fiber Delay Line Filters)
b) Implementation of OPRFS using fiber gratings as delay lines and/or weighting elements in conjuntion with single or multiple tunable source arrays. We will refer to these as FGDLFs (Fiber Grating Delay Line Filters).

The main activities in FDLFs have been carried during the period 1980-1994, while those of FDLFs have been relevant since 1994 and extend to the present time. In the following we briefly outline the work on FDLFs, focusing more effort in describing the main results obtained in FGDLFs.

2.4.3. Fiber Delay Line Filters

Intense research work on passive FDLFs was carried during the period between 1980 and 1990, including the development of special purpose components such as variable 2x2 couplers, star couplers etc. and the experimental demonstration of simple passive structures performing basic signal processing operations such as transversal and notch filtering, correlation, data storage etc [45]-[47]. Both noise and signal analysis methods were developed for these structures. The advent of optical amplifiers (OAs) opened the possibility of overcoming the limitations imposed by the static nature of passive structures both in terms of reconfiguration and loss compensation. The inclusion of OAs provided the structures with enhanced flexibility as sample weighting could be altered. Several contributions proved these advantages both theoretically as well as experimentally [48]-[64]. Synthesis methods where developed [66],[67] and filters with negative coefficients resulting from the application of some of the above methods were experimentally demonstrated [64]. By the middle of the 90s it was thus clear that the main restriction faced by FDLFs was that related to resonance tunability. In 1994 a solution for this problem was proposed by implementing a tunable delay combining of a tunable source and high dispersion fiber delay lines [70]. The concept was extended by Frankel and Esman [71] who demostrated the implementation of a transversal filter with continously tunable unit time delays consisting of 8 taps with progressively longer segments of high dispersion fiber, but completed with dispersion-shifted fiber to nominally identical overall lengths. The time delay tuning at each tap was achieved by tuning the wavelength of the optical carrier. A Q=30 bandpass Rf filter tunable over 1 octave was demonstrated.

2.4.4. Fiber Grating Delay Line Filters

In 1994 Ball and co-workers [76] proposed the combined use of an externally modulated tunable fiber laser and a 6 element wavelength multiplexed uniform fiber Bragg grating array with the grating spacing set to yield the desired delay to implement a programmable delay line capable of generating 50 ns true time delay in discrete 10 ns intervals. This fueled the research toward the application of the recently available fiber gratings in the implementation of OPRFSs. Uniform Fiber Bragg gratings can be employed both as weighting and delay line elements, since their reflectivity changes with the signal wavelengths and their Bragg wavelength is adjustable. When used as delay elements they can only provide discrete changes in the value of the sampling period. This limitation is removed by the use of linearly chirped gratings.

A simple discretely tunable notch filter was demonstrated [79] using a michelson interferometer with two uniform fiber gratings placed in series in one of its arms and subsquently continous tuning was demonstrated replacing the uniform with chirped gratings. In a further step [80], multitap (29 taps) transversal bandpass filter was demonstrated by spectrally slicing a broadband source with wavelength multiplexed Bragg grating arrays equispaced in time. The reflectivity of the gratings in the array were apodised according to a Kaiser window.

High Q filters have also been demonstrated using fiber Bragg gratings both in fixed and tunable operation by sandwiching an active fiber between to gratings, one of which is partially reflecting while the second one is 100% reflecting. The presence of the internal optical amplifiers allow for the existence of a very high number of signal samples which results in resonances with high selectivity (Q=325) [81], [85]. The main drawback is that although the weight of the samples can be reconfigured by changing the amplifier gain, it cannot be done independently sample by sample. Furthermore the length of the active medium severely restricts the filter FSR. A solution has however been proposed [93] for this last inconvenient by placing the former filter in tanden with a Mach-Zehnder lattice filter the period of which is N times larger than that of the former. Q factors over 800 have been demonstrated.

The approaches described above do not usually address dynamic tunability and reconfiguration simultaneously. The dispersive nature of Linearly Chirped Fiber Bragg Gratings (LCFBGs) can be employed to obtain programable RF transversal filters by means of feeding the RF modulated output of an array of sources to the device [87], [88], [98], [100]. The layout of the filter for a specific case of a laser array of 5 elements is shown in figure 41, although in general it is composed of N sources. The advantage of using a laser array as a feeding element to the delay line is twofold: On one hand the wavelengths of the lasers can be independently adjusted. Thus spectrally equispaced signals representing RF signal samples can be feeded to the fiber grating suffering different delays, but keeping constant the incremental delay $\Delta\tau$ between two adjacent wavelengths emitted by the array if the delay line is implemented by means of a linearly chirped fiber grating. This means for instance and referring to figure 43 that the delay between the signals at λ_1 and λ_2, λ_3, λ_4, $\lambda_{5,}$ .....λ_N is repectively $\Delta\tau$, $2\Delta\tau$, $3\Delta\tau$, $4\Delta\tau$ and $(N-1)\Delta\tau$. Hence the configuration can act as a transversal filter, where the basic delay is given by $\Delta\tau$: Futhermore $\Delta\tau$ can be changed by proper variation of the laser central wavelengths in the array. Thus these structure provides the potential for implementing tunable RF filters.

The second advantage stems from the fact that the output powers of the lasers can be adjusted independently and at high speed. This means that the time response of the filter can be apodised or in other words, temporal windowing can be easily implemented and therefore the filter transfer function can be reconfigured at high speed. The RF transfer functions for the case of AM and SSB RF modulation are given respectively by [100]:

$$\left|H_{RF}(\Omega)\right| = R\cos\left(\frac{\beta_2\Omega^2}{2}\right)\left|\sum_{k=1}^{N} P_k \cos^2\theta_{km} e^{-j[\Omega(k-1)\Delta\tau]}\right| \qquad (34)$$

And

$$\left|H_{RF}(\Omega)\right| = R\left|\sum_{k=1}^{N} P_k \cos^2\theta_{km} e^{-j[\Omega(k-1)\Delta\tau]}\right| \qquad (35)$$

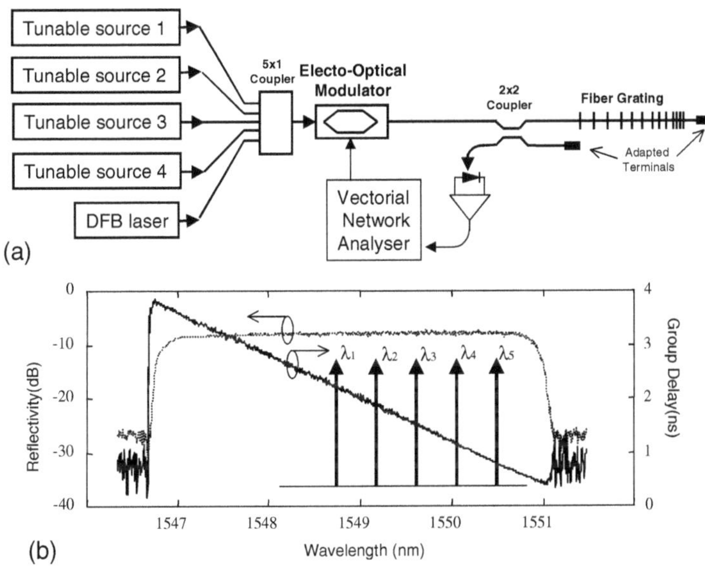

Figure 43 : Layout of an RF transversal filter composed of a LCFBG feeded by a laser array [59].

In the above expressions, P_k represents the output power from the kth source of the array, R is the receiver responsivity, Ω the RF frequency, $cos^2\theta_{km}$, represents the possible mismatch between the linearly polarised output from the kth laser of the array and the optimum input polarisation to the external modulator, and finally $\Delta\tau = \beta_2\Delta w$ represents the incremental differential delay experienced by two adjacent carriers of the laser array with β_2 representing the group delay slope of the linearly chirped grating. Note that the carrier supression effect that appears as a factor in the spectral response under AM modulation, disappears when SSB modulation is employed.

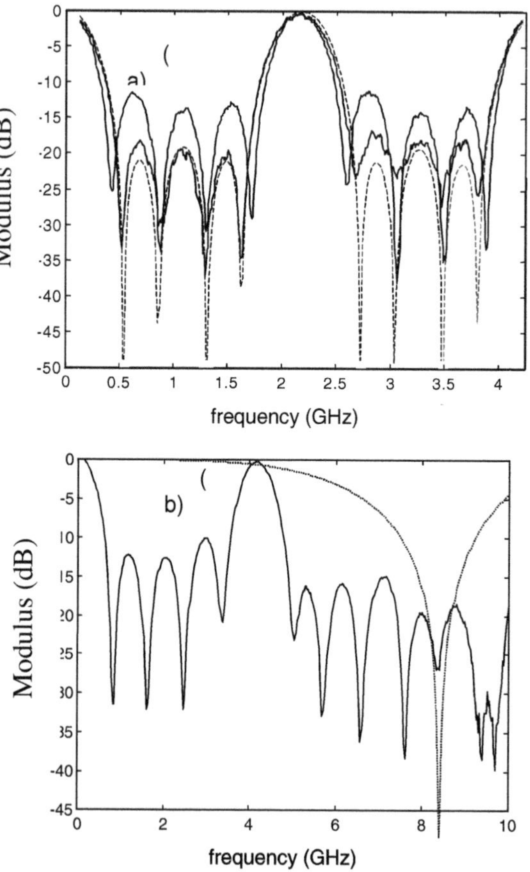

Figure 44 : a Filter reconfigurability demonstration. b Filter tunability demonstration [100].

We have experimentally suceeded in the demonstration of both tunability and reconfigurability. For instance, figures 42a and 42b show the results when the samples of the 5 stage uniform filter where weighted using a truncated Gaussian window. Trace (a) in figure 44a shows for the sake of reference the spectrum corresponding to the uniform filter (i.e unapodised) where the normalised output powers from the lasers in the array is [1 1 1 1 1]. Trace (b) corresponds to a 5 stage Gaussian windowed filter where the normalised output powers from the lasers in the array is given by [0.46 0.81 1 0.81 0.46]. For the sake of comparison, the broken trace shows the theoretical results as expected from (31). Figure 44b demonstrates the resonance tunability.

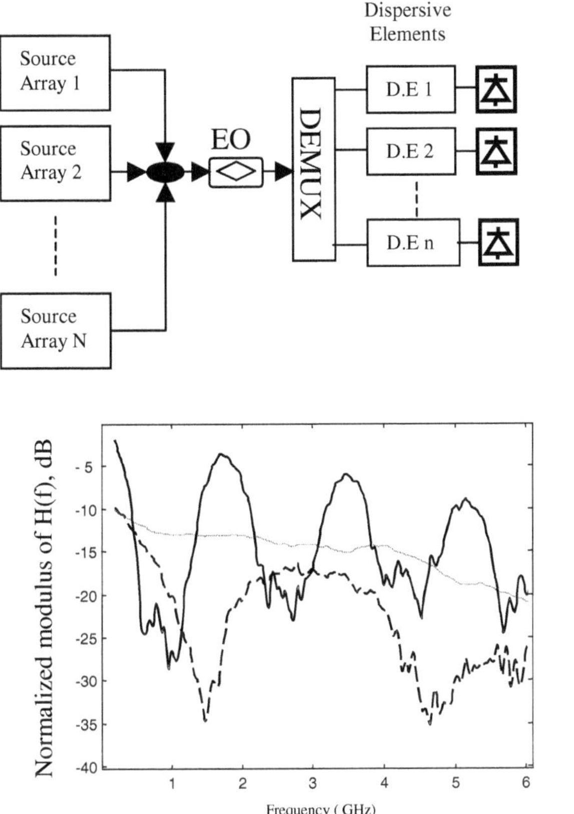

Figure 45 : Main results from [102]

An additional advantage of employing laser arrays is the possibility of exploiting WDM techniques for parallel signal processing [101] and to provide both a a large number of taps and arbitrary coefficients [92]. In the first case, the possibility of implementing a bank of parallel transversal filters is feasible by extending the concept of a single fiber-optic RF transversal filter based on multiple linearly chirped fiber Bragg gratings and dispersive elements into the implementation of a bank of transversal filters, by means of utilising, wavelength division multiplexing techniques. Using this technique allows for the simultaneous processing of a single RF signal by various filters. Figures 45a and 45b represent the concept and the transfer functions resulting from the implementation of a bank of two filters (bandpass and notch).

In the second case, WDM combined with the regularity characteristics of the sampling process allows for the implementation of a high number of taps by exploiting the concept of spectral mapping and partitioning.

2.4.5. Summary and Conclusions

In this section we have revised the fundamental concepts and limitations of optical filters for the processing of Rf signals. The main achievements both as far as theoretical as well as practical and experimental work carried by different research groups working on the field have been reviewed considering the two main approaches followed in their implementation; those based on the use fiber coils as delay lines, single source illumination and signal tapping combination and weighting by means of discrete fiber or integrated optics components, and those based on fiber gratings as delay lines and/or weighting elements in conjunction with single or multiple tunable sources.

2.4.6. Acknowledgements

The authors wish to acknowledge the financial support through Spanish government CICYT projects TIC98-0346 and TEL99-0437.

2.5. Signal Processing Methods for Subcarrier Optical Transmission

T. Berceli
Technical University of Budapest, György ter 3, Goldmann 1111, Budapest, HUNGARY
berceli@mht.bme.hu

Abstract

The subcarrier transmission of microwave signals is a perspective method for optical fiber links. Several approaches are presented covering the generation and reception of subcarrier optical signals. The transmission problems like linearity, distortion, chromatic dispersion, etc. are also discussed. The optical-microwave mixing process is utilized for an improved reception.

2.5.1. Introduction

There is an increasing demand for better and more communication services all over the world. In this progress the main transmission medium is the optical fiber offering an enormous bandwidth along with low attenuation and light weight. However, this huge transmission capacity cannot be fruitfully exploited due to the capacity limitations of the photonic and electronic components of the fiber optic link [112-115].

For increasing the transmission capacity two approaches are at our disposal. One method applies higher and higher bit rates like 2.5-10-40 Gbit/s to provide increased capacity. This method is well applicable for the backbones of the communication networks which are mainly used as point-to-point connections. However, this approach is not well suited to distribution/collection systems or local area networks where point-to-multipoint or multipoint-to-multipoint connections are required.

In these applications the problem arises from the fact that a high speed system requires high bit rate components everywhere in the system what is a big drawback if the number of terminals is high. Another disadvantage is that every signal processing unit of the system should have the highest forseen capacity in the time of installation. All of these units have to be replaced if the capacity has to be increased or parallel transmission channels have to be installed utilizing the wavelength division multiplexing (WDM) principle. However for the WDM system more sophisticated photonic components (tunable lasers, optical filters, wavelength converters, selective receivers, etc.) are to be applied making the system more expensive.

In the other method the multiplexing is performed in the electronic region instead of the optical region. Therefore it applies subcarriers with different frequencies and low bit rate (2, 8, 34 or 140 Mbit/s) channels on each subcarrier. The subcarrier multiplexed (SCM) system offers many advantages: inserting or dropping a channel is easy, therefore point-to-multipoint or multipoint-to-multipoint connections can be established using inexpensive electronic components. The capacity of the system is enchanced by introducing new subcarrier frequencies. The bit rate is relatively low therefore the system is well suited to the distribution and collection of information and for local area networks (LANs). Multiplexing the channels is accomplished in the electronic region what is less expensive than optical multiplexing.

In this paper the SCM optical system is discussed in detail. Its main application fields are :

- distribution of entertainment programs (TV, radio, music, etc.),
- broadcasting public information (teleeducation, journals, announcements, traffic timetables, weather forecasts, etc.),
- collection of data (telemetering, telecontrol, etc.),
- multipoint-to-multipoint communications (voice, data, picture etc.),
- cellular mobile networks,

- indoor communications,
- integrated services systems.

For these applications different system architectures are preferable, and different realization problems are encountered which are overviewed in detail.

2.5.2. Distribution System

In a distribution system a center station distributes information for the terminals of many subscribers. The reception can be full or selective. Therefore the different channels are put on different subcarriers and this way the subscribers can receive any or all of the channels.

The multi-subcarrier transmitter has to meet some special requirements like: high linearity, broad bandwidth, low noise, high dynamic range and low distortion. As intensity modulation is applied the wavelength stability is not critical because at the reception only the intensity of the optical beam is detected.

2.5.2.1. Linearity

The linearity of a direct modulated laser diode is usually not high enough for a multi-carrier modulation. To improve the modulation linearity there are two main methods: the laser diode inner construction can be properly designed to get a higher linearity or the nonlinearity of the laser can be equalized utilizing different compensation approaches.

As the availability of high linearity laser diodes is limited, in many cases a compensation method is to be applied. The best results are obtained by the active matching techniques which offers an adaptive behaviour as well [116].

In figure 46 the modulation characteristics of a direct modulated laser diode are presented for two cases. The upper curve of the figure shows the modulation characteristics using passive matching and its lower curve presents the result of the linearization applying the active matching method. As seen a very high linearity is achieved in a wide modulation range.

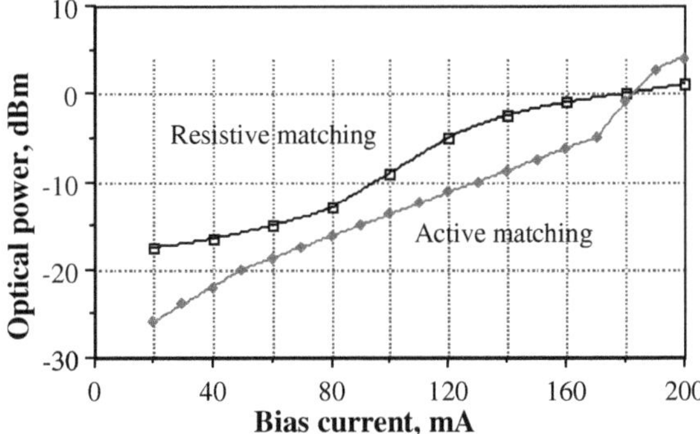

Figure 46 : The optical output power versus driving current.

2.5.2.2. Distortion

The linearity is also checked by measuring the harmonic distortion. Figure 47 shows the fundamental, second and third harmonics as a function of the modulation signal power. The achieved linearity is high enough, if the power of the modulation signal is below a certain level. This power level is dependent on the specific application.

For a higher number of subcarriers a higher linearity is required to keep the third order intermodulation product below the specification. The third order intermodulation distortion has been tested using 4 subcarriers to modulate the intensity of Fabry-Perot (FP) and distributed feedback (DFB) lasers.

The transmission band of the system is around 1 GHz. However, the band is rather narrow. Four channels are transmitted with 1 MHz separations. Therefore, the harmonics, sum and difference frequency mixing products are out of the band. Only the third order intermodulation products are in the band.

The results are presented in table 2. The measurement is performed with and without an optical isolator.

There is a very interesting phenomena: the intermodulation is dependent on the optical reflection. As seen the intermodulation product is reduced when an optical isolator is inserted at the output of the laser. To achieve a very low intermodulation product the optical reflections should be very low, below –60 dB. Another curiosity is that the intermodulation is practically the same for Fabry-Perot and distributed feedback (DFB) lasers.

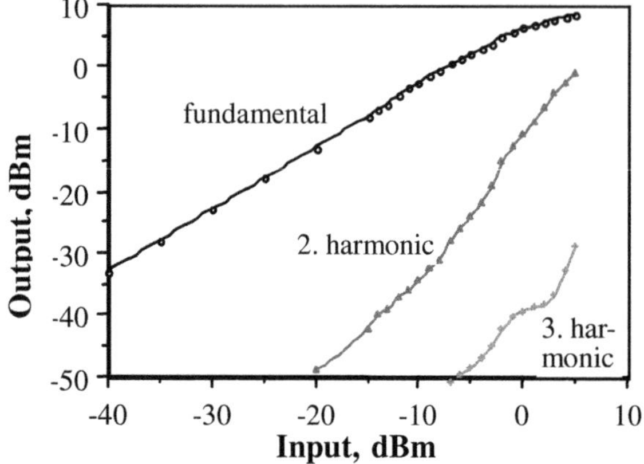

Figure 47 : The output level of the fundamental, second and third harmonic frequency signals as functions of the fundamental frequency input level.

Laser type	DFB	Fabry-Perot
With isolator	79 dBc	79 dBc
Without isolator	71 dBc	71 dBc

Table 2 : Third order intermodulation products.

In case of a large number of subcarriers another test method is used: the matrix generator. This generator provides a large number of signals except two: one close to the lower edge of the total band and one close to the upper edge of the total band. This composite signal is used to modulate the laser. In the receiver a special filter is applied which has stop band for every signals of the matrix generator and two pass-bands where there are no modulating signals.

The matrix generator and the reception bands are shown in figure 48. For the measurements first the matrix generator is switched off. Then the receiver gets only noise in the two pass-bands. In the next step the matrix generator is switched on. In this case the total distortion is measured in the pass-bands which is the result of every intermodulation of all signals. During the test the total modulation power is varied, and thus the optical modulation depth is adjusted considering the allowable total distortion.

The distortion products are dependent on the linearity of the system and on the frequency allocation of the channels. Thus the distortion can be improved by a higher linearity and/or by a better frequency allocation.

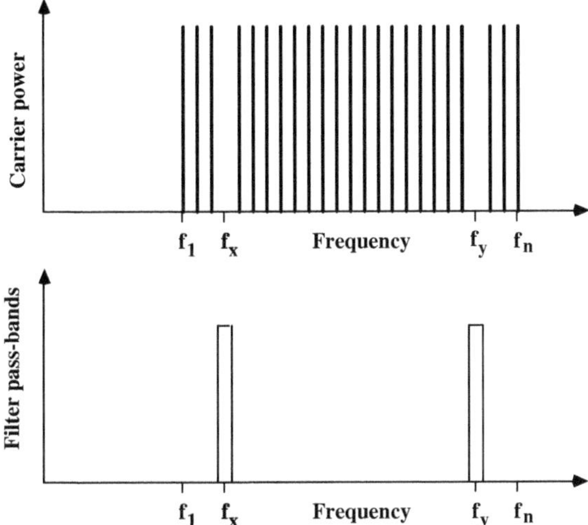

Figure 48 : The matrix generator spectrum (upper figure), and the reception bands (lower figure).

2.5.3. Multipoint-to-Multipoint System

In a multipoint-to-multipoint system each terminal wants to communicate with any other terminal in the network or via a gate-way with an external terminal. The principle of this type of system is shown in figure 49. The optical transmitters and receivers of the terminals are connected by a passive optical network (PON) composed of optical hybrids or couplers. The network contains a control unit to establish the requested connection. Each optical transmitter has its own subcarrier frequency and in the optical receivers the proper subcarrier is selected to establish the connection. A common channel is used for signaling and control.

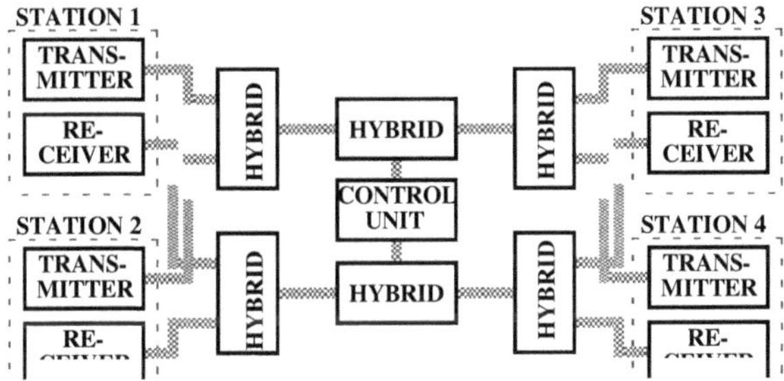

Figure 49 : Optical local area network.

Each optical terminal can serve as a traffic concentrator. In that case a TDMA (time division multiple access) or CDMA method is used to connect several subscribers to the optical terminal.

2.5.3.1. Combined Optical- Wireless System

A simple and economical system can be established applying subcarrier multiplexing (SCM) and time division multiplexing (TDM) simultaneously.

In that case the center station is only a controlling and signaling unit in the network and switching is actually done by the selection of the subcarrier frequency and the time slot.

This system offers optimum solution for the local area network of an office building. Wireless communications is used in the office rooms serving the different terminals. Each large room or a group of neighbouring small rooms has its own wireless network and thus the transmitter power can be small enough because there is no need to establish connections by radio waves through many walls. The individual wireless networks use TDMA or CDMA (code division multiple access) techniques. They are interconnected via optical fibers applying the subcarrier multiplexing method.

The combined optical- wireless communications offers several further benefits for the customers. The optical networking is expensive if it connects all the terminals. However, in this application its cost is low because it is applied only in the highways of the network. On the other hand the wireless section becomes more economic due to the small area of a picocell. The system is very flexible, it can easily be extended to serve more terminals including mobiles as well.

The new architecture takes advantage of the very wide transmission band offered by optical fibers. Therefore a huge number of subcarriers can be accommodated providing a high traffic capacity. The subcarrier multiplexed transmission also offers a high flexibility for changing traffic conditions.

The block diagram of the new system is presented in figure 50. Each large office room or a group of adjacent small rooms has a specific carrier frequency for the terminal radio transmitters while the terminal radio receivers are tunable.

The modulation methods can be FSK (frequency shift keying), BPSK (binary phase shift keying), QPSK (quadrature phase shift keying), 16 QAM (16 state quadrature amplitude modulation), etc. depending on the wireless links because the optical part of the system is transparent. The transmitter signals of the terminals are collected by the receiver of the radio node and they are transmitted via the optical fiber. This way the radio carriers are used as subcarriers in the optical region.

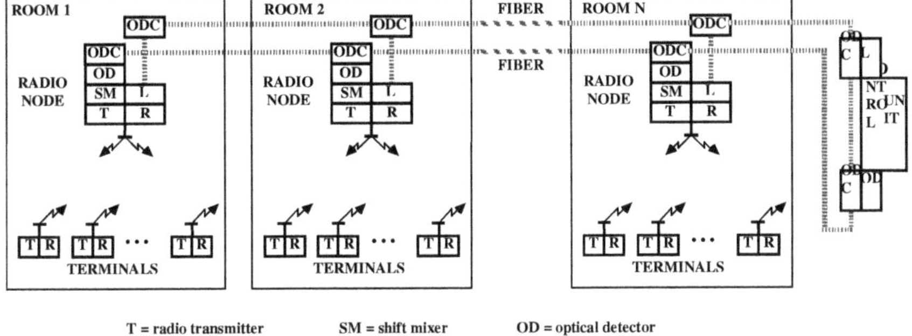

Figure 50 : Block diagram of the combined system.

The optical fiber operates as a one-way bus using the principle of collection and distribution. When all the subcarriers are transposed into the optical region the fiber is routed back to the radio nodes where all of the channels are converted back into the radio frequency region. However, before radiation their frequencies should be shifted to use a different frequency band for the radio up-link and down-link. The route from the terminal to the radio node is called up-link, and the route from the radio node to the terminal is called down-link.

The number of the terminals is determined basically by the available radio frequency bandwidth. Assuming 2 Mbit/s bit rate for every subcarrier and FSK (frequency shift keying) modulation of the radio waves, 90 subcarriers can be accommodated in a 200 MHz frequency band keeping 10 % bandwidth for the separation of the channels. That means the number of the simultaneously operated simplex channels is 2700 if their bit rate is 64 kbit/s. This way the network can provide a high quality service for at least 10000 terminals with 64 kbit/s bit rate assuming 13.5 % simultaneous traffic (or availability) in the network.

Naturally, some channels are used for connections to the public switched network and to other local area networks. In many cases the bit rate can be smaller resulting in a higher number of simultaneously operating channels. Utilizing the total available bandwidth, i.e. 2 GHz the number of channels can be almost 10 times higher. This very high capacity is usually not needed, however, it can be utilized for broadband communications services.

2.5.4. Cellular Mobile Networks Utilizing Fiber Optic Links

In a cellular mobile network the fiber connection can be applied in two different ways :

- the information channels are transmitted over the fiber and the carrier frequency is generated locally at the radio base stations,
- the information channels and the carrier frequencies are transmitted together over the fiber to the radio base stations (radio over fiber system).

In the first case the optical transmission has less troubles however, the carrier frequencies are not synchronized. In the second case the radio base station has less functions, the carrier frequencies are synchronous, but the fiber transmission is more complicated. Nevertheless, the second method offers many advantages mainly in the millimeter wave region. Here we discuss this approach in more detail. There are several methods for a radio over fiber system.

In one approach two lasers are used with off-set frequency stabilization. Their frequency difference is kept constant utilizing a millimeter wave signal as a reference. For the stabilization one of the lasers is tuned by a phase locked loop. This way the frequency difference between the two laser beams is in the millimeter wave region. These two beams are transmitted via a fiber to the radio base stations where the millimeter wave signal is regained by optical detection.

In another approach a single laser operating in two modes is applied. The frequency difference between the two laser modes is kept constant by injection locking techniques utilizing a millimeter wave signal.

In a third approach a single mode laser beam is modulated by the millimeter wave signal. This method seems to be simpler than the previous two ones, however, it needs a high frequency external modulator what is rather expensive. A further problem arises in the transmission of the optical wave carrying a millimeter wave signal. Due to the chromatic dispersion of the fiber transmission minima are obtained for longer fiber lengths. This problem may be overcome by the use of several modulation techniques at the transmitter end which effectively mitigate the effect of the fiber chromatic dispersion, such as single-side-band modulation [117,118], minimum transmission bias or maximum transmission bias of the MZ modulator [119]. However, the single-side-band modulation is more complex while at the minimum or maximum transmission bias the modulation linearity is poor.

For the optical generation of a stable, low noise signal based on the first approach DFB lasers are used which have a low relative intensity noise (RIN) and can be tuned to accomplish the off-set frequency stabilization. For the second approach a two-mode laser with a low RIN is needed along with a high mode purity and stability. In the third approach a high frequency external modulator is necessary.

Beside these requirements the millimeter wave signal – used as a reference of the phase locked loop for off-set frequency stabilization in the first approach, or for injection locking of the two-mode laser in the second approach, or for external modulation of the single-mode laser in the third approach - has to be stable and of very low noise as well. Further at the reception side a high-speed photo-diode is to be applied [120]. Therefore, these methods are very complex and expensive.

Nevertheless, there is an increasing need for higher frequencies and thus carrier frequencies in the millimeter wave band are to be used in cellular mobile communication systems as well. The optical transmission of millimeter waves faces many obstacles, thus the optical transmission of signals is more and more lossy when the frequency is increased.

2.5.4.1. Novel Optical Signal Generation Methods

The common basic principle of these methods is that a low frequency reference signal is transmitted to the radio base stations instead of the millimeter wave signal and utilizing this low frequency reference the millimeter wave signal is generated at the radio base stations [121,122].

In this approach a single mode laser is intensity modulated by the sub-harmonic reference signal. The detected signal is used to stabilize the VCO frequency of the radio base station by a phase locked loop [123,124]. Beside the reference signal subcarriers are used for the optical transmission of the information channels. The block diagram of the system is shown in figure 51 as it is applied in a cellular mobile network.

The main task is to ensure the low noise property of the millimeter wave signal. Comparing the well-known methods and the present method it is obvious that the electronic system part producing the millimeter wave signal provides the same stability and noise performance when it is applied either in the optical transmitter or in the optical receiver. Therefore it is very important to use a low noise, high stability quartz crystal oscillator as the basic source for the reference signal in every case.

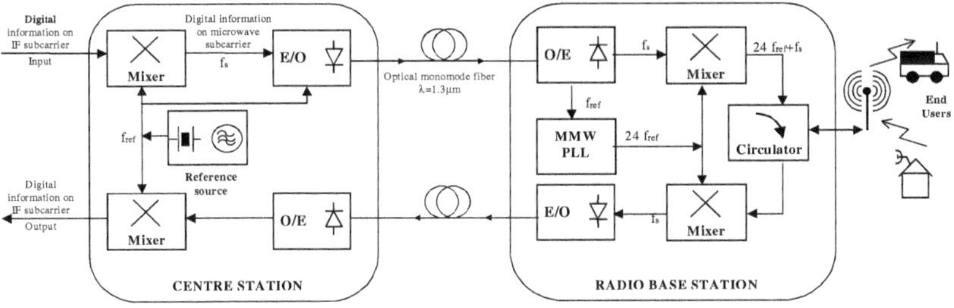

Figure 51 : The block diagram of the optically fed cellular radio system.

2.5.4.2. Phase Jitter Measurements

An IQ demodulator operating at 885 MHz was used for the phase jitter measurements. The millimeter wave signal was down-converted into this band by a mixer. A measurement setup has been developed to eliminate the phase jitters originating from the generators used for down-converting the millimeter wave signal. The I and Q signals were displayed on an oscilloscope and the phase jitter distribution was recorded and calculated by a computer.

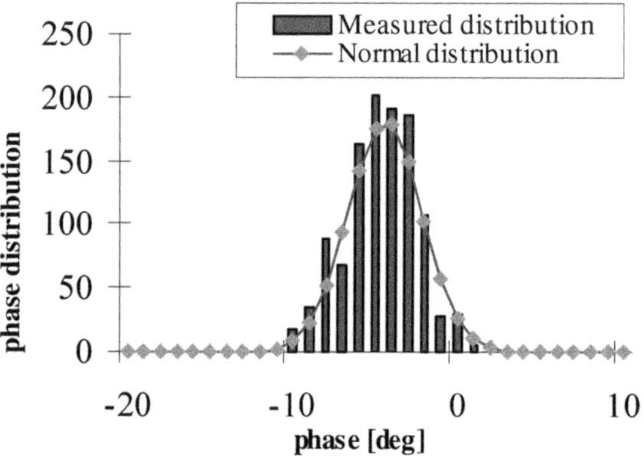

Figure 52 : Phase distribution of I/Q signal.

The distribution has 2.25° standard deviation. As the reference signal has a phase jitter of ≈0.09° and the multiplication number of the frequency is 24, the noise contribution of the system is negligible. Figure 52 shows the phase histogram compared to the Gaussian distribution.

2.5.4.3. Bit Error Rate Measurements

In these tests the bit error rate (BER) of the whole system was measured in case of different modulations. The system performance has been evaluated with changing signal to noise ratio (S/N) of the radio frequency signal (see figure 53). The curve for MODEM refers to the back-to-back MODEM measurement. The curve of ELECTRICAL test gives the data for the case when direct electrical connection was between the center station and the radio base station. Finally, the curve of OPTICAL transmission shows a very small degradation compared to the direct ELECTRICAL connection.

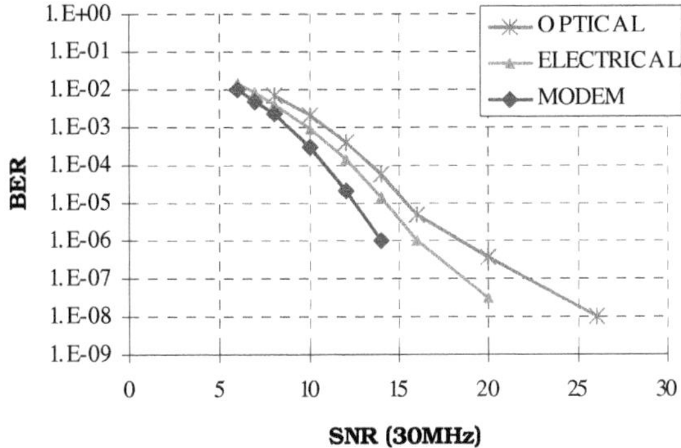

Figure 53 : BER measurements vs. S/N ratio with QPSK modulation.

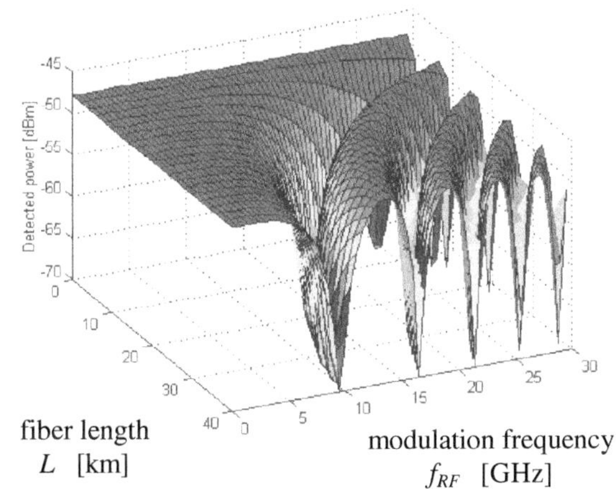

Figure 54 :Detected power level of MW signal transmitted optically via a dispersive fiber. (Linear modulator bias of γ=0.5, α=0.25, D=17ps/km/nm, R_{PD}=0.35 A/W).

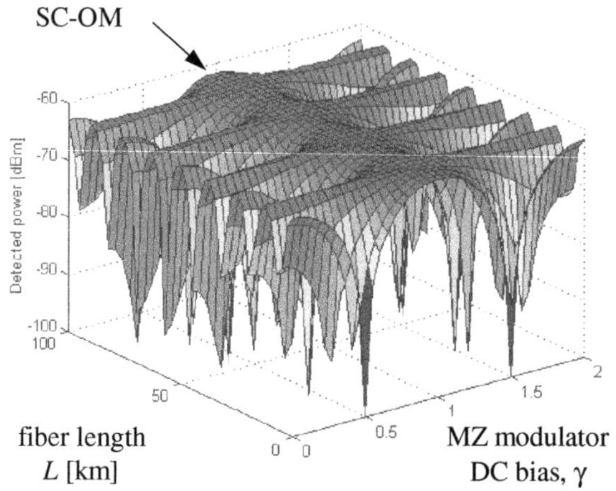

Figure 55 : Detected 2nd harmonic level after transmission on dispersive fiber, (f_{RF} *=12 GHz,* α*=0.25,* D=*17ps/km/nm,* R_{PD}*=0.35 A/W).*

2.5.5. Chromatic Dispersion Effects

Based on a coherent model of the fiber-optical link a general simulation tool has been developed for studying different optical link architectures. Figure 54 shows the detected power of the optically transmitted signal using externally modulated optical link. The effect of chromatic dispersion is clearly seen resulting in periodic rejections as a function of fiber length L and modulation frequency f_{RF}.

To avoid chromatic dispersion, several proposals have been reported [125,126]. Figure 55 plots suppressed carrier optical modulation (SC-OM) achieved by a normalized modulator bias of $\gamma = V_{DC}/V_{\pi} = 1$. At this special bias the second harmonic of the modulation signal is generated. Advantageously, SC-OM is unaffected by chromatic dispersion.

Single sideband optical modulation (SSB-OM) offers a further perspective solution of dispersion-free optical transmission of MW/MMW signals. One SSB-OM method filters out optically one of the sidebands [127].

2.5.6. Selection of the Subcarrier Signals

The reception of the modulated subcarrier signals are performed using a photo-detector and thus all of the subcarriers are regained. Then the wanted subcarrier can be obtained in two different ways :

- all of the subcarriers are separated by a series of fixed frequency filters, the wanted subcarrier is selected by a switch and it is down-converted to a fixed intermediate frequency,

- a specific subcarrier is selected by a tuned filter and it is down-converted to a fixed intermediate frequency. However, keeping the performance of a filter unchanged when it is tuned is very difficult because several resonators and their couplings are controlled.

The block diagrams of these receivers are shown in figure 56 and 57. The calling and controlling signals are received in the signaling channel which is transmitted via a fixed frequency subcarrier, and therefore a fixed filter is used for the signaling channel.

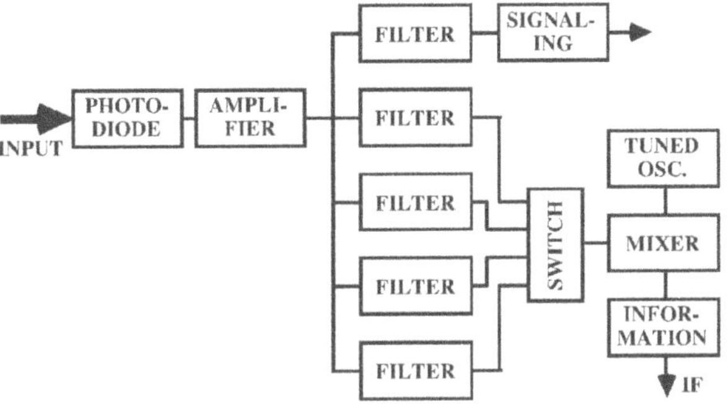

Figure 56 : Selection by series of filters.

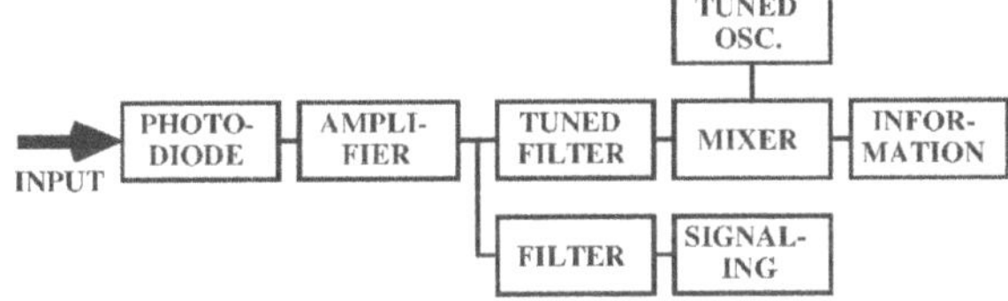

Figure 57 : Selection by a tuned filter

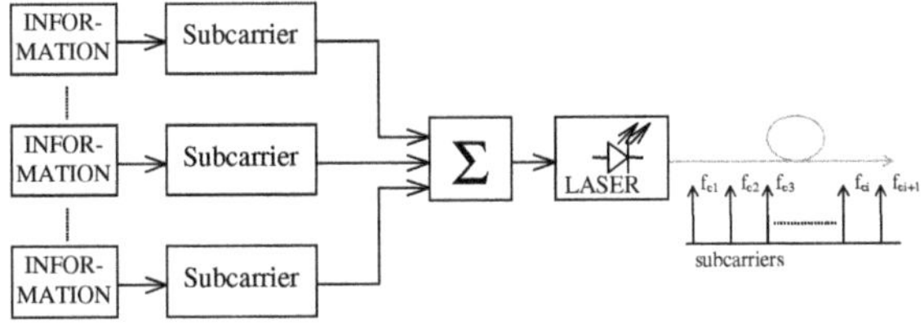

Figure 58 : Subcarrier signal transmission (transmitter).

Mixing of optical waves and microwaves offers new perspectives for the reception of subcarrier multiplexed optical signals [128-131]. In the subcarrier type optical communications each transmitter has its own subcarrier frequency as it is shown in figure 58. The transmission capacity of the network can be increased by applying new subcarriers, and thus the digital processing rate per subcarrier remains fixed.

Utilizing the optical-microwave double mixing method in a receiver of a subcarrier multiplexed signal transmission (figure 59) a simple receiver structure and channel selection can be achieved. By tuning the local oscillator the mixing product is obtained always at the same intermediate frequency. That is a big advantage because tuning an oscillator is much easier than tuning a filter.

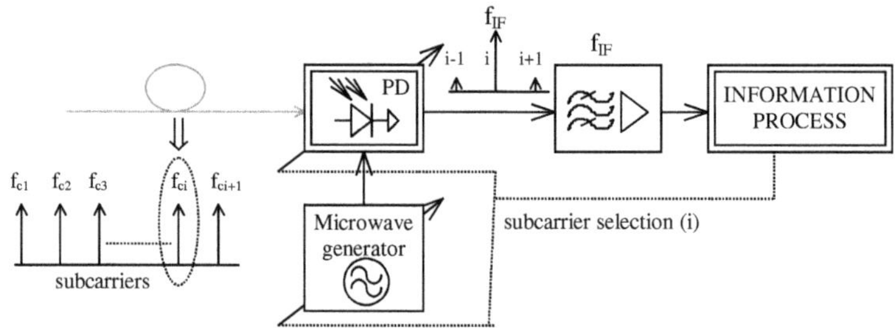

Figure 59 : Subcarrier signal transmission (receiver with optical-microwave mixing).

2.5.7. Optical-Microwave Mixing by a Photo-Diode

2.5.7.1. Photo-Detection Investigation

Before the mixing experiments, the optical detection was characterized. The investigated photodiode was an 1A358 type CATV PIN photo-diode. The detection response was used as a reference. In this arrangement the dynamic behavior of the photodiode (PD) was investigated in its detection mode of operation. The intensity modulated optical signal is generated by a HP 83424A 1550 nm DFB laser source and a HP 83422A external Mach-Zehnder optical modulator, with a typical optical modulation depth (OMD) of 25 %.

The detected intensity modulated optical signal was measured by a spectrum analyzer. The response of the diode was flat up to 3 GHz and the detected signal level at high reverse bias voltage was ≈ -36 dBm at 200 MHz.

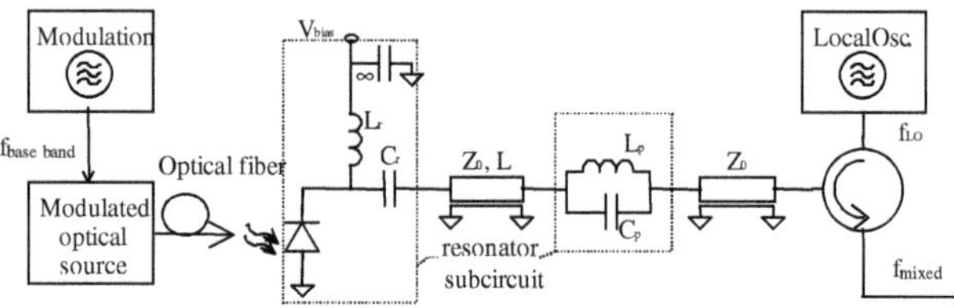

Figure 60 : Mixing measurement setup with electrical LC resonators.

2.5.7.2. Optical-Microwave Double Mixer

The developed new optical-microwave mixing setup is shown in figure 60. The local oscillator (LO) signal is fed to the photodiode via a circulator [132]. A wideband circulator is used to separate the input LO signal and the output mixing product.

The high impedance in the baseband is generated by a resonator circuit constructed by inductors and capacitors (L_r, C_r, L_p, C_p). The series resonator inductance is L_r. The parallel resonator (L_p, C_p) in the series branch constitutes a branching filter and it is tuned in the base-band. In a narrow band determined by the quality factor of the resonator it shows a high input impedance thus it separates the photodiode from the Z_0 system impedance. This high impedance and the remaining resonator elements with the photodiode capacitance can produce the desired high impedance. With proper values of the resonators the attenuation of the embedding circuit in the LO and up-converted signal band can be negligible.

This peak overcomes the normal detected level by more than 30 dB.

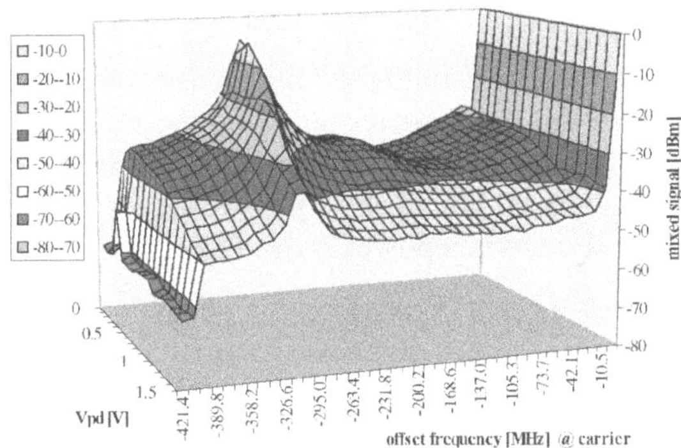

Figure 61: Lower sideband of the upconverted signal using series and parallel resonance., P_{LO} = 5 dBm (the maximum is –2.89 dBm @ 289 MHz).

The measured lower sideband of the upconverted signal is shown in figure 61. The used circulator has a bandwidth of 2-4 GHz. The optical carrier was modulated by a f_{mod} = 10 - 410 MHz signal. The local oscillator signal has a 5 dBm power at 2.5 GHz frequency. The reverse bias of the photodiode was varied 0 - 1.5 V in 0.1 V steps. The horizontal axis of the surface plot in figure 61 is the frequency offset from the 2.5 GHz carrier, the perpendicular axis shows the reverse bias of the photodiode and the vertical scale is the power level of the up-converted signal. The up-converted spectrum has a –2.89 dBm peak at the modulation frequency of 289 MHz with about 0.2V reverse bias voltage.The noise properties of the optical-microwave mixing effect were also investigated. The theoretical investigation shows that the equivalent input noise density of the mixer is 0.15 $pA/Hz^{1/2}$. The measured output noise spectral density of the mixer is -141.7 dBm/Hz at 270 μW illumination without modulation.

2.5.8. Nonlinear Distortion

In case of multi-carrier transmission the nonlinearity of the system causes intermodulation distortion. The optical link consists of a laser transmitter, a connecting fiber and an optical receiver. In this system the main nonlinear components are the laser driver, the laser itself and the transimpedance amplifier in the receiver.

When many carriers are present in a specific band the distortion in a channel comes simultaneously from many other channels. The harmonic distortion is increasing toward the higher frequencies in the band. A similar trend is with the sum frequency mixing products. However the difference frequency mixing products are increasing toward the lower frequencies of the band. These relations are presented by means of an example when only two signals with different frequencies are applied at the input of the optical system.

The transfer of the link is determined by the relationship between the output current of the receiver and the input current of the transmitter. That relationship is usually nonlinear. For the investigation of the distortion the nonlinearity is described in a power series form:

$$I_{out} = a_1 I_{in} + a_2 I_{in}^2 + a_3 I_{in}^3 + \dots$$

In general the input current contains several components with different frequencies. For simplicity now only two components are considered at the input of the system:

$$I_{in} = I_1 \cos \omega_1 t + I_2 \cos \omega_2 t$$

The output current is obtained by substituting the expression of the input current into the nonlinear relationship for the output current :

$$
\begin{aligned}
I_{out} &= a_1\left[I_1\cos\omega_1 t + I_2\cos\omega_2 t\right] + \\
&+ a_2\left[I_1^2\left(\frac{1}{2}+\frac{1}{2}\cos 2\omega_1 t\right) + I_1 I_2\left(\cos(\omega_1+\omega_2)t + \cos(\omega_1-\omega_2)t\right) + I_2^2\left(\frac{1}{2}+\frac{1}{2}\cos 2\omega_2 t\right)\right] + \\
&+ a_3\left[I_1^3\left(\frac{3}{4}\cos\omega_1 t + \frac{1}{4}\cos 3\omega_1 t\right) + + 3I_1^2 I_2\left(\frac{1}{2}\cdot\cos\omega_2 t + \frac{1}{4}\cos(2\omega_1+\omega_2)t + \frac{1}{4}\cos(2\omega_1 - -\omega_2)t\right) + \right. \\
&\left. 3I_1 I_2^2\left(\frac{1}{2}\cos\omega_1 t + \frac{1}{4}\cos(2\omega_2+\omega_1)t + \frac{1}{4}\cos(2\omega_2-\omega_1)t\right) + I_2^3\left(\frac{3}{4}\cos\omega_2 t + \frac{1}{4}\cos 3\omega_2 t\right)\right]
\end{aligned}
$$

As seen due to the nonlinearity several mixing products appear at the output of the fiber link even in this two-frequency case.

Based on this simple example the distortion is composed of the harmonic frequencies as well as the sum and difference frequency mixing products. There are two mixing products which attract a special interest. These are composed from the second harmonic frequency of one signal deducted the frequency of the other signal. That mixing process is called the third order intermodulation.

2.5.9. Conclusions

The subcarrier transmission of microwave signals is a perspective method for optical fiber links. Several approaches hare been presented covering the generation and reception of subcarrier optical signals. The transmission properties like noise, nonlinearity, distortion, chromatic dispersion, etc. have also been discussed including the problems of multi-carrier transmission. The optical-microwave mixing process has been utilized for an improved reception.

Acknowledgements

The authors wish to acknowledge the Commission of the European Union and the Hungarian National Scientific Research Foundation (OTKA T030148, T017295, T026557, F024113) for their continuous support to their research work.

2.6. Photonic Processing of Microwave Signals

D. Dolfi, S. Tonda-Goldstein, J.P. Huignard
Thomson-CSF, Laboratoire Central de Recherches, Domaine de Corbeville, 91404 Orsay, FRANCE
Daniel.Dolfi@thomson-lcr.fr

2.6.1. Introduction

Future active radar systems, based on solid state T/R modules will provide new capabilities, especially in terms of angular coverage, reliability, jamming robustness and complete flexibility. For the future, airborne antennas are planned to be distributed over the entire aircraft while ground based antennas will be remoted from the processing unit. Signal distributions will have to fulfill strict performance criteria including high isolation from both electromagnetic interference and cross-talk between module or subarray feeds with increased instantaneous bandwidths; dramatic reduction in size and weight regarding present fielded radars; and performance compatible with growing requirements such as low phase noise and high dynamic range. Moreover future multifunction phased array antennas will require frequency bandwidth which largely exceed those of existing radars. Such wide instantaneous bandwidths (up to 30 %) lead to the definition of advanced concepts and technologies.

The availability of optoelectronic components operating up to 20 GHz brings attractive perspectives for optical processing of microwave signals. Furthermore optically carried microwave signals can experience large time delays, especially in fiber based systems, providing time-frequency products in the range between 10^2-10^3. Owing to their inherent parallel processing capabilities, optoelectronic architectures are well suited for the implementation in radar and electronic warfare systems of basic functions such as spectrum analysis, time-delay beamforming, adaptive and programmable filtering, correlation and waveform generation.

This paper presents, only as examples, a review of some architectures developed within Thomson-CSF for transit/receive phased-array time-delay beam-forming (TDBF), programmable filtering and waveform generation.

2.6.2. Optically Controlled Phased Array Antennas – 2D Free Space Polarization Switching Approach

The operating principle of our 2D optical architecture[133] is detailed in figure 62. A dual frequency expanded laser beam is the optical carrier of the microwave signal. It travels through a set of SLMs (spatial light modulators) whose number of pixels (pxp) is the number of elements of the antenna. M_O is a parallely aligned liquid crystal SLM. It controls the phase of the microwave signals by changing the relative optical phase of the cross polarized components of the dual frequency beam.

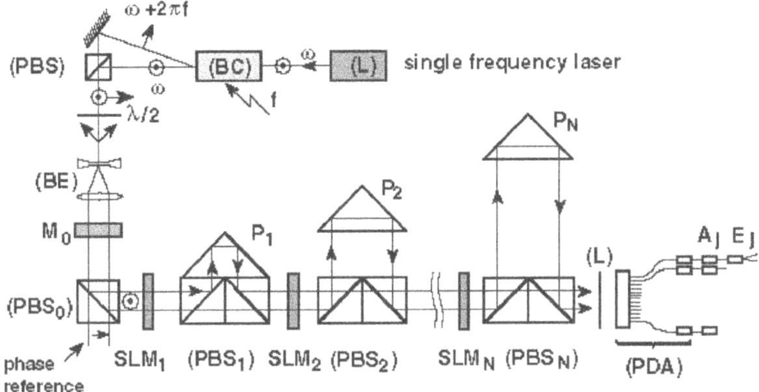

Figure 62 : Time and phase delays for phased array antenna: (BC) frequency shifter Bragg cell. (Aj): microwave amplifiers - (Ej): radiating elements.

At the output of M_O, the now linearly polarized optical carrier intercepts N spatial light modulators SLM_i, polarizing beam splitters PBS_i and prisms P_i. They provide the parallel control of the time delays assigned to the antenna. On each pixel the beam polarization is rotated by 0° or 90°. That is, PBS_i is transparent (and the light beam intercepts the next SLM_{i+1}) or reflective (and the microwave signal is delayed). The channelized beam is then detected by an array of pxp fiber pigtailed photodiodes (PDA). The phase of the microwave signal delivered by each photodiode is determined by the applied voltage on the corresponding pixel of M_O and by the choice of the PBS_i on which the reflections occur. Since time delay values are set according to a geometric progression (τ, 2τ, 4τ,.), the beating signal can be delayed from 0 to $(2^N-1)\tau$ with step τ.

An experimental proof of concept, implemented with 6 SLMs of 4 x 4 pixels, feeds a 16 element antenna with 32 delay values and an analogue control of the phase. The far field pattern of this antenna was characterized, using optical phase and time delay switching, with no beam squint between 2.5 and 3.5 GHz [133].

The receive mode can be performed by reversing the previous architecture. For a radar detection in the same direction as for emission, the optically carried transmitted and received signals, have to travel through the same TDBF network. It permits in-phase addition, over a large frequency bandwidth, of the microwave signals received by the antenna. In this case, these signals are spread over a dynamic range as large as 120 dB which is still difficult to obtain, over large bandwidth, with currently available optoelectronic components.

In order to overcome this limitation, we proposed an original architecture in which a channelized microwave local oscillator (LO), optically carried, is used for mixing with the received microwave signals. There is no more microwave-to-optic conversion of the received signals. The transmit/receive architecture which operates in a way similar to optical phase conjugation is shown on figure 63. Two cross-polarized optical beams at wavelengths λ_1 and λ_2 are modulated by microwave signals at frequencies f_{tr} (signal to be transmitted) and f_{lo} (local oscillator). They travel through a 2D switching network similar to the one of figure 62. The delays experienced by signals at f_{tr} permit the control of the emitted far field pattern. The delays experienced by LOs at f_{lo} are chosen to be complementary to the one experienced by f_{tr}, using a remarkable property of an optical architecture based on polarization switching. When two cross-polarized beams travel along the same channel, their polarizations remain orthogonal and they experience complementary paths. In this case, when the carrier of the frequency f_{tr} is delayed by τ_k along channel k, the cross-polarized carrier of frequency f_{lo} is delayed by τ_{max}-τ_k, where τ_{max} is the maximum available time-delay. On each channel, a dichroïc mirror switches the carriers on two different photodiodes (figure 63) which provides the signal to be emitted and a perfectly matched microwave LO, respectively.

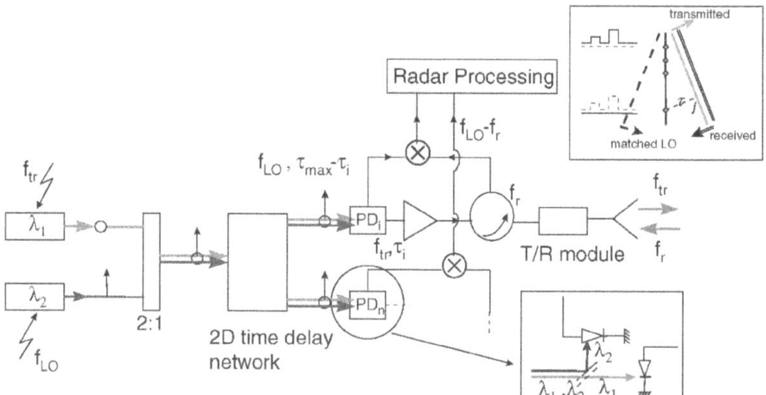

Figure 63: Principle of the Matched Local Oscillator Architecture.

Because of jamming, frequency f_{lo} must stand out of the radar bandwidth Δf. The ideal homodyne processing must be replaced by an heterodyne detection, where $f_{lo} \neq f_r$ (f_r is the frequency of the received signals). It can results, for large Δf, in prohibitive phase errors. In order to minimize these errors, the LO has to be generated in two successive steps. The first step provides a channelized LO with $f_{lo}=f_{tr}$. Then, those signals are mixed with in-phase microwave signals at an intermediate frequency $f_i>\Delta f$. It provides, on each channel k, a LO with phase $\varphi_k^{LO}(t)=2\pi(f_{tr}-f_i)t-2\pi f_{tr}(\tau_{max}-\tau_k)$, which is mixed with the corresponding received signal.

A proof of concept was recently completed [134], which consists of two transmit/receive modules and two delay blocks. On each channel, a laser beam is modulated at frequency f_r or f_{lo}. Phase differences equivalent to 10 ps delays at $f_i = 700$ MHz were measured that permits in-phase addition of the received signals (received signals were generated using time delays that simulate reflection from the target).

2.6.3. Programmable Transversal Filtering

A transversal filter optimizes the detection in a signal x(t) = S(t) + N(t) of a given signal S(t) with duration T, in presence of a stationary noise N(t) or permits jammers rejection from detected signals. These signals are often processed in a sampled form using digital electronic delay lines, allowing a rather large number of sampling points (up to 10^2 - 10^3), but with a frequency bandwidth limited to the low and intermediate frequencies (100 MHz-1 GHz). This frequency bandwidth can be extended up to the 10 GHz region using optoelectronic architectures [135-143], especially in fiber based systems, but with the limitation of a number of sampling points in the range $10\text{-}10^2$.

Therefore, we propose a free space optical architecture of a programmable filter which could provide a large number of samples of about 10^3 and which may process signals over a frequency bandwidth as large as 20 GHz. The operating principle of this programmable filter [144] is shown in figure 64.

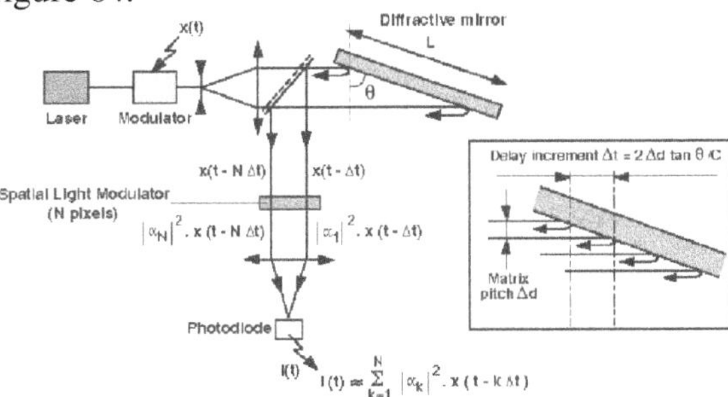

Figure 64 : Programmable transversal optical filter.

A C.W. laser diode is coupled into an integrated optic amplitude modulator, excited by a microwave signal x(t). This provides an optical carrier of this signal x(t). This optical carrier is expanded and reflects off a diffractive mirror, which operates in the Littrow geometry and provides the necessary time delays. This reflected beam, extracted with a beam-splitter, passes through a one dimensional LCSLM of N pixels providing parallel weighting. Finally the channelized beam is focused onto a photodiode. The response time of the proposed programmable filter is mainly determined by the response time of the liquid crystal SLM (in the range 10-100 μs for ferroelectric or chiral smectic LC). In order to meet radar or EW systems requirements in term of adaptive processing speed in the range 10 ns - 10 μs, it would be necessary to use multiple quantum well SLMs or to take advantage of the high resolution of LCSLMs. Furthermore, it is possible to extend the concept to a 2D geometry including a photodiode array in order to provide high speed processing and large time delays, i.e time-frequency products up to 10^3-10^4.

As a proof of concept a simple rejection filter is implemented using a CW fiber laser (40 mW at 1550 nm). It is coupled to an integrated optic Mach-Zehnder modulator, excited by a CW microwave signal. An image of two slits is displayed onto the SLM, providing two out-of-phase signals. The diffractive mirror operates in the double pass geometry, providing a measured maximum delay of 750 ps. In these conditions, using a multimode-fiber pigtailed photodiode, it is possible to measure at 1.3 GHz a 52 dB signal rejection as shown in figure 65.

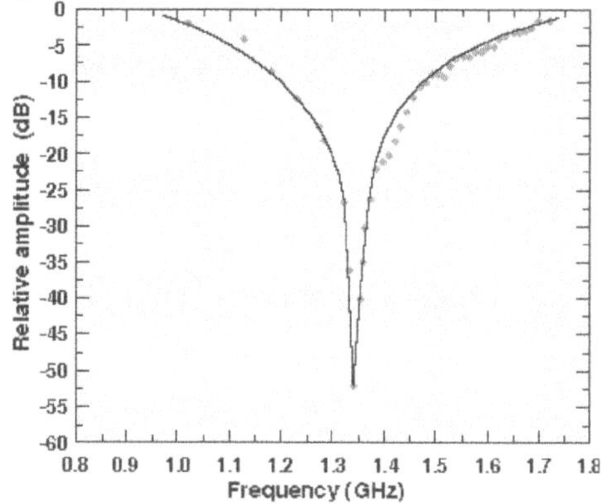

Figure 65 : 52 dB stop-band filter at 1.3 GHz.

2.6.4. Optical Waveform Generation

In order to increase the resolution and the jamming robustness of radar systems, highly complex synthetic waveforms are needed to perform sophisticated signal processing functions at high speed. Typical nowadays solutions are based on digital processing techniques requiring high speed sampling of the signals. Hundreds of thousands sampling points with 8 to

10 bits coding are required because the radar pulse typically lasts 1 to 10μs for radar frequency of 10GHz in 1GHz-bandwidth. A new realistic solution would be to exploit and combine the flexible numerical processing strengths of electronics with the communication and parallel processing strengths of optics to accomplish computationally intensive tasks with high processing speed [145]. We propose and demonstrate an arbitrary waveform generator based on the heterodyne detection of optically carried microwave signals whose phase and amplitude is optically controlled through the use of LCSLMs.

The operating principle of controlling the amplitudes and phases of optical carriers of microwave signals using heterodyne detection is depicted in figure 66. It is based on a combination of the basic principles already illustrated in the previous applications. A single-frequency laser beam (ω) is focused through an anisotropic acousto-optic Bragg cell (BC), excited by a continuous microwave signal at pulsation 2πf.

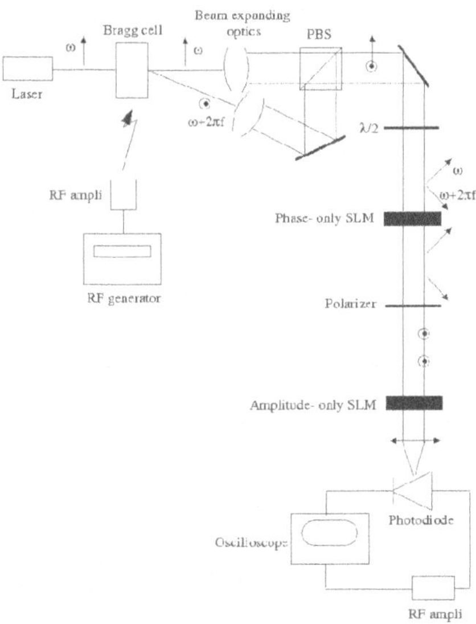

Figure 66 : Experimental set-up of the control of amplitudes and phase of optically carried microwave signals.

The transmitted beam (ω) and the diffracted beam (ω+2πf) at the output of the Bragg cell are cross-polarized. They are recombined without loss on a polarizing beam splitter (PBS) to get a dual-frequency optical carrier for the microwave signal. When a photodiode detects this dual-frequency beam through a 45°-oriented polarizer, a microwave beating signal at frequency f is observed. The dual-frequency beam intercepts a first nematic liquid crystal (NLC) spatial light modulator (SLM1) of P pixels providing P channelized beams. This phase-only SLM permits, independently on each channel, an analog phase control of the optically carried microwave signal by changing the relative optical phase of the cross-polarized components of the dual-frequency beam. Next, a twisted nematic liquid crystal (TNLC) SLM (SLM2) of P pixels, placed between two crossed polarizers (polarizing beam splitter PBS at input and exit polarizer at output of SLM2), provides the amplitude control of the microwave signals. Images up to 8 to 10 bits are displayed on the TNLC cell, so that the transmission of each pixel of SLM2 is changed. By control of the displayed grey levels onto the TNLC cell, the amount of light passing through the exit polarizer is controlled, giving an amplitude-controlled output microwave signals. At the output of SLM2, the beating signal is detected at the photodiode. According to the set-up features (photodiode area, optical wavelength, number and size of the SLM pixels and focal length), the amplitudes of the optical signals are coherently summed onto the photodiode[144], thus also generating coherent summation in the electrical domain of the optically carried microwave signals.

For microwave signals composed of several frequencies, as it is the case for radar pulses, the acousto-optic Bragg cell has to be excited by several continuous microwave signals at different frequencies f_k, (k=1 to N). Each diffracted beam, frequency-shifted by f_k passes through one given pixel i (i=1 to P) of the SLMs'. We control independently the features of each microwave frequency component f_k (phase and amplitude) by the control of the transmission law of the SLMs'. When attributing one frequency to a given pixel i and when doing so for all the frequencies of a given radar spectrum, one can generate any arbitrary waveform. In practical case, a radar antenna transmits a coherent pulse train e(t) that has a pulsewidth T and an interpulse period PRI=1/PRF. The Fourier transform of the pulse train has a sinc(u) envelope, modulating a series of spectral lines f_k spaced by the PRF of the radar ($f_k=f_a+k$PRF, $k=-\infty..+\infty$). The peak of this response is at the radar center frequency f_a, and the zeros of the sinc(u) envelope are located at frequencies $f_a\pm j/T$ ($j=-\infty..+\infty$) where T is the pulse width. At the output of the SLMs', and on channel i, the expression of the modulated microwave signal is :

$$s(t)=\frac{1}{2}\sum_{k=0}^{+\infty}a_{k,i}E_k\left(\cos(2\pi f_{k,i}t+\phi_{k,i})+\cos(2\pi f_{-k,i}t+\phi_{-k,i})\right)$$

where the coefficients $a_{k,i}$ are the attenuations of the frequencies $f_{k,i}$ due to the pixel i of the amplitude-only SLM2 and $\Phi_{k,i}$ the phase shifts introduced by the pixel i of the phase-only SLM1. The E_k coefficients are the Fourier coefficients of the coherent pulse train. The frequency $f_{k,i}$ is associated to the pixel i and $f_{-k}=f_a-kPRF$ and $f_{+k}=f_a+kPRF$. We decompose the output signal s(t) onto the kPRF frequencies of the envelope in order to be able to modify s(t) by arbitrary changing the parameters (a_k,Φ_{+k} and Φ_{-k}).

According to the Fourier Transform of the radar pulse train, two consecutive frequencies f_k and f_{k+1} are spaced from only PRF=1/PRI, that is in the range of a few tens of kHz. In order to cope with this low resolution, we propose an alternative optical architecture (Figure 67) that combines an acousto-optic Bragg cell (BC) with a moving grating that diffracts multiple orders [146]. The moving enforced to the grating is driven by applying a periodical signal on the grating. We consider as for the frequency of the applied signal, the frequency of the radar pulse train: f_s=PRF.

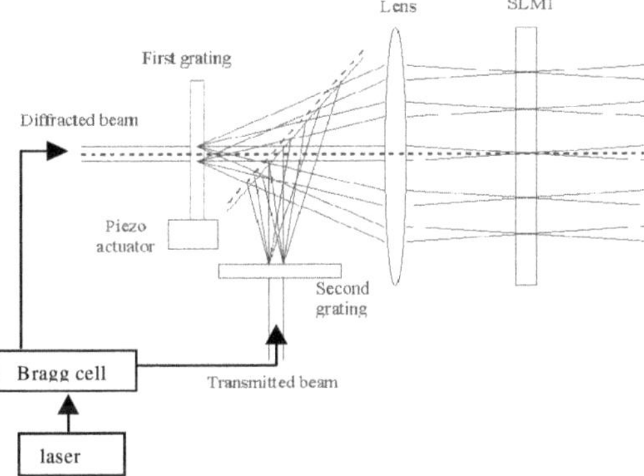

Figure 67 : kHz rate architecture with two gratings, that drives each frequency component at kPRF towards each pixel.

In order to modulate the maximum number of frequencies, the grating should diffract many orders. The second grating in the figure 67 is a fixed one that diffracts the transmitted beam coming from BC. This architecture has the advantage to discriminate the optical beams on the assumption of SLMs' with classical sizes of pitches (larger than 40 μm). According to the required kHz frequency shift, the Raman-Nath grating (first grating) is moved using a piezoelectric actuator, excited with a sawtooth signal of pulsation $2\pi f_s$. This moving grating creates a Doppler shift resulting in diffracted orders at pulsation $\omega+2\pi kf_s$. Owing to the capabilities of the Bragg cell combined with the moving grating, the optical architecture enables us to drive each frequency f_k shifted by kPRF towards each pixel i of the SLMs'. The microwave carrier f_a and the frequency spectrum of the radar f_s are respectively transposed on the optical carrier by the Bragg cell and the moving grating. Figure 68 presents the complete optical architecture of the proposed arbitrary waveform generator.

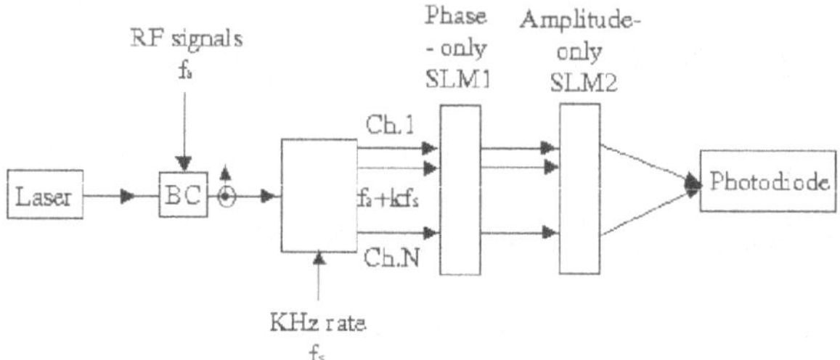

Figure 68 : Complete optical architecture of the proposed arbitrary waveform generator.

A proof of concept is experimentally demonstrated [147] using two computer-controlled SLMs of 4x4 pixels. Each pixel is 3.5x3.5mm^2 and exhibits an optical transmission of 95%. It permits to explore amplitude and phase ranges of 30dB and 2π rad respectively. The Bragg cell central frequency is 2GHz with bandwidth of 2GHz. Our aim was to control independently the features of the two frequencies (phase and amplitude). We used two RF generators and a combiner to sum the two RF signals.

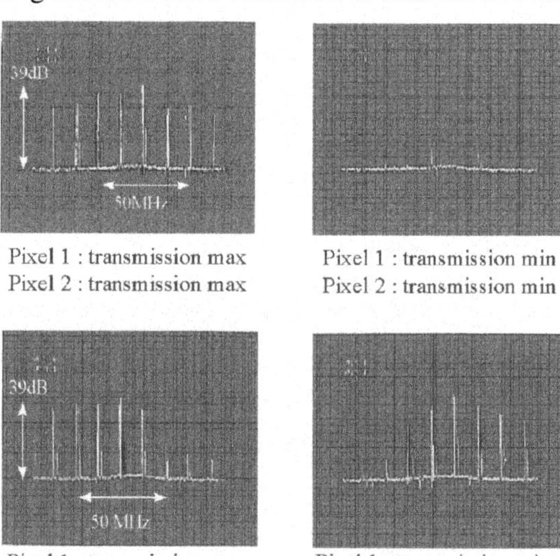

Figure 69 : Amplitude modulation in case of a pulse train (T=12ns, PRI=77ns).

Owing to an appropriate set-up, the two diffracted beams at the output of the Bragg cell were made to travel in parallel through two different pixels of the 4x4 pixel SLM. The signal detected at the photodiode was composed of a carrier at frequency $(f_1+f_2)/2$ and an envelope at frequency $(f_1-f_2)/2$. By switching off independently pixel 1 or pixel 2 of the amplitude-only SLM, we were able to attenuate of 25dB either the spectrum line 1 or the spectrum line 2 of a dual-frequency signal (Δf=45MHz).

We also demonstrated the control of a pulse train (pulsewidth T=12ns and interpulse period PRI=77ns, PRF=13MHz). Because the pixel sizes of the SLM we used were quite large (3.5x3.5mm^2), the signal modulation could only be realized with two pixels. Figure 69 shows the sinc(u) envelope when the light beams travel through the two pixels. The spectrum lines are spaced of 13MHz. We first verified that switching off all the pixels causes all the spectrum lines to disappear (top of figure 69), next, that switching off one pixel causes a part of the spectrum to disappear (bottom of figure 69). For example, the extinction of pixel 1 leads to the attenuation of the left lines with respect to the central line. And the extinction of pixel 2 leads to the attenuation of the right lines. The attenuation of the lines was about 25dB, which is close to the SLM maximum attenuation.

We have finally demonstrated with the experimental set-up of figure 67 that the moving grating controlled by a piezoelectric material permits us to increase the resolution of the architecture. A 1μm-step grating was used that combine the transmitted zero-order beam with the 1-order and 2-order beams. The periodic electrical signal was at 10kHz. The beating signals were detected at frequencies 10kHz for the 1-order beam and 20kHz for the 2-order beam.

2.6.5. Conclusion

The development of the optoelectronic technology offers new opportunities for introducing optical RF distribution in airborne phased array radars. It has been shown a large considerable interest for such technologies as soon as complex architectures are considered. In these cases, more technology advances are needed to give all flexibility to the system, with the help of digital beamforming. On more simple implementations, an analysis must be done in each case to balance this technology with more classic ones. In addition, one can notice that 2D optical TDBFN approaches could greatly benefit from a holographic backplane scheme, in order to solve the problems of compactness, reliability and scaling up to 10^3 transmit/receive modules.

Due to a great improvement in the performances of optoelectronic components over the last ten years, photonics becomes one of the major technology for advanced telecommunication, radar and EW systems. Further progress in the near future will have a very significant impact on the design of new microwave system architectures. As examples, architectures for optically controlled phase array antennas, programmable filters and waveform generation were demonstrated.

2.6.6. Acknowledgments

The authors would like to thank the DGA/STTC (Service des Techniques et Technologies Communes) for partial support of this work. Ph. Richin, D. Mongardien and E. Goutain (Thomson-CSF/LCR), T. Merlet, O. Maas (Thomson-CSF/Airsys), N.Breuil, G. Granger and J. Chazelas (Thomson-CSF/Detexis) are acknowledged for their contributions.

3. SEMICONDUCTOR DEVICES FOR ALL OPTICAL PROCESSING

3.1. Optical Processing with Semiconductors

P. Spano
Fondazione Ugo Bordoni, Via B. Castiglione, 59, I00142 Rome, ITALY
Pspano@fub.it

Abstract

The evolution of the optical networks towards very high bit rate (10-40 Gbit/s or more) and time domain or wavelength domain multiplexing, led to the development of devices for all-optical processing of the signals propagating in the optical fibers. This is due to the possibility of implementing at the optical level some functions which are quite difficult or impossible to obtain with the electronics. A general description of the functions which can be implemented at the optical level will be given, with a particular emphasis on those functions not attainable using the electronics. The physical processes responsible for the optical non-linearities, which underlie the possibility of attaining all-optical processing will be analyzed in semiconductor optical amplifiers, which are the most promising devices for this goal.

3.1.1. Introduction

In the last years, the capacity of the optical transmission systems based on optical fibers, is dramatically increasing thanks to the adoption of time domain multiplexing (TDM) and wavelength domain multiplexing (WDM) techniques. A capacity of 3.28 Tbit/s over a fiber length of 300 km [148] and 3.2 Tbit/s over a fiber length of 1500 km has been obtained using a WDM approach [149] in which 82 channels at 40 Gbit/s each and 160 channels at 20 Gbit/s, respectively, have been transmitted. TDM has been proven to be feasible at 320 Gbit/s [150] using 1.8 ps pulsewidth over 200 km fiber length, while an hybrid WDM/TDM technique, in which 19 TDM channels each at 160 Gbit/s have been multiplexed in the wavelength domain [151], showed a total capacity of 3 Tbit/s. In the mean time, the complexity of the optical networks is also increasing as well as the need for processing the optical signals crossing the network. This need seemed to be unnecessary when the erbium doped fiber amplifiers (EDFA) where first introduced in the Eighties. In fact EDFAs permit the re-amplification of the optical signals in the point-to-point transmissions, allowing a transparent transmission of TDM or WDM signals with minimum degradation [152]. However, the above mentioned increase of the transmission bit rate and the related distortion of the signal, as well as the complexity of the optical networks, brought again to the attention the processing both along the transmission line (regenerators) and in the nodes, to permit routing, add and drop, demultiplexing and other functions.

The standard processing obtained through electronics suffers for at least four orders of problems.

1. Electronics has a limited maximum speed.
2. It is not convenient from a cost/power consumption point of view, it requires two passages (optical to electrical and electrical to optical) that are avoided in the all optical processing.
3. In general the electronic processes are bit rate specific, while all optical processing can be made transparent to the bit rate and also to the signal format (intensity, phase or frequency modulation).
4. Electronics can process only one channel of a WDM signal at a time, while a simultaneous processing of all the WDM channels is feasible using all optical processing [153].

On the other side, all optical processing can be used only for simple functions, while more complex functions should be performed by electronics.

3.1.2. Optical Nonlinear Processes

The devices feasible for all optical processing in fiber transmission systems are based on the optical nonlinearities in both active and passive optical waveguides. In linear optics the electrical field Eout at the output of a generic device is linked to the input field by

$$E_{out} = CE_{in} \tag{36}$$

where C is a linear operator, while in the non linear optics the field at the out is, in general a function of the input

$$E_{out} = f(E_{in}) \tag{37}$$

where the particular function f depends on the medium itself.

In fact the linear relationship between the polarization of the medium and the field, which in linear media reads

$$P = \chi^{(1)} E \tag{38}$$

is substituted in the non linear media by the more general expression

$$P = \chi^{(1)} E + \chi^{(2)} EE + \chi^{(3)} EEE + \ldots \tag{39}$$

where $\chi^{(n)}$ is the n-th order susceptibility. In particular, the polarization at the optical frequency ω_1 is given by

$$\begin{aligned} P_i(\omega_1) = {} & \chi_{ij}^{(1)}(\omega_1) E_j(\omega_1) + \chi_{ijk}^{(2)}(\omega_1;\omega_2,\omega_3) E_j(\omega_2) E_k(\omega_3) \\ & + \chi_{ijkl}^{(3)}(\omega_1;\omega_4,\omega_5,\omega_6) E_j(\omega_4) E_k(\omega_5) E_l(\omega_6) + \ldots \end{aligned} \tag{40}$$

where the relations

$$\begin{cases} \omega_1 = \omega_2 + \omega_3 \\ \omega_1 = \omega_4 + \omega_5 + \omega_6 \end{cases} \tag{41}$$

hold and the subscripts $i,j,\ldots$ indicate the polarization states.

The second order term is responsible for the three wave mixing, which takes place when $\omega_2 \neq \omega_3$, and for the second harmonic generation [154] when $\omega_2 = \omega_3$. This is a coherent process and will give rise to a macroscopic field at frequency ω_1 only in the case in which the wave-vectors of the three waves are linked by the phase matching condition

$$\underline{k}_1 = \underline{k}_2 + \underline{k}_3 \tag{42}$$

The second order non-linearity is not the most exploited non-linearity in devices for all optical processes, nevertheless very interesting results have been obtained in the field of wavelength translators [155], both using semiconductor devices based on the GaAs and $LiNbO_3$.

The third order term is responsible for the Kerr effect, in the case in which $\omega_4 = \omega_5 = \omega_1$ and $\omega_6 = -\omega_1$. In this case the non-linear polarization becomes

$$P^{nl}(\omega_1) = \chi^{(3)}(\omega_1) |E(\omega_1)|^2 E(\omega_1) \tag{43}$$

This is an incoherent effect, in the sense that the field interacts directly with the matter and no phase matching conditions are required. In this case the gain/absorption and the refractive index of the media are a function of the field intensity: $\alpha = \alpha_l + \alpha_{nl} I$, $\eta = \eta_l + \eta_{nl} I$.

The third order term is responsible also for four-wave mixing (FWM) when $\omega_4 = \omega_6 \approx -\omega_5 \approx \omega_1$.

In this case

$$P^{nl}(\omega_1)=\chi^{(3)}(\omega_1;\omega_4,\omega_4,-\omega_5)E(\omega_4)E(\omega_4)E^*(\omega_5) \quad (44).$$

This is a coherent process, but the phase matching conditions are much more relaxed with respect to the second harmonic generation, because, in the cases we will analyze in below, the involved wave-vectors are always very similar [156] due to the similarity of the involved optical frequency values.

The two third order non-linear effects are widely used, both in fibers [157], and in semiconductor devices, to get devices for all optical processing in fiber transmission networks. Here we will consider only the semiconductor devices because, presently they are the preferred devices due to their small dimensions, their capability to be integrated and their possibility to present gain which can compensate for the unavoidable loss and the possible low efficiency of the non-linear optical processes.

3.1.3. Nonlinear Processes in Ative Semiconductor Devices

In this section we will show why active semiconductor devices, both lasers and optical amplifiers (SOA), have a third order optical non-linearity and how it can be exploited to make devices for all optical processing in optical fiber transmission systems.

The non-linear optical interaction in active semiconductor devices can be traced back to three different processes occurring when an optical field interacts with an electron-hole gas, as shown in figure 70. The first and second processes are due to a redistribution of the carriers inside the bands, hence they are intra-band effects. The third one is due to carrier density variation, so it is a inter-band effect. Let us examine these processes in some details :

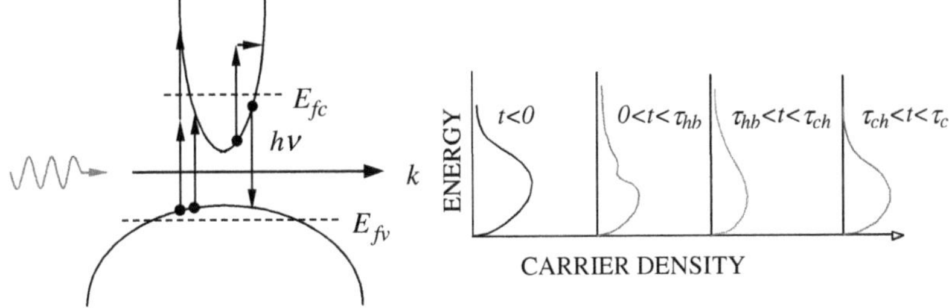

Figure 70a) : Schematic picture of the interaction processes in inverted semiconductors.
Figure 70b) : Evolution of the carrier distribution in the conduction band in the presence of an ideal short optical pulse injected at the time t=0.

- The first effect is due to a variation of the carrier distribution inside the semiconductor bands. It is due to the burning of a hole in the spectral distribution of the carrier energy caused by the stimulated emission. The process which tends to restore the original carrier distribution is the fast carrier-carrier scattering, which occurs on a time scale ~ 50-100 fs.
- The second effect is the variation of the carrier temperature induced by the stimulated transitions and the free carrier absorption of the electromagnetic field while it propagates through the active device. After the above-mentioned fast thermalization of the electron gas, in fact, the Fermi distribution of the carriers will assume, in general, a temperature higher than in the absence of the field. This is due to the excess of high energy carriers caused by free carrier absorption and the reduction of low energy carriers due to the stimulated emission [158]. The time needed to the carrier gas to recover the original lattice temperature is the electron-phonon scattering time, typically ≤1 ps.
- The third effect is connected with the variation of the carrier density induced by the increase of the stimulated emission rate in the presence of the injected optical field. The variation of the carrier density induces a change in both gain and refractive index, the latter through the linewidth enhancement factor α [159]. The characteristic time for the field-induced carrier density variation is the spontaneous lifetime, which is ~ 0.1-1 ns depending upon the values of the carrier density.

Due to these effects, the gain and the refractive index of the device is dependent upon the injected field intensity. This dependence can be exploited for all optical processing both using incoherent and coherent effects. The major differences between the two approaches will be better clarified in the following, but we can anticipate here that the incoherent processes permit a larger variety of applications, but they exploit all the non-linear interaction processes at once, so that the most efficient one determines the device performance. Such process is that related to the variation of the total carrier density, so that the speed of the devices, in general, is limited by the carrier lifetime. The coherent processes, on the contrary, can exploit the different interaction processes and could be used to process signals up to some hundred GHz [160], though with lower efficiency.

3.1.4. Devices Based on Incoherent Processes

The two main possible ways to exploit the incoherent non-linear processes in SOAs and semiconductor lasers, for getting all optical processing, are based on cross-gain [161] or cross-phase [162] effects. The former effect is the variation of the gain of an optical field crossing the device, induced by the presence (or absence) of another field which acts as a saturation beam. In figure 71 one can see a typical plot of gain vs. input power for a SOA, and the standard basic device which exploits the cross gain effect for all optical processing.

The CW field at wavelength λ_2 undergoes a modulation because of the gain saturation induced by the saturation field at wavelength λ_1. At the output of the device, after filtering out the signal at λ_1, we have a translation of the signal from λ_1 to λ_2 with mark-space inversion.

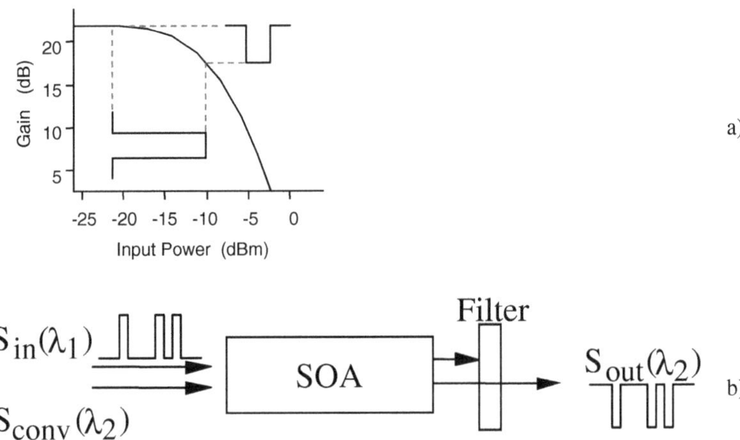

Figure 71 : Typical gain vs. input power in a SOA,
a). Building block of a device for all optical process based on cross gain
b). The signal is converted with mark-space inversion from wavelength λ_1 to wavelength λ_2 because of the gain saturation induced by the field at λ_1.

Similarly one can exploit the refractive index variation which is linked to the gain variation in an interferometric configuration. In figure 72 a Mach-Zehnder interferometer, in which the arms contain two SOAs, is reported.

In this configuration a CW field at a wavelength λ_1 is equally split between the two arms of the interferometer. A phase shifter sets the interferometer to obtain a zero output. The injection of a field at a wavelength λ_2 in one of the two arms changes the optical length of that arm and unbalances the interferometer thus causing the emission of the field at λ_1. In the case in which the field at λ_2 is a pulse signal, the recovery time of the interferometer is given by the carrier lifetime.

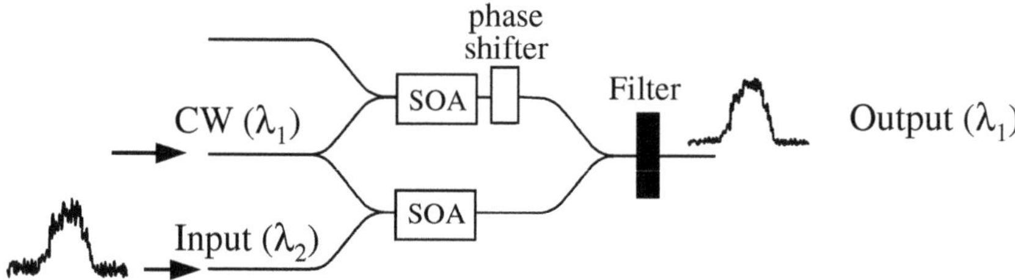

Figure 72 : Building block of a device for all optical processing based on cross phase modulation. The refractive index variation, induced by the input signal (λ_2) in the bottom arm of the interferometer unbalances the interferometer itself and permits the CW field (λ_1) to exit the interferometer.

This scheme is very powerful because the transfer function of the interferometer is strongly non-linear and, for this reason it can be used to reshape a signal corrupted by crossing an optical fiber. It is also able to supply amplification, which is the second need of a complete 3R [163] (reshaping, reamplification, retiming) optical regenerator. The interferometer has been also integrated in order to give a real usable device. The performance of this interferometer is very interesting and show, for example that a 3R all optical regenerator permits to obtain a penalty lower than 1 dB after 15,000 km and 300 regenerations at 10 Gbit/s [164]. In figure 73 the eye diagram after 1, 10, 100 and 300 regenerations is reported. Other interferometric configurations, which show some better performance of that reported in figure 72 have been proposed and realized. They basically permit to avoid the frequency translation from λ_2 to λ_1 [165] or present a steeper transfer function [166].

Among the limitations of all the devices based on cross phase modulation, one must underline the gain variation which is linked by the linewidth enhancement factor α, to any variation of the SOA refractive index, and tends to reduce the extinction ratio of the output signal. It has been shown [167,168] that this effect can be reduced or even suppressed using a CW field with a wavelength larger than that corresponding to the semiconductor band-gap, while maintaining the wavelength of the control pulse at the SOA gain peak.

Due to the spectral extension of the refractive index variation for any change of the carrier density, which is much broader than that of the gain (α tends to increase for a photon energy close or even lower than the semiconductor band-gap [159]), in this configuration the CW field still experiences a phase variation while the gain variation is negligible.

The last function (retiming) to obtain a 3R all optical regenerator has been implemented using unstable devices. A nice configuration [169] consists of a controlled self pulsing laser. In the case of ref. [169], self pulsing is obtained in a DFB/DBR laser in which the Q factor of the cavity is self switched by the field in the cavity, as shown in figure 74, where a sketch of the device is reported. When the field intensity in the active region increases the refractive index variation tends to match the reflectivity peaks of the DFB and DBR sections, thus reducing the lasing threshold. This gives rise to the emission of a pulse as in standard Q-switching. After the emission the Q factor decreases and the process starts again. The repetition frequency of the pulses is determined by the DFB pump current and, in the case of ref. [169] can reach values up to 45 GHz. When the signal to be retimed is injected in the device, it locks the repetition frequency of the self Q-switched laser, thus giving the retiming. It has been shown that a few bits are sufficient to lock the self-pulsing frequency, while the locking is maintained for very long sequences of zeros.

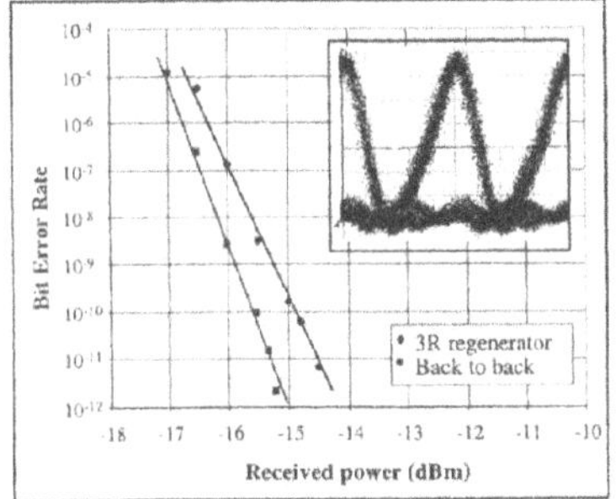

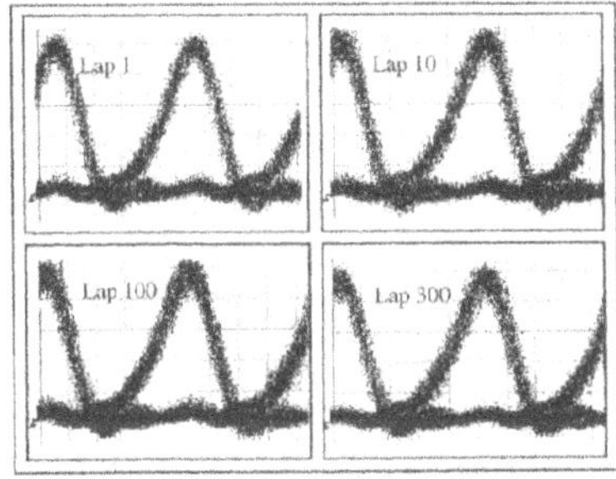

Figure 73: BER curves back to back and after all-optical regeneration for a 10 Gbit/s transmission system (a). Evolution of the eye diagram after 1, 10, 100 and 300 laps. Total length 15,000 km (b). (After [164]).

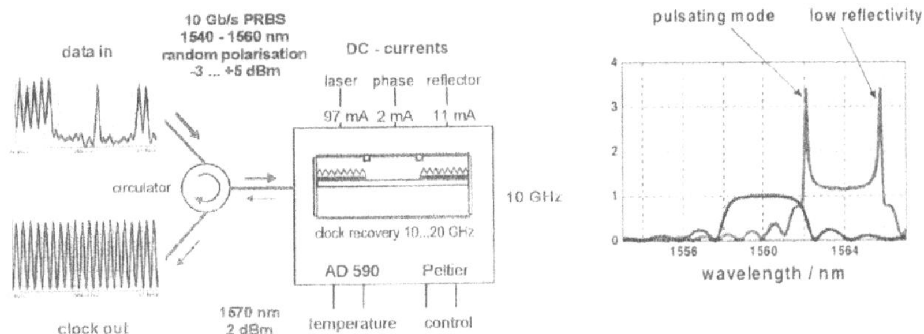

Figure 74 : Scheme of the self Q-switching laser in the configuration for retiming (a) and transfer function of the DFB and DBR sections (b). (After [169]).

The main limitation of the devices based on the incoherent effects is the maximum operating speed, which is limited by the spontaneous lifetime of the carriers. It limits the maximum speed to about 20 Gbit/s, although processing at 40 Gbit/s has been shown [170]. In order to overcome this limit, Mach-Zehnder interferometers based on a differential time delay have been implemented [171]. In figure 75 the configuration which can be used for regeneration or demultiplexing [172] of TDM optical signals is reported. Similar configurations have been used also for add and drop multiplexing.

In the first case two replicas of the signal, degraded by the transmission along the fiber, are sent delayed by a time τ, to the two amplifiers of the interferometer. The clock, obtained from the signal e.g. by the technique reported above, is sent to the third input, with a delay $<\tau$ with respect to the signal 1. When a pulse of the signal 1 crosses the SOA, due to the induced refractive index change, the interferometer switches to the on state and the clock pulse is transmitted through the interferometer. When the corresponding pulse of the delayed signal 2 crosses the second SOA, the interferometer is switched off because now both SOAs undergo the same optical length variation, almost independently of the carrier lifetime. The device is able to transmit the clean clock pulses with no limitations on the maximum speed if it arrives to the interferometer within a time window τ after a pulse of signal 1. Inverting the role of the clock and the signal one can obtain a demultiplexer as shown in figure 75 b). In this configuration the clock pulses open the transmission window of the interferometer. The main limitation of this approach is the maximum length τ of the window. If τ is comparable with the carrier lifetime, in fact, the recovery in the first SOA prevents the perfect balance of the interferometer after the pulse in the SOA number 2 and affects the performance of the device. Moreover a bit interval much shorter than the carrier lifetime causes a behavior dependent on the particular bit sequence. In this condition a long sequence of ones is responsible for a refractive index variation in the SOAs lower than in the case of a one following a long sequence of zeros.

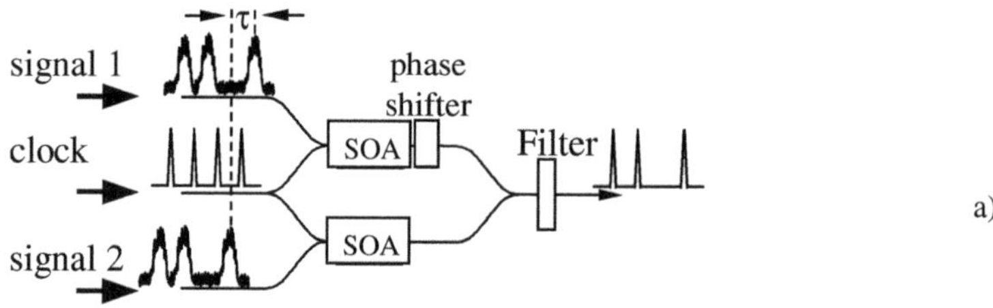

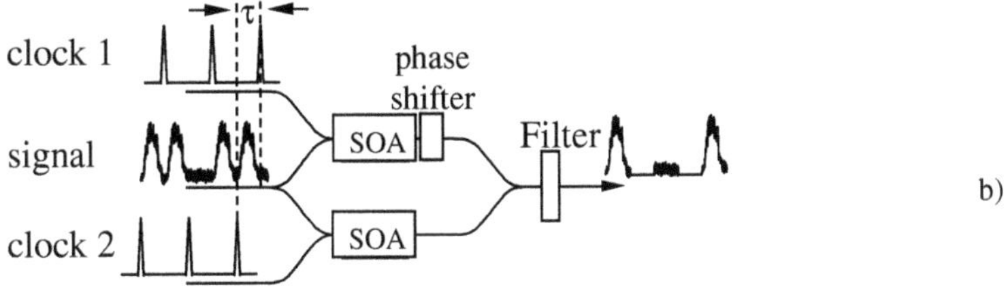

Figure 75 : Scheme of a differential delay Mach-Zehnder interferometer used to regenerate the signal (a), and to demultiplex TDM signals (b).

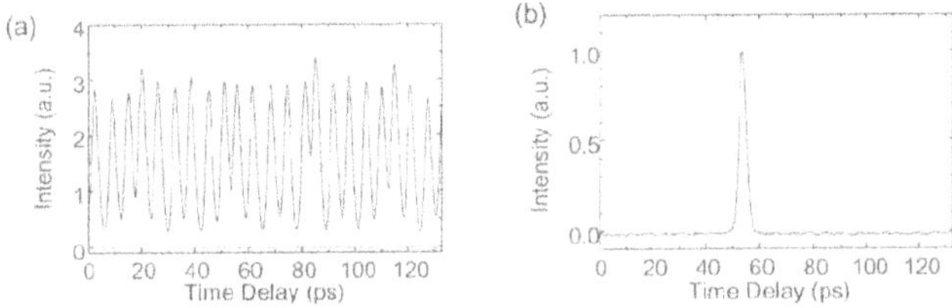

Figure 76 : Bit sequence at 168 Gbit/s and a demultiplexed bit at 10.5 GHz. (After ref. [172]).

In spite of these constraints, good performance have been obtained as reported for instance in figure 76, where a signal at 10.5 GHz is demultiplexed from a sequence at 168 Gbit/s [172].

3.1.5. Devices Based on Coherent Processes

As stated in one of the previous sections the coherent non-linear third order process is the FWM. It takes place when a pump field at the optical frequency ν_p is injected in the SOA together with the signal at frequency ν_s. The beating between the two fields at the frequency $\Delta\nu=|\nu_p-\nu_s|$ generates a modulation of the gain and refractive index if the medium is able to follow the modulation. This modulation, applied to the pump field produces two modulation sidebands. One of them at the signal frequency, the second one at a new frequency $\nu_c = 2\nu_p-\nu_s$. If the signal is not monochromatic, the spectrum of the new generated signal, centered at ν_c is reversed with respect to the original one. By filtering the output signal one gets a wavelength converter which, besides, offers the optical phase conjugation, as shown in figure 77.

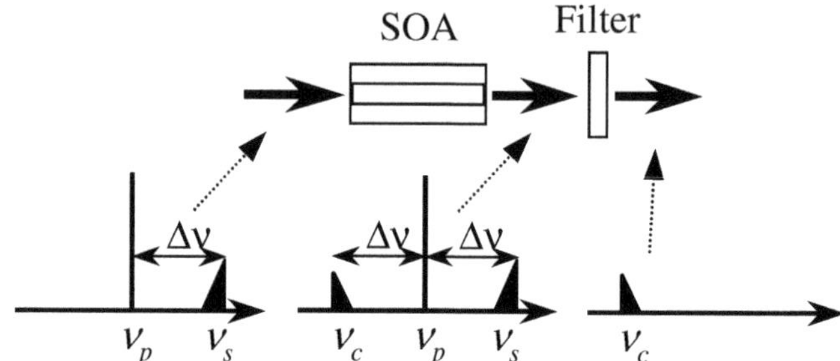

Figure 77 : Basic configuration of a wavelength converter based on FWM in SOAs and optical spectra in different points.

The presence of different non-linear processes in SOAs with different interaction coefficients, reflects into a dependence of the conversion efficiency $\eta=|E_{c,out}|^2/|E_{s,in}|^2$ on the conversion interval $2\Delta\nu$. In opposition to devices based on the incoherent effects, where the carrier lifetime limits the maximum response velocity, here one can exploit also the fast non-linear processes. The lower non-linearity associated with the faster non-linear processes is responsible only for a reduction of η. The intensity of the conjugate field, in fact, can be written as [173]:

$$|E_c|^2 \propto |E_s^*|^2 |E_p|^4 |H(\Omega)|^2 \tag{45}$$

Ω being equal to $2\Delta\nu$ and

$$H(\Omega)=\frac{1}{\left(1+\Omega^2\tau_{dp}^2\right)}\left[-\frac{(1-i\alpha)}{(1-i\Omega\tau_{hb})(1-i\Omega\tau_c)}\frac{1}{P_s}-\frac{(1-i\beta_{ch})\varepsilon_{ch}}{(1-i\Omega\tau_{hb})(1-i\Omega\tau_{ch})}-\frac{\varepsilon_{hb}}{1-i\Omega\tau_{hb}}\right]$$
$$-(1-i\beta_{inst})\varepsilon_{inst} \tag{46}$$

where the subscripts *dp, ch, hb, inst* stand for carrier density population, carrier heating, spectral hole burning and instantaneous processes (two photon absorption and Kerr effect), respectively.

The expression $\varepsilon=g^{-1}\partial g/\partial I$ is the gain compression coefficient due to a particular non-linear process; τ is the relaxation time of the process, P_s is the SOA saturation power and α, β are the linear and non-linear linewidth enhancement factors.

In figure 78 the plot of the conversion efficiency vs. the pump signal detuning $\Delta\nu$, is reported for a standard 0.5 mm long SOA [174]. In the figure, the dashed lines represent the conversion efficiency one could expect in the case in which only one non-linear process is present at a time. The two solid lines represent the efficiency for positive frequency conversion (*$\Delta\Omega>0$, $v_p>v_c$*) and negative frequency conversion (*$\Delta\Omega<0$, $v_p<v_c$*). It is evident the decrease of the efficiency when faster and less efficient non-linear processes are involved. The difference of the conversion efficiency for positive and negative detunings is due to the relative phase of the terms present in eq. (46). By using the basic frequency-converter building-block, one is able to make, beyond the frequency converters to be used in WDM networks for routing or add/drop [175], demultiplexers for TDM signals [176] and fiber-dispersion compensators [177, 178]. For a demultiplexer both the pump and the signal are pulsed. The pump acts as the clock field. Only when both fields are present the FWM process takes place and the frequency translated field is generated. A 6.3 Gbit/s signal, demultiplexed from a 100Gbit/s has been successfully obtained with a maximum penalty of less than 1 dB [179]. For dispersion compensation, the device is based on the spectral inversion operated on the conjugate field by the FWM process (see figure 79). The dispersion is, in fact, a linear process which does not affect the optical spectrum of the signal, but only its time shape.

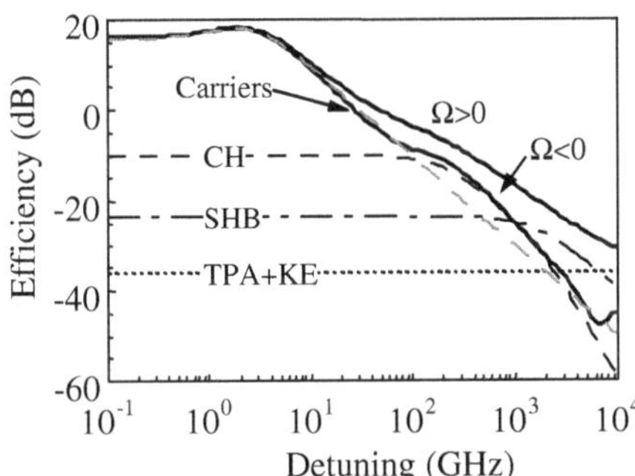

Figure 78: Conversion efficiency vs. frequency detuning for a FWM converter based on a 500 μm long SOA. The dashed lines represent the conversion efficiency due to the single non-linear processes. $\Delta\Omega>0$, means $v_p>v_c$; $\Delta\Omega<0$ means $v_p<v_c$ (after [174]).

This happens because the spectral components of the signal cross the fiber at a different velocity. If in the middle of the fiber a FWM converter is used, the spectrum of the signal is reversed and the fastest spectral components become the slowest ones, thus re-compressing the broadened pulse in the following part of the fiber span.

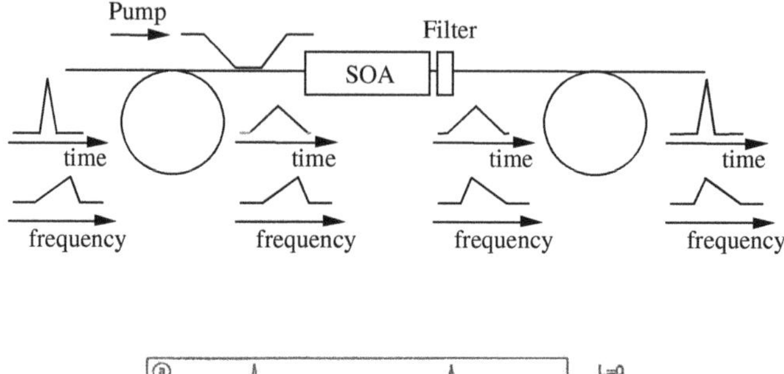

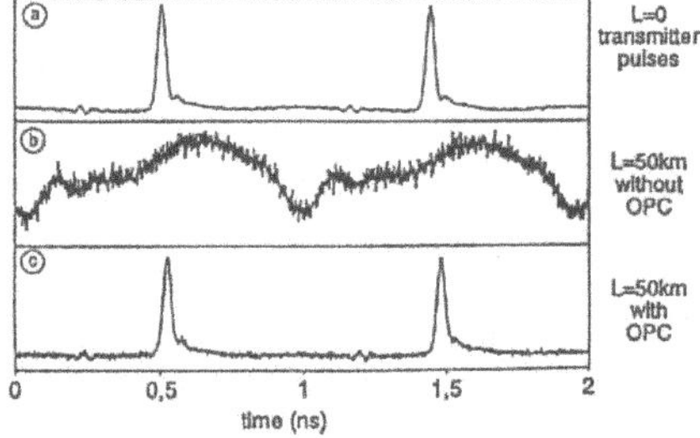

Figure 79 : Basic scheme of a mid-span fiber dispersion compensator and pulse-shape at the beginning of a fiber and after 50 km without and with dispersion compensation by optical phase conjugation (after [177]).

3.1.6. Coherent VS Incoherent Processes

The main question, after the simple explanation of the working principles of the two main classes of semiconductor devices for all-optical processing, is which is the best. As one can easily understand the answer is not unique and depends in a large extent on the frame in which the devices are used. As a guideline in table 3 one can find the main advantages and drawbacks of the two approaches.

The basic difference between the interferometers based on cross-phase modulations and the frequency converters based on FWM lies in the transfer function of the two devices. The transfer function of the interferometer is strongly non-linear, so it can be easily exploited to remove the noise affecting the signal and, hence, to make the regenerators which act better and better as the non-linearity of the transfer function increases [163]. On the contrary FWM has a linear transfer function and can be used for regeneration only in an indirect way [180]. However in many other application FWM is better than the inchoerent techniques. For instance it permits to convert a whole comb of WDM channels simultaneously [181], while using the inchoerent techniques every channel must be converted individually. This implies a much more complex architecture of the converter. Again, due to its linear transfer function, FWM permits a transparent processing, i.e. it can be used not only to process intensity modulated signals, but also phase or frequency modulated signals as well as analog signals. However the best feature of FWM is to permit the processing of signals at a much higher speed than the incoherent techniques. The limitation to the maximum speed of the signals to be processed by FWM lies in the slope of the conversion efficiency vs $\Delta\nu$. Fast signals correspond to a large modulation bandwidth. In order to avoid a spectral deformation of the converted field, the conversion efficiency must remain constant on the modulation bandwidth, and this requires a larger value of $\Delta\nu$ (see fig. 78). Large values of $\Delta\nu$, on the other hand, correspond to low efficiency and a low signal-to-noise-ratio (SNR) because of the amplified spontaneous emission (ASE).

Incoherent techniques	Coherent techniques
Nonlinear transfer function	*Linear transfer function*
☺ Regeneration possible	☺ No limitation to the bit rate
☺ Improvement of SNR	☺ No limitation to the signal format
☹ Limited bit rate	☺ Directly usable with WDM systems
☹ Possible only with intensity modulated signals	☺ Fiber dispersion compensation possible
	☹ Regeneration difficult
	☹ Degradation of SNR (ASE of the SOA)
	☹ Low efficiency
	☹ Efficiency dependent upon the conversion interval
	☹ Polarization sensitivity

Table 3 : Pros and cons of coherent and incoherent techniques to get all-optical processing.

Among the main drawbacks of FWM one can consider the low efficiency of the process, the associated degradation of the SNR, the reduction of the conversion efficiency for large detunings and the polarization sensitivity, being FWM a coherent process, in fact, it takes place only for co-polarized pump and signal. However, after a long fiber link the polarization of the signal cannot be easily controlled and tends to change with time. Some FWM configurations, more complicated than the simplest one shown in figure 77, have been proposed and realized [182,183] to solve the last two problems, while the efficiency can be dramatically increased by using long SOAs [184,185].

3.1.7. Conclusions

In this paper we tried to answer the general questions concerning the all-optical processing in high capacity optical networks: Why, Where and How. The two main techniques able to get all-optical processing using semiconductors have been discussed and the performance of the devices made according with these techniques have been compared.

3.2. The Use of InGaAs / In Photo – HBT's in Optical / Microwave Processing

G. Eisenstein
Electrical Engineering Department, Technion, Haifa 32000, ISRAEL
gad@ee.technion.ac.il

Abstract

This paper describes the use of InGaAs / InP heterojunction bipolar photo transistors in optical / microwave processing. The basic physical process we use is the optoelectronic mixing process which is described in detail. Applications of optoelectronic mixing in these photo transistors are surveyed.

3.2.1. Introduction

Heterojunction bipolar photo – transistors (photo – HBT's) combine the advantages of an HBT with the possibility of integration with optical control signals. Photo – HBT's constructed in the InGaAs / InP material system have the added advantage of operating in the same wavelength regime (near 1550 nm) as the optical fiber network.

The basic idea of an HBT was conceived early in the history of bipolar transistors [186]. The different band gap in the three transistor regions enables high current gains and a wide modulation bandwidth [187]. The addition of an optical port, obtained by an opening in the metallic base contact transforms the device into a four terminal device with different combinations of electrical and optical feed configurations. These configurations enable a variety of functions to be performed with large possible implications in the field of RF photonics communication and instrumentation.

The combined electrical and optical feeds lead to the process of optoelectronic mixing (OEM) which is the basis for many of the advanced applications described hereon. The OEM process in a photo – HBT has been studied extensively [188,189] and the results are summarized in section II. Section III describes several applications of optoelectronic mixing such as diode laser locking for WDM systems, remote millimeter wave signal generation and its analog and digital modulation, self oscillation in a photo HBT including injection locking to reduce its noise and modulation by optical means.

3.2.2. Optoelectronic Mixing in a Photo HBT

3.2.2.1. Model

The most common model of an HBT operating in the active mode is the so called large signal π-model shown schematically in figure 80. We employed that model for treating frequency mixing and conversion assuming that the HBT is subject to a modulation at two different RF frequencies: the LO frequency ω_{LO} and the signal frequency ω_S.

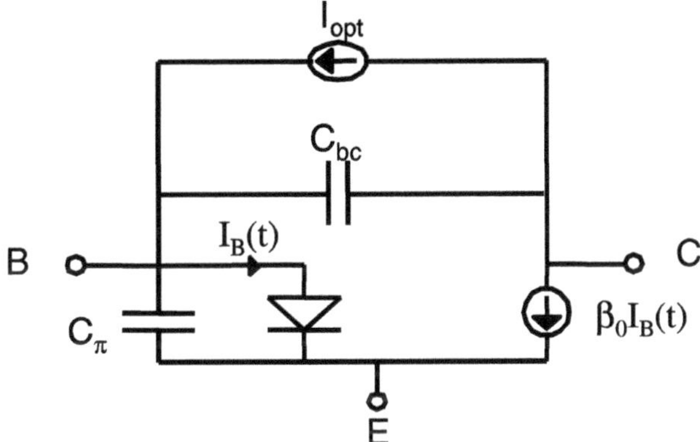

Figure 80 : The HBT π-model.

The nonlinear input capacitance C_π is

$C_\pi = C_{be} + \frac{\beta_0 + 1}{\beta_0} \frac{\tau I_S \exp(V_{be}(t)/V_T)}{V_T}$, where β_0 is the low frequency AC current gain of the HBT, C_{bc} and C_{be} are the base-collector and the base-emitter junction capacitance, τ is the emitter to collector delay time, I_S is the saturation current, $V_{be}(t)$ is the time dependent base-emitter voltage and $V_T = nKT$ is the thermal voltage, with n being the ideality factor of the base-emitter junction, K is the Boltzman constant and T the temperature .

In order to calculate the performance of the HBT as an OEM, we considered the circuit diagram of figure 80(b). The LO signal, connected to the base, was represented as a voltage source: $V_{LO} = V_{LO} \sin(\omega_{LO} t)$. The input optical signal which impinged on the base served as a current source: $I_{opt} = I_{opt}\sin(\omega_S t)$. The 50 Ω resistors connected to the LO source and the collector output, represented the LO output impedance and the input impedance of the measurement apparatus (spectrum analyzer), respectively. The capacitance and inductance of the Bias-T are C_{BT}=1mF and L_{BT}=1mH. Other symbols represent the following : V_1 - the voltage at the output of the LO source (after the 50Ω resistor), V_b - voltage at the base port, V_{bDC} - the base DC bias, I_{bDC} - the base DC current, V_c - the voltage at the collector port and V_{cDC} - the collector DC bias.

Note that the π-model does not take into account saturation effects due to a voltage drop on the 50Ω load at high collector currents, which results in forward biasing of the base collector junction. This model simulates the HBT in the active mode regime – the regime for the best mixing performance [188]. Therefore, only large V_{ce} values and relatively small LO power will be used in order to keep the HBT in the active mode regime (V_{ce}-V_b>0.3V).

The differential equations describing the model are:

$$C_{BT}\frac{dV_1}{dt} = C_{BT}\frac{dV_b}{dt} + \frac{V_{LO} - V_1}{R} \tag{47a}$$

$$(C_\pi + C_{bc})\frac{dV_b}{dt} = I_{opt} + \frac{V_{LO} - V_1}{R} + I_{DC} + C_{bc}\frac{dV_c}{dt} - \frac{I_s}{\beta_0}\left[\exp\left(\frac{V_b}{V_T}\right) - 1\right] \tag{47b}$$

$$C_{bc}\frac{dV_c}{dt} = C_{bc}\frac{dV_b}{dt} + \frac{V_{cDC} - V_c}{R} - I_s\left[\exp\left(\frac{V_b}{V_T}\right) - 1\right] \tag{47c}$$

$$\frac{dI_{DC}}{dt} = \frac{V_{bDC} - V_b}{L_{BT}} \tag{47d}$$

The output signal of the OEM is the current on the 50Ω input impedance of the spectrum analyzer. When neglecting the base-collector capacitance, this current is equal to the collector current, I_c, and is given by

$$I_c = I_S\exp\left(V_{be}(t)/V_T\right) \tag{48}$$

Solving numerically the differential equations (47), substituting the solution for $V_{be}(t)$ in (48) and Fourier transforming the result, yields the output spectrum. This spectrum contains the LO component at ω_{LO}, the signal component at ω_S, the down-conversion product at $\omega_{down} = \omega_{LO} - \omega_S$, the up-conversion product at $\omega_{up} = \omega_{LO} + \omega_S$ and additional mixing harmonics which are not considered in the present analysis.

$$V_{LO} = v_{LO}\exp(i\omega_{LO}t) + c.c \tag{49a}$$

$$I_{opt} = i_{opt} \exp(i\omega_S t) + c.c \tag{49b}$$

Applying the small modulation approximation, equation (3) can be simplified by using Taylor's expansion for the exponent. Keeping terms quadratic in V_{be}, we obtain:

$$C_\pi \frac{dV_{be}}{dt} \approx I_{opt} + \frac{V_{LO} - V_1}{R} + I_{DC} - \frac{I_s}{\beta_0}\left[1 + \frac{V_{be}}{V_T} + \frac{1}{2}\left(\frac{V_{be}}{V_T}\right)^2\right] \tag{50}$$

With the solution of the form:

$$V_{be} = Vb_{DC} + \upsilon_{be_{LO}} \exp(i\omega_{LO} t) + \upsilon_{be_S} \exp(i\omega_S t) + \sum_{\omega_2} \upsilon_{be_2} \exp(i\omega_2 t) + c.c \tag{51}$$

where $\omega_2 = \omega_{LO} \pm \omega_S,\ 2\omega_{LO},\ 2\omega_S$ and the spectral coefficients are:

$$\upsilon_{be_{LO}} \approx \upsilon_{LO} \cdot \frac{1}{1 + i\omega_{LO}/\omega_0} \tag{52a}$$

$$\upsilon_{be_S} \approx R \cdot i_{opt} \cdot \frac{1}{1 + i\omega_S/\omega_0} \tag{52b}$$

$$\upsilon_{be_{down}} \approx \frac{\upsilon_{be_{LO}} \cdot [\upsilon_{be_S}]^*}{\tilde{\upsilon}} \cdot \frac{1}{1 + i\omega_{down}/\omega_0} \tag{52c}$$

$$\upsilon_{be_{up}} \approx \frac{\upsilon_{be_{LO}} \cdot \upsilon_{be_S}}{\tilde{\upsilon}} \cdot \frac{1}{1 + i\omega_{up}/\omega_0} \tag{52d}$$

where $\omega_0 = 1/(R \cdot \overline{C}_\pi)$, $\overline{C}_\pi = C_{be} + \frac{\beta_0 + 1}{\beta_0} \frac{\tau I_S \exp(Vb_{DC}/V_T)}{V_T}$ and

$$\tilde{\upsilon} = \frac{V_T}{1 + i(\omega_{LO} + \omega_S)/\omega_0}.$$

Substituting equations (51) and (52) into equation (48) yields expressions for the amplitudes of the output signal at the up and down converted frequencies:

$$I_{C_{down}} \approx \frac{I_s}{V_T} \cdot \exp(Vb_{DC}/V_T) \cdot \frac{\upsilon_{be_{LO}} \cdot [\upsilon_{be_S}]^*}{\tilde{\upsilon}} \cdot \frac{1}{1 + i\omega_{down}/\omega_0} \tag{53a}$$

$$I_{C_{up}} \approx \frac{I_s}{V_T} \cdot \exp(Vb_{DC}/V_T) \cdot \frac{\upsilon_{be_{LO}} \cdot \upsilon_{be_S}}{\tilde{\upsilon}} \cdot \frac{1}{1 + i\omega_{up}/\omega_0} \tag{53b}$$

Equations (53a) and (53b) are the most important results of the small signal analysis.

An informative characterization of the high-frequency behavior of the mixing scheme is an asymptotic scenario, when the signal and the local oscillator are close in frequency, $\omega_{LO} \sim \omega_S = \omega > \omega_0$. The frequency ω_0 is associated with the RC delay time related to the 50Ω resistor connected to the LO source and the sum of the diffusion capacitance and the base emitter capacitance. In this asymptotic case, we can describe the HBT OEM performance as a cascade of two transfer functions, $H_1(\omega)$ which depends on the input signal frequency ω and $H_2(\omega_2)$ which depends on the mixing-product frequency ω_2. Both $H_1(\omega)$ and $H_2(\omega_2)$ depend on the frequency ω_0, i.e. - the input impedance and the nonlinear input capacitance. Cascading $H_1(\omega)$ and $H_2(\omega_2)$ to form the input mixing network - results in different currents values for the down and up conversion processes. The input signal passes through H_1, which is a frequency dependent amplification function. Note: this function is not the current gain amplification function of the transistor due to the pole at ω_0. Then, the "amplified signal" passes through the H_2 function which is responsible for the different conversion gains of the various mixing products. The later implies that we can model the operation of the HBT OEM as and 'special' amplification stage followed by a mixing stage, however it should be noted that both of these processes are actually simultaneous and taking place at the input network of figure 80. This is consistent with one of our previous results [189], related to large signal operation, which states that the HBT OEM cannot be separated into an ideal input mixing stage followed by an amplifying frequency dependent output stage. Rather, mixing and amplification take place simultaneously and the voltage dependence of the input network is the main nonlinear effect in the HBT OEM.

In the asymptotic case, the frequency of the down-conversion process is constant at ω_{LO} - ω_s and the frequency of the up-conversion process ($\omega_{LO} + \omega_s$) varies as ω. Therefore, in this case - equations (53) become:

$$I_{C_{down}} = H_1(\omega)\cdot H_2(\omega_{down}) \xrightarrow[\omega>>\omega_0]{} c\cdot\frac{1}{\omega/\omega_0}\cdot\frac{1}{\omega_{down}/\omega_0} \xrightarrow[\omega>>\omega_0]{} \frac{1}{\omega} \tag{54a}$$

$$I_{C_{up}} = H_1(\omega)\cdot H_2(\omega_{up}) \xrightarrow[\omega>>\omega_0]{} c\cdot\frac{1}{\omega/\omega_0}\cdot\frac{1}{\omega_{up}/\omega_0} \xrightarrow[\omega>>\omega_0]{} \frac{1}{\omega^2} \tag{54b}$$

The result implies that in the small signal regime, the amplitude of the intermediate-frequency output signal decays as $1/\omega$ for the down-conversion process and as $1/\omega^2$ for the up-conversion process. This frequency behavior makes the use of up-conversion process at low LO power levels (small signal regime) very inefficient at high frequencies.

3.2.2.2. Experiments

The experiments were carried out using an HBT grown on a semi-insulating InP substrate by a compact metalorganic molecular beam epitaxy system [189]. The layer structure of the HBT is shown schematically in figure 81 and contained: 400nm GaInAs ($n=3x10^{19}$ cm^{-3}) and 250nm InP ($n=2x10^{18}$ cm^{-3}) sub collector, 750nm undoped GaInAs collector, 50nm GaInAs ($p=3x10^{19}$ cm^{-3}) base, 150nm InP ($n=2x10^{17}$ cm^{-3}) emitter and 200nm ($n=3x10^{19}$ cm^{-3}) GaInAs contact layer.

Conventional wet etching and a self-aligned Pt/Ti/Pt/Au one step metalization process were employed to fabricate the devices. Polymide passivation and Ti/Au pads completed the fabrication process. A 5x6 micron opening in the base metalization served as an optical window. Small signal of the HBT yielded an F_t and an F_{max} of 70GHz and 40GHz, respectively, at Ic=20mA and Vce=2V.

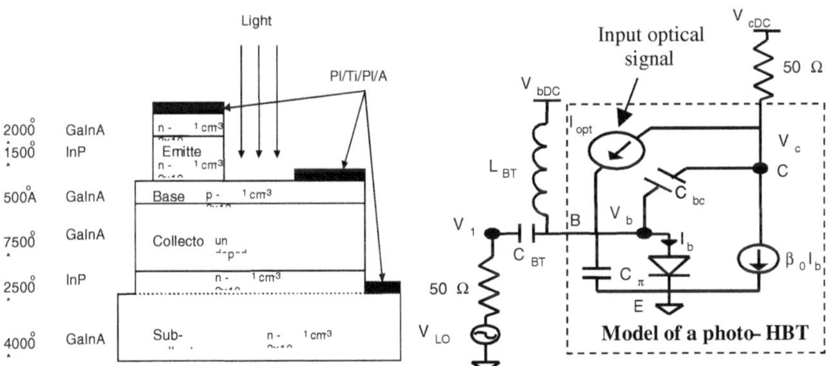

Figure 81 : The layer structure of the HBT.

Figure 82 : Schematic of the experimental optoelectronic mixing setup.

The schematics of the experimental setup is shown in figure 82. An LO source and a DC voltage source were connected via a Bias-T to the base. A DFB laser operating at 1.55μm was externally modulated by an RF source and the optical modulated signal was amplified by an Erbium doped fiber amplifier (EDFA) before being focused onto the optical window of the HBT. The EDFA serves to compensate for various coupling losses in the experimental arrangement. A spectrum analyzer was connected via a Bias-T to the collector which served as the output port.

Conversion Gain measurements

Both the intrinsic and extrinsic conversion gains are useful figures of merit for OEMs [190]. The intrinsic conversion gain, G_{int}, is defined as the ratio of the output power (P_{down}, P_{up} for down- , up- conversion) to P_{prime}, the primary photo detected RF power. P_{prime} is the photo induced RF electrical power detected by the base collector junction without amplification. It was measured by shorting the base emitter junction of the photo-detector transistor. The extrinsic conversion gain, G_{ext}, is defined as the ratio of the output power of the up or down converted signal to the equivalent electrical RF power, P_{in}, that would have been detected by an ideal photo-diode with an equal load resistance. This ideal power is related to the peak power of the modulated component of the incident optical signal by $P_{in} = \left(\frac{qP_{\text{mod}}}{h\upsilon}\right)^2 \frac{R_{Load}}{2}$ [191], where q is the electron charge, $h\upsilon$ is the photon energy and $R_{LOAD} = 50\Omega$.

P_{prime} and P_{in} are related by the external quantum efficiency, η, of the base-collector photo-diode, thus $G_{ext} = \eta^2 G_{\text{int}}$. The external quantum efficiency was measured from the DC photo-response of the base-collector junction and was $\eta = 29\%$ or $\eta^2 = -10.8\ dB$. This result agrees with a calculation assuming an absorption depth of 1.5 μm, and 30% reflection.

The electrical response of the HBT to the modulated optical signal was first measured as a function of the frequency modulation of the optical signal. The electrical frequency response of the base-collector PIN photo-diode served as a reference for calculating the intrinsic signal gain, excluding the effects of the external quantum efficiency of the HBT. Next, the HBT was driven at its optimum bias [188] while the LO power level and the collector emitter bias were held constant at -10 dBm and 2V, respectively.

In both down- and up- conversion experiments, we applied two RF signals (one to the external OE modulator and the other to the HBT base) in the range of 0.5 - 20 GHz, keeping the separation between them at 500MHz. For each conversion experiment the applied frequency $\omega = \omega_{LO} \sim \omega_S$ was varied while measuring the amplitude of the converted product. Note that the down-converted product was at a constant 500MHz and the up-converted product was varied as ω. The results of the intrinsic down and up conversion gain responses are shown in figure 83.

The results show that the conversion gain of the down-conversion process is larger than that of the up-conversion case for all frequencies.

The conversion cut-off frequency (defined as the frequency where the intrinsic conversion gain is 0dB) was 9 GHz for the down-conversion process and ~ 4GHz for up-conversion process.

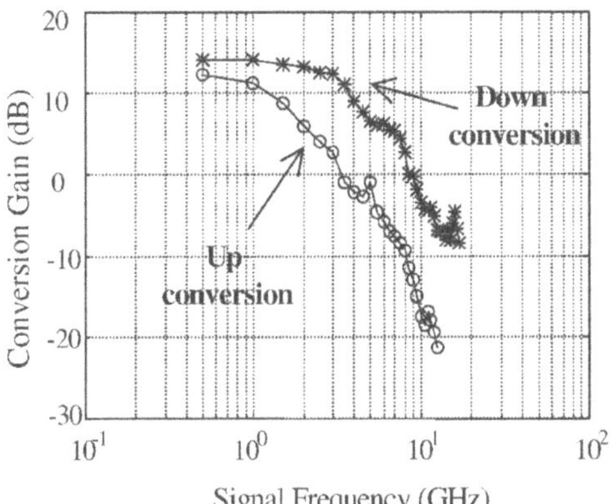

Figure 83 : Measured intrinsic up and down conversion gain.

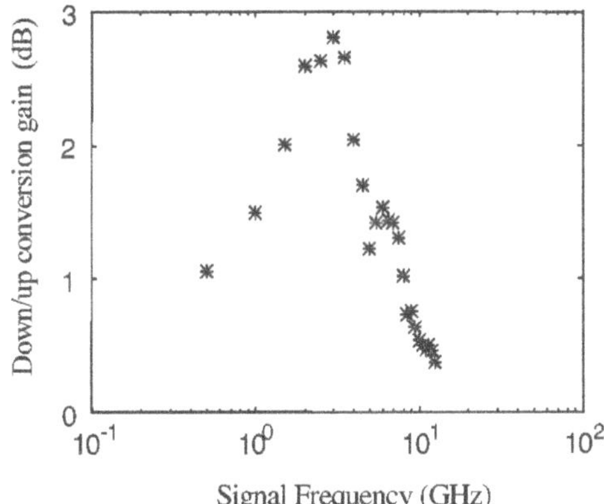

Figure 84 : Ratio between down and up conversion products.

The measured ratio between the down- and up- conversion gain can be used to identify the dominant nonlinear effects in the OEM operating in the large signal regime. Using the PSPICE model of the HBT, we showed previously [188] that the main nonlinear effect is the exponential dependence of the dynamic emitter resistance, r_e, on the base emitter time dependent bias voltage $V_{be}(t)$ (i.e, the input network). The amplitudes and phases of the down- and up- conversion components of $V_{be}(t)$ differ due to the effect of the input network, and thus, we expect to have a difference between down- and up- conversion efficiencies. Figure 84 shows the frequency dependence of the measured ratio between the down-conversion and the up-conversion products. One obvious observation is that the absolute efficiency of the down-conversion process is higher than that of the up-conversion process by as much as 3 dB at $\omega/2\pi$=3 GHz. The applied LO power of –10dBm was optimized for the 3 GHz RF frequency. This is consistent with LO powers of similar transistors.

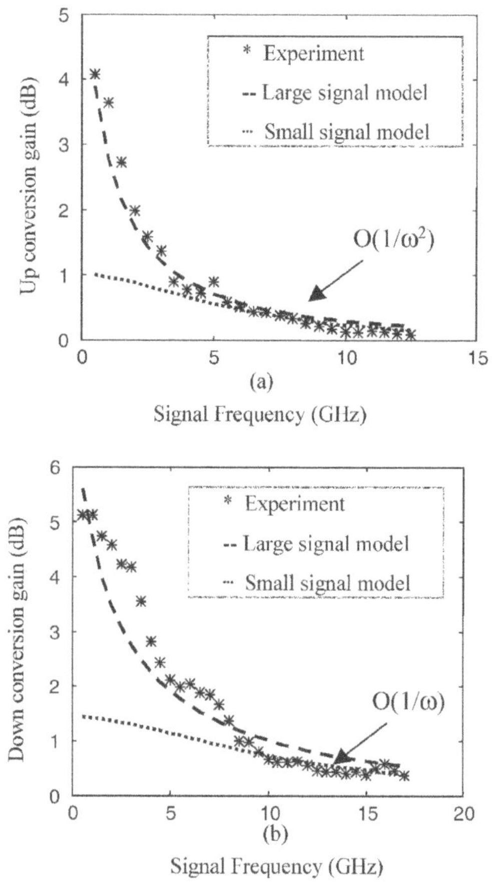

Figure 85 : Comparison between experimental data and calculated values for (a) up conversion process. (b) down conversion process.

This 3 dB difference is a further proof of the statement made previously regarding the inability to separate the HBT OEM into an ideal mixing stage followed by a frequency dependent amplifying stage. Would this separation be valid, and if we assuming a frequency response of the amplifying stage to be the same as that of the current gain of the HBT, the difference between down- and up- conversion products would be larger than 20dB for a 3GHz RF signal and 3.5GHz LO. Since the measured difference is only 3dB we conclude that this separation is invalid and that the HBT is a distributed mixer and a frequency dependent current amplifier operating simultaneously.

Comparison to the Model

The differential equations (47), representing the large signal model, were solved numerically using the MATLAB/SIMULINK software package for the same biasing and power level conditions used in the experiments. In addition, the conversion gain of the small signal model was calculated from the analytic terms of equation (53) for large signal conditions. For both models, the values of the HBT parameters used for the calculations were extracted from small-signal s-parameter measurements using the technique described in [189]. A comparison between the experimental data and the calculated values is shown in figure 85.

In general, the large signal simulation results are in good agreement with the experimental data with only minor deviations. The reason for the deviations is mainly due to calibration errors and to non-ideal behavior of the Bias-T we used. Regarding the small signal calculated results, a good match to the experimental data was obtained only at high frequencies where the LO power levels (–10dBm) are in fact in the small signal modulation range. Indeed, in the high frequency regime, the dependence of the experimental data on frequency agrees with the small signal theoretical predictions of equation (154). In the asymptotic case, $\omega > \omega_0$ (which is about 10 GHz), the efficiency dependence on frequency varies as $1/\omega$ for the down-conversion process and as $1/\omega^2$ for the up-conversion process.

3.2.3. Applications of OEM in Photo - HBT's

3.2.3.1. Diode Laser Frequency Locking for DWDM Applications

Dense DWM fiber optics systems require laser sources with precise frequencies set to the standards of the ITU and detuned by 50 GHz from each other. To achieve this, one of the lasers must be locked to a frequency standard such as a temperature stabilized fiber grating. Opto electronic mixing in a photo HBT is used then to lock the separation between the two at 50 GHz. Additional lasers can be added making a series of frequency locked sources.

Figure 86 shows the schematic of the wavelength stabilizing and refernce locking system of two DFB lasers detuned by 50 GHz. First, in order to stabilize the wavelength of laser 1 and use it as an absolute source, the output of the laser feeds a temperature stabilized fiber grating (Bragg filter) and its wavelength, λ_1, is tuned to the negative slope of the grating transmission function. The Bragg filter acts as a frequency discriminator whose output feeds a control circuit which generates an error current $I_{error\ 1}$. This error current is combined with the DC bias and fed back into laser 1 in order to correct for the deviations from λ_1. The frequency stability of the unlocked laser was measured using a conventional wave meter having a resolution of 1 pm (125 MHz) and was found to be ± 1 GHz. For the locked case the stability can not be measured with the present resolution. It is estimated to be about 10 MHz and may be improved with a faster control circuit. Next, laser 2 is temperature tuned to $\lambda_2=\lambda_1+0.4$nm (a detuning of 50 GHz) and a part of both laser output power is coupled to the optical window of the photo HBT. The base electrode is fed by a 46 GHz electrical signal, resulting in the collector current having a component at the down converted frequency of 4 GHz. The down converted signal feeds a frequency discriminator which yields an error current $I_{error\ 2}$ that is combined with the DC bias of laser 2 and hence corrects for the deviations from the 50 GHz. detuning.

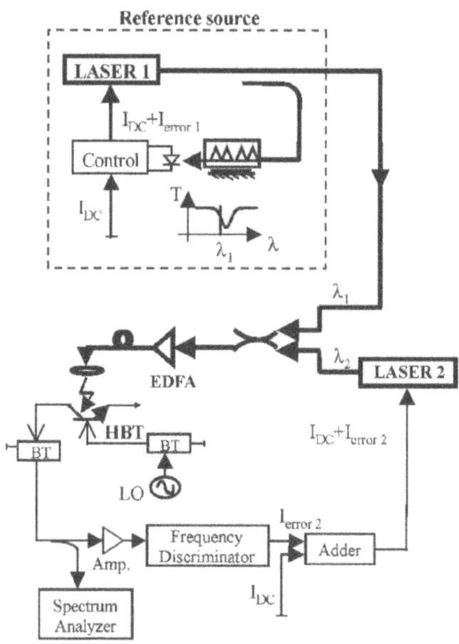

Figure 86 : Experimental set up for the wavelength locking of two DFB lasers detuned by 50 GHz.

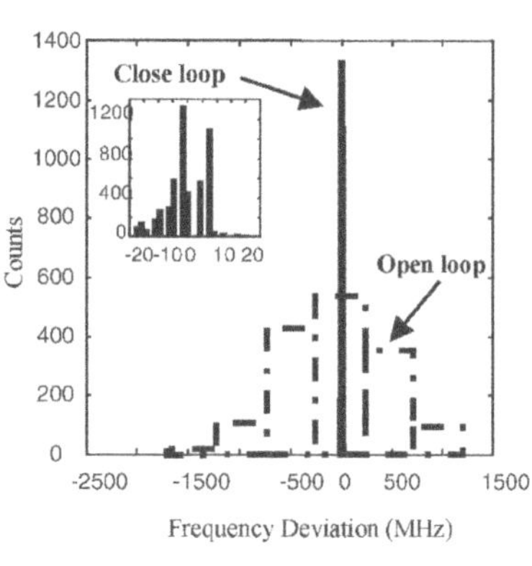

Figure 87 : Distribution of frequency deviations from 50 GHz detuning. Both the open loop and the locked cases are shown. The insert shows a zoomed picture of a closed loop case.

The accuracy of the mutual locking scheme was characterized by sampling the long term frequency differences and extracting the distribution of their deviations from the desired 50 GHz. A narrow distribution of only a few megahertz was achieved, as shown in figure 87, more than two orders of magnitude narower than the unlocked distribution, also shown in figure 87.

3.2.3.2. Optoelectronic Generation and Modulation of Millimeter Waves

The use of millimeter waves requires optical means for the signal generation with OEM mixing being an efficient technique to achieve it. The optical signals can be fed from a remote site to a photo – HBT which can in tern be also modulated electronically.

The scheme we use is described schematically in figure 88. A single frequency (1550 nm) laser is externally modulated at 22.5 GHz with the modulator bias adjusted so as to maximize the two side bands and minimize the optical carrier [191]. At the optimum operating point, the carrier is suppressed by more than 20 dB relative to the side bands. The modulated optical signal is amplified by an Erbium doped fiber amplifier (EDFA) before being focused on the optical port at the base of the photo HBT. The optical detection process generates the frequency difference signal at 45 GHz which serves now as the millimeter wave carrier. At the same time, an analog or digital modulation current drives the base yielding a collector current which consists of a modulated 45 GHz signal. The photo-HBT is used here simultaneously as a fast detector, mixer / modulator and amplifier.

The analog modulation capabilities of the photo HBT were characterized for a 300 MHz sinusoidal drive with a typical output spectrum shown in figure 89. The results shown were obtained for an average optical power of + 0.6 dBm and a modulating signal power of – 28 dBm. The modulation capabilities were quantified in terms of the amplitude modulation index and its dependence on the input modulating signal power.

Characterization of the digital modulation capabilities requires down conversion to base band of the bit pattern imposed on the millimeter wave carrier as described in figure 88. The down conversion was achieved by means of homodyne detection with a harmonic mixer, the output of which was amplified and fed back to the receiving part of the bit error rate (BER) test set. The IF bandwidth of the harmonic mixer limited the modulation rates to a maximum of 300 Mb/s. Measured bit error rates as a function of the optical power coupled to the photo HBT are shown in figure 90 for a data rate of 300 Mb/s. The modulated signal was error free for optical powers larger than +1.7 dBm. The inset of the figure shows an open and relatively noise free eye pattern of the recovered 300Mb/s data.

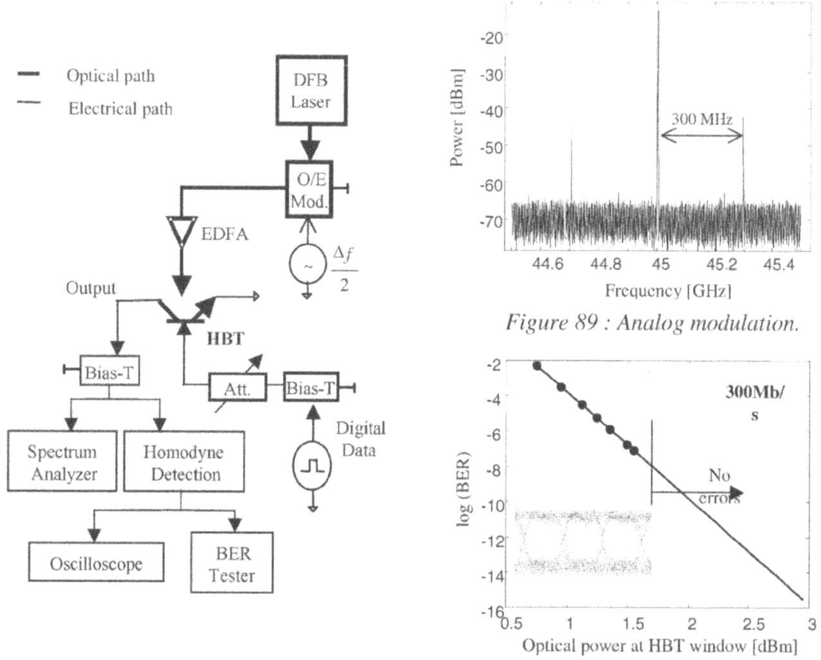

Figure 89 : Analog modulation.

Figure 88 : Experimental set up for optoelectronic generation and modulation of mm-waves.

Figure 90 : Digital modulation.

3.2.3.3. Self Oscillating Photo HBT

The future use of broad band wireless communication services will require massive integration between the network distributing microwave and millimeter wave radio signals and the optical fiber network which will carry broad band data and control signals. One potentially useful configuration places a simple millimeter waver source at remote sites and optical fibers feed those sites with broad band data signals that get imprinted on the millimeter wave carriers. The remote oscillator may be a self oscillating photo – HBT [192] as shown schematically in figure 91.

The collector electrode of the photo HBT is fed back to the base via a narrow band pass filter centered at 2, 10 or 30 GHz, an attenuator and a microwave tuner to form the oscillator. The non linear HBT generates several harmonics of the fundamental self oscillating frequency which are extracted simultaneously via a broad band directional coupler at the collector. A modulated optical carrier is coupled to an optical window at the base region of the photo-HBT and the modulation is imprinted simultaneously on all harmonics. Figure 92 shows a spectrum consisting of the first four harmonics in the 10 GHz oscillator with the optical signal being turned off.

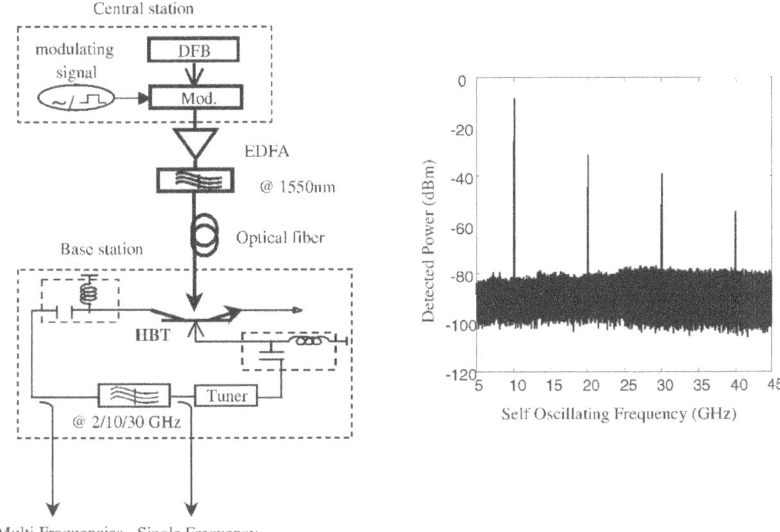

Figure 91 : Experimental set up. of the self oscillator

Figure 92: Spectrum of the first four harmonics of the 10 GHz oscillator with the optical signal being tuned off.

Optical Injection Locking

Optical injection locking is achieved with the injection of an optical signal modulated at the oscillation frequency yielding a reduction in phase noise. The phase noise of the free running oscillator is determined by the Q value of the microwave filter and the loop losses. For the 10 GHz oscillator, the free running phase noise measured at a 10 KHz offset was –81 dBc / Hz. For the 30 GHz oscillator, the loop losses were larger resulting in an increased phase noise of –68 dBc / Hz. These phase noise values, while rather low, may cause difficulties for phase sensitive modulation schemes. With optical injection locking under optimum conditions, the phase noise of the 10 GHz oscillator was reduced to below –100 dBc / Hz and in the corresponding 30 GHz case it was – 80 dBc / Hz.

In figure 93a we display the phase noise dependence on the detuning between the free running 10 GHz oscillator and the injection locking input frequency. The injection level was low, (less than -10dBm, as compared with to a few mW of the free running oscillator) measured using the HBT as a photodetector (namely, with zero bias to the base). Operation with a low optical power is important for applications since it enables the multicasting the optical locking signal between many base stations. The significant effect of the injection locking process is clearly seen in the figure with the phase noise reaching a minimum below –100 dBc / Hz. Figure 93b shows the phase noise as a function of the average optical power under optimum detuning and modulation depth conditions. We note that at injected powers above –13 dBm the phase noise saturates at the low level of ~ -100 dBc / Hz.

The locking range is defined as the spectral region within which the external signal of power P_1 at frequency ω_1 can be tuned relative to the frequency ω_0 of the free running oscillator so that the amplified output power at ω_1 is larger than the free running oscillation output P_0.

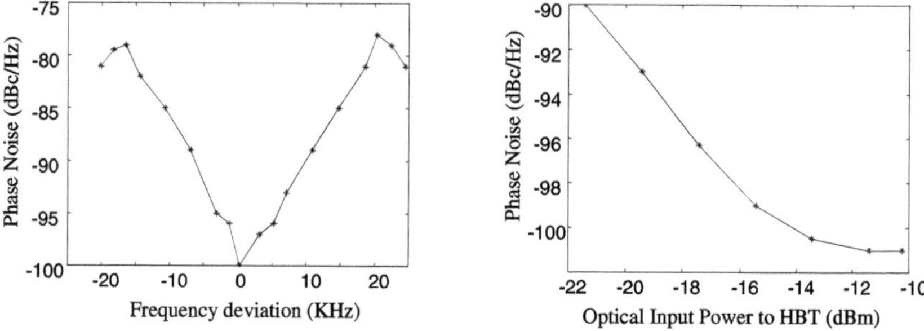

Figure 93 : Phase noise dependence on (a) the detuning, (b) average optical power.

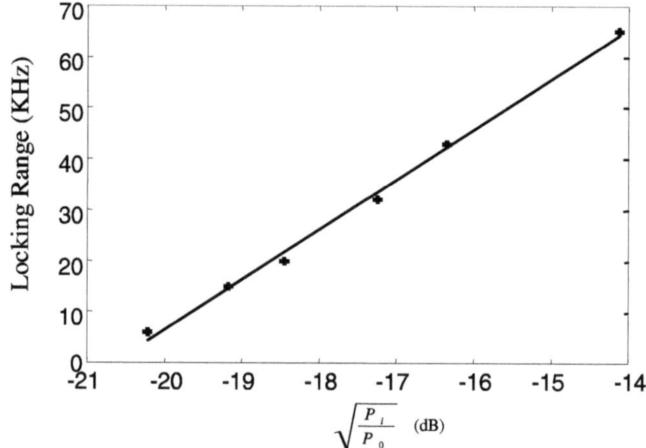

Figure 94 : Locking range as a function of the injection power.

The locking range is quantified by the so-called Adler equation [193], resulting in:

$$\Delta\omega_{lock} = \frac{2\omega_0}{Q}\sqrt{\frac{P_1}{P_0}}$$, where ω_0 / Q is the cold cavity bandwidth.

The results shown in figure 94 validate square root dependence of the locking range on the injection power, and from the experimental data we obtained a cold cavity bandwidth of ~ 1MHz, implying that the Q factor of the suggested oscillator is on the order of 10000. We note that the bandwidth of the bandpass filter was 10 MHz.

Analog modulation

Figure 95 shows a measured spectrum with the first few harmonics of a 10 GHz oscillator. Every spectral component is accompanied by side bands due to the 300 MHz optical modulation. The conversion gain of photo - HBT's operating as optoelectronic mixers was previously characterized using a constant IF frequency, namely, the local oscillator and the modulating signal frequencies were tuned together at a constant difference. The present experiment employed a constant modulating frequency (200 MHz) and the up conversion gain for all spectral lines was characterized at once. For relatively narrow band modulation, the up and down conversion gains are equal. The HBT non linearity generates in this case harmonics of the optically carried modulating signal in addition to mediating the up conversion process.

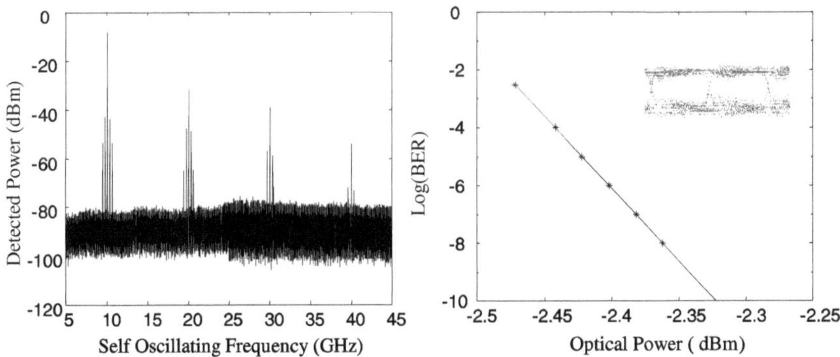

Figure 95 : Measured spectrum of with 300 MHz analog modulation.

Figure 96 : Measured spectrum of with 300 MB/s digital modulation.

Digital modulation

Digital modulation was characterized in both the frequency and time domains. A coherent homodyne receiver detected the digitally modulated signal. The base band signal was amplified and fed to the bit error rate receiver. Bit error rate measurements as a function of average optical power together with the detected eye pattern are shown in figure 96. The results demonstrate the ability to generate a high quality digitally modulated signal for error free transmission.

3.2.3.4. A Two HBT Configuration for Millimeter Wave Generation and Modulation

An advanced millimeter wave source employing a two photo – HBT's configuration is described in figure 97. One transistor is used as a 30 GHz self oscillator which can be, if needed, injection locked by an optical signal to improve its spectral characteristics. The 30 GHz signal is coupled to a second photo - HBT serving as an optoelectronic mixer/modulator with the data being fed to its optical port at the base via an optical fiber. The two optical signals, one for injection locking and the second for modulation can be carried by different wavelengths and separated by a standard WDM demultiplexing scheme.

The self oscillator is constructed by connecting the collector to the base via a 30 GHz narrow band pass filter, an attenuator and a tuner. In a free running mode, the 30 GHz output exhibits – 68 dBc/Hz of phase noise of at 10KHz offset. A sinusoidally modulated optical input signal was used to injection-lock the oscillator and under optimum conditions, the phase noise was reduced to -80 dBc/Hz. A typical analog modulated spectrum (at 300 MHz) is shown in figure 98.

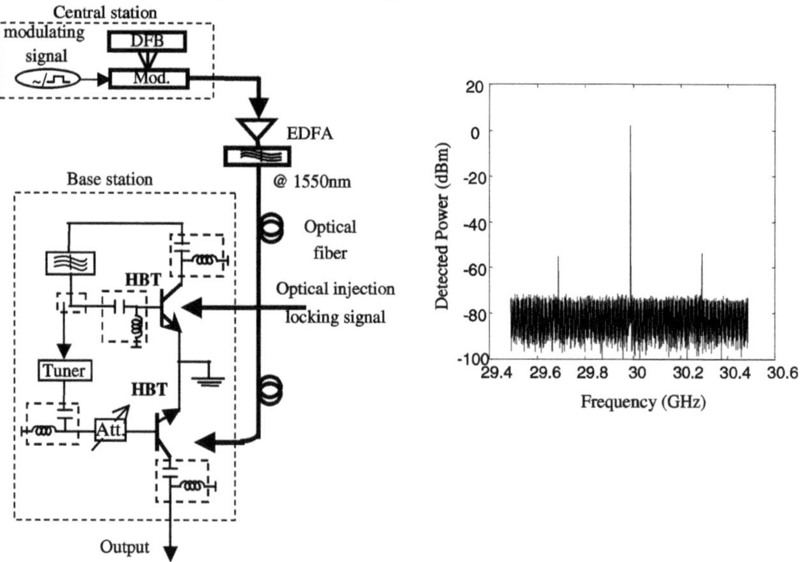

Figure 97 : Experimental arrangement employing a two photo-HBTs configuration.

Figure 98 : Analog modulated spectrum. at 300 MHz

The digital modulation capabilities were tested at a bit rate of 500 Mb/s. The collector output was detected using a coherent homodyne receiver and the resulting base band signal was characterized in both frequency and time domains. Figure 99 shows the base band spectrum with the typical sync envelope and no distortions. The bit error rate (BER) curve exhibited in figure 100 was obtained by feeding the base band signal to the receiver of a BER test set and changing the optical power impinging on the HBT. The results show that the 30 GHz signal can be properly modulated with no errors.

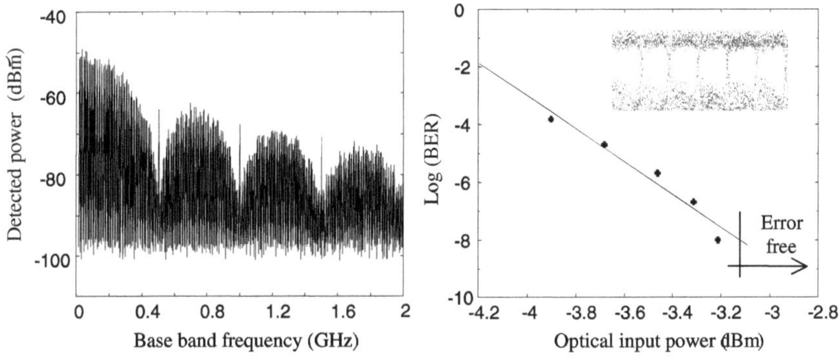

Figure 99 : Base band spectrum of 500 Mb/s digital modulation.

Figure 100 : Digital modulation at 500 Mb/s of the 30 GHz source.

3.2.5. Acknowledgement

The research surveyed in this paper was performed by the graduate students Jacob Lasri and Alberto Bilenca in collaboration with my colleges Prof. Dan Ritter and Meir Orenstein, all from the Technion – Israel Institute of Technology.

4. DIGITAL CONVERTORS, OPTOELECTRONIC PROCESSORS

4.1. Optical and Microwave Signal Processing Applied to A to D Converters

Y. Langard
THOMSON-CSF DETEXIS, 1 bd Jean Moulin, 78 852 Elancourt, FRANCE
yves.langard@detexis.thomson-csf.com

Abstract

Analogue to Digital Conversion represents one of the main limitating factor in linearity for many receiving architectures such as communications satallites, wireless, and multi-purpose radars. In this tutorial, we present a state-of-the-art of analogue to digital converters, relatively to their architectures and technologies. The requirements of the different applications are analysed and point out the limitations of current technologies to achieve high resolution multi-GSPS analogue to digital converters. Major analogue to digital conversion parameters and definitions are mentioned. A review on the current photonics architectures is detailed showing the respective advantages and drawbacks. Benefits of photonics technologies such as low phase noise generation are analysed. Future trends are discussed as a conclusion.

4.1.1. Review of ADC Performance Parameters

Definitions are presented for the terms and parameters that are relevant to the specification and assessment of an ADC. Where appropriate, these are based on those in common use [194-198].

4.1.1.1. Basic Quantiser Structure

Suppose that the quantiser is intended to provide N-bit, two's complement encoding with a uniform input step size. Then the coding map for a device with no imperfections is:

$$Q(V) \equiv \begin{cases} 2^{N-1}-1 & \text{for} \quad 2^{N-1}-1 < \lfloor v \rfloor \\ \lfloor v \rfloor & \text{for} \quad 0 \leq \lfloor v \rfloor \leq 2^{N-1}-1 \\ 2^{N}+\lfloor v \rfloor & \text{for} \quad -2^{N-1} \leq \lfloor v \rfloor < 0 \\ 2^{N-1} & \text{for} \quad \lfloor v \rfloor < -2^{N-1} \end{cases}$$

where V is the input signal voltage, $v = (V - \delta/2)/\delta$, and δ is the input voltage range associated with a single code level – often referred to as the 'input quantum' or 'least significant bit voltage' LSB (see below).

It is also useful to have a quasi-inverse that maps each code k to the mid point of the corresponding input range:

$$W(k) = \begin{cases} k\delta & \text{for } \; 0 \le k < 2^{N-1} \\ \left(k - 2^N\right)\delta & \text{for } \; 2^{N-1} \le k < 2^N \end{cases}$$

4.1.1.2. Input Thresholds, Operating Range, and Basic Linearity

Basic Quantiser Parameters and Figures-of-Merit		
Symbol	**Name**	**Definition**
$W(k)$	Ideal mid-range input voltage for output code *k*.	See above
$V_L(k)$	Lower input voltage threshold for output code *k*.	The minimum input voltage that will produce output code *k*.
$V_U(k)$	Upper input voltage threshold for output code *k*.	The maximum input voltage that will produce output code *k*.
$\Delta(k)$	Input step size for code *k*	The difference between the upper and lower input voltage thresholds for output code *k*: $\Delta(k) = V_U(k) - V_L(k)$
LSB	The input step size δ for an ideal quantiser	In an ideal quantiser, LSB and Full Scale input amplitude are related by $\text{LSB} = \dfrac{2\text{FS}}{(2^N - 1)}$
FS	Full-Scale input amplitude	The maximum amplitude for a sinewave input that does not overload the quantiser. For an ideal quantiser, $\text{FS} = \dfrac{(2^N - 1)\text{LSB}}{2}$.
DNL	Differential non-linearity	The differential non-linearity DNL(*k*), for an output code *k* is defined by: $\text{DNL}(k) = \dfrac{\Delta(k) - \text{LSB}}{\text{LSB}}$, where $\Delta(k)$ is the measured step size for code *k*. The overall differential non-linearity DNL is:

		$\mathrm{DNL} = \max_k \lvert \mathrm{DNL}(k) \rvert$ DNL is expressed in units of LSB
INL	Integral non-linearity	The integral non-linearity INL(k), for an output code k is defined by: $\mathrm{INL}(k) = \frac{V_L(k) - W(k) - \mathrm{LSB}/2}{\mathrm{LSB}}$ where $V_L(k)$ is the measured lower input threshold and $W(k)$ the ideal mid-range input, both for code k. The overall integral non-linearity INL is: $\mathrm{INL} = \max_k \lvert \mathrm{INL}(k) \rvert$ INL is expressed in units of LSB

4.1.1.3. Internal Random Noise and Sampling Jitter

Internal Noise and Sampling Jitter		
Symbol	**Name**	**Definition**
T_A	Aperture delay	The delay between the rising edge of the clock (zero-crossing point) and the time at which the input signal is sampled
Jitter	Aperture uncertainty	The sample-to-sample variation in aperture delay. Normally referred to the sampling epochs determined by the average clock rate, and modelled as a zero-mean random variable characterised by its variance.

4.1.1.4. Tests of Non-linearity with Single Sinewave Input

All the following tests are carried out as frequency-domain assessments, with an input signal that is a pure sinewave of amplitude a and frequency f. Both parameters can be varied, so each figure-of-merit should be condsidered to be a function of frequency and sinewave amplitude. A convenient reference level is Full-Scale input, and the corresponding relative sinewave level is then A dB, where:

$$A = 20\log_{10}(a / FS)$$

Conventional ADC characterisation often concentrates on results for a particular sinewave level, namely 1 dB below Full-Scale amplitude ($A = -1$).

Sinewave Testing – Figures of Merit		
Symbol	**Name**	**Definition**
SINAD	Signal to noise and distortion ratio	The ratio of the power in the wanted signal to the total power in all other spectral components, including harmonics, except DC. Expressed in dB.
SNR	Signal to noise ratio	The ratio of the power in the wanted signal to the total power in all other spectral components, excluding the first five harmonics and DC. Expressed in dB. For an ideal quantiser, $\text{SNR} \approx 10\log_{10}\left(\left(\frac{a^2}{2}\right)\Big/\left(\frac{\delta^2}{12}\right)\right)$ $= 20\log_{10}\left(2^N - 1\right) + 10\log_{10}(3/2) + A$
THD	Total harmonic distortion	The ratio of the total power in the first five harmonic components to the power in the wanted signal. Expressed in dB.
ENOB	Effective number of bits	$\text{ENOB} = (\text{SINAD} - 1.76 - A)/6.02$
SFDR	Spurious-free dynamic range	The ratio of the power in the wanted signal to the power of the next highest spectral component (i.e. the peak spurious spectral component). Expressed in dB. $\text{SFDR(dB)} = 20\log_{10}\frac{\lvert X_{avg}(f_r)\rvert}{\max_{f_{SP},f_H}\left\{X_{avg}(f_H)\rvert \text{ or } \lvert X\right.}$ where: X_{avg} is the averaged spectrum of the ADC output, f_r is the input signal frequency, f_H and f_{SP} are the frequencies of the set of harmonic and spurious spectral components.

SFDR is generally a function of both the amplitude, $X_{avm}(f_r)$, and the frequency, f_r, of the input sinewave, and possibly the ADC sample frequency f_S. Thus, the amplitude and frequency of the input, and the sample frequency, for which SFDR measurement(s) are made shall be specified.

The term "spurious free dynamic range" (SFDR) is often used for this measure where both harmonic distortion and spurious signals are considered to be undesirable "spurs" in the spectrum of a sampled pure sine wave. SFDR is used to indicate the ADC usable dynamic range beyond which special detection and thresholding problems occur in spectral analysis.

4.1.1.5. Tests of Intermodulation

Two-tone test

To avoid the shortcomings of single sinewave testing as a method of assessing low-order non-linearities – in particular for band-pass applications – two-tone testing is recommended. The test signal comprises a pair of equal-amplitude sinewaves with different frequencies, approximately equally spaced from the centre of the band of interest. Intermodulation distortion spectral components may occur at sum and difference frequencies for all possible integer multiples of the input frequency tones or signal group frequencies. The frequency separation f_d from band centre is chosen so that intermodulation products up to the maximum order of interest, *m,* (at frequency separation $(2m-1)f_d$) will fall within the system pass-band. If the input test frequencies f_{R1} and f_{R2} are set to values which are an odd number of DFT bins away from band-centre then the difference f_d between f_{R1} and f_{R2} is then always an even number of DFT bins.

Table 4. Intermodulation

Symbol	**Name**	**Definition**
IMD	Intermodulation distortion	The two-tone intermodulation distortion is the ratio of either input tone to the worst third-order intermodulation product. To avoid overload, the (equal) input tone levels must each be no greater than -6 dB relative to Full-Scale (-7 dB is often used).

4.1.2. State-of-the-Art High Speed Electronic ADCs

The state-of-the art of conventional high speed electronic ADCs is reviewed. Firstly, what is meant by 'high-speed' must be defined for analogue to digital conversion. The definition of high speed for ADCs is dependant on the number of bits of resolution. A 10 MSPS sampling frequency represents a high speed frequency for very high resolution 16 bits ADC. For 12 to 14 bits ADC, 10 to 100MSPS represent high speed frequencies while 8 and 10 bit ADCs, which represent relatively low resolution, will be considered at high frequencies for sampling rate above 100MSPS. Figure 101 gives an overview of the global state-of-the art of electronic ADC.

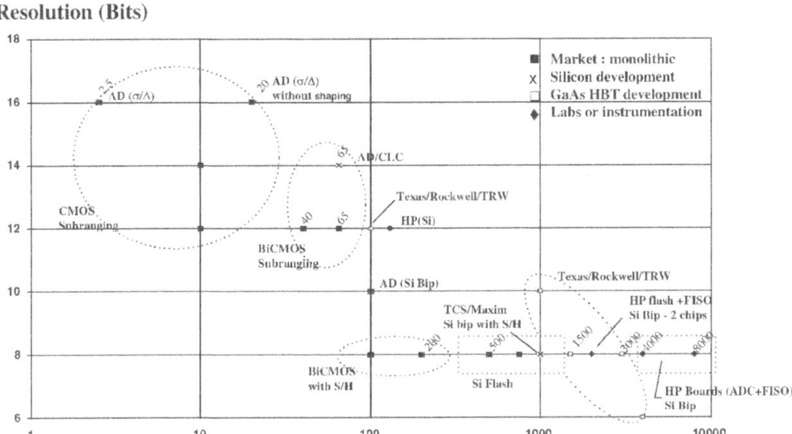

Figure 101 : State of the art of electronics ADCs.

However sample rate should not be the sole consideration as a criteria for the definition of a high speed ADC. Input frequency (analogue bandwidth) should also be considered. For example in an undersampling scheme, for an 8-bit ADC, a 10MHz instantaneous bandwidth at a 100MHz input carrier frequency, which is sampled at 20MHz should be considered as a high speed A-to-D conversion process.

The state-of-the art of 8 to 10 bits high speed electronic ADC will now be focussed upon. Three classes of products can be discerned :

- Commercially available products
- Existing products in specific systems but which are not commercially available
- Devices demonstrated in a laboratory environment

4.1.2.1. Commercially Available Products

4.1.2.1.1. Monolithic Parts.

The materials technology used is primarily silicon. Research is also ongoing into devices based on GaAs technology but this will be described later. The TS8388 from THOMSON-CSF TCS is representative of the state-of-the-art, allowing a sampling up to 1GSPS with 50dB linearity up to 500MHz input frequency and 40 dB at 1GHz. TCS use an innovative architecture, including an on chip sample-and-hold, fabricated with an advanced high speed bipolar process (B6HF from Siemens). The ADC allows a low 500mVpp differential or single ended analog inputs and provides either Gray or binary output data format. It present a 1.8GHz full power input bandwidth. The power consumption of the product is a quite low 3W. The product is currently packaged in a Ceramic Quad Flat Pack package (CQFP 68).

Rockwell Science Center offer a preliminary data sheet for a nominal 8-bit, 2 GS/s device [199] (RSC-ADC080E/S) with an analogue bandwidth to 3 GHz. The SINAD performance of this device drops quite rapidly to around 30 dB as input frequencies approach 1 GHz and this for on-wafer probing. The performance and availability of appropriately packaged devices is still to be determined.

Maxim is trying with the MAX104 to achieve 1 GSPS sampling rate with an amplitude interleaving technique. It seems however, that they have some difficulties since the product is still not available, several months after its nominal launch date.

4.1.2.1.2. Hybrid Parts

Before the availability of high speed technologies allowing the realization of high speed monolithic products, Tektronix used a well known method to achieve high speed sampling, known as the interleaving technique. The TKAD10C, originally developed by Tektronix, is now available from Maxim as the MAX101 device. It uses a bipolar technology from Tektronix to perform a high-speed Sample-and-Hold on a single chip and also includes two high speed ADCs interleaved together on another chip. Each of these ADCs samples at 250MSPS to achieve a global 500MSPS sampling rate. The device offers a 1.2GHz full power input bandwidth.

4.1.2.2. Existing Products in Specific System but not Commercially Available

High speed electronic ADC have been widely used for the last tens years to realize digitising oscilloscopes. Mainly Tektronix and Hewlett-Packard have been working in this domain. Both have used interleaving technique to increase the sampling frequency .

HP has pushed the ultimate to realize an 8 GSPS sampling rate. The converter is realized using 4 boards working each one at 2 GSPS to achieve a global 8GSPS. Each board is build up with four 7 bits ADCs interleaved together, working each one at 500MHz. A dithering technique allows an additional 8th bit to be obtained. A lot of calibration is necessary to adjust different levels of the boards. Such a complicated technique is only achievable in an instrument such as an oscilloscope where calibration before measurement is more easily realisable.

All of these products are generally not available for customers due to the fact that they have not been designed for that, and even some of them would be available, they would be too difficult to implement and too expensive.

4.1.2.3. Demonstrated in Research Laboratories

Some industrials have focussed on other technologies trying to achieve very high speed devices. In the U.S. the DARPA organisation has financed Companies such as Hughes, TRW, Rockwell and Texas Instruments over a number of years to develop very high-speed ADCs. Hughes have reported [200] on a Flash device developed in an AlInAs/GaInAs HBT process on InP substrate which offered 8 GS/s but only 3-bit resolution. The analogue bandwidth of this device however was implied to be 3.6 GHz.

Rockwell Science Centre have also reported [201] on a 1 GS/s, 11-bit Track-and-Hold amplifier implemented in a production AlGaAs/GaAs HBT process. Experimental results for a two-stage pipelined 10-bit ADC which is presumed to make use of the T&H amplifier have also been reported [202]. Such a device would certainly seem to be of interest to a future full implementation of the HELIOS sampler system.

4.1.3. Photonic ADC Developments

This section presents a summary review of other reported photonic based approaches to high speed Analogue-to-Digital Conversion.

4.1.3.1. Photoconductive Switch

4.1.3.1.1. Auston Optoelectronic Sample-and-Hold

The arrangement described by [197] and shown in figure 102 describes the most simple technique to sample electrical signals with an optically activated photoconductor switch. The photoconductor switch is used as a series sampling switch. It is turned on when illuminated by a pulse of laser light and charges the capacitor C, which is read out by a buffer.

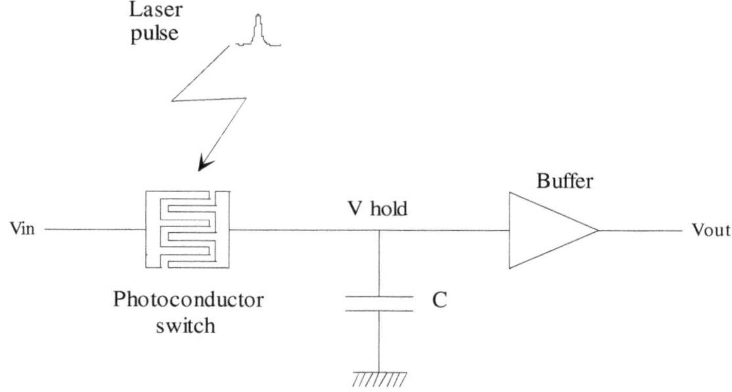

Figure 102 : Auston switch.

In reference [197], the photoconductor used consists of a thin slab of high-resistivity silicon, the switch active area width and length are equal to 340μm. The gap resistance of the switch is more than 100kΩ, and its gap capacitance is 10^{-14}F. A rise time of approximately 10ps and a fall time of approximately 15ps have been measured for this switch.

The switch realised by Sun [198] is an interdigited metal-semiconductor-metal (MSM) structure fabricated on a semi-insulating (Fe doped) InP substrate. Each switch has an active area of 50x50 μm^2, a 2μm electrode spacing and a 2μm electrode width. The switch is activated by a 8ns train of optical pulses from a GaAs/AlGaAs laser, the repetition rate is 20MHz and each train consists of eight 1ns duration optical pulses. The measured optical energy of the optical pulses train is 8nJ. For this sample and hold (S/H) circuit, a R_{on} value of 1.5kΩ and a R_{off} value of 57kΩ have been measured.

4.1.3.1.2. Monolithic Sample-and-Hold Circuit

The principle of this monolithic hybrid optoelectronic S/H circuit [199] is presented in figure 103. The switching device is a GaAs MESFET which is turned on and off by the photoconductor. The MESFET control is done by varying its gate-source potential by passing the photocurrent I_p through the combination of a current source I_{source} and a shunt conductance G. When the photoconductor switch is not illuminated ($I_p < I_{source}$), the gate voltage is close to V_{ss} and the MESFET is off. When the switch is under illumination ($I_p > I_{source}$), the gate voltage rises until the MESFET is on.

With this architecture, the voltage transition at the gate of transistor can be set by a small change in the photoconductor current. Because of that, the circuit is immune to variations of the extinction ratio of the laser source. Moreover, the sampling instant can be adjusted by varying the current I_{source}.

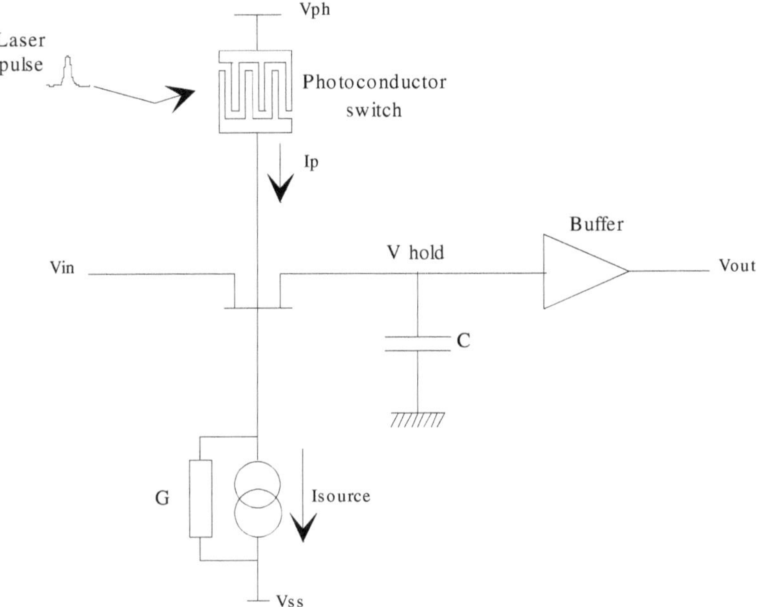

Figure 103 : Monolithic sample and hold circuit.

Mason and Taylor [199] have employed a MSM photoconductor with an active area of 60x60 μm^2 and an electrode spacing and width of 4μm. An integrated circuit has been realised using the standard GaAs MESFET foundry process. The photoconductor switch is activated by a on-off modulation of an 830nm laser diode. They have measured samples rates approaching 500Ms/s (sample per second) for a 400mV r.m.s. input sinusoid with a resolution of 7.6 bits.

4.1.3.1.3. Low Temperature GaAs Photoconductor Switch

Unlike photodiodes whose bandwidths are determined by the electrode spacing, the response times of photoconductive detectors are determined by the carrier lifetime of the material that is used. Using material with ultrashort carrier lifetime (about 1ps), ultrafast and high-power optical switches can be achieved. The influence of the physical properties of the photoconductive material on the photoconductor performances are presented in the following table.

Parameter	**Symbol/Units**	**Photoconductor Performance Influenced**
Mobility	μ (cm^2/Vs)	Sensitivity
Resistivity (dark)	ρ (Ω.cm)	Dynamic Range
Lifetime	τ (ps)	Bandwidth

Influence of the physical properties of the photoconductive material on the photoconductor performance.

Usually when the life time of a material is shortened, the mobility is also lowered. In 1990 Smith *et al.* [201] discovered that MBE GaAs grown at low temperature (LT-GaAs) (200-300°C) exhibited new and interesting properties. This material exhibits, after annealing, semi-insulating properties [202], and ultra short carrier lifetimes [203] (about 1ps) have been measured in this material. Unlike other materials used to make fast photoconductors (implanted, highly doped or polycrystalline materials), in LT-GaAs, the properties needed to make fast photoconductors (ultrashort carrier lifetime, high dark R_{off} resistance and high breakdown field) can be obtained together which a good mobility (>500cm^2/Vs). Low R_{on} resistance can therefore be obtained when the photoconductor is illuminated. These properties make LT-GaAs ideal for electrical pulse generation and gating [204].

4.1.3.2. HELIOS Architecture

4.1.3.2.1. Partners

This work is carried out by two partners, BAE SYSTEMS Research Centre (with support and participation of BAES Defense Systems Division) and THOMSON-CSF DETEXIS and LCR.

4.1.3.2.2. Demonstrator

The architecture of the proposed system is shown in figure 104. The RF signal is converted to an optical signal using a (linearised) modulator at the array face. The modulator is fed by a high power laser which will probably be located remotely from the antenna. This aspect of the demonstrator is essential for any real application, since a prime aim of this R&D is to remove mass and volume from the array face.

The rf-on-fibre signal is then fed to an optical commutator switch, driven at some sampling frequency and its (phase-locked) sub-harmonics. The output from this is a pulsed sample of contiguous sections of the rf signal from the antenna, fed sequentially into the output channels. Each of these are then detected using photodiodes and preamplified at an appropriate bandwidth for the succeeding channel. A delay network to synchronise all channels and to bring them to their photodiodes at exactly the same time may be optionally insterted at this stage.

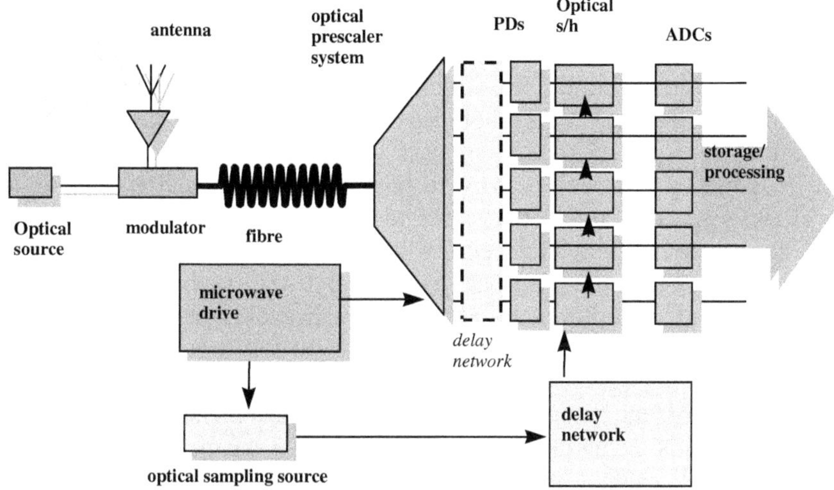

Figure 104 : Demonstrator architecture.

Phase-locked to the optical switch drive signals is a pulsed laser source which gates a series of photoconductors, used as sample-and-hold devices. Beyond this an array of electrical analogue to digital convertors (ADCs) will digitize the signals for storage or onward processing.

4.1.3.3. Time-Stretched Analogue-to-Digital Approach

The approach proposed and demonstrated by Jalali *et al* [210] and Esman *et al* [211], is to time-stretch the electrical signal in the optical domain before sampling and digital conversion. Performing time-stretch in the optical domain is equivalent to slowing down the microwave carrier and its modulation.

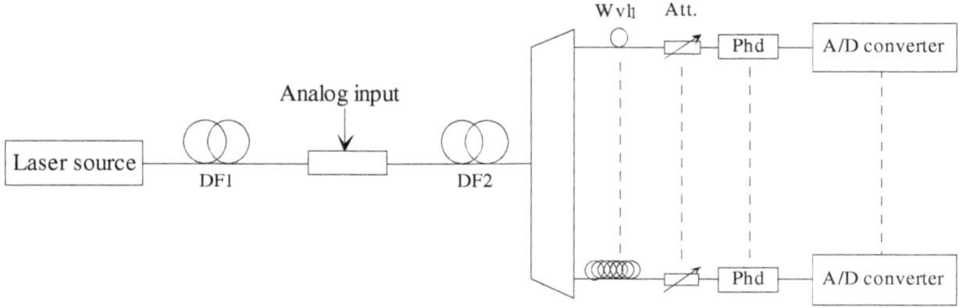

Figure 105 : Time stretched general approach.

The implementation of the time-stretched analogue to digital converter (ADC) is depicted in figure 105. It uses a modelocked erbium–doped fibre ring laser, the pulse generated by the laser is dispersed in a highly-dispersive fibre. The intensity of this chirped pulse is modulated by the RF signal to be converted. The amplitude modulated pulse is then dispersed in a second fibre, which stretched the modulation envelope. The envelope detected by a photodetector is then digitised with a familiar ADC.

This approach can be used to both finite-time signal and continuous-time signal. In this latter case, the signal has to be segmented and interleaved into *n* channels, where *n* is the stretched factor. The segmentation could be done by optical filters such as a wavelength-division demultiplexer to separate the wavelength-coded signal into parallel channels. In this case, each channel following the demultiplexer must have a length of fibre to synchronise the arrival time of the signal on the detector, and an optical attenuator to equalise interchannel amplitude.

With this technique, a stretched factor of 8 has been demonstrated by Jalali et al., making it possible to obtain an overall sample rate of 8Gs/s with eight 1Gs/s interleaved ADC. Esman et al. have demonstrated sampling of a 10GHz narrow band signal, the dynamic range for this signal is 6 bits. The optoelectronic ADC has been characterised with frequencies up to 18GHz. The effective number of bits approximates to 6 bits at 2GHz, and 5 bits at 18GHz.

4.1.4. Conclusion

Test parameters and techniques associated with the characterisation of ADC's in general have been summarised to provide a framework around which the test schedule of the experimental demonstrator can be defined. The current status of the optical components which are central to the HELIOS architecture, i.e. photoconductive switch elements, have been described together with a summary of the state-of-the-art of electronic and other photonic approaches to Analogue-to-Digital Conversion.

4.2. Optoelectronic Processors : an Overview

P. Chavel
Laboratoire Charles Fabry de l'Institut d'Optique (CNRS), BP 147, 91403 Orsay cedex, FRANCE
Pierre.Chavel@iota.u-psud.fr

Abstract

It has been known for a long time that imaging in optics is to a very good approximation a correlation between the input image and some impulse response describing diffraction and other effects in the optical setup. Consequently, image correlation can be implemented by a suitably designed optical setup, where in particular diffractive optical elements may be used to tailor the impulse response. This approach, while intrinsically operating in parallel over all pixels of the input image, has suffered for many years from the slow response of existing input and output devices. Some improvement has been witnessed recently and optical image correlators can sometimes offer performance levels comparable to dedicated electronic hardware. After a review of the basics, the major algorithmic approaches that advocate the use of correlation for applications are described, performances are given, and optoelectronic cellular automata, an extension of the concept of an optical correlator, will be mentioned.

4.2.1. Optical Imaging and Convolution: basics

In this introductory section, we review the fundamentals of Fourier Optics, i.e., the description of light propagation in terms of linear operators. We concentrate on imaging systems, mentioning the main concepts and warning the reader about the main pitfalls of the subject.

4.2.1.1. Optical Convolution : a Systems Approach to Imaging

One way to look at imaging, the most standard operation of optics, is to adopt a systems point of view derived from electronics. Essential part of electrical and electronic circuits are of the linear, time–shift invariant category. From elementary algebraic properties of the Fourier Transform, the output of such systems is known to be described in the time domain by a convolution between the input and the impulse response proper to the system, in the frequency domain by a product between the input spectrum and the frequency dependant gain function. Denoting by O the input function, by I the output function (note that, intentionally, letter I does not denote the input in this case) and by the tilde operator ~ the Fourier transform, the input–output relationship in a linear, shift–invariant system is described by the following well–known equations:

$$I = O * R \tag{55}$$

$$\tilde{I} = \tilde{O}G \tag{56}$$

where R is the impulse response and G, its Fourier transform, is the gain function. The star denotes convolution. Intentionally, these equations are given as relations between functions, without explicitly mentioning the variables. For electronic systems, the variable is time, t, for equation (55), and ν, frequency, for equation (56): it is notorious that a vast class of common devices including resistors, capacitors, inductances, are linear in terms of input–output relationship for voltages (or currents) – in short, twice the input gives rise to twice the output – and shift–invariant – if the input is delayed by t_o, so is the output.

That this approach is to a large extent valid for optics, and in particular for optical imaging, has been recognized about fifty years ago [220] and has become the basis of the branch of optics known as Fourier Optics, of which reference [221] is probably the most classical textbook. In Namely, optical imaging is essentially a linear phenomenon in space between the input, called the "object", and the output, called the "image": if the object is multiplied by two, so is the image, and if the object is shifted by an amount x_o, y_o, so is the image. In this case, the proper variables are the two space coordinates that describe the object (or image) plane x,y. The Fourier conjugate space μ,ν is known as the spatial frequency plane. Optical imaging systems are therefore essentially linear, shift invariant systems operating on a two–dimensional space. The impulse response, also termed "point spread function", R(x,y) is the image of a single object point, which, both because of aberrations and of diffraction, is not a point. Its Fourier transform, G(μ,ν), is usually known as the Transfer Function. A few other striking differences between linear optical imaging systems and linear electrical circuits may deserve mentioning here.

4.2.1.2. The Optical Fourier Transform

One wonderful property of optical imaging systems is that they seem to be the only case where, at least in some situations, physical access the Fourier plane is possible: this is simply due to the fact that Fraunhofer diffraction (also sometimes called far field diffraction, although this expression is not exempt of ambiguity) is expressed analytically as a Fourier transform. Specifically, consider the system sketched in figure 106, where in the left hand part an object O(x,y) is illuminated by a spherical wave originating from a laser of wavelength λ and converging to point Ω located in plane ξ, η on the optical axis z. The ξ and η axes are parallel to the x and y axes and are used simply to distinguish between the two parallel planes involved. In the limit of a set of standard approximations known as the Fresnel approximations, which in particular imply low aperture of the beams, the wave in plane ξ, η is expressed as

$$A\tilde{O}\left(\frac{\xi}{\lambda d}, \frac{\eta}{\lambda d}\right) \tag{57}$$

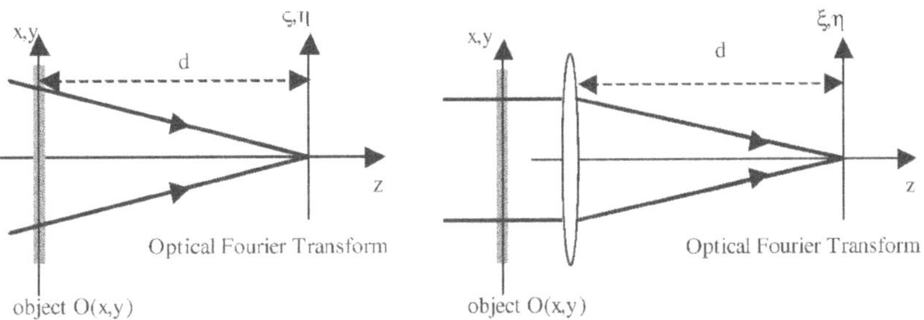

Figure 106 : Two standard optical Fourier transform situations.

In this equation, A is a constant of no further relevance to this discussion and d is the distance shown on the figure. Indeed, the optical spectral plane is displayed in scaled units, whereby spatial frequency coordinate μ,ν appears at point ξ/λd, η/λd in the Fraunhofer diffraction plane, also known as the optical Fourier plane. Equation (57) is a straightforward extension of elementary textbook treatments of diffraction by slits, disks, diffraction gratings and similar simple objects. The right hand part of figure 106 is, to a good first approximation, an alternative to the same operation where a lens transforms the plane wave illuminating the object into a spherical wave. In this case, d is just the focal length of the lens. This latter setup is often sketched in the case where the object is located in the front focal plane, a distance d in front of the lens, because this slightly simplifies the mathematical analysis of the system in terms of diffraction, it is not an essential requirement.

From above, it follows that two cascaded optical Fourier transforms constitute an archetypal optical case of linear, shift invariant optical imaging system. The system is sketched in figure 107.

The left–hand side of both parts of figure 107 are identical with the two parts of figure 106, respectively. They are complemented with a second Fraunhofer diffraction on their right–hand side, which according to a well–known property of Fourier transformation yields a copy of the object in the image plane. The interesting point is that because the Fourier transform of object O is physically available in plane ξ,η, it can be manipulated by inserting any kind of mask that will change the modulus or the phase of any arbitrary part of the spatial frequency plane: this is the initial principle of spatial frequency filtering, that was already recognized in rudimentary form by Ernst Abbe in 1873 [222]. The upper and lower part of figure 107, again, do not essentially differ by their function but the second, also known as the 4f optical system, is simpler to analyse for a first introduction in Fourier Optics courses; this is why it is more common. In both cases, the ξ,η plane, where the input Fourier transform appears, is known as the "pupil plane".

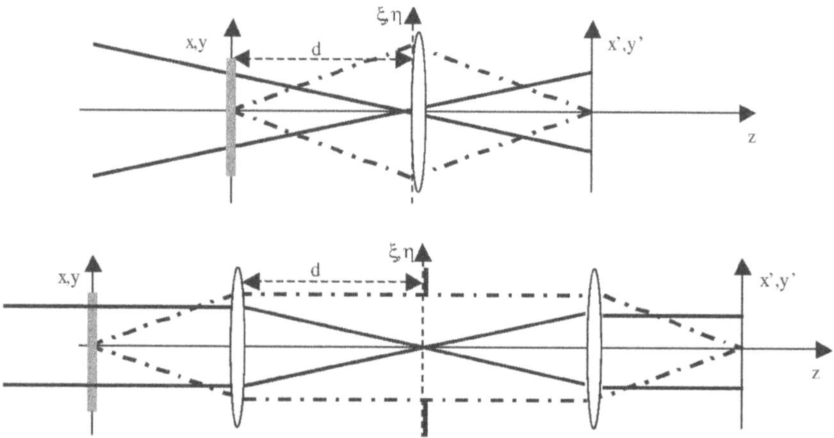

Figure 107 : The basic double diffraction imaging setup for spatial frequency filtering. Bold dash–dotted lines indicate optical conjugation.

The principles of spatial filtery filtering best apply if it corresponds to the pupil plane of the theory of fields and stops in optical imaging, i.e., the aperture of beams emitted by object points considered individually should be limited by a stop in that plane: this stop is just the diaphragm of the lens in the upper part of figure 107, while in the lower part it is represented in cross section as two segments.

4.2.1.3. Some Particularities of Optical Convolution

While the above consideration have shown that imaging is a two–dimensional analog of linear electrical circuits, it is appropriate to point out four differences that probably explain why it took relatively so long to analyse optics in terms of linear systems theory.

Magnification:

While time is a uniquely defined parameter for the input and the output space of an electrical linear system, the same does not straightforwardly apply to object and image coordinates in optics. Because of magnification, linear system theory applies only between the object at input and a scaled version of the image at output (or conversely, between a scaled version of the object and the image). Using the normalized coordinates system already by Seidel and Schwarzschild in the nineteenth century [223] yields a formal solution to the difficulty at the expense of some loss in immediate intuitive appeal of the parameters.

Field and aberrations:

More fundamental is the fact that space invariance in optical systems is limited by two factors: an optical instrument has a field of view, which is the portion of space that sends light into the system. In addition, most aberrations vary inside the field of view, which is equivalent to saying that the impulse response R(x,y) is not shift invariant. Linear system theory therefore only applies inside the field of view of optical instruments and over a spatial extent where aberrations are essentially constant.

Coherence:

The third factor with linearity in optics is the optical quantity of interest. So far we have not clearly mentioned what physical quantity is denoted by O(x,y) and by I(x,y), except that we called it "wave" in connection with equation (57). Indeed, the appropriate quantity depends on the state of coherence. In a completely coherent situation, such as usually with laser illumination, O and I denote so–called light "disturbance", i.e., the complex amplitude of one arbitrary component of the electromagnetic field (extension to vector coordinates in polarization sensitive systems is possible). In a completely incoherent situation, O and I denote photometric quantities: typically, O is object brightness and I is image illumination. Partial coherence requires a more complex treatment, where bilinear rather than linear relations come into play. To summarize:

- in coherent illumination, light amplitudes are related linearly. The impulse response R(x,y) is the complex amplitude of the "light disturbance" generated in the image plane by one single source point. Because of the optical Fourier transforming properties mentioned above, the transfer function $G(\mu,\nu)$ is just a scaled version of the pupil transmittance. In particular, the field stop sketched in figure 107 automatically acts as a low pass filter.
- Under incoherent illumination, photometric quantities related with the power carried by light are related linearly. The impulse response R(x,y) is the flux per unit area in the image of one single source point. The Fourier transforming properties do not exist in this regime, and the transfer function $G(\mu,\nu)$ can be shown to be the autocorrelation of the pupil transmittance.

Transmittance:

One final point worth commenting on is the following: how is an optical "object" defined? Self–luminous objects are mostly incoherent sources that, as just pointed out, are represented by their brightness. But most objects of interest in signal handling and processing are not self–luminous. Instead, they are illuminated by some primary source such as a laser. One then has to enquire whether the object can be characterized in an unambiguous manner as a mere function of two space coordinates or whether they are directly affected by the illuminating field. Of course, spectral properties – object colour – constitute by themselves a separate dimension, wavelength (or frequency). Let us consider here monochromatic illumination. The very idea, already used above, that coherently illuminated objects possess a "transmittance" implicitly relies of the following assumption: if A(x,y) be the light disturbance impinging on the object, the disturbance leaving the object is

$$O(x,y)=A(x,y)t(x,y) \tag{58}$$

where the "complex amplitude transmittance" t depends on x and y but is independent of A(x,y). This is true only for objects that are sufficiently thin. Thickness effects arise from three phenomena

- oblique incidence increases the path in the object substrate, which affects the phase delay and therefore changes the phase of t(x,y);
- shadowing effects between various planes inside the object thickness are strongly dependent on illumination,
- and the Bragg effect, a multiple interference effect between the various layers inside a thick periodic object that arise as soon as object thickness e and period Λ obey the following equation, where λ is the wavelength in the medium:

$$e\Lambda \geq \lambda^2 \tag{59}$$

To conclude this short discussion, we have to admit that the concept of transmittance usually depends on illumination, i.e., t(x,y) depends on A(x,y). If the quantity of interest is t and the illumination has significant variations in space, for example like a spherical wave, this is one further restriction to linearity. What has just been said about the object transmittance applies exactly as well to the pupil transmittance.

4.2.2. Optical Correlation Processors

4.2.2.1. Shaping the MTF/Impulse Response

While spatial frequency analysis has first been applied to optical systems to develop an analytical model of the imaging process, a straightforward extension is to investigate the possibility to arbitrarily shape the impulse response, or equivalently the transfer function, thus transforming optical imaging systems into linear filters that operate on two dimensional objects, as first pointed out by Maréchal et Croce [224]. This might at first sight appear an easy task since the transfer function is simply an appropriately scaled version of the pupil transmittance: if, in addition to the lenses sketched in figure 107, a thin plate with transmittance $t_p(\xi,\eta)$, the filter, is introduced in the pupil plane, then the transfer function is simply

$$G(\mu,\nu)=t_p(\lambda d\mu,\lambda d\nu) \tag{60}$$

Cancelling spatial frequency domains by blocking some regions of the pupil plane is therefore the easiest optical processing operation, also known as Abbe's experiments. However, it is a technically demanding operation to implement more general transfer functions, i.e. to shape at random the filter phase and modulus so that each point in the mask can be modulated in transmittance to reach any t_p value inside the unit disk in the complex plane. In the following subsection, we discuss four possible options.

4.2.2.2. Four Basic Correlation Setups

Direct modulation

The most straightforward approach is to use existing technology to fabricate the best possible approximation of the desired complex amplitude function $G(\mu,\nu)$. Some liquid crystal devices allow to access, using as control parameter the voltage at each pixel, a curve that spans a reasonably extended part of the complex unit disk, such as that presented in figure 108.

Dammann gratings

In the early days of computer generated holography, Tribillon and Dammann [225] recognized the need for methods to use binary masks to generate a class of intensity distribution patterns of special interests: arrays of regularly spaced spots, usually all of the same brightness although the method is straightforwardly extended to arrays of weighted spots. The references indicate extensive work on the calculation of binary masks whose Fourier transforms belong to that class: essentially, these are a special set of diffraction gratings. Efficiencies up to 80% are obtained. Extensions to two dimensions have been proposed.

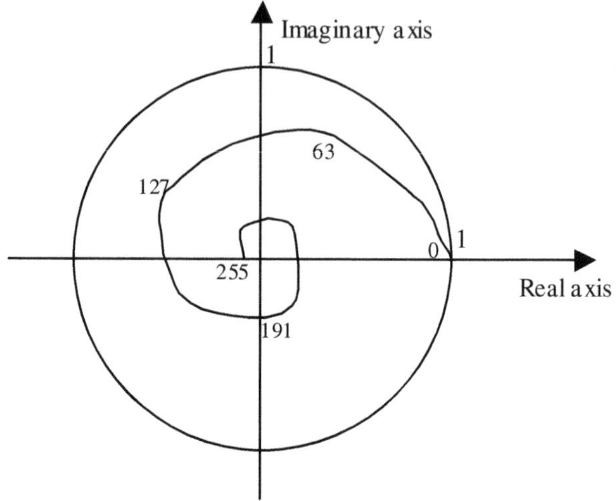

Figure 108 : A typical complex unit disk path of a liquid crystal device. Figures indicate eight–bit coded voltage.

With current technology, these two approaches remain either limited to a fairly sparse subset of the complex unit disk or fairly costly. This is why holographic methods, where field of view is traded for filter fabrication flexibility, have been introduced as explained below.

The Vander Lugt correlator:

Figure 109 schematically depicts the two–step process involved in a Vander Lugt correlator [226], which is directly derived from holography. The filter is recorded as the hologram of the impulse response, as shown in the left part of the figure. The impulse response can either be available physically for direct recording, or available in numerical form for the fabrication of a computer generated hologram, in which case the filter is plotted or displayed using some computer controlled device rather than optically recorded.

In the right part of the figure, the filter is "reconstructed" using the setup of figure 107. Each point P of the object reconstructs the hologram, thus generating in the first diffracted order of the hologram a copy of the desired impulse response, shifted according to the position of P. All impulse responses add up and generate the desired result. However, the other diffracted orders of the hologram, in particular orders 0 and 2, limit the space available for the correlator output.

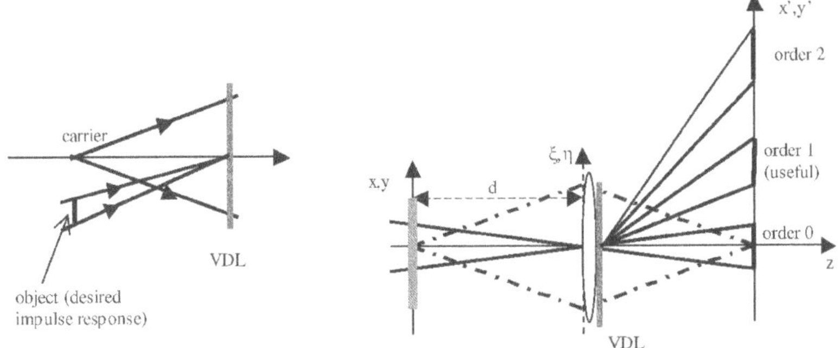

Figure 109 : The Vander Lugt correlator. The holographic Vander Lugt hologram recorded in the right hand side part of the figure is used as the filter in the left hand side.

While obviously recording the hologram requires coherent light, in a Vander Lugt correlator the reconstruction part, which is the correlation itself, may involve coherent or spatially incoherent illumination.

The joint transform correlator:

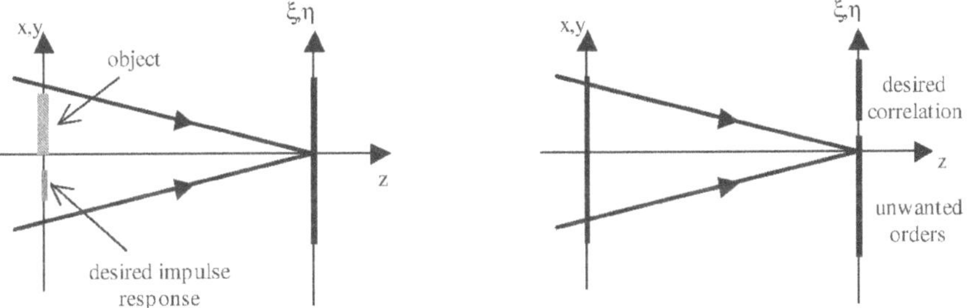

Figure 110 : The joint transform correlator relies on two successive optical Fourier transforms.

Figure 110 schematically depicts the "joint transform" correlator, a two–step process as well.

The object and the impulse response are placed side to side in an input plane and their Fourier transform is recorded as mentioned in figure 106. The recorded pattern is similar to a hologram, except that the carrier is replaced by the impulse response. Upon reconstruction in a further optical Fourier transform setup, one can show that because image detectors detect energy, i.e. they are quadratic in the light disturbance, the desired correlation is obtained. Here again, space is limited by unwanted diffraction orders. Because the joint transform correlator consists of two Optical Fourier Transform, it is limited to coherent light.

4.2.2.3. Algorithmics

Convolution, the combination of shifted versions of an image, has a variety of applications in signal processing. If the signals at all object pixels are modulated in time, as might be appropriate in the context of this Optics and Microwaves school, each image pixel may be used to combine input channels. By far the most common domain of application however, is pattern recognition and its correlates, tracking, target localisation, motion detection, pattern classification. Initial work on optical pattern recognition motivated Vander Lugt (ref. [226]) when he invented his filter: the idea was to extend to pattern recognition the concept of matched filtering that had been known for some time in one–dimensional signal processing. If an object O(x,y) is searched for the presence of some target pattern s(x,y), using s(x,y) as the impulse response of a correlator will produce sharp peaks in the output that is known to be the optimal correlator in the case of white, gaussian, object independent noise, thus indicating at the same time the presence and the position of the target in the object.

Noise in images, however, is usually not white, gaussian and object independent: instead, it is composed of series of patterns in the scene that are often quite similar to the target and generate false alarms with matched filters. A whole new direction of research, that has proved very fruitful, was therefore derived from the idea of using statistical estimation theory to derive filters that are optimal in some specified conditions, taking into account the specification of the filter fabrication technology available. Two approaches have been followed.

- In the so called “heuristic” filters, the approach is to maximize some criterion under a given statistical model of the complete object (target and noise). Criteria include maximizing the correlation peak energy, or maximizing noise robustness in the presence of the given noise statistics, or maximizing the total energy in the image. Trading off among these criteria has proved to be a good way to stabilize the filters [227].
- “Statistical” filters optimise the cost of a given decision given the model. They include “non linear correlations” where the filter depends not only on the target but also on the scene itself. Impressive detection results have been obtained in the case of objects buried in strong noise when the noise statistics and the objects statistics clearly differ [228].
-

4.2.2.4. Devices and Systems

While optical processing is known to be fast inasmuch as it intrinsically operates in parallel over the whole image being processed, the most critical aspect of optical correlator performance is the speed and resolution of input and output devices. Image input devices are needed for the filter plane also in the fairly frequent case where the filter needs to be updated during the processor operation, e.g. to adjust to changes in the target or the noise, or to sequentially implement several filters on one given scene. While fast cameras of various kinds are available, the more critical factor is fast, high resolution image input: this function is provided by components called "spatial light modulators", or SLM, that have progressed significantly in recent years.

While many physical principles have been tried for SLMs, we shall restrict this discussion to the three SLM categories that appear most relevant in the present context: acousto–optic devices, liquid crystal devices and "micro–electro–mechanical systems" (MEMS), or more precisely "micro-opto–electro–mechanical systems" (MOEMS). In the future, all solid state SLMs based on electro–absorption devices might lead to the best performance but so far they have not yet met commercial success.

Acousto–optics:

While acousto–optics is an obvious solution for one–dimensional SLMs and have been discussed elsewhere in this book, they are not directly adapted to two–dimensional signals. Nevertheless, stacks of one–dimensional acousto–optic crystals can be used. While the number lines in the image is still fairly limited by the lack of convenience of stacking crystals, their obvious advantage is high modulation bandwidth, in the gigahertz range.

Liquid crystals

The advantage with liquid crystal displays is that optical processing application can ripe the benefits of developments in the display domain. However, a liquid crystal device optimised for display is not always optimised for image correlation. Let us mention some characteristics relevant to optical processing.

Typical nematic displays are very well known to be useable at video frame refresh frequencies, typically 25 or 30 Hz, and to be available in sizes up to about 1000*1000 pixels. Video projector type displays are best suited for optical processing because they are compact in size, which is convenient for a compact setup with moderate size, high quality lenses, and they are used in the transmittive rather than reflective mode, which is appropriate for transmitted beam phase control. Uniformity, optical quality, time response, dynamic range of the accessible modulus and phase excursion and polarization behaviour are the typical factors that need to be adjusted for optimal application to optical processing [230].

Ferroelectric liquid crystal present a faster switching time. The so–called smectic C* phase shows a binary, bistable behaviour. They are usually produced in the form of a matrix of birefringent half–wave plates that can be rotated in plane by $\pi/4$. The switching time is on the order of 10μs.

MOEMS

Microtechnology–based SLMs include [231].

- switchable micromirror arrays composed of typically 1000*1000 smallplane mirrors (down to 10μm on a side) that can be electrostatically tilted by a few degrees at frequencies up to a few kilohertz,
- and deformable mirror membranes that are driven by an underlying electrode structure under a continuous flexible surface.
-

4.2.2.5. Performance

With the above figures, it is possible to derive orders of magnitude for the number of correlations per second that can be implemented by a dedicated optical system [232]. Dedicated application to target tracking operating at several hundred hertz on relatively small images (typically 200*200 pixels) have been demonstrated, while correlators operating at video rate on 1000*1000 pixel arrays are commercially available. The net result is that these specialized processors easily outperform general purpose electronic processors as well as digital signal processor (DSP) arrays and compare favourably with specialized electronics. The limitation to their development is the size of the market.

4.2.3. Extension: Optical Cellular Automata

Based on the above comment that processors dedicated to image correlation have so far failed to open a seizable market, research on more powerful optical processors has been going on for years. In the 1980ies in particular [233], general purpose processors were designed and in the early 1993 one design by Opticomp corporation in the US met or slightly exceeded the performance of a workstation of that time. This was not enough of a performance difference, however, to justify commercial development in a situation where conventional microprocessors were still gaining in power following Moore's exponential law, as they still are nowadays.

An intermediate track of research is that of Optical Cellular Automata, that aim at providing, rather than just correlation, a more complete set of functions adapted to the need of image processing, in particular low level vision, without pretending to compete with general purpose electronic processors. In the following, we summarize the approach and the results of a recent project implemented by A. Cassinelli during his doctoral thesis work using a dedicated chip designed by colleagues from Institut d'Electronique Fondamentale, CNRS/Université Paris–Sud, Orsay, France.

4.2.3.1. Shift Invariance and Colouring in Parallel Processors

It is commonplace that computer architectures are mostly organized in relatively large sections composed of identical units. This is most evident when looking at the layout of a memory chip or a cache memory section in a processor, but can be found also in other parts of computer integrated circuits where fine grain parallelism at the level of registers and logic units is best implemented using a high degree of symmetry and more specifically, quite often, a high degree of shift invariance. The same property of local shift invariance appears at higher levels in the architecture hierarchy as well, where for example random access memory circuits are arranged in regular arrays with regular interconnection paths.

Because of the suitability of optics for implemented convolutions, a suitable point non linear operation NL acting on every pixel P of the image plane resulting from an optical convolution C is a good way to implement cellular automata with a degree of parallelism equal to the number N of image pixels [234], which can be a very large number compared to figures usual in parallel computers. The time sequential operation of the automaton should then be implemented by feeding the result of convolution C and non linearity NL from the image back to the object plane and starting the next cycle (see figure 111).

Our driving idea here is therefore to efficiently implement more general classes of parallel machines by arranging the computing sites in a parallel computer into patches, each patch performing a shift-invariant operation that can be implemented with the support of optical convolution in an efficient way [235] (see figure 112).

There is no reason, however, that the patches form compact areae. On the contrary, it turns out that fairly important image processing tasks can be performed by a semi-shift invariant layout where the invariant parts interlaced into an array of processor cells. While, with suitable spatial sampling, the whole array is globally shift-invariant, at a finer sampling its constituent processor cells are not. Instead, processor cells can be grouped into so-called " colors " of the processor and colors are interlaced in a fine, regular fabric. For example, all " blue " processor cells of the array perform one given operation at the same time and all " yellow " processor cells perform another given operation at the same time. Here we use the terms "blue" and "yellow" in a completely arbitrary manner, without reference to the true colors. We refrain from using "black" and "white" to avoid confusion with binary states of a one-bit processor.

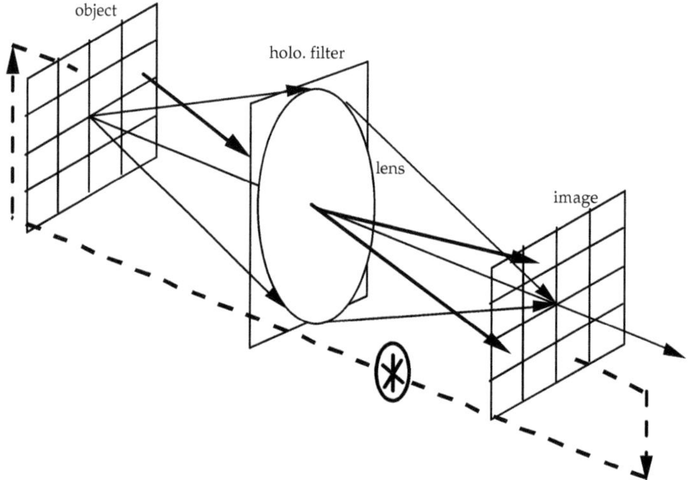

Figure 111 : An optical cellular automaton, based on optical convolution using an holographic filter and electronic feedback from the image to the object. The star denotes a point non linear operation performed on every pixel, the plain arrows exemplify light rays and the dotted lines arrows denote electronic interconnects. Every image pixel together with its associated electronic feedback and the corresponding object pixel constitute a processor cell, the light rays interconnect the processor cells with each other.

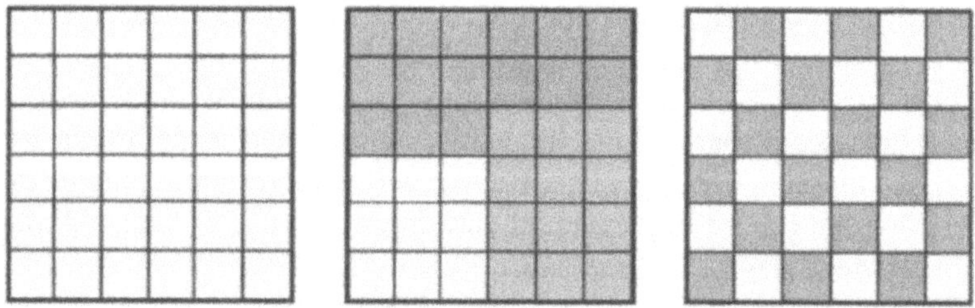

Figure 112 : Shift-invariant (left), semi-shift invariant (center) and interlaced semi-shift invariant (right) processor arrays. All processor cells of the same shade perform the same operation at the same time. In the illustration, there are four patches in the semi-invariant array and two "colors" (distributed patches) in the interlaced semi-shift invariant array.

4.2.3.2. Massively Parallel Stochastic Approaches to Image Optimisation

Image optimization is an approach to low level image processing where the quality of the processed image is expressed with respect to a set of desirable criteria in the form of an energy (or cost) function. The energy function should be minimized. It is a challenge to find the global minimum of a problem that depends on a number of variable equal to the number of pixels in a realistic image, therefore sub optimal approaches need to be adopted. We identified stochastic approaches to image optimization as one powerful algorithmic tool to reach good sub optimal solutions in a massively parallel way. Specifically, every pixel in the energy function affects the energy through its interplay with a limited number of other pixels, called its neighbors ; parallel implementation must make sure that two pixels that are neighbors to each other do not evolve in parallel. Therefore, the image array should be " colored ", in the meaning of the word " color " that was just explained, in such a way that two pixels of the same colors are not neighbors to each other. A typical number of colors is a few units, so that massively parallel image processing can be envisioned : the degree of parallelism is linear with the number of pixels in the image.

In our work, we used simulated annealing, a stochastic procedure that was theoretically and experimentally shown to provide excellent results in various optimization problems (see references [236,237]for a discussion of this issue) and applied it to motion detection in a gray level image sequence, adding to every gray level pixel a binary value that gives its state of motion. It is appropriate here to mention the difference between motion detection and the simple subtraction of successive images in a time sequence : a non zero difference between two such images may indicate motion, but it may as well be due to time variable noise or a change in illumination conditions ; the purpose of motion detection is to give a complete estimate of all pixels in the scene that are moving at a given time and follow their motion in terms of compact, but deformable, moving objects.

Specifically, we started from a class of motion detection algorithms based on Markov Random Fields and first proposed by Bouthemy and Lalande [236] and simplified it, at the expense of some loss in performance, to suit the possibilities of our hardware. The algorithm proceeds by compromising at every pixel between :

- the magnitude of the gradient
- the previous state of motion
- the present state of motion of its neighbors.

4.2.3.3. Optoelectronic Implementation

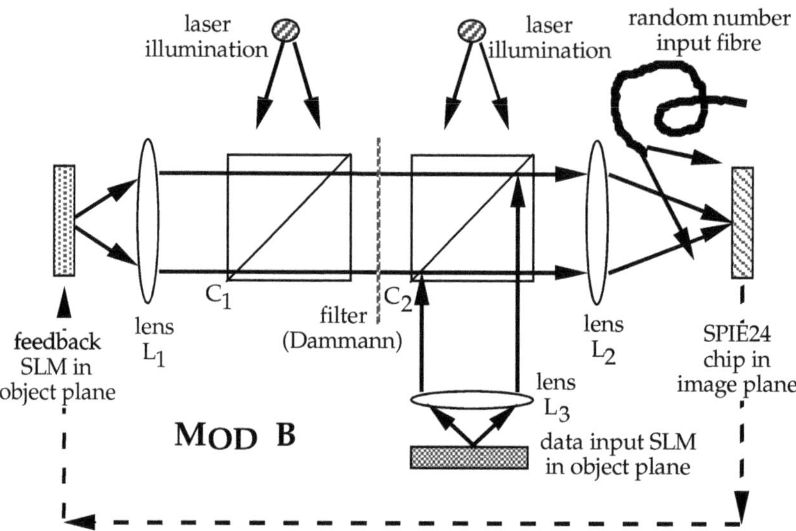

Figure 113 : General overview of the system. This mode of operation includes optical input, optical random number generation and optical convolution.

Every parallel optoelectronic implementation of a stochastic image optimization processor relies on three ingredients : an optical imaging or convolution setup, an optical random number generator and an optoelectronic device to implement the required nonlinear operations. For the random number generator, based on extensive previous experience [237], we selected moving laser speckle. A dedicated CMOS chip was designed and fabricated for the purpose of this and similar demonstrators. Dammann gratings and a set of SLMs and lenses, mirrors and beamsplitters were used to complete the setup depicted in figure 113, that apart from communication delays with the host computer provided video–real time estimation of motion on a small image (24*24 pixels, limited only by the cost of dedicated circuit fabrication in small quantities).

4.2.4. Conclusion

We have reviewed the basics of optical correlation, which presently is capable of providing state of the art performance on this specialized operation. As a basic processing function, it can stimulate new applications in domains other than the traditional field of pattern recognition. More complex dedicated processors such as optoelectronic cellular automata have followed this approach. Other, perhaps in the domain of optics for high frequency signal transmission and telecommunications, may follow.

The author wishes to acknowledge the essential contribution of his graduate student A. Cassinelli to the results mentioned in part III. Early work on the subject has benefited from contributions from his colleagues Ph. Lalanne, J.C. Rodier, L. Garnero, D. Prévost, from I. Glaser of MMRI, Israel, and from F. Devos, P. Garda, E. Belhaire and A. Dupret of Institut d'Electronique Fondamentale, CNRS/Université Paris–Sud, Orsay, France. Part of this work was funded by the European Commission under contract ERBCI1*CT93-0004.

5. OPTOELECTRONICS IN THE TERAHERTZ FREQUENCY RANGE

J.L. Coutaz
LAHC, Université de Savoie, 73 376 Le Bourget du Lac Cedex, FRANCE
and Department of Physics-Optics, Royal Institute of Technology, S-100 44 Stockholm, SWEDEN
coutaz@univ-savoie.fr

5.1. Introduction

The emergence of the "information society" has, during the last years, demanded strong improvements of the performances of electronics devices and systems, and of telecommunication means. This intense research and development activity will certainly stand for several years. Today, commercially available processors have reached the GHz clock frequency domain ; the 10 GHz limit is foreseen for the years 2010's by pushing the silicon technology towards its ultimate limit. Telecommunication technology has also been much improved in terms of information rate by the use of both optical fiber systems and hertzian satellite links. 10 Gbits/s is the common performance for installed high-rate optical systems. 40-60 Gbits/s rates have been demonstrated in laboratories. The 100 Gbits/s range is expected soon. Simultaneously, using frequency multiplexing, the rate of information carried by a single fiber could reach the Tbits/s level: in this case, tens of signals, each modulated at 10-20 GHz, are carried at different wavelengths.

Other applications of high frequency signals can be found in radar technology (70-90 GHz anti-collision radar equipment for cars and vehicles), in local communications (60 GHz antennae), in environmental studies (detection of atmospheric pollution), in the medical domain (tomography)... Applications in the scientific domain for research purposes are extremely numerous: investigation of solids (excitation of phonons, free carrier plasma, intraband transitions in quantum wells...), of liquids and gas (vibrations and rotations of molecules, *Debye* relaxation time, dipolar interaction ...), radioastronomy in the millimeter range (3 K black body background radiation of the Universe, detection of interstellar molecules ...), and so on.

Therefore, a strong research effort is carried out towards components and devices able to operate at frequencies higher than 100 GHz, up to the THz range. This domain of frequencies, spreading from 100 GHz or less, up to several THz, was not strongly studied before the 80's, because on one hand there was at these times a lack of practical sources, and, on the other hand, technological applications did not exist. Moreover, this domain of frequencies (see Fig. 114) lies in between the more classical electromagnetism areas, that is, infrared optics on the short wavelength side, and microwaves and high-frequency electronics on the long wavelength side: for the study of THz signals, one has to deal with techniques of both infrared optics and microwaves.

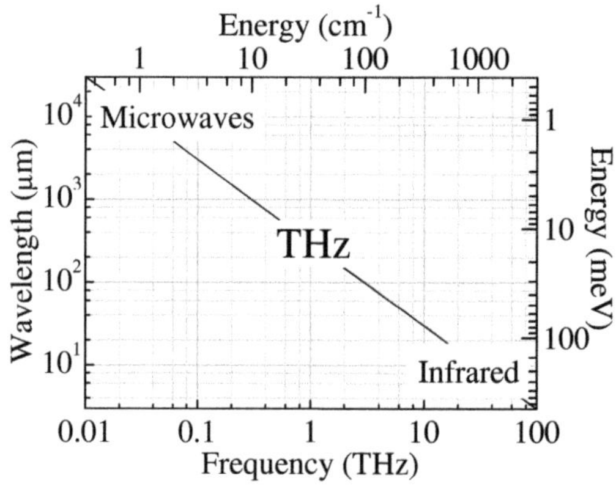

Figure 114 : The THz domain.

In the 80's, the situation concerning the THz studies has strongly involved. As already explained, technological applications were demanding components working at higher and higher frequencies. Besides, new methods of generation and detection of THz signals appeared: among them, optoelectronic techniques based on the use of femtosecond lasers were made possible because of the development of such reliable lasers. Simultaneously, classical techniques and devices, like bolometers, have been much improved.

This lecture will be devoted to the description of optoelectronic devices and methods to investigate the THz frequency domain. Most of them are based on temporal techniques, through the use of ultrashort optical pulses. These pulses serve both to generate and to detect, after propagation in materials or through devices, electrical transients. Frequency data are numerically obtained by *Fourier* transformation of the temporal signals. Such optoelectronic methods allow us to spread the entire range from some tens of GHz up to several tens of THz, that is wavelengths from centimeters down to a few microns. In order to present the advantages and drawbacks of the optoelectronic way, we will also briefly present the other main techniques in competition: emitters such as *Gunn* diodes and related devices, detectors like bolometers, and so on. Useful basic notions, concerning the time-frequency duality, the sampling technique and some aspects of nonlinear optics, are given in the appendixes.

5.2. Sources for THz Radiation

Both frequency-domain, using CW waves at THz frequencies, and time-domain, using picosecond electrical bursts, are in competition and each shows advantages and drawbacks. The frequency-domain allows high frequency resolution studies, but the useful power is either high with a very narrow bandwidth, or rather weak with an extremely wide bandwidth. CW-operating optoelectronic devices, that are both tunable and powerful, are still under study. The time-domain permits wide bandwidth and time-resolved experiments: in addition, switching the detector during a picosecond duration eliminates most of the room thermal noise at THz frequencies and the resulting data could be promptly recorded at room temperature with high dynamic. As a drawback, the frequency resolution is bad.

5.2.1. Frequency Domain

5.2.1.1. Blackbodies

Any body at temperature T radiates an infinitely wide spectrum of electromagnetic waves, which results from the thermal equilibrium of all the sources of radiation in the body. An ideal blackbody, that perfectly absorbs incoming electromagnetic waves, radiates according to *Planck's* law :

$$P(\nu) = \frac{2\pi h}{c^2} \frac{\nu^3}{e^{h\nu/k_B T} - 1}, \qquad (61)$$

where ν is the frequency. The emission spectrum of commercially available blackbodies obeys fairly well the *Planck's* law. Figure 115 shows the spectrum of commercial blackbody radiators. Even for low-temperature blackbodies, the power at ~1 THz remains weak. Moreover, the blackbody radiation is incoherent. Nevertheless, most of the far infrared physics up the years of the 60's was performed with blackbody sources.

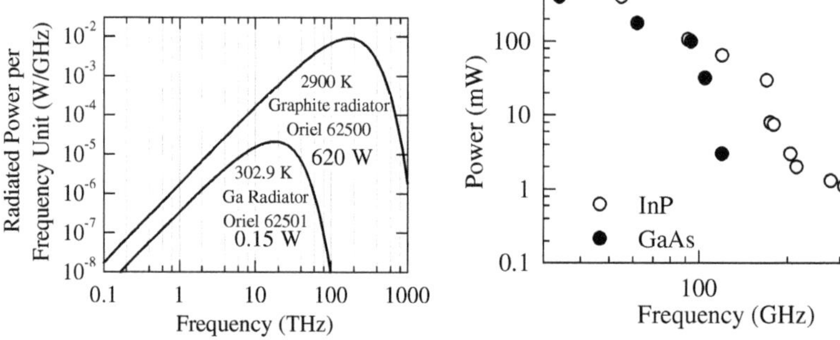

Figure 115 : Blackbody radiation of commercial devices (adapted from Oriel catalogue).

Figure 116 : State-of-the-art published results for CW high frequency Gunn-diodes (from H. Eisele et al. [239]).

5.2.1.2. Gunn Diodes and Related Devices

The *Gunn* effect was discovered in 1962 [238]. It appears in doped semiconductors whose conduction band exhibits a satellite valley, such as in GaAs or InP. Carrier velocity $\vec{v}$ increases with the applied electrical field $\vec{E}$ up to a maximum and then it decreases to reach a saturation plateau. This is caused by the transfer of the electrons from high- to low-mobility valleys. The negative slope region in the $v(E)$ curve gives rise to a negative differential resistance (NDR). When the semiconductor sample is biased in the NDR region, small fluctuations of the electron density near the cathode induce a local perturbation of the electrical field. Due to the NDR regime, the electrons in the perturbated region are gathered together, leading to an increase of the field perturbation. The effect grows up while the electrons move to the anode. Reduction of the electrical field strength in the rest of the sample prevents the simultaneous formation of other pulses. Thus, the repetition rate of the pulses is roughly equal to the length of the sample divided by the carrier velocity. Typically, a 5 µm-thick GaAs diode biased under 3 V delivers a signal at 25 GHz. Figure 116 shows the published state-of-the-art results for such diodes [239]. More than 100 mW CW are available under 100 GHz. Similar performances are obtained with other fast diodes (IMPATT, TUNNETT)

[239]. The THz regime is reached by means of harmonic generation, with a rather low efficiency (≈ 0.1 %) [240].

5.2.1.3. FIR Lasers

Far infrared (FIR) molecular gas lasers were the first source of coherent light radiation ever built [241]. The stimulated recombination occurs between rotation levels of molecules. Although the stimulated emission strongly prevails over the spontaneous emission in the far infrared, the molecular lasers are difficult to handle and show a lack of stability. Table 5 gives several strong FIR laser wavelengths and the corresponding molecules.

	wavelength (μm)	frequency (THz)	pump
CH_3OH [242]	119, 171, 369, 392, 570 ...	2.52, 1.75, 0.81, 0.76, 0.53 ...	CO_2 laser
HCN [242]	99, 126, 129, 131, 311 ...	3.03, 2.38, 2.33, 2.29, 0.96 ...	electrical discharge
CD_3Cl [243]	246, 286, 349, 443 ...	1.22, 1.05, 0.86, 0.68 ...	CO_2 laser

Table 5 : Selected lines of some FIR lasers.

Nowadays, new concepts are proposed to fabricate compact semiconductor FIR laser diodes. Intraband transitions in multi-quantum wells devices seem very promising, even if the laser emission has not yet been demonstrated. In a quantum well, the energy levels in the valence and the conduction bands are discrete. The difference between two neighboring energy levels can be tuned to several meV by adjusting the wells width and choosing the right device materials. THz luminescence in such structures has already been demonstrated [244]. A *pin* junction design, including a superlattice, has been proposed recently by Soref *et al.* [245] for laser emission in the THz domain.

5.2.1.4. Other Techniques

Optical beating corresponds to interferences in the time-domain. Let us consider two optical beams whose frequencies are ω and $\omega_2=\omega+\Omega$, respectively, in which Ω belonging to the THz range, and which are phase-locked. When the two beams are superimposed, the total E-field is :

$$E(t) = E_1 \sin(\omega t) + E_2 \sin(\omega_2 t + \varphi) = E_1 \sin(\omega t) + E_2 \sin(\omega t + \Omega t + \varphi) \tag{62}$$

This total field is then rectified (see the Appendix) in a nonlinear device or in an ultrafast detector (see 5.2.2.1.). The rectified signal is proportional to $E^2(t)$:

$$E^2(t) = \frac{1}{2}\left[E_1^2 + E_2^2 + 2E_1E_2\cos(\Omega t + \varphi) + \text{terms}(2\omega)\right] \quad (63)$$

Using a "slow detector", the terms vibrating at optical frequencies are removed and the rectified signal exhibits mostly a component at THz frequency. If the principle of optical beating is quite simple, the practical realization is somehow difficult, because: 1) one needs to phase-lock two optical beams with a THz spectral difference, and 2) an efficient and fast conversion device is necessary. CW THz generation has only been achieved in LT-GaAs photoswitches [246,247]. Two laser diodes, extremely well regulated in temperature, deliver the light beams: the THz frequency could be tuned by changing the operation temperature of one laser diode [246]. Signals up to 3.5 THz have been produced in this way [246]. Narrow frequency signals can reach the μW CW-power range [247]. Another scheme consists in using a multi-mode laser and their mixing two of these modes: this requires a drastic control of the temperature of the *Fabry-Perot* etalons that select the two modes [248]. Related works performed at *Hsinchu* University with a two-wavelength laser-diode array have produced signals of up to 7 THz [249].

In contrast to CW single-longitudinal mode (or bi-mode) lasers, mode-locked lasers will generate light pulses. The duration of the pulses is inversely proportional to the number of locked modes, while the time delay between two consecutive pulses is equal to the back and forth light propagation time in the cavity. Using short cavity lengths (mm), the pulse repetition rate could be as high as several tens of GHz. Selecting the lasing modes with an intra-cavity *Bragg* reflector, S. Arahari *et al.* have observed signals of up to 3.4 THz [250].

5.2.2. Time Domain

Here, the goal is to produce very short voltage pulses, which can be either guided along microwave lines, or radiated into free space.

5.2.2.1. Photoconducting Devices

Today, they are the most popular devices for the generation of ultrashort electrical pulses. Free carriers are photogenerated by absorption in a biased semiconductor. They give rise to a current under the applied electric field. In order to get pulse durations of the order of picoseconds (THz frequencies), the following rules should apply :

- excitation with a femtosecond pulsed laser,
- ultrafast semiconductor, which means that the free carrier lifetime should be as short as possible, and,

- the whole electrical circuit, including the photoswitch and the microwave line, should exhibit a sub-picosecond RC response time.

Moreover, for efficient opto-electrical conversion, the illuminated device area must be large and the semiconductor must present a high carrier mobility and a high electrical field breakdown. We will now first describe the semiconductor materials used for ultrafast photoswitching, then the design of the devices and related performances in both cases of guided pulses (detectors) and radiated pulses (THz antennae).

5.2.2.1.1. Ultrafast Semiconductors

The semiconductors should exhibit an ultrashort carrier lifetime in such a way that the current will flow only during a time comparable with the duration of light excitation. Moreover, the semiconductor should be highly resistive to avoid dark current, when the device is not illuminated. The current density in the semiconductor is given by :

$$\vec{j}(t) = n(t)\, q\, \mu\, \vec{E}(t), \qquad (64)$$

where $n(t)$ is the photogenerated carrier density, μ is the carrier mobility, and $\vec{E}(t)$ is the total field in the semiconductor. For achieving a good conversion efficiency, one desires to obtain a high current density. This imposes the requirement of high carrier mobility, high quantum efficiency, and as large as possible applied field.

semiconductor		lifetime (fs)	Hall mobility ($cm^2/V/s$)	transient mobility ($cm^2/V/s$)
LT-GaAs (low-temperature grown)		150-600 [251,252]	1650 [253]	120-150 [254]
Be:LT-GaAs		<100 [255,256]		
ii-GaAs	(ion-implanted)	200-5000 [257]	0.5-2000 [257]	
rd-SOS	(radiation damaged)	600 [258]	360 [259]	29 [260]
p-CdTe	(polycrystalline)	450 [261]		180 [261]

Table 6 : Ultrafast semiconductors for optoelectronic applications. See [262] for the definition of the transient mobility.

In order to reduce the carrier lifetime in the semiconductor, efficient traps have to be introduced in the material. This could be done by irradiation with particle beams, which induces structural defects in the crystal, by ion implantation, or by epitaxial growth at low-temperature (LT). In the latter case, the LT-grown material is no more stoichiometric and it presents an excess of species that form antisite defects. Post-grown annealing at high temperature precipitates the defect. Polycrystalline layers show also high density of recombination centers located at the grain boundaries. Table 6 presents a selection of ultrafast semiconductors for photoconducting applications.

LT-GaAs is certainly the most employed material as it exhibits very nice photoconduction and other related properties [251]. Doping the LT-GaAs with beryllium increases the proportion of ionized As antisites that are extremely efficient electron traps. With such a material, and with improved structural quality induced by Be-doping, lifetime shorter than 100 fs could be observed [255,256].

5.2.2.1.2. Ultrafast Optical Detectors

In such devices, metallic electrodes are generally deposited over the ultrafast semiconductor substrate. The laser beam generates photocarriers in the bare area of the semiconductor located in between the electrodes. The detector-rise time is mainly proportional to the laser pulse duration, and also to the RC time constant of the circuit. The detector decay time is governed by the RC time constant, and by the shortest among the carrier transit time and the carrier lifetime. Most of the photogenerated carriers should reach the electrodes to give rise to a strong signal in the detection circuit. Otherwise, this signal mostly originates in the displacement current [263] and, therefore, it remains weak. This means that the transit time should be longer than the carrier lifetime.

The efficiency of such photoswitches is improved by designing an interdigitated electrode pattern (Fig. 117), in order to enlarge the illuminated area. Even for sub-μm inter-finger distance, the switch capacity remains small, allowing ultrafast photoresponse and possible ballistic transport [264]. Kordos *et al.* [265] have reported a 0.58 ps FWHM time response for MSM LT-GaAs detectors, corresponding to a 550 GHz 3 dB-bandwidth (Fig.118).

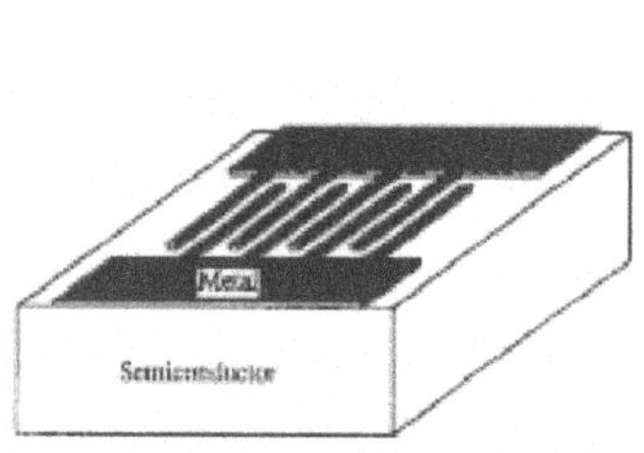

Figure 117 : Typical MSM photodetector design [264]

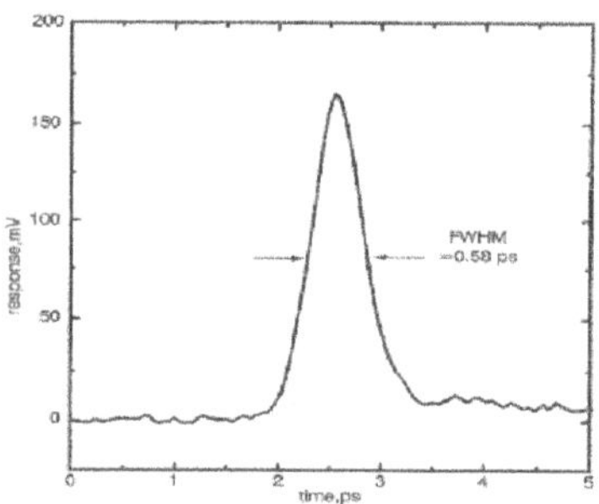

Figure 118: Ultrashort response time measured by Kordos [265].

Traveling-wave devices also constitute efficient and ultrafast detectors. Their electrodes are patterned to form coplanar waveguides in such a way that edge-coupled light and electrical signals propagate at the same speed. A 370 GHz LT-GaAs device, with 8% efficiency, has been reported by Chiu *et al.* [266].

5.2.2.1.3. THz Antennae

Narrow-gap biased photoswitches look very much like ideal dipoles (Fig. 121). When the device is not photo excited, similar charges, but opposite in sign, are piled up at the facing electrodes forming the switch. The dipole moment $\vec{p}$ is equal to the accumulated charges times the gap width. When the switch is illuminated, charges move to the opposite electrodes because of the induced conductivity in the semiconductor. Thus, the dipole moment is strongly varying. We know from basic physics that such a time-varying dipole is efficiently radiating electromagnetic fields. The far field amplitude of the dipolar field is given by :

$$E(t) \approx \frac{\partial^2 \vec{p}(t)}{\partial t^2} \approx \frac{\partial \vec{j}(t)}{\partial t}, \tag{65}$$

where $\vec{j}(t)$ is the current density flowing in the photoswitch. Using short-lifetime semiconductors, the current pulse duration could be shorter than 1 picosecond. Therefore, the far field is also a transient of picosecond duration and its spectrum reaches the THz frequency range.

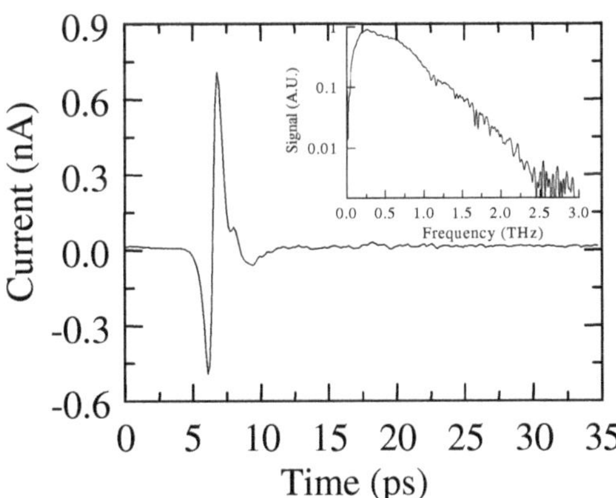

Figure 119 : Typical THz temporal pulse and the corresponding frequency spectrum [267].

Let us notice that the photoswitch acts as a nonlinear rectifying device: indeed, the incoming laser pulse is composed of a sinusoidal wave limited by a time-limited envelope. Its spectrum is a broad signal centered at the optical frequency.

On the other hand, the THz field corresponds only to the time-derivative of the envelope: its spectrum is now centered at zero. Figure 119 shows a typical THz temporal signal generated by a photoconducting switch, together with its spectrum.

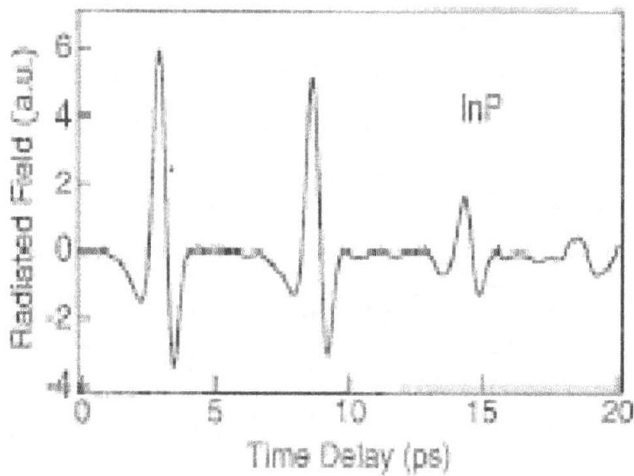

Figure 120 : THz temporal pulses radiated from semi-insulating InP [268].

For wide-band spectrum generation, the *Grischkowsky* dipole geometry [269] is widely used (Fig. 121), with LT-GaAs as semiconductor material, together with a quasi-optical focusing system (hyper-hemispheric lens and parabolic mirrors). Such devices will radiate signals up to 7 GHz [270]. Using amplified femtosecond lasers and wide-gap geometries, such photoswitches will deliver energies in the THz range up to 0.4 μJ per pulse under 45 kV bias [271]. When a signal with an enhanced power at a given frequency is preferred, efficient antennae with narrower bandwidth are chosen (Fig. 122).

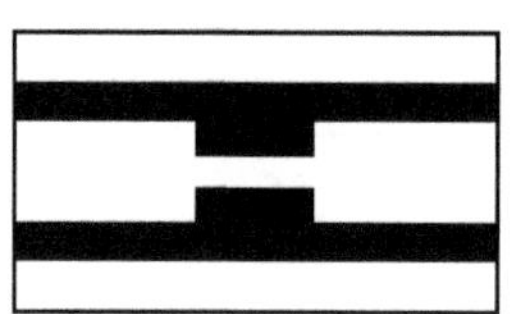

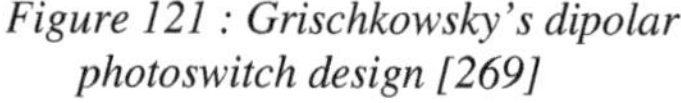

Figure 121 : Grischkowsky's dipolar photoswitch design [269]

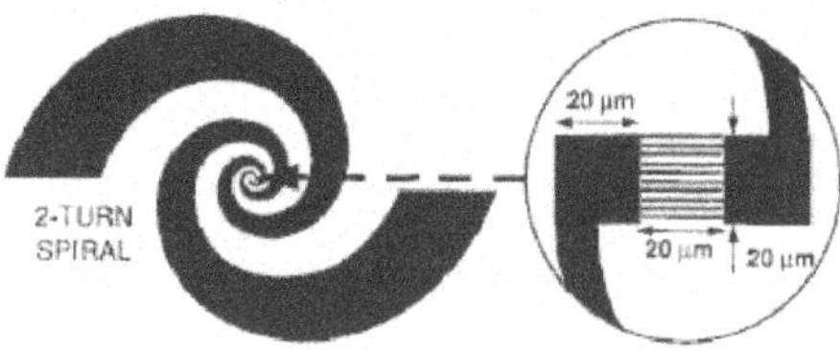

Figure 122 : Spiral antenna for narrow band terahertz emission[272]

5.2.2.1.4. Other Semiconductor Devices

Semiconductor surface generation: Due to the bending of the energy bands at the surface of a semiconductor, there exists a native surface field normal the surface. When carriers are photoexcited at the surface, they are accelerated perpendicularly to the surface by the native field. This current gives rise to a strong THz radiation (Fig. 122). Such a technique is very simple and cheap, as it requires only a bare semiconductor wafer. Unfortunately, the intense part of the THz radiation propagates mostly along the wafer surface, except if extended laser beams are employed to excite the wafer. In such a case, generalized *Fresnel's* laws apply and the THz beam can be used in applications such as THz time-domain spectroscopy. Table 7 gives the efficiency of different semiconductors, relative to InP [268]. Signals generated through the surface effects exhibit rather similar characteristics as those radiated by photoswitches.

	InP	GaAs	CdTe	CdSe	Ge	Si
THz signal	1	0.71	0.33	0.11	0.07	0.005

Table 7 : Surface-effect THz emission efficiency of several semiconductors relative to InP (from [268]).

When ultrahigh temporal resolution (~10 ps) is used, we may observe the very first events of the carrier acceleration [273] : first an instantaneous polarization due to electron-hole pairs formation appears, then the carriers undergo ballistic acceleration, and, finally, they move at an average drift velocity mostly limited by phonon scattering. All these stages participate in the THz radiation emission [273].

Bloch oscillations: Under an applied field E, the electrons in a semiconductor move, in the k-space, towards the border of the first *Brillouin* zone ($k=\pi/d$, d is the lattice periodicity). When they reach the second *Brillouin* zone, their energy state is equivalent to the one at the other border of the first *Brillouin* zone [274]. Therefore, the electrons are *Bragg*-reflected and they undergo periodic motion both in the k-space and in the real space, namely, the *Bloch* oscillations [275]. The period of this motion is the time to go from $k=-\pi/d$ to $k=+\pi/d$. For bulk semiconductors, d is of the order of a few Å, and the period is shorter than the electron dephasing time, making it impossible to observe any of the *Bloch* oscillations. In superlattices, d is the periodicity of the quantum wells and it is much longer than in bulk materials.

The *Bloch* oscillation frequency ν is given by [275] :

$$\nu = \frac{e}{h} d\,E\,. \tag{66}$$

Using superlattices, however, with $d \approx 100$ Å and $E \approx 1$ V/μm gives rise to a signal at ~2.4 THz. To avoid fast relaxation through phonon coupling, experiments are generally performed at low temperatures. However,

Dekorsy *et al.* [276] have demonstrated *Bloch* oscillations even at room temperature, with frequency tunable from 4.5 to 8 THz.

<u>*p-i-n* diodes:</u> In the depletion region of a reverse-biased *p-i-n* diode, the built-in electrical field could be as high as 100 kV/cm. Photogenerated carriers are thus strongly accelerated through the intrinsic region. As the electrons and holes move in opposite directions, a space-charge field is created that screens the built-in field. As a consequence, the carrier plasma oscillates due to this restoring force at the plasma frequency, ν_p, given by :

$$\nu_p^2 = \frac{n e^2}{4\pi^2 \varepsilon m^*} . \tag{67}$$

Sha *et al.* [277] have first reported THz plasma oscillations in a vertical GaAs *p-i-n* diode. Recently, Leitenstorfer *et al.* [278] have demonstrated emission up to 70 THz with a GaAs *p-i-n* diode excited by 12 fs duration laser pulses (Fig. 124).

5.2.2.2. Nonlinear Optical Methods

Nonlinear optics is certainly the best way to produce tens of THz signals [279], because it implies no free carriers generation and drift. Here, the nonlinear phenomenon which produces the THz signal is frequency difference generation in a nonlinear crystal showing no center of symmetry. The origin of this phenomenon is linked to the nonparabolic molecular potential wells experienced by the electrons. This means that the effect is almost instantaneous, at least shorter than 10 femtoseconds, as it corresponds to a reshaping of the molecular electronic cloud. When using femtosecond pulses with a wide associated spectrum, the effect will mix two different frequencies of the spectrum to produce the frequency difference. As for typical 100 fs optical pulses, the –3 dB frequency limit of the associated spectrum is 5.3 THz, it reaches, however, 7.5 THz for the rectified signal. Using detectors with a large dynamic range, together with sampling techniques, frequencies of over 30 THz have been measured [280]. Recently, using a GaSe crystal, Huber *et al.* [281] have produced signals of up to 60 THz, whose central frequency is tunable up to 41 THz by varying the phase-matching conditions.

The main limitation of the technique is due to the weak efficiency of optical rectification. Getting useful THz power requires us to employ thick crystals. Unfortunately, there exists a large refractive index dispersion in most of the nonlinear crystals, due to material resonances in between the optical and the THz ranges. Thus, the accumulation effect saturates with the crystal thickness due to velocity mismatch. Table 7 gives the main crystals used for THz generation through optical rectification [282], together with relevant parameters. As compared to photoconducting antennae, the spectra are much wider, but the efficiencies are much smaller.

Material	EO coefficient (pm/V) at λ in μm	Index of refraction at λ in μm	Dielectric constant	OR Figure of merit (pm/V) $2rn^3/(1+\sqrt{\varepsilon})$	GVM (ps/mm)	EOS V_π (kV)
ZnTe	r_{41}= 4.04 at 0.633	n=2.853 at 0.800	ε=10.1	~ 44.9	1.1	6.75
CdTe	r_{41}= 4.5 at 1.000	n=2.84 at 0.800	ε=9.4	~ 50.7	0.75	9.70
GaAs	r_{41}= 1.43 at 1.150	n=3.61 at 0.886	ε=13	~ 24.2	0.015	17.09
$LiTaO_3$	r_{33}= 30.5 at 0.633 r_{13}= 8.4 at 0.633	n_0=2.176 at 0.633 n_e=2.180 at 0.633	$\varepsilon_{1,2}$=41 ε_3=43	up to 84.9	14.1	2.01 7.27
$LiNbO_3$	r_{33}= 30.9 at 0.633 r_{51}= 32.6 at 0.633	n_0=2.286 at 0.633 n_e=2.200 at 0.633	$\varepsilon_{1,2}$=43 ε_3=28	up to 110.3	14.2	1.71 1.82
ADP	r_{41}= 26.8 at 0.546 r_{41}= 8.56 at 0.546	n_0=1.527 at 0.546 n_e=1.481 at 0.546	$\varepsilon_{1,2}$=58 ε_3=14	up to 22.2	20.28	5.72 19.64
KDP	r_{41}= 8.77 at 0.546 r_{63}= 10.3 at 0.546	n_0=1.512 at 0.546 n_e=1.470 at 0.546	$\varepsilon_{1,2}$=44 ε_3=21	up to 11.7	17.1	18.01 16.69
KTP [283]	r_{33}= 36.3 at 0.633 r_{23}= 15.7 at 0.633	n_e=1.866 at 0.633	$\varepsilon_{1,2}$=11 ε_3=15	up to 109.2	5.4	2.68
DAST	r_{11}= 160 at 0.820	n_0=2.46 at 0.820 n_e=1.70 at 0.820	ε=8.0 ε=2.9	up to 1244.3	1.22	0.34 1.04

Table 7 : Efficient crystals for optical rectification and electrooptic sampling (from [282] and [283]).

Parametric THz generation through stimulated scattering of polaritons can occur in polar crystals, such as $LiNbO_3$. The crystal is excited in an upper virtual state by the laser pulse and it relaxes down to a polariton state, which in the far infrared behaves like a photon. In stimulated processes, both the energy and the wavevector are conserved, thus an idler wave participates in the phenomenon :

$$\nu_{laser} = \nu_{THz} + \nu_{idler} \qquad \vec{k}_{laser} = \vec{k}_{THz} + \vec{k}_{idler} \,. \tag{68}$$

The THz signal frequency is tuned by adjusting the angular phase-matching conditions (68). Efficient THz generation have been reported by

Morikawa *et al.* [285], who used a $LiNbO_3$ nonlinear crystal located in a *Fabry-Perot* cavity. The cavity is resonant for the idler wave and the THz beam is efficiently coupled out of the crystal by using a trapezoidal design. Pumped with 25 ns-38 mJ pulses from a Nd-YAG laser, 45 pJ pulses are generated around 1.7 THz.

5.2.2.3. Other Methods

The high conductivity of superconductors is related to the pairing of free electrons to form *Cooper* pairs. The binding energy of a *Cooper* pair is of the order of meV. When excited with photons of higher energy, the *Cooper* pairs are broken. The lifetime of separated electrons is sub-picosecond. Therefore, when a superconductor sample is illuminated by a femtosecond optical pulse, its resistivity increases during a time shorter than 1 ps. In current-biased samples, the current shows an as-short decay transient, that is then a source for THz radiation. M. Hangyo *et al.* [286] have produced a sub-picosecond electromagnetic pulse, whose spectrum spreads up to 2 THz: contrary to the *Grischkowsky* scheme, here, the dipole is a conductive one (YBaCuO bridge), whose resistance increases under illumination.

Finally, let us mention a fully-electronic, time-domain source based on temporal compression of electrical pulses along a nonlinear microwave transmission line. 1.4 ps fall time for step-function voltage transients has been reported by Rodwell *et al.* [287], showing frequency components of up to several hundreds of GHz. This technique cannot compete with the tens-of-THz optoelectronic ones, but it is a very competitive method, in terms of simplicity and low price, to study devices, materials and circuits in the 10-100 GHz range [288].

5.3. Detectors of THz Radiation

5.3.1. Frequency Domain

Bolometers are widely used as broadband detectors in the infrared and far-infrared ranges. Typically, the bolometer consists of an absorber connected to a thermal bath. The temperature of the absorber is increased by the absorption of the incident wave and it is measured with a very precise thermometer (thermistor, superconductor film with ammeter). Because of the strong variation of the conductivity properties of low-temperature superconductors at the superconduting transition and of their weak Cooper pair energy, these materials are very suitable to fabricate ultra-sensible thermal sensors in the far infrared. In the millimeter domain, low-temperature superconductor bolometers exhibit sensitivities (generated current/incident power) bigger than 10^5 A/W and noise equivalent power down to 10^{-17} W/√Hz [289]. To allow frequency selection, filters should be placed in front of the bolometer. Jointly, the main drawbacks of the bolometers are their long time-response, their working operation at low-temperature and in some cases, their too-high sensitivity: they could be damaged by any strong variation of the incident signal. Nevertheless, their unique properties make them ideal millimeter-waves detectors for radio astronomy.

5.3.2. Time Domain

5.3.2.1. Photoconducting Devices

Antennae similar to the emitters (see 5.2.2.1.3.) permit us to detect THz pulses and to resolve their temporal shape using sampling techniques (see Appendix 5.7.3.). The detection photoswitch, now unbiased, is linked to a slow response-time amperemeter, which reads and integrates the current flowing through the photoswitch. Let's imagine that the THz signal is photogenerated by a part of a pulsed laser beam (pump beam). The other part of the laser beam (probe beam), delayed (time delay τ) as compared to the pump beam, excites carriers in the photoswitch. As no bias is applied, these photocarriers do not move and they recombine after a duration corresponding to their lifetime: therefore, no current is read. On the other hand, if the THz field, E_{THz}, illuminates simultaneously the photoswitch, the photocarriers are accelerated by this field. They give rise to a current whose amplitude is :

$$i(\tau) = \int_{-\infty}^{+\infty} \int_{-\infty}^{+\infty} E_{THz}(t-\tau)\, I(t')\, \eta(t'-t)\, dt'\, dt\,, \tag{69}$$

where $I(t)$ is the laser intensity and $\eta(t)$ is the temporal behavior of the photocarrier population in the photoswitch. Usually, the laser pulses are very short and, thus, information on the temporal shape of E_{THz} is obtained when ultrafast semiconductors are used. Therefore, these detectors should be fabricated with the same semiconductors as the emitters (see 5.2.2.1.1.). Their bandwidths typically reach 3~7 THz. They are employed in association with quasi-optical focusing systems. The main advantage of the photoconducting detector is its great sensitivity and dynamics as compared to other sampling detection schemes. Signals as small as 0.1 pA are detected. The dynamics in the field amplitude can be larger than 50 dB. Let us finally notice that the current delivered by the photoswitch is proportional to the THz field amplitude and not to its intensity.

5.3.2.2. Electro-Optic Sampling [290]

The electrooptic (EO) effect is very similar to optical rectification described in 5.2.2.2. When an EO crystal is located in an electrical field, it turns out that a birefringence is induced in the crystal. This effect corresponds to a small modification of the index ellipsoid, mostly of the length of the axes of the ellipsoid. This perturbation is described through the following expression [291] :

$$(x,y,z)\begin{pmatrix} \frac{1}{n_x^2}+\sum_i r_{1i}E_i & \sum_i r_{6i}E_i & \sum_i r_{5i}E_i \\ \sum_i r_{6i}E_i & \frac{1}{n_y^2}+\sum_i r_{2i}E_i & \sum_i r_{4i}E_i \\ \sum_i r_{5i}E_i & \sum_i r_{4i}E_i & \frac{1}{n_z^2}+\sum_i r_{3i}E_i \end{pmatrix}\begin{pmatrix} x \\ y \\ z \end{pmatrix}=1, \tag{70}$$

where x,y,z are the principal axes directions of the crystal and the index i takes the values *1=x*, *2=y*, *3=z*. r_{ij} represents the elements of the EO tensor and E_i the components of the electrical field. Solving this equation means to diagonalize the matrix, in order to find the new principal directions and the length of the axes, that is the principal indices. The calculation is quite tedious in the general case and the solution must be computed. However, for many realistic crystals, most of the tensor elements are null and, moreover, the single orientation of the electrical field in the crystal considerably simplifies the problem. The solution is given is many textbooks [291] : when a polarized light-beam propagates through the crystal, its polarization direction varies under the effect of the field-induced perturbation. Usually, a linear polarization is slightly rotated. However, this situation is not the best one as the rotation does not vary proportionally to the applied field [290] : it is preferable to use a circularly polarized input beam. At the output, then, the beam polarization is slightly elliptic. A polarizing prism (*Glan*, *Wollaston*) separates the two perpendicular components of the polarization. The EO effect adds a positive part to one of the components and a negative part to the other one. Recording the two signals with two detectors and subtracting them electronically removes the main part of the light beam which is not affected by the EO effect. Simultaneously, this procedure doubles the amplitude of the recorded signal. This leads to a strong reduction of the noise and it allows us to measure very weak fields. Typically, measurements of light polarization rotation down to 10^{-7} rad are reachable, corresponding to fields as weak as some V/m [292]. The parameters of the principal EO crystals are given in Table 7.

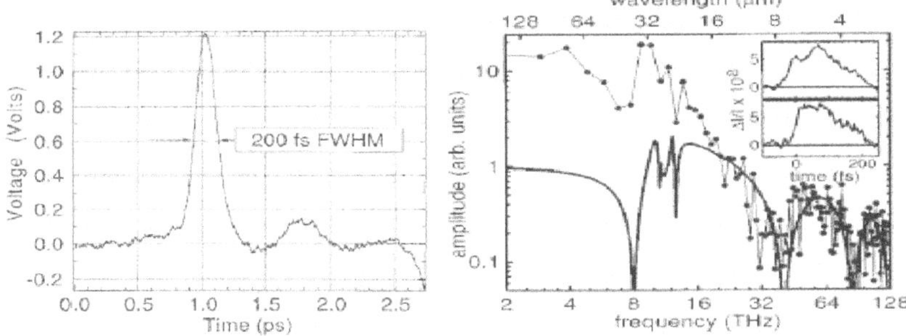

Figure 123 : Ultrashort electrical pulse generated at a coplanar line and measured by EOS [293].

Figure 124 : Measured EO THz amplitude spectrum radiated by a GaAs p–i–n diode. The EO crystal is a 13 mm thick GaP slab [278].

Such an amazing sensitivity is necessary in order to measure weak THz pulses. Here again, the temporal shape of these pulses is obtained by sampling. The probe laser pulse is time-delayed as compared to the electrical pulse. This requires a perfect synchronization between both, which is generally achieved through the optical generation or triggering of the electrical signal using a delayed part of the probe beam.

The bandwidth limitation of such a method is mostly due to the difference of the velocities of the waves in the optical and THz ranges, when the two signals propagate in the same direction. When the light beam goes through the crystal perpendicularly to the THz pulse, the time-resolution is more or less given by the time of flight of the light pulse in the crystal. Sub-picosecond resolution times have been achieved [293] (Fig. 123), which means frequencies up to the THz range, for the characterization of fast electrical signals guided by microwave lines. For freely propagated THz pulses, records at more than 50 THz have been reported [278,281](Fig. 124).

Taking the benefit of the vanishing elements of the EO tensor (Eq. 70), and optimizing both the orientation of the crystal and the probe beam direction, makes it possible to map the three components of the electrical field [294]. Strong efforts are done at the present time to build-up practical EO systems: fibered devices [295], pulsed laser diode sources [296], noise-reduced systems [297], free running systems [298] and quantitative measurements [299]. Developments of commercial instruments are now under way [300].

5.3.2.3. Single-Shot EO Measurement

Single-shot measurements of freely propagating THz fields are possible when the EO signal is strong enough. Two methods have been proposed. The principle of the first one [301] (Fig. 124) is to enlarge the lateral spatial profile of the optical probe-beam and to make it impinge at an oblique incidence on the EO crystal, while the THz signal is illuminating the crystal at normal incidence. The time-delay between both the optical and the THz beams is thus continuously varying along the crystal length : therefore, the time variable is transformed into a spatial one. Imaging the EO signal onto a CCD array allows us to single-shot record the THz pulse's temporal shape. The time resolution in this experiment was limited by the crystal thickness to 29 fs.

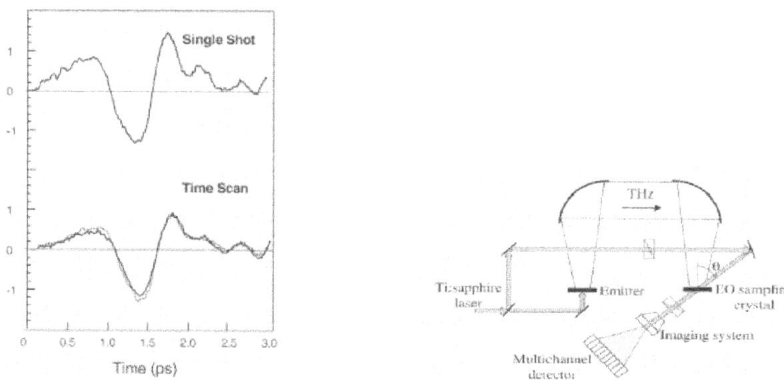

Figure 125 : Single shot EO measurement method from T. Heinz's group [301].

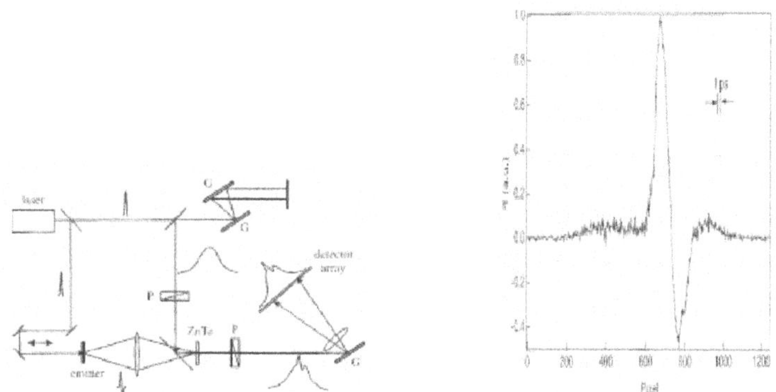

Figure 126 : Single shot EO measurement, X.-C. Zhang's group method [302].

In the second method [302], the laser pulse is linearly chirped: the time variable is thus transformed into a wavelength variable. Both the THz and the optical pulses interact in the EO crystal. The temporal shape of the EO signal, carried by the transmitted optical beam, is single-shot analyzed using a spectrometer to separate the wavelengths (Fig. 125). As compared to the first method [301], this one is indirect and its time resolution, given by the temporal linearity of the chirp, is certainly much weaker.

5.4. Applications

5.4.1. THz Time-Domain Spectroscopy (THz-TDS)

The emission of THz pulses and the detection and measurement of their temporal shape make it possible to perform spectroscopy in the far-infrared range. This technique is called THz time-domain spectroscopy. The principle is to put a sample of the material in between the antennae, and to measure the THz signal transmitted by the sample. A second measurement without the sample will serve as reference. Then, both signal and reference are *Fourier*-transformed. The ratio of both the *Fourier* transforms is equal to the transmission coefficient of the sample in the frequency domain. For homogeneous slides with parallel faces, the theoretical plane-wave transmission coefficient is known from the electromagnetic theory [303] :

$$T(\omega) = \frac{4\tilde{n}\, e^{-j(\tilde{n}-1)\omega d/c}}{(1+\tilde{n})^2 - (1-\tilde{n})^2 e^{-2j\tilde{n}\,\omega d/c}}, \qquad \text{with } \tilde{n} = n - j\kappa = n - j\frac{c\alpha}{2\omega}, \tag{71}$$

when n is the refractive index and α is the absorption coefficient. Fitting the experimental data to the expression (11) allows us to determine n and α (Fig. 127). Usually, the fitting is done using a *Newton-Raphson* method or derived numerical codes [303]. Let us notice two important points: 1) the temporal measurement gives us two informations, i.e. the strength of the signal and the relative time delay between the signal and the reference. In the frequency domain, however, these two informations correspond to the amplitude and the phase. From them, it is possible to derive the two material parameters n and α ; 2) the expression (71) is given for the amplitude and not for the intensity, as the detected current is proportional to the THz field amplitude. The precision on the determination of n and α could be as good as a few percent. It is mostly limited [304] by the precision on the sample thickness, and, at a second order level, by the experimental noise and the plane-wave model accuracy as regards to the experimental beam shape.

Due to their great emissivity and sensitivity, photoswitches are widely employed in the THz-TDS set-ups. Figure 127 shows some examples of far infrared dispersion curves measured with this technique. Despite being less sensitive, THz-TDS with EO detectors seems achievable [305,306].

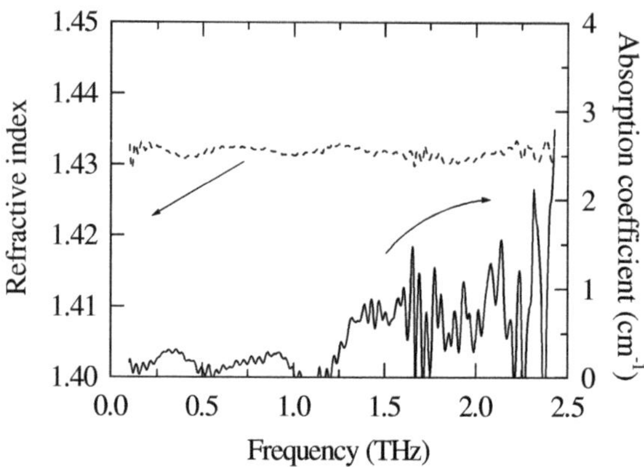

Figure 127 : Refractive index and absorption of Teflon measured by THz time domain spectroscopy [267].

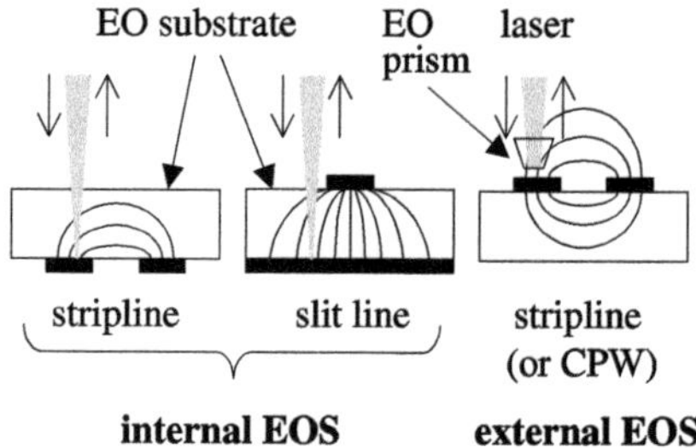

Figure 128 : External and internal EOS schemes.

5.4.2. Imaging

Performing imaging in the far infrared seems to be a promising non-invasing and non-ionizing tool, mostly for the examination of biological tissues, as the THz waves are strongly absorbed by water. There are two ways of imaging with time domain THz beams: 1) The first one consists in focusing the THz beam in between the emitter and the detector and to translate a sample in the focal plane. For each *x-y* position of the sample, the THz signal is recorded. The spatial resolution of the set-up is limited by the THz-beam waist diameter at the focus. As this waist diameter varies as the wavelength, high-frequency spectral distribution is necessary to obtain a fairly good spatial resolution. With an all-EO set-up and using 15 fs laser pulses, Yan *et al.* [306] have recorded THz images with a ~50 μm spatial resolution. They applied this technique to image biological tissues. Similar resolution has been reached by Hunsche *et al.* [307], but now with a THz beam centered at 1.5 THz, using a near-field technique; 2) The principle of the second method [308] is to use wide-section THz and optical probe beams which interact in a big-size EO crystal. Then the EO information is extracted from the transmitted optical probe-beam with a polarization analyzer and directly imaged by a CCD camera. This method requires intense THz radiation in order not to use differential detection as explained previously (5.3.2.2.).

5.4.3. High Frequency Electronic Circuits Characterization

5.4.3.1. EOS [309]

EO sampling (EOS) constitutes a high-performance technique to characterize electronic circuits. As compared to classical electronics techniques, EOS is almost not invasive, it exhibits a much superior bandwidth and it also allows us to map the electrical signal everywhere over the circuit (Fig. 129). The electrical signal in the circuit is detected through its electrical field by EOS. Two cases are of interest (Fig. 128). When the substrate of the circuit is EO (eg, GaAs, InP), the laser probe-beam measures the induced birefringence in the substrate. No probing device should be brought in the vicinity of the circuit and, thus, the measurement perturbation on the circuit running behavior is almost null. On the other hand, the substrate should exhibit optically flat surfaces, the laser wavelength should lie in the transparency range of the substrate, and these methods cannot be applied to centro-symmetric substrates (Si, Ge, amorphous wafers...). For such components, an EO component should be brought close to the circuit, in the evanescent part of the electrical field (Fig. 128). Small-size EO prisms ($LiTaO_3$, GaAs (see table 7)) are used [290,294,309]. The presence of such a prism over the circuit induces a small perturbation that can slightly modify the operation of the circuit at high frequencies [310]. In practical EOS, the major difficulties are connected to: 1) the time jitter between the electrical signal to be measured and the optical probe pulse. Free running systems, avoiding the optical triggering of the circuit, have already been proposed [298]. 2) The size and the price of THz laboratory equipments. Practical applications, limited to a few tens of GHz, will need compact systems, based on fibered laser diodes.

5.4.3.2. Photoconductive (PC) Sampling

Using a microprobe to pick up a small part of the electrical signal and sampling it with a photoswitch constitutes an alternative technique to EOS. The microprobe could be either a metallic tip in contact with the circuit [311], or a tunneling-effect probe [312]. The detected current is converted into a voltage, using, for example, a JFET source-follower amplifier [313]. A photoconducting switch, located on the tip holder (Fig. 130), permits us to sample the signal. This method is very sensitive and the induced perturbation remains weak as the tip size is small. Picosecond time resolution can be achieved with LT-GaAs photoswitches. The spatial resolution could be submicrometer when the probe is installed in an STM microscope. However, the measurement is limited to the current, and no field mapping is possible.

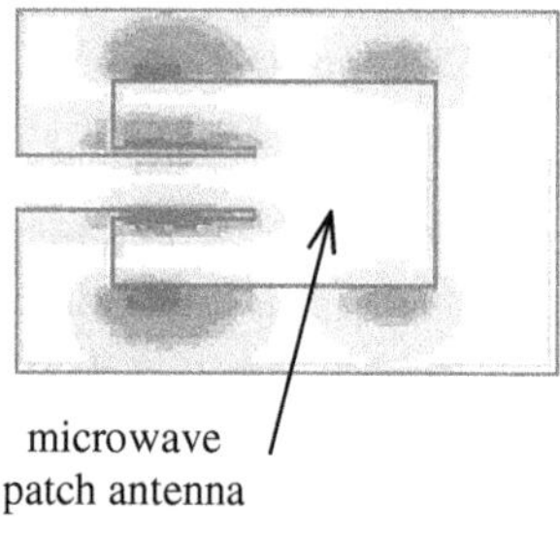

Figure 129 : EO mapping of the x component of the electrical field radiated at 100 GHz by a patch antenna [294].

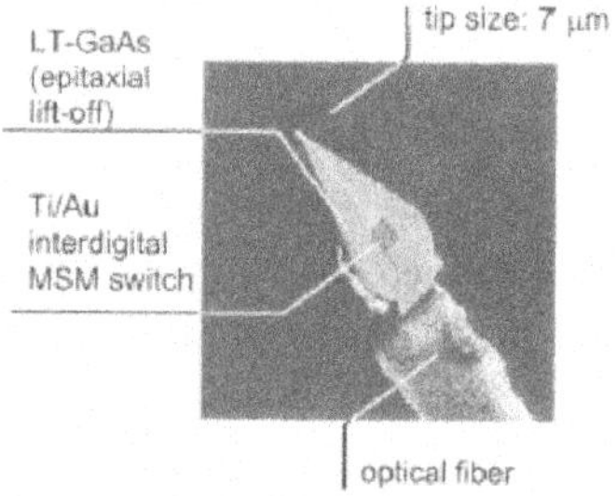

LT-GaAs (epitaxial lift-off)
Ti/Au interdigital MSM switch
tip size: 7 μm
optical fiber

Figure 130: Picture of a PC sampling probe [311].

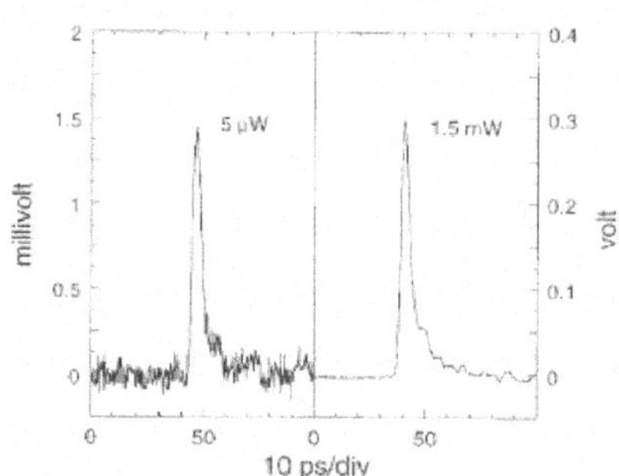

Figure 131 : Example of PC sampling of temporal records [313].

5.5. Conclusion

The optical generation of sub-picosecond electrical signals has opened the field to numerous applications, from fundamental solid state physics, to ultrafast electronics and biological studies. These works have been strongly facilitated by the development of reliable femtosecond lasers. The progress of THz optoelectronics will certainly follow three complementary ways: 1) fabrication of compact systems using laser diodes (THz spectroscopy, EOS), 2) development of electronic systems working at more than 100 GHz, and 3) academic researches in order to definitively go under the 100 femtosecond threshold, that is for delivering electrical transient of a few tens of femtosecond duration. No doubt that THz optoelectronics will continue to produce amazing results for several years !

5.6. Acknowledgements

These lecture notes were written while the author was visiting the Department of Physics-Optics, Royal Institute of Technology, Stockholm, Sweden. The author would like to acknowledge a visiting grant from the *Werner Gren* Foundation (Stockholm-Sweden), and critical reading of the manuscript by Prof. Jens A. Tellefsen. Jr.

5.7. Appendix

5.7.1. Juggling with Time and Frequency

Time and frequency are conjugated variables. The frequency signature, $S(\omega)$, of a temporal signal, $S(t)$, is thus simply its *Fourier* transform :

$$S(\omega)=\frac{1}{\sqrt{2\pi}}\int_{-\infty}^{+\infty} S(t)e^{-j\omega t}dt\,. \tag{72}$$

It follows that for a monochromatic electromagnetic wave, the spectrum exhibits a single peak, while for time-limited electromagnetic pulses (oscillating signal limited by an envelope), the spectrum is also frequency-limited, centred at the carrier frequency (see Fig. 132). In the case of gaussian electromagnetic pulse, one writes :

$$S(t)=e^{-\frac{t^2}{\tau^2}}e^{j\omega_o t} \qquad \Rightarrow \qquad S(\omega)=\frac{1}{\sqrt{2}}e^{-\frac{(\omega-\omega_o)^2\tau^2}{4}}\,. \tag{73}$$

If the pulse does not exhibit fast oscillations, its spectrum is maximum at zero as deduced from the relation (A2) with ω_o=0.

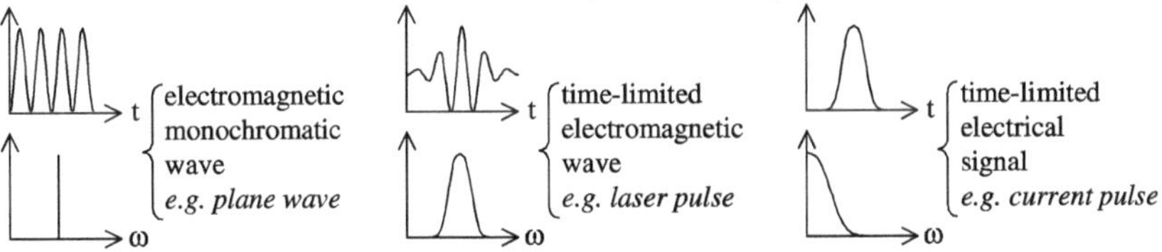

Figure 132 : Temporal pulses and the corresponding frequency spectra.

Therefore, when dealing with THz frequencies, we will either look at monochromatic electromagnetic waves oscillating at around 10^{12} Hz, or at electrical pulses with picosecond duration. In THz optoelectronics, the question to be answered will be how to transform optical signals whose central frequency is in the 10^{15} Hz into THz signals.

5.7.2. Nonlinear Optics

One way of answering the preceding question is to make use of nonlinear optics. Let us recall some basic ideas and definitions of nonlinear optical phenomena. The response of a dielectric medium to an electromagnetic excitation (with electric field $\vec{E}$) is depicted by the displacement field $\vec{D}$:

$$\vec{D} = \varepsilon_o \vec{\vec{\varepsilon}} \vec{E} = \varepsilon_o \vec{E} + \vec{P}, \tag{74}$$

where $\vec{\vec{\varepsilon}}$ is the relative permittivity tensor, ε_o is the vacuum permittivity, and $\vec{P}$ is the polarization of the medium. The polarization $\vec{P}$ arises from the dipoles induced at each atom or molecule. For commonly applied fields, the dipole moment is proportional to $\vec{E}$. For strong applied fields, each dipole moment saturates and, thus, $\vec{P}$ is no more a linear function of $\vec{E}$. Nevertheless, the nonlinear terms are small and $\vec{P}$ can be written as a development in the power of $\vec{E}$:

$$\vec{P} = \varepsilon_o (\vec{\vec{\chi}}^L \cdot \vec{E} + \vec{\vec{\chi}}_2^{NL} \cdot \vec{E} \cdot \vec{E} + \vec{\vec{\chi}}_3^{NL} \cdot \vec{E} \cdot \vec{E} \cdot \vec{E} + ...), \tag{75}$$

where $\vec{\vec{\chi}}_i^{NL}$ is the nonlinear susceptibility tensor of order i. The first right-hand side term in the expression (75) corresponds to linear optics. Here, we are only interested in second order phenomena. For a general treatment, let us imagine that the field is the sum of two components, $\vec{E}_1$ and $\vec{E}_2$, oscillating respectively at frequencies ω_1 and ω_2. It is easy to show that the second-order term of $\vec{P}$ is oscillating at frequency $\omega_3 = \omega_1 \pm \omega_2$. The following cases are of interest for THz optoelectronics :

$$\begin{cases} \omega_3 = \omega_1 + \omega_2 & \Rightarrow & \text{sum - frequency generation} \\ \omega_3 = \omega_1 - \omega_2 & \Rightarrow & \text{difference - frequency generation} \\ \omega_3 = \omega + \omega = 0 & \Rightarrow & \text{optical rectification} \\ \omega_3 = \omega + 0 = \omega & \Rightarrow & \text{electro - optic effect or Pockels effect} \end{cases} \tag{76}$$

In the three first cases, a beam is generated at a frequency different from the incident beams. In the last case (EO effect), the frequency of the incoming beam is conserved: the effect results in a variation of the refractive index of the medium (ellipsoid of index for anisotropic crystals), which will induce a modification of the polarization of the light beam propagating through the crystal. Let us notice that all the effects (76) are related to a deformation of the electronic clouds of atoms and molecules: therefore, they can be considered as instantaneous since their risetime is shorter than 10 fs. However, other contributions (ionic, dipole orientation...) could be observed for the Pockels effect, with much longer response times. The amplitude, χ_2, of the nonlinear susceptibility is weak ($\approx 10^{-12}$ m/V), making the efficiency of the frequency generation low, unless a high amplitude of the E-field is used: that is the case with femtosecond laser pulses, for which the peak power is usually higher than 1 kW.

5.7.3. Sampling Techniques

Ultrashort pulse-durations are far beyond the temporal performance of any present electronic system. Best streak cameras show a subpicosecond resolution in the visible range, but this type of equipment is only at the stage of development for far-infrared radiation [314]. Thus, it is impossible to directly observe the temporal shape of ultrashort pulses. The way to overcome this problem is to take the benefit of the repetition of the laser pulses by making use of sampling techniques. The signal, $S(t)$, to be measured is mixed with another signal, $S_{probe}(t)$, shorter than $S(t)$ if possible, in a nonlinear device. The response of the nonlinear device, recorded with a slow detector, corresponds to the overlap integral of both signals.

This response, $R(\tau)$, is measured as a function of the time-delay, τ, separating the two signals peaks :

$$R(\tau) \propto \int_{-\infty}^{+\infty} S(t-\tau)\, S_{probe}(t)\, dt \,. \tag{77}$$

$R(\tau)$ corresponds to the convolution product of $S(t)$ and $S_{probe}(t)$. If $S_{probe}(t)$ is much shorter than the signal, $S(\tau) \approx R(\tau)$. This technique requires very reproducible pulses, as each measurement, for a given delay τ, is performed using pulses different from the previous measurement. In ultrafast optoelectronics, the ultimate time limit of this technique is usually given by the autocorrelation curve of the laser pulse and/or by the time response of the nonlinear device.

6. REFERENCES

[1] K.E. Razavi, and P. A. Davis
Semiconductor Laser Sources for the Generation of Millimeterwave Signals"
IEE Proc.-Optoelectronics, Vol. 145, pp. 159-163, 1998.

[2] Goldberg L. et. al.
Electronics Letters, Vol.19, No.13, p.491, 1983.

[3] Gliese U. et al.
IEEE Photonic Technology Letters, Vol.4, No.8, p.936, 1995.
[4] Bordonalli, A.C. et. Al
IEEE Photonic Technology Letters, Vol.8, No.9, p.1217, 1996.
[5] G. P. Olbright et al.
"Linwidth, Tunability, and VHF- Millimeter Wave Frequency Synthesis of Vertical-Cavity GaAs Quantum-Well Surface-Emitting Laser Diode Array", IEEE Photon. Technol. Lett., Vol.3, pp. 799781-599, 1991.
[6] Dowd P. et al.
IEEE Journ. Selected Topics in Quantum Electronics, Vol.3, No2, p. 405, 1997.
[7] S. Gevorgian, N. Calander, J. Een, M. Karlsson, M. Nilsson, M. Ghisoni L. Pendrill, and M. Tan
Characterisation of VCSELs for Applications in Industrial Microwave Photonic Devices
Proc. EuMC, 99.
[8] L R Pendrill 1988
Laser measurement station for the maintenance of the national unit of length (in Swedish), Technical Report SP-RAPP 1988:01, National Testing Institute, Borås, Sweden.
[9] T. Milster et al.
A single-mode high-power vertical cavity surface emitting laser
Applied Physical Letters, Vol. 72, No. 26, pp 3425-3427, June 1988, ISBN: 0003 6951L.
[10] A. Coldren, S. W. Corzine
Diode Lasers and Photonic Integrated Circuits
John Wiley & Sons Inc., New York, USA, 1995.
[11] F. Sjöström, A Ölme
Spectral characterization of Vertical Cavity Surface Emitting Lasers (VCSELs)
Swedish National Testing and Research Institute, SP Report 1998:14.
[12] C. Carlsson, M. Ghisoni, A. Larsson, and A. Alping
Analog Modulation Performance of Single and multimode Vertical Cavity Surface Emitting Lasers
International Topical Meeting on Microwave Photonics –99 (MWP-99), Nov.17-19, 1999, Melbourne, Australia.
[13] Shuji s. et al.
A Tunable Cavity-Locked Diode Laser for Terahertz Photomixing
IEEE Trans. Microwave Theory Techn., Vol. 48, pp. 380-387, 2000.
[14] C. Walton et. al.
High-Performance Heterodyne Optical Injection Phase-Lock Loop Using Wide Linewidth Semiconductor Lasers
IEEE Phot technol. Letters, Vol.10, pp.427429, 1998.
[15] M. Ohtsu et al.
FM Noise Reduction and Subkilohertz Linewidth of an AlGa As Lasre by Negative Electrical Feedback
IEEE J. Quant. Electron., Vol.26, pp.231-241, 1990.
[16] J. Lasri et al.
Wavelength Locking of DFB lasers at 50 GHz Spacing Using a Single Fiber Grating reference and a Photo-Heterojunction Bipolar Transistor Optoelectronic Mixer
Proc. IEEE Top. Meeting MWP'99, pp.1316.
[17] T. Hoshida et al.

Generation of 33 GHz Stable pulse trains by Subharmonic Electrical Modulation of a monolithic Passively mode locked semiconductor Laser
El. Lett., Vol.32, pp. 572-573, 1996.

[18] D. Wake et al.
Optical generation of Millimeter-Wave Signals for Fiber-radio Systems Using a Dual-Mode DFB Semiconductor Laser
IEEE Trans. Microwave Theory Techn., Vol. 43, pp.2270-2276.

[19] D. Decoster et al.
InP Photodetectors for Millimeter Wave Applications
SPIE Proc. Photonic West, Photon. Materials and Devices, Vol. 3948, Jan. 2000.

[20] S.M. Sze
Physiscs of Semiconductor Devices, p. 751, 1981.

[21] H. Jacobsson et al.
Low Phase Noise, Low Power IC VCOs for 5-8 GHz Wireless Applications
Proc. IEEE MTT-S'2000.

[22] C. Gonzalez et al.
In/InGaAs Bipolar Phototransistors as a Front End Photoreceever for HFR Distribution Networks
Proc. MWP'99, pp. 35-38.

[23] H. Kamitsuna
Monolithically Integrated High Gain and High Sensitive Photoreceivers with Tunable Filtering Functions for Subcarrier Multiplexed Optical/microwave Systems
IEEE Trans. Micr. Theory Techn., Vol.43, pp.2351-2356, 1995.

[24] R. A. Suematsu, et al.
Noise Performance of MMIC HBTs as Photodetectors
proc. EuMC'93, pp.311-313.

[25] Y. Umezawa et al.
Frequency onverters Using an Illuminated PIN-PD
Proc., WMP'98, pp.1063-1066.

[26] S. G. Ayling
First Demonstration of a High Power, Wide Band Microwave Aplifier Based Upon an Optically Coupled Transistor
Proc. MWP'99, pp.3942.

[27] M. S. Islam et al.
High power Distributed Balanced Photodetectors with High Linearity
Proc. MWP'99, pp. 79-82.

[28] O. Sjölund et al.
Technique for Integration of Vertical Cavity Lasers and Resonant Photodetectors
Appl. Phys. Letters., Vol.73, pp.1-3, 1998.

[29] Y. C. Chung and Y.H. Lee
Spectral Characteristics of Vertical Cavity Surface Emitting Lasers with External Feedback
IEEE Photon. Technol. Lett., Vol.3, pp597-599, 1991.

[30] S. Gevorgian et al.
Swedish patent application.

[31] V.M. Hietala, K.L. Lear, M.G. Armendariz, C.P. Tigges, H.Q. Hou and J.C. Zolper
Electrical characterization and application of very high speed vertical cavity surface emitting lasers (VCSELs)
Sandia National Laboratories.

[32] S. Matsuura et al.
A Tunable Cavity-Locked Diode Laser Source for terahertz Photomixing

IEEE Trans. Micr. Theory Techn., Vol. 48, pp. 380387, 2000.

[33] An Ho Quoc
Etude et Réalisation de Dispositifs Tout-optiques pour le Traitement de Signaux Rapides
PhD thesis, INPG Grenoble, France, July 1996.

[34] T.A. Cusik, S. Iezekiel, R.E. Miles, S. Sales, J. Capmany
Synthesis of All Optical Microwave Filters Using Mach Zehnder Lattices
IEEE Trans. on Microwave Theory Tech., vol. MTT-45, pp. 1458-1461, August 1997.

[35] G. K. Gopalakrishnan, R.P. Moeller, M.M. Howerton, W.K. Burns, K.J. Williams, R.D. Esman
A Low-Loss Downconverting Analog Fiber-OpticLink
IEEE Trans. on Microwave Theory Tech., vol. MTT-43, pp. 2318-2323, September 021995.

[36] H. Ogawa
Fiber Optic Microwave Transmission using Harmonic Laser Mixing, Optoelectronic Mixing, and Optically Pumped Mixing
IEEE Trans. on Microwave Theory Tech., vol. MTT-39, pp. 2045-2051, December 1991.

[37] C. K. Sun, R.J. Orazi, S.A.Pappert, W.K. Burns
Efficient Microwave Frequency Conversion Using Photonic Link Signal Mixing
IEEE Photonics Tech. Lett., Vol. 8, pp. 1166-1168, September 1996.

[38] G. Maury, B. Cabon
New Technique for Optical Generation of Microwave Mixing
Microwave and Optical Technology letters, Vol. 16, N°5, pp. 275-280, December 1997.

[39] C. Gonzalez
HBT Phototransistor as an Optic/Millimetre-Wave Converted
Proceedings of the First International Summer School, OMW'98, p. 63, Autrans-France, August 1998.

[40] G. Maury, A. Hilt, T. Berceli, B. Cabon, A. Vilcot
Microwave Frequency Conversion Methods by Optical Interferometer and Photodiode
Special issue "Microwave and Millimeter-Wave Photonics II", IEEE Trans. on Microwave Theory and Techniques, Vol. 45, pp. 1481-1485, August 1997.

[41] V. Girod, G. Maury, B. Cabon
Microwave Mixing on Multiplexed Optical Carriers
Proceedings of the 3rd International Summer School, OMW'2000, Autrans-France, August 2000.

[42] K. Wilner and A.P. Van Den Heuvel
Fiber-optic delay lines for microwave signal processing
Proc. IEEE, vol 64, pp. 805-807, 1976.

[43] C. Chang, J.A. Cassaboom and H.F. Taylor
Fibre optic delay line devices for RF signal processing
Electron. Lett., vol 13, pp. 678-680, 1977.

[44] H.F. Taylor
Fiber and integrated optical devices for signal processing
SPIE, vol 176, pp. 17-27, 1979.

[45] K. Jackson, S. Newton, B. Moslehi, M. Tur , C. Cutler, J. Goodman and H.J. Shaw
Optical fiber dely-line signal processing
IEEE Trans. Microwave Theory Tech, vol 33, pp. 193-204, 1985.

[46] B. Moslehi, J. Goodman, M. Tur and H.J. Shaw
Fiber-optic lattice signal processing
Proc. IEEE, vol 72, pp. 909-930.1984

[47] K.P. Jackson, and H.J. Shaw
Fiber-optic delay-line signal processors
in Optical Signal Processing, chapter 7, Academic Press, San Diego, 1987.

[48] B. Moslehi
Fibre-optic filters employing optical amplifiers to provide design flexibility
Electron. Lett, vol. 28, pp. 226-228, 1992.

[49] M.C. Vázquez, B. Vizoso, M.López-Amo and M.A.Muriel
Single and double amplified recirculating delay lines as fibre-optic filters
Electron. Lett., vol 28, pp. 1017-1019.

[50] J. Capmany and J. Cascón
Optical programmable transversal filters using fibre amplifiers
Electronics Letters, pp. 1245-1246, 1992.

[51] B. Moslehi and j.W. Goodman
Novel amplified fiber-optic recirculating delay line processor
IEEE J. Lightwave Technol, vol 10, pp. 1142-1146, 1992.

[52] B. Moslehi, K. Chau and J.W. Goodman
Optical amplifiers and liquid-crystal shutters applied to electrically reconfigurable fiber optic signal processors
Opt. Eng., vol 32, pp. 974-981. 1993.

[53] J. Capmany and J. Cascón
Direct form I fiber-optic discrete-time signal processors using optical amplifiers and embedded Mach-Zehnder structures
IEEE Photonics Technology Letters, pp. 842-844, 1993.

[54] J. Capmany
Investigation on phase induced intensity noise in amplified fibre-optic recirculating delay line
Electronics Letters, pp 346-347, 1993.

[55] J. Capmany and J. Cascón
Discrete-time fiber-optic signal processors using optical amplifiers
IEEE Journal of Lightwave Technology, pp. 106-117, 1994.

[56] M.C. Vazquez, R. Civera, M. López-Amo and M.A.Muriel
Analysis of double-parallel amplified recirculating optical delay lines
Appl. Opt., vol 33, pp. 1015-1021, 1994.

[57] B. Vizoso, C. Vazquez, R. Civera, M. Lopez-Amo and M.A. Muriel
Amplified fiber-optic recirculating delay lines
IEEE J. Lightwave Technol., vol 12, pp. 294-305. 1994.

[58] J. Capmany
Amplified double recirculating delay line using a 3x3 coupler
IEEE Journal of Lightwave Technology. pp. 1136-1143, 1994.

[59] A. Ho Quoc and S. Tedjini
Experimental investigation on the optical unbalanced Mach-Zehnder interferometers as microwave filters
IEEE Microwave and Guided Wave Lett., vol 4., pp 183-185, 1994.

[60] D. Pastor, S. Sales, J. Capmany, J. Martí and J. Cascón
Amplified double-coupler fiber-optic delay line filter
IEEE Photonics Technology Letters, pp. 75-77, 1995.

[61] M.C. Vázquez, M. López-Amo, M.A. Muriel and J. Capmany
Performance parameters and applications of a modified amplified recirculating delay line

[62] J. Capmany, J. Cascón, J.L. Marín and J. Martí
Fibre-optic delay line filter synthesis employing a modified PADE method
Electronics Letters, pp. 479-480, 1995.

[63] J. Capmany, J. Martí, S. Sales, D. Pastor and J. Cascón
Comment on "New topologies of fiber-optic delay line filters" by K.K. Goel
IEEE Photonics Technology Letters, pp. 822-823, 1995.

[64] S. Sales, J. Capmany, J. Martí and D. Pastor
Experimental demonstration of fibre-optic delay line filters with negative coefficients
Electronics Letters, pp. 1095-1096, 1995.

[65] S. Sales, J. Capmany, D. Pastor and J. Martí
Fiber-optic delay line filters employing fiber loops: Signal and noise analysis and experimental characterization
Journal of the Optical Society of America A, JOSA A, pp. 2129-2135, 1995.

[66] J. Capmany, J. Cascón, J.L. Marín, S. Sales, D. Pastor and J. Martí
Synthesis of Fiber-optic delay line filters
IEEE Journal of Lightwave Technology, pp. 2003-2012, 1995.

[67] E. Heyde and R.A. Minasian
A solution to the synthesis problem of recirculating delay line filters
IEEE Photon. Technol. Lett., vol 6, pp. 833-835, 1995.

[68] S. Sales, J. Capmany, J. Martí and D. Pastor
Novel and significant results on the nonrecirculating delay line with a fibre loop
IEEE Photonics Technology Letters, pp. 1439-1440, 1995.

[69] S. Sales, J. Capmany, J. Martí and D. Pastor
Solutions to the synthesis problem of optical delay line filter
Optics Letters, pp. 2438-2440, 1995.

[70] D. Norton, S. Johns, C. Keefer and R. Soref
Tunable microwave filtering using high dispersion fiber time delays
IEEE Photon. Tech. Lett., vol 6, pp. 831-832, 1994.

[71] M.Y. Frankel and R.D. Esman
Fiber-optic tunable transversal filter
IEEE Photon. Tech. Lett., vol 7, pp. 191-193, 1995.

[72] D. Dolfi et al
Optical architectures for programmable filtering and correlation of microwave signals
IEEE Trans microwave Theory Tech. Vol 45, pp. 1467-1472, 1997.

[73] A.P. Foord, P.A. Davies and P.A. Greenhalgh
Synthesis of microwave and millimetre-wave filters using optical spectrum-slicing
Electron. Lett., vol 32, pp. 390-391, 1996.

[74] J. Capmany, D. Pastor and B. Ortega
Fibre-optic microwave and millimetre wave filter with high density sampling and very high sidelobe supression using subnanometre optical spectrum slicing
Electronics Letters, 1999, pp. 494-496.

[75] J. Capmany, J. Cascón, D. Pastor y B. Ortega
Reconfigurable fiber-optic delay line filters incorporating electrooptic and electroabsorption modulators
IEEE Photonics Technology Letters, 1999, pp. 1174-1176

[76] G.A. Ball, W.H. Glenn and W.W. Morey
Programmable fiber optic delay line
IEEE Photon. Technol. Lett., vol 6, pp. 741-743, 1994.

[77] A. Molony, C. Edge and I. Bennion
Fibre grating time delay for phased array antennas

Electron. Lett., vol 31, pp. 1485-1486, 1995.

[78] D.B. Hunter and R.A. Minasian
Reflectivity tapped fibre-optic transversal filter using in-fibre Bragg gratings
Electron Lett., vol 31, pp. 1010-1012, 1995.

[79] D.B. Hunter and R.A. Minasian
Tunable transversal filter based on chirped gratings
Electron. Lett., vol 31, pp. 2205-2207, 1995.

[80] D.B. Hunter and R.A. Minasian
Microwave optical filters using in-fiber Bragg grating arrays
IEEE Microwave and Guided Wave Lett., vol 6, pp. 103-105, 1996.

[81] D.B. Hunter and R.A. Minasian
Photonic signal processing of microwave signals using active-fiber Bragg-grating pair structure
IEEE Trans. Microwave Theory and Techn, vol 8, pp. 1463-1466, 1997.

[82] J.L. Cruz, B. Ortega, M.V. Andres, B. Gimeno, D. Pastor, J. Capmany and L. Dong
Chirped fibre Bragg gratings for phased array antennas
Electron. Lett., vol 33, pp. 545-546, 1997.

[83] D.B. Hunter and R.A. Minasian
Microwave optical filters based on fibre Bragg grating in a loop structure
Proc OFC '97, paper TH2-2, pp. 273-275.

[84] D.B. Hunter and R.A. Minasian
Microwave optical filters based on fibre Bragg grating in a loop structure
Proc OFC '97, paper TH2-2, pp. 273-275.

[85] D.B. Hunter and R.A. Minasian
Photonic signal processing of microwave signals using active-fiber Bragg-grating pair structure
IEEE Trans. Microwave Theory and Techn, vol 8, pp. 1463-1466, 1997.

[86] S.C. Kim, et al
Recirculating fiber delay line filter using a fiber Bragg grating
IEEE LEOS '98 Digest., pp. 328-329, 1998.

[87] D. Pastor and J. Capmany
Fiber optic tunable transversal filter using laser array and linearly chirped fibre grating
Electronics Letters, 1998, pp. 1684-1685.

[88] J. Capmany, D. Pastor and B. Ortega
Experimental demonstration of tunability and transfer function reconfiguration in fibre-optic microwave filters composed of a linearly chirped fibre grating fed by a laser array
Electronics Letters, 1998, pp.2262-2264

[89] W. Zhang, J.A.R. Williams, L.A. Everall and I. Bennion
Fibre-optic radio frequency notch filter with linear and continous tuning by using a chirped fibre grating
Electron. Lett., vol 34, pp. 1770-1772.

[90] W. Zhang, J.A.R. Williams, L.A. Everall and I. Bennion
Tuneable radio frequency filtering using linearly chirped fibre grating
proc ECOC '98, pp. 611-612, 1998.

[91] D.B. Hunter and R.A. Minasian
Tunable microwave fiber-optic bandpass filters
IEEE Photon. Technol. Lett, vol 11, pp. 874-876, 1999.

[92] N. You and R.A. Minasian

Synthesis of WDM grating based optical microwave filter with arbitrary impulse response
Int. Top. Meeting MWP '99 Digest, pp. 223-226, 1999.

[93] N. You and R.A. Minasian
A novel high-Q optical microwave processor using hybrid delay line filters
IEEE Trans. Microwave Theory Tech., vol 47, pp. 1304-1308, 1999.

[94] R. Minasian
Photonic signal processing of high speed signals using fiber gratings
Int. Top. Meeting MWP '99 Digest, pp. 219-223, 1999.

[95] W. Zhang, J.A.R. Williams, L.A. Everall and I. Bennion
Tuneable radio frequency filtering using linearly chirped fibre grating
proc ECOC '98, pp. 611-612, 1998.

[96] K.J. Williams, J.L. Dexter and R.D. Esman
Photonic microwave signal processing
Int. Top. Meeting MWP '98 Digest, pp. 187-189. 1998.

[97] W. Zhang, J.A.R. Williams and I. Bennion
Recirculating fiber-optic notch filter employing fiber gratings
IEEE Photon. Technol. Lett., vol 11, pp. 836-838, 1999.

[98] J. Capmany, D. Pastor and B. Ortega
Efficient sidelobe supression by source power apodisation on fibre-optic microwave filters composed of linearly chirped fibre grating fed by laser array
Electronics Letters, 1999, pp. 640-642.

[99] J. Capmany, D. Pastor, B. Ortega
Applications of fibre Bragg Gratings to Microwave Photonics
IEE Colloquium on fibre gratings 1999, Birmingham, UK.

[100] J. Capmany, D. Pastor and B. Ortega
New and flexible fiber-optic delay line filters using chirped Bragg gratings and laser arrays
IEEE Trans. Microwave Theory Tech., vol 47, pp. 1321-1327, 1999.

[101] P.J. Mathews and P.D. Biernacki
Photonic signal processing for microwave applications
IEEE MTT-S Digest, paper WE1B-1, pp. 877-880, 1999.

[102] D. Pastor, J. Capmany and B. Ortega
Experimental demonstration of parallel fiber-optic-based RF filtering using WDM technique
IEEE Photonics Technology Letters, 2000, pp. 77-79.

[103] R.A. Minasian
Photonic signal processing of high-speed signals using fiber gratings
Opt. Fib. Tech., vol 6, pp. 91-108, 2000.

[104] W. Zhang, J.A.R. Williams and I. Bennion
Optical fibre delay line filter free of limitation imposed by optical coherence
Electron.Lett., vol 35, pp. 2133-2134, 1999.

[105] A. Ho-Quoc, S. Tedjini and A. Hilt
Optical polarization effect in discrete time fiber-optic structures for microwave signal processing
IEEE MTT- Symposium Digest, pp. 907-910, 1996.

[106] F. Coppinger, S. Yegnanarayanan, P.D. Trinh and B. Jalali
All-optical incoherent negative taps for photonic signal processing
Electron. Lett., vol 33, pp. 973-975, 1995.

[107] M. Tur and B. Moslehi
Laser phase noise effects in fiber-optic signal processors with recirculating loops
Opt. Lett., vol 8, pp. 229-231, 1983.

[108] M. Tue, B. Moslehi and J.W. Goodman
Theory of laser phase noise in recirculating fiber optic delay lines
IEEE J. Lightwave Technol, vol 3, pp. 20-30, 1985.

[109] M. Tur and A.Arie
Phase induced intensity noise in concatenated fiber-optic delay lines
IEEE J. Lightwave Technol., vol 6, pp. 120-130, 1988.

[110] B. Moslehi
Analysis of optical phase noise in fiber-optic systems employing a laser source with arbitrary coherence time
IEEE J. Lightwave Technol., vol 4, pp. 1334-1351, 1986.

[111] J.T. Kringlebotn and K. Blotekjaer
Noise analysis of an amplified fiber-optic recirculating delay line
IEEE J. Lightwave Technol, vol 12, pp. 573-581, 1994.

[112] Westbrook, D. Nesset
Novel Techniques for High-Capacity 60-GHz Fiber-Radio Transmission Systems
IEEE Transactions on Microwave Theory and Techniques, Vol.45, pp.1416-1423, August 1997.

[113] R.-P. Braun et al.
Optical Microwave Generation and Transmission Experiments in the 12- and 60 GHz Region for Wireless Communications
IEEE Trans-actions on Microwave Theory and Techniques, Vol.46, No.4, pp.320 330, 1998.

[114] N. Imai, H. Kawamura, K. Inagaki, Y. Karasawa
Wide-Band Millimeter-Wave/Optical-Network Appli-cations in Japan
IEEE Transactions on Microwave Theory and Techniques, Vol.45, No.12, pp.2197-2207, 1997.

[115] D. Wake, D.G. Moodie
Passive Picocell - prospects for increasing the radio range
IEEE Topical Meeting on Microwave Photonics, MWP'97 Digest, pp.269-271, Duisburg, Germany, September 1997.

[116] T. Berceli, I. Frigyes, P. Gottwald, P.R. Herczfeld, F. Mernyei
Improvements in Fiber - Optic Transmis-sion of Multi-Carrier TV Signals
IEEE Transactions on Microwave Theory and Techniques, Vol.40, pp.910-915, May 1992.

[117] G.H. Smith, D. Novak, Z. Ahmed,
Novel Technique For Generation of Optical SSB With Carrier Using A Single MZM To Overcome Fiber Chromatic Dispersion
Microwave Photonics Workshop, MWP'96, pp.5-8 (PDP-2), Kyoto, Japan, 1996.

[118] G.H. Smith, D. Novak
Broad-Band Millimeter-Wave (38GHz) Fiber-Wireless Transmission System Using Electrical and Optical SSB Modulation to Overcome Dispersion Effects
IEEE PTL, Vol.10, No.1, pp.141-143, January 1998.

[119] J.M. Fuster, J. Marti, V. Polo, F. Ramos, J.L. Corral
Mitigation of dispersion-induced power penalty in millimetre-wave fibre optic links
Electronics Letters, Vol.34, pp.1869-1871.

[120] M. Tsuchiya, and T. Hoshida
Nonlinear Photodetection Scheme and Its System Applications to Fiber-Optic Millimeter-Wave Wireless Down-Links
IEEE Trans. on MTT, Vol.47, No.7, pp.1342-1350, July 1999.

[121] T. Berceli

A new approach for optical millimeter wave generation utilizing locking techniques
IEEE MTT-S, International Microwave Symposium Digest, Vol.III, pp.1721-1724, Denver, USA, June 1997.

[122] T. Marozsák, T. Berceli, G. Járó, A. Zólomy, A. Hilt, S. Mihály, E. Udvary, Z. Varga
A New Optical Distribution Approach for Millimeter Wave Radio
MWP'98 Digest, pp.63-66, Princeton, New Jersey, USA, October, 1998.

[123] A. Zólomy, V. Bíró, T. Berceli, G. Járó, A. Hilt
Design Of Harmonic Oscillators For Millimeter Wave Signal Generation In Optical Systems
Proc. of the 28th EuMC, pp.75-80, Amsterdam, The Netherlands, October 1998.

[124] A. Hilt, A. Zólomy, T. Berceli, G. Járó, E. Udvary
Millimeter Wave Synthesizer Locked to an Optically Transmitted Reference Using Harmonic Mixing
Digest of the IEEE Topical Meeting on Microwave Photonics, MWP'97, pp.91-94, Duisburg, Germany, September 1997.

[125] H. Schmuck
Comparison of optical millimetre-wave system concepts with regard to chromatic dispersion
Electronics Letters, Vol.31, No.21, pp.1848-1849, 12th October 1995.

[126] A. Hilt, A. Vilcot, T. Berceli, T. Marozsák, B. Cabon
New Carrier Generation Approach for Fiber-Radio Systems to Overcome Chromatic Dispersion Problems
IEEE MTT-S Digest, pp.1525-1528, Baltimore, USA, June 1998.

[127] J. Park, W.V.S orin, K.Y. Lau
Elimination of the Fibre Chromatic Dispersion Penalty on 1550 nm Millimetre Wave Optical Transmission
Electronics Letters, Vol.33, No.6, pp.512-513, 13th March 1997.

[128] T. Berceli
A new optical reception method using microwave subcarriers
Proc. of the 24th EuMC, pp.1679-1684, Cannes, France, September 1994.

[129] R. Helkey
Advances in frequency conversion optical links
Proc. of 10th MICROCOLL, pp.365-369, Budapest, Hungary, 1999.

[130] T. Berceli, B. Cabon, A. Hilt, G.Járó
Improved optical-microwave mixing process utilizing high-speed photo-diodes
26th EuMC Proc., Vol.1, pp.125-129, Prague, Czech Republic, September 1996.

[131] Q.Z. Liu, R. Davies, R.I. MacDonald
Experimental Investigation of Fiber Optic Microwave Link with Monolithic Integrated Optoelectonic Mixing Receiver
IEEE Trans. on MTT, Vol.43, No.9, pp. 2357-2360, September 1995.

[132] T. Berceli, G. Járó
A new optical-microwave double mixing method
IEEE MTT-S, International Microwave Symposium Digest, pp. 1003-1006, Baltimore, Maryland, USA, June 7-12, 1998.

[133] D. Dolfi et al.
Applied Optics 35, 5293, 1996.

[134] L. Pastur et al
Applied Optics 38, 3105,1999.

[135] M.Y. Frankel et al
IEEE Trans. on Microwave Theory and Techniques, 43, 2387, 1995.

[136] S.H. Song et al
Applied Optics, 34, 5913, 1995.

[137] F.Xu et al
Applied Optics 34, 256, 1995.

[138] Capmany et al.
IEEE Trans. on Microwave Theory and Techniques, 47, 1321,1999.

[139] D. Norton et al
IEEE Photonic Technology Letters, 6, 831, 1994.

[140] M.Y. Frankel et al
IEEE Photon. Technol. Lett.,7, 191, 1995.

[141] R.M. Montgomery et al
Applied Optics, 30, 2844, 1991.

[142] B. Moslehi et al
Optical Engineering, 947, 1993.

[143] D.B. Hunter et al.
IEEE Trans. on Microwave Theory and Techniques, 8, 1463, 1997.

[144] D. Dolfi et al
IEEE Trans. on Microwave Theory and Techn. 45, 1467, 1997.

[145] Optical Processing and Computing, Special Issue, N.A. Riza, ed.
SPIE's International Techn. Group Newsletter 10 (April 1999).

[146] M. Born and E. Wolf
Principles of Optics
Pergamon Press, New York, 1980

[147] S. Tonda et al.
Optical Engineering.

[148] T. N. Nielsen et al.
Proceedings of "OFC 2000", Baltimore, March 7-10, 2000, paper PD23.

[149] T. Ito et al.
Proceedings of "OFC 2000", Baltimore, March 7-10, 2000, paper PD24.

[150] G. Raybon et al.
Proceedings of "OFC 2000", Baltimore, March 7-10, 2000, paper PD29.

[151] S. Kawanishi et al,
Proceedings of "1999 OFC/IOOC 1999", San Diego, February 21-26, 1999, PD1.

[152] K. Hagimoto et al.
J. Lightwave Technol., 8, pp. 1387-1392, Sept. 1990.

[153] R. Shnabel et al.
IEEE Photon. Technol. Lett., vol. 6, pp. 56-58, January 1994.

[154] A. Yariv
Quantum Electronics
John Wiley & Sons 1988, Chapt. 16.

[155] S. J. B. Yoo et al.
Appl. Phys. Lett., vol. 66, pp. 3410-3412, June 1995.

[156] A. Mecozzi et al.
Proceedings of SPIE "Physics and simulation of optoelectronic devices IV"
San Jose, Jan. 29-Feb. 2, 1996, pp. 288-302.

[157] S. Watanabe et al.
Proceedings of "ECOC'98", Madrid, September 20-24, 1998, PD paper, p. 85.

[158] C. T. Hultgren et al.
Proceedings of "Ultrafast electronics and opto-electronics, Topical meeting", San Francisco, 1993.

[159] K. Vahala et al.
Appl. Phys. Lett., vol. 42, pp. 631-633, April 1983.

[160] S. Kawanishi et al.
Electron. Lett., vol. 30, pp. 981-982, June 1994.

[161] T. Durhuus et al.
Proceedings of "CLEO'92", Anaheim, May 1992, paper CThS4.

[162] C. Jorgensen et al.
Proceedings of "Optical amplifiers and their applications", Yokohama, 1993, Paper MD2.

[163] C. Janz
Proceedings of "OFC 2000", Baltimore, March 7-10, 2000, paper ThF7.

[164] B. Lavigne et al.
Proceedings of "ECOC'99", Nice, September 26-30, pp. 262-263.

[165] D. Wolson et al.
Electron. Lett., vol. 35, pp. 59-60, Jan. 1999.

[166] B. Lavigne et al.
Proceedings of "1999 OFC/IOOC 1999", San Diego, February 21-26, 1999, paper TuJ3.

[167] R. J. Manning et al.
Electron. Lett., vol. 34, pp. 916-917, April 1998.

[168] S. Diez et al.
Electron. Lett., vol. 34, pp. 803-804, April 1998.

[169] S. Bauer et al.
Proceedings of "OFC 2000", Baltimore, March 7-10, 2000, paper TuF5.

[170] I. D. Phillips et al.
Electron. Lett., vol. 34, pp. 2340-2341, 1998.

[171] G. Raybon et al.
Proceedings of "1999 OFC/IOOC 1999", San Diego, February 21-26, 1999, paper FB2.

[172] S. Nakamura et al.
"OFC 2000", Baltimore, March 7-10, 2000, paper ThF3.

[173] S. Scotti et al.
IEEE J. of Selected Topics on Quantum Electron., 3, pp.1156- 1161, Oct. 1997.

[174] A. Mecozzi et al.
Appl. Phys. Lett., vol. 66, pp. 1184-1186, March 1995.

[175] A. D'Ottavi et al,
IEEE Photon. Technol. Lett., vol. 10, pp. 952-954, July 1994

[176] W. Pieper et al.
Electron. Lett., vol. 30, pp. 724-725, 1994.

[177] R. Ludwig et al.
Fiber and Integrated Optics, vol. 15, pp.211-223, 1996.

[178] A. Corchia et al.
IEEE Photon. Technol. Lett., vol. 11, pp. 275-277, Feb. 1999.

[179] T. Morioka et al.
Electron. Lett., vol. 32, pp. 840-841, April 1996.

[180] E. Ciaramella and S. Trillo
IEEE Photon. Technol. Lett., vol. 12, pp. 849-851, July 2000.

[181] J. P. R. Lacey et al.
IEEE Photon. Technol. Lett., vol. 9, pp. 1355-1357, Oct.. 1997.

[182] G. Contestabile et al.
IEEE Photon. Technol. Lett., vol. 10, pp. 1398-1400, Oct.. 1998.

[183] A. Mecozzi et al.
Appl. Phys. Lett., vol. 72, pp. 2651-2652, May 1998.

[184] A. D'Ottavi et al.

IEEE J. of Selected Topics on Quantum Electron., vol. 3, pp.522- 528, April 1997.
[185] D. Nesset et al.
Electron. Lett., vol. 34, pp. 107-109, 1998.
[186] J. Bardeen and W.H. Brattain
Phys. Rev. 74 p. 230, 1948.
[187] W. Shockley
Bell Sys. Tech. J. 28, p. 435, 1949.
[188] Y. Betser, et. Al
IEEE Journal of Lightwave Technol., vol.16, no. 4, p. 605, 1998.
[189] J. Lasri, et. al.
IEEE Journal of Lightwave Technol., vol. 17, no. 8, p. 1423, 1999.
[190] S. Chandrasekhar, et al
IEEE Photonic. Tech. Lett., vol. 5, p. 1316, 1993.
[191] A. J. Seeds et al
Technical Digest of Microwave Photonics , p. 217. 1996.
[192] [H. Sawada et al
IEEE Trans. on Microwave Theory and Tech., 47, p. 1515, 1999.
[193] R. Adler
IRE 34 p. 351, 1946.
[194] IEEE MTT-S International Microwave Symposium & Exhibition
Workshop on ADC's for Digital Receiver Systems, 8 June 1998, Baltimore, U.S.
[195] R.G. Vaughan, N.L. Scott, D.R. White
The Theory of Bandpass Sampling
IEEE. Trans. Sig. Proc., Vol. 39, No. 9, Sept 1991, pp73-84
[196] J. U. Kang, M. Y. Frankel, and R. D. Esman
Highly parallel pulsed optoelectronic analog-digital converter
IEEE Photonics Technology Lett., vol. 10, 11, 1998, pp1626-1628
[197] D.H. Auston
Picosecond optoelectronic switching and gating in silicon
Applied Physics Letters, 1975, Vol.26 (3), pp. 101-103.
[198] C.K. Sun, C.-C. Wu, et al.
A bridge type optoelectronic sample and hold circuit
Journal of Lightwave Technology, 1991, Vol. 9 (3), pp. 341-345.
[199] R. Mason and J.T. Taylor
Design and evaluation of an optically triggered monolithic sample and hold circuit using GaAs MESFET technology
Journal of Lightwave Technology, 1995, Vol. 13 (3), pp. 422-429.
[200] K. Poulton, J.J. Corcoran, and T. Hornak
A 1-GHz 6-bit ADC system
IEEE Journal of Solid-state Circuits, 1987, Vol. 22 (6), pp. 962-970.
[201] F.W. Smith, H.R. Calawa, C.L. Chen, M.J. Mantra et L.J. Mahoney
IEEE Electronics Developement Letters, 1988, Vol. 9, p. 77.
[202] D.C. Look, D.C. Walters, M.O. Manasreh, J.R. Sizelove, C.E. Stutz, and K.R. Evans
Physics Review B, 1990, Vol. 42, p. 3578.
[203] Gupta, M.Y. Frankel, J.A. Valdmanis, J.F. Whitaker, F.W. Smith and A.R. Calawa
Applied Physics Letters, 1991, Vol. 59, p. 3276.
[204] [Y. Chen, S. Williamson, T. Brock, F.W. Smith, and A.R. Calama
Applied Physics Letters, 1991, Vol. 59, p. 1984.
[205] C.T. Chang, C.K. Sun, D.J. Albares, and E.W. Jacobs
High-energy (59pJ) and low-jitter (250fs) picosecond pulses from gain-switching of a tapered-stripe laser diode via resonant driving

IEEE Photonics Technology Letters, Vol.8 (9), 1996, pp.1157-1559.

[206] P.P. Vasil'ev
High-power high frequency picosecond pulse generation by passively Q-switched 1.55μm diode lasers
IEEE Journal of Quantum Electronics, Vol. 29 (6), 1993, pp. 1687-1692.

[207] Y. Arakawa, A. Larsson, J. Paslaski, A. Yariv
Active Q-switching in GaAs/AlGaAs multiquantum well laser with an intracavity monolithic loss modulator
Applied Physics Letters, Vol 48 (9), 1986, pp. 561-563.

[208] T. Schrans, R.A. Salvatore, S. Sanders, A. Yariv
Subpicosecond pulses from CW passively mode locked external cavity two sections multiquantum well lasers
Electronics Letters, Vol. 18 (16), 1992, pp. 1480-1482.

[209] A.D. Carr
Advanced Electro-Optic Pre-Processor for DRFM
RP 413-278 Final Technical Report, Marconi Research Centre Document MTR 97/46D, November 1997.

[210] F. Coppinger, A.S. Bhushan, and B. Jalali
Time magnification of electrical signals using chirped optical pulses
Electronics Letters, 1998, Vol. 34 (1), pp. 399-400.

[211] M.Y. Frankel, Jin U. Kang, and R.D. Esman
High performance photonic analogue-digital converter
Electronics Letters, 1997, Vol. 33(25), pp 2096-2097.

[212] H. F. Taylor
An electro-optic analog-to-digital converter
Proceedings of the IEEE, 1975, Vol. 63, pp 1524-1525.

[213] R.A. Becker, C.E. Woodward, F.J. Leonberger, and R.C. Williamson
Wide-band electro-optic guided-wave analog-to-digital converters
Proceedings of the IEEE, 1984, Vol. 72 (7), pp. 802-819.

[214] B. Jalali and Y.M. Xie
Optical folding-flash analog-to-digital converter with analog encoding
Optics Letters, 1997, Vol. 20 (18), pp. 1901-1903.

[215] R.G. Walker, I. Bennion, A.C. Carter
Novel GaAs/AlGaAs Guided Wave Analogue-to-Digital Converter
Elect. Lett. Vol. 25, No. 21, pp. 1443-1444, October 1989.

[216] R. Mason and J.T. Taylor
High-speed electro-optic analogue to digital converters
Proceedings of IEEE, 1993, pp. 1081-1084.

[217] P.J. Delfyett, D.H. Hartman, and S.Z. Ahmad
Optical clock distribution using a mode-locked semiconductor laser diode system
Journal of Lightwave Technology, 1991, Vol. 9 (12), pp. 1646-1649.

[218] J.A. Bell, M.C. Hamilton, D.A. Leep, H.F. Taylor, Y. Lee
A/D Conversion of Microwave Signals Using a Hybrid Optical/Electronic Technique
SPIE Vol 1476, 1991, pp 326-329.

[219] A. Johnstone, M.F. Lewis, J. Hares
Optical Replication Technique for Wideband Transient Waveform Digitisation
Proc. SPIE Vol. 3285, Fabrication, Testing, Reliability of Semiconductor Lasers III, San Jose, January 1998, Paper No. 3285-38.

[220] P.M. Duffieux
L'intégrale de Fourier et son application à l'Optique

first edition, chez l'auteur, Besançon, 1946. Later published by Masson and also translated into English.

[221] J.W. Goodman
Introduction to Fourier Optics
second, revised edition, Mc Graw Hill,

[222] E. Abbe
Archiv für mikroskopische Anatomie, 9 (1873) 413.

[223] See M. Born and E. Wolf
Principles of Optics
Pergamon Press, New York (first edition, 1959), chapter V.

[224] A. Maréchal, P. Croce
Compt. Rend. Acad. Sci., 1953.

[225] J.L. Tribillon
PhD thesis, Besançon, France, 197
H. Dammann and K. Görtler
Opt. Commun. 3, pp. 321-323, 1971
J.L. Tribillon
Pure Appl. Opt. 3, pp. 389-404, 1994
H. Dammann, E. Klotz
Opt. Acta 24, pp. 505-515, 1977
U. Krackhardt, N. Streibl
Opt. Commun. 74, pp. 31-36, 1989.

[226] A. Vander Lugt
IEEE Trans. Inform. Theory, IT10 (1964) 139–145.

[227] Ph. Réfrégier
Opt. Lett. 16 (1991) 829–831 ; J. Opt. Soc. Am. A11 (1994) 1243–1251.

[228] F. Goudail and Ph. Réfrégier
Opt. Commun. 125 (1996) 211–216
Opt. Lett. 21 (1996) 495–497.

[229] See contribution by D. Dolfi
OMW2000.

[230] K.H. Lu and B.E.A. Saleh, Opt. Engin. 29 (1990) 240–246; C. Soutar et al., Opt. Engin. 33 (1994) 1061–1068 ; V. Laude, S. Mazé, P. Chavel, P. Réfrégier, Opt. Commun. 103 (1993) 33-38.

[231] A. O'Hara et al., Appl. Opt. 32 (1993) 5549–5556; M.C. Roggeman et al., Opt. Engin. 36 (1997) 1326–1328; T.G. Bifano et al., Opt. Engin. 36 (1997) 1354–1359 ; G. Vdovin et al., Opt. Engin. 36 (1997) 1382–1390.

[232] J. Colin, PhD Thesis, université Paris Pierre et Marie Curie, France, novembre 1998.

[233] For a review, see P. Chavel, "Computing : A joint venture for light and electricity?" in "International Trends in Optics", J.W. Goodman, editor, Academic Press (1990) 499-519.

[234] A. Huang, IEEE 10th Optical Computing Conf. pp. 13-17, 1980 ; J. Taboury, J.M. Wang, P. Chavel, F. Devos and P. Garda, Appl. Opt. 27, pp. 1643-1650, 1988.

[235] B.K. Jenkins, P. Chavel, R. Forchheimer, A.A. Sawchuk and T.C. Strand, Appl. Opt. 23, pp. 3465-3474,1984.

[236] Ph. Bouthemy and P. Lalande, Opt. Engin. 32, pp. 1205-1212, 1993.

[237] Ph. Lalanne, D. Prévost, P. Chavel, L. Garnero, "Stochastic Artificial Retinae", Ann. Phys., 1999.

[238] For a tutorial on Gunn diodes, see R. van Zyl, W. Perold, R. Botha, IEEE Proceedings of the 1998 South African Symposium on Communications and Signal Processing, COMSIG'98, p. 407 (1998)

[239] H. Eisele, A. Rydberg, G. Haddad, IEEE Trans. MTT vol. 48, 626 (2000)

[240] T. Crowe, T. Grein, R. Zimmermann, P. Zimmermann, IEEE Microwaves Guided Wave Lett. vol. 6, 207 (1996)

[241] J.P. Gordon, H. Zeiger, C.H. Townes, Phys. Rev. vol. 99, 1264 (1955)

[242] M. Pollack, in "Handbook of Lasers", p. 298, R. Pressley ed., Chemical Rubber Co., Cleveland (1971)

[243] G. Carelli, A. Moretti, D. Pereira, F. Strumia, IEEE J. Quant. Electron. vol. 31, 144 (1995)

[244] B. S. Williams, B. Xu, Q. Hu, M.R. Melloch, Appl. Phys. Lett. vol. 75, 2927 (1999)

[245] R. Soref, L. Friedman, G. Sun, Gregory, M.J. Noble, L.R. Ram-Mohan, Proc. SPIE vol. 3795, 516 (1999)

[246] S. Matsuura, M. Tani, K. Sakai, Appl. Phys. Lett. vol. 70, 559 (1997)

[247] S. Verghese, E.K. Duerr, K.A. McIntosh, S.M. Duffy, S.D. Calawa, C.-Y. Tong, R. Kimberk, R. Blundell, IEEE Microwaves Guided Wave Lett. vol. 9, 245 (1999)

[248] M.D. Pelusi, H. F. Liu, D. Novak, Y. Ogawa, Appl. Phys. Lett. vol. 71, 449 (1997)

[249] C.-L. Wang and C.- L. Pan, Opt. Lett. vol. 20, 1292 (1995)

[250] Y. Matsui, M.D. Pelusi, S. Arahari, Y. Ogawa' Electron. Lett 35, 472 (1999)

[251] J.F. Whitaker, Mater. Sci. Eng. vol. B22, 61 (1993)

[252] Proc. of the 2nd Symposium on Non-Stoichiometric III-V Compounds, ed. by T. Marek, S. Malzer and P. Kiesel, Physik Mikrostrukturierter Hableiter vol. 10, Erlangen (1999)

[253] P. Benicewicz, J.P. Roberts, A.J. Taylor
J. Opt. Soc. Amer. vol. B11, 2533 (1994)
J.-F. Roux, F. Garet, J.-L. Coutaz, J.F. Whitaker
PIERS'96, Innsbruck (1996)

[254] S. Gupta, J.F. Whitaker, G.A. Mourou
IEEE J. Quantum Electron. vol. 28, 2464 (1992)

[255] M. Hailm, U. Siegner, F. Morier-Genoud, U. Keller, M. Luysberg, P. Specht, E.R Weber
Appl. Phys. Lett. vol. 74, 1269 (1999)

[256] A. Krotkus, K. Bertulis, L. Dapkus, U. Olin, S. Marcinkevicius
Appl. Phys. Lett. 75, 3336 (1999)

[257] A. Krotkus, S. Marcinkevicius, J. Jasinski, M. Kaminska, H.H. Tan, C. Jagadish
Appl. Phys. Lett. vol. 66, 3304 (1995)

[258] F.E.Doanay, D. Grischowsky,C.C. Chi
Appl. Phys. Lett. vol. 50, 460 (1987)

[259] M. Roser, S.R. Clayton, P.R. de la Houssaye, G.A. Garcia
IEEE Trans. Electron Devices vol. 39, 2665 (1992)

[260] P.R. Smith, D.H. Auston, M.C. Nuss
IEEE J. Quant. Electron. vol. 24, 255 (1988)

[261] M.C. Nuss, D.W. Kisker, P.R. Smith, T.E. Harvey
Appl. Phys. Lett. vol. 54, 57 (1989)

[262] It has been observed in many semiconductors that the mobility of ultrashort-lifetime carriers measured in sub-picosecond time-resolved experiments is much weaker than the DC mobility, measured using Hall-effect techniques. The slowing down of the mobility is attributed to the non-thermal behavior of the free carrier at very short time after excitation. Also, relaxation to the L valley (small mobility) and then to the Γ valley (high mobility) could be an explanation

See J.T. Darrow, X.-C. Zhang, D.H. Auston and J.D. Morse
IEEE J. Quantum Electron. vol. 28, 1607 (1992)

[263] M. Klingenstein, J. Kuhl, R. Notzel, K. Ploog, J. Rosenzweig, C. Moglestue, A. Hulsmann, J. Schneider, K. Kohler
Appl. Phys. Lett. vol. 60, 627 (1992)

[264] S.Y. Chou and M.Y. Liu
IEEE J. Quantum Electron. vol. 28, 2358 (1992)

[265] P. Kordos, A. Förster, M. Marso, F. Rüders
Electron. Lett. vol. 34, 119 (1998)

[266] Y.-J. Chiu, S.B. Fleischer, D. Lasaosa, J.E. Bowers
Appl. Phys. Lett. vol. 71, 2508 (1997)

[267] F. Garet, L. Duvillaret, L. Noel, J.M. Munier, J.-L. Coutaz, P. Febvre
ESA workshop on mm waves and applications, Espoo, Finland, 27-29 may 1998

[268] X.-C. Xhang and D.H. Auston
J. Appl. Phys. vol. 71, 326 (1992)

[269] M. van Exter and D. Grischkowsky
Appl. Phys. Lett. vol. 56, 1694 (1990)

[270] N. Katzenllenbogen, H. Chan, D. Grischkowsky
Quantum Electronics and Laser Science Conference
OSA Technical Digest Series vol. 3, p. 155, Washington, DC (1993)

[271] E. Budiarto, J. Margolies, S. Jeong, J. Son, J. Bokor
IEEE J. Quant. Electron. vol. 32, 1839 (1996)

[272] S. Verghese, K.A. McIntsoh and E.R. Brown
IEEE Trans. MTT vol. 45, 1301 (1997)

[273] B. Hu, E. de Souza, W. Knox, J. Cunningham, M. Nuss, A.V. Kuznetsov, S. Chuang
Phys. Rev. Lett. vol. 74, 1689 (1995)

[274] Ch. Kittel
"Introduction to Solid State Physics"
Wiley, New York (1956)

[275] J. Shah
"Ultrafast Spectroscopy of Semiconductors and Semiconductor Nanostructures"
Springer Series in Solid State Physics vol. 115, Berlin (1996)

[276] T. Dekorsy, R. Ott, H. Kurz, K. Kohler
Phys. Rev. B vol. 51, 17275 (1995)

[277] W. Sha, A.L. Smirl, W.F. Tseng
Phys. Rev. Lett. vol. 74, 4273 (1995)

[278] A. Leitenstorfer, S. Hunsche, J. Shah, M.C. Nuss, W.H. Knox
Appl. Phys. Lett. vol. 74, 1516 (1999)

[279] A. Bonvalet, M. Joffre, J.L. Martin, A. Migus
Appl. Phys. Lett. vol. 67, 2907 (1995);
M. Joffre, A. Bonvalet, A. Migus, J.L. Martin
Opt. Lett. vol. 21, 964 (1996)

[280] P. Y. Han and X-C. Zhang
Appl. Phys. Lett. vol. 73, 3049 (1998)

[281] R. Huber, A. Brodschelm, F. Tauser, A. Leitenstorfer
Appl. Phys. Lett. vol. 76, 3191 (2000)

[282] Q. Wu and X.-C. Zhang
Appl. Phys. Lett. vol. 68, 1604 (1996)

[283] H. Karlsson
PhD Thesis, TRITA-FYS 2197, KTH Stockholm (1999)

[284] I. Kaminow and E. Turner

in "Handbook of Lasers", p. 447, ed. by R. Pressley, Chemical Rubber Co., Cleveland (1971)

[285] A. Morikawa, K. Kawase, J.-i. Shikata, T. Taniuchi, H. Ito
SPIE vol. 3828, 302 (1999)

[286] M. Hangyo, S. Tomozawa, Y. Murakami, M. Tonouchi, M. Tani, Z. Wang, K. Sakai, S. Nakashima
Appl. Phys. Lett. vol. 69, 2122 (1996)

[287] M.J.W. Rodwell, M. Kamegawa, R. Yu, M. Case, E. Carman, K.S. Giboney
IEEE MTT vol. 39, 1194 (1991)

[288] O. Wohlgemuth, M.J.W. Rodwell, R. Reuter, J. Braunstein, M. Schlechtweg
IEEE MTT vol. 47, 2591 (1999)

[289] A.T. Lee, P.L. Richards, S.W. Nam, B. Cabrera, K.D. Irwin
Appl. Phys. Lett. vol. 69, 1801 (1996)

[290] J.A. Valdmanis
chapter 4, in "Semiconductors and Semimetals"
Academic Press, New York (1989)

[291] A. Yariv
"Optical Electronics"
Saunders College Pub. (1991)
R. Guenther, "Modern Optics", J. Wiley & Sons (1990)

[292] F.A. Hegmann, D. Jacobs-Perkins, C.-C. Wang, S.H. Moffat, R.A. Hughes, J.S. Preston, M. Currie, P.M. Fauchet, T.Y. Hsiang, R. Sobolewski
Appl. Phys. Lett. vol. 67, 285 (1995)

[293] U.D. Keil and D.R. Dykaar
OSA Proc. Ultrafast Electronics and Optoelectronics 14, 189 (1993)

[294] K. Yang, L. P. B. Katehi, J. F. Whitaker
Appl. Phys. Lett. vol. 77, 486 (2000)

[295] V.N. Filippov, A.N. Starodumov, Y.O. Barmenkov, V.V. Makarov
Appl. Opt. vol. 39, 1389 (2000)

[296] O. Reimann, D. Huhse, E. Dröge, E. Böttcher, D. Bimberg, H. Stahlmann
IEEE Photonics Tech. Lett. 11, 1024 (1999)

[297] J.M. Zhang, R.L. Ruo, M.K. Jackson
Appl. Phys. Lett. vol. 75, 3446 (1999)

[298] R. Hofmann and H.-J. Pfleiderer
J. Lightwave Technol. vol. 14, 1788 (1996)

[299] J. Allam, C.L. Yuca, J.R.A. Cleaver,
EEE Proceedings of the Symposium EDMO, p. 20 (1999)

[300] M. Shinagawa, T. Nagatsuma, K. Ohno, Y. Jin,
Proceed. of Instrumentation and Measurement Technology Conference vol. 3, 1541, (2000)
M. Shinagawa, T. Nagatsuma, S. Miyazawa, IEEE Trans. Instrum. Meas. vol. 47, 235 (1998)

[301] J. Shan, A.S. Weling, E. Knoesel, L. Bartels, M. Bonn, A. Nahata, G.A. Reider, T.F. Heinz
Opt. Lett. vol. 25, 426 (2000)

[302] J. Ziang and X.-C. Zhang
Appl. Phys. Lett. vol. 72, 1945 (1998)

[303] L. Duvillaret, F. Garet, J.-L. Coutaz
IEEE J. Select. Topics Quant. Electron. vol. 2, 739 (1996)

[304] L. Duvillaret, F. Garet, J.-L. Coutaz
Appl. Opt. vol. 38, 409 (1999)

[305] C. Gallot and D. Grischkowsky

J. Opt. Soc. Am. B vol. 16, 1204 (1999)
[306] P.Y. Han, G.C. Cho, X.-C. Zhang
Opt. Lett. vol. 25, 242 (2000)
[307] S. Hunsche, M. Koch, I. Brener, M.C. Nuss
Opt. Commun. vol. 150, 22 (1998)
[308] Q. Wu, T.D. Hewitt, X.-C. Zhang
Appl. Phys. Lett. vol. 69, 1026, 2268 (1996)
[309] Special Issue on "Optical Probing of Ultrafast Devices"
Optical and Quantum Electronics vol. 28 (1996)
[310] H.-J. Cheng, J.F. Whitaker, K.J. Herrick, N. Dib, L.P.B. Katehi, J.-L. Coutaz,
Ultrafast Electronics and Optoelectronics
OSA Technical Digest Series vol. 13, 128 (1995)
[311] J. Kim, S. Williamson, J. Nees, S. Wakana, J. Whitaker
Appl. Phys. Lett. vol. 62, 2268 (1993)
M.D. Weiss, M.H. Crites, E.W. Bryerton, J.F. Whitaker, Z. Popovi´c
IEEE MTT vol. 47, 2599 (1999)
[312] S. Weiss, D.F. Ogletree, D. Botkin, M. Salmeron, D.S. Chemla
Appl. Phys. Lett. vol. 63, 2567 (1993)
[313] J.-R. Hwang, H.-J. Cheng, J.F. Whitaker, J.V. Rudd
Appl. Phys. Lett. vol. 68, 1464 (1996)
[314] M. Drabbels, G. M. Lankhuijzen, L. D. Noordam
IEEE J. Quant. Electron. vol. 34, 2138 (1998)

GPSR Compliance
The European Union's (EU) General Product Safety Regulation (GPSR) is a set of rules that requires consumer products to be safe and our obligations to ensure this.

If you have any concerns about our products, you can contact us on

ProductSafety@springernature.com

In case Publisher is established outside the EU, the EU authorized representative is:

Springer Nature Customer Service Center GmbH
Europaplatz 3
69115 Heidelberg, Germany

www.ingramcontent.com/pod-product-compliance
Ingram Content Group UK Ltd.
Pitfield, Milton Keynes, MK11 3LW, UK
UKHW021939200726
13856UKWH00005B/97

* 9 7 8 1 4 7 5 7 8 4 4 6 6 *